T0319230

PEDIATRIC BRAIN STIMULATION

Companion Web Site:

http://booksite.elsevier.com/9780128020012/

Pediatric Brain Stimulation: Mapping and Modulating the Developing Brain
Adam Kirton and Donald L. Gilbert, Editors

ACADEMIC
PRESS

PEDIATRIC BRAIN STIMULATION

MAPPING AND MODULATING THE DEVELOPING BRAIN

Edited by

ADAM KIRTON
University of Calgary, Calgary, AB, Canada

DONALD L. GILBERT
Cincinnati Children's Hospital Medical Center, Cincinnati, OH, United States

AMSTERDAM • BOSTON • HEIDELBERG • LONDON
NEW YORK • OXFORD • PARIS • SAN DIEGO
SAN FRANCISCO • SINGAPORE • SYDNEY • TOKYO
Academic Press is an imprint of Elsevier

Academic Press is an imprint of Elsevier
125 London Wall, London EC2Y 5AS, UK
525 B Street, Suite 1800, San Diego, CA 92101-4495, USA
50 Hampshire Street, 5th Floor, Cambridge, MA 02139, USA
The Boulevard, Langford Lane, Kidlington, Oxford OX5 1GB, UK

British Library Cataloguing-in-Publication Data
A catalogue record for this book is available from the British Library

Library of Congress Cataloging-in-Publication Data
A catalog record for this book is available from the Library of Congress

ISBN: 978-0-12-802001-2

For Information on all Academic Press publications
visit our website at https://www.elsevier.com/

  Working together
to grow libraries in
developing countries

www.elsevier.com • www.bookaid.org

Typeset by TNQ Books and Journals
www.tnq.co.in

Contents

I

FUNDAMENTALS OF NIBS IN CHILDREN

1. TMS Basics: Single and Paired Pulse Neurophysiology

E. ZEWDIE AND A. KIRTON

2. Assessing Normal Developmental Neurobiology With Brain Stimulation

K. LIMBURG, N.H. JUNG, AND V. MALL

3. Neuroplasticity Protocols: Inducing and Measuring Change

S.W. WU AND E.V. PEDAPATI

4. Therapeutic rTMS in Children
D.L. GILBERT

5. Transcranial Direct-Current Stimulation (tDCS): Principles and Emerging Applications in Children
P. CIECHANSKI AND A. KIRTON

6. Insights Into Pediatric Brain Stimulation Protocols From Preclinical Research
M.Q. HAMEED, M.J. SANCHEZ, R. GERSNER, AND A. ROTENBERG

7. Pediatric Issues in Neuromodulation: Safety, Tolerability and Ethical Considerations

K.M. FRIEL, A.M. GORDON, J.B. CARMEL, A. KIRTON, AND B.T. GILLICK

II

NIBS IN PEDIATRIC NEUROLOGICAL CONDITIONS

8. TMS Applications in ADHD and Developmental Disorders

D.L. GILBERT

9. TMS Mapping of Motor Development After Perinatal Brain Injury

M. STAUDT

10. The Right Stimulation of the Right Circuits: Merging Understanding of Brain Stimulation Mechanisms and Systems Neuroscience for Effective Neuromodulation in Children

J.B. CARMEL AND K.M. FRIEL

11. Therapeutic Brain Stimulation Trials in Children With Cerebral Palsy

B.T. GILLICK, K.M. FRIEL, J. MENK, AND K. RUDSER

12. Brain Stimulation in Children Born Preterm—Promises and Pitfalls

J.B. PITCHER

13. Brain Stimulation to Understand and Modulate the Autism Spectrum

L.M. OBERMAN, A. PASCUAL-LEONE, AND A. ROTENBERG

14. Non-Invasive Brain Stimulation in Pediatric Epilepsy: Diagnostic and Therapeutic Uses

S.K. KESSLER

15. Brain Stimulation in Pediatric Depression: Biological Mechanisms

P. CROARKIN, S.H. AMEIS, AND F.P. MACMASTER

16. Brain Stimulation in Childhood Mental Health: Therapeutic Applications

F.P. MACMASTER, M. SEMBO, K. MA, AND P. CROARKIN

17. Transcranial Magnetic Stimulation Neurophysiology of Pediatric Traumatic Brain Injury

K.M. BARLOW AND T.A. SEEGER

18. Brain Stimulation Applications in Pediatric Headache and Pain Disorders

T. RAJAPAKSE AND A. KIRTON

III

INVASIVE BRAIN STIMULATION IN CHILDREN

19. Deep Brain Stimulation in Children: Clinical Considerations

J.-P. LIN

20. Deep Brain Stimulation Children – Surgical Considerations
R.D. BHARDWAJ

21. Invasive Neuromodulation in Pediatric Epilepsy: VNS and Emerging Technologies
M. RANJAN AND W.J. HADER

22. Emerging Applications and Future Directions in Pediatric Neurostimulation
D.L. GILBERT AND A. KIRTON

List of Contributors

S.H. Ameis Campbell Family Mental Health Research Institute, CAMH; University of Toronto, Toronto, ON, Canada

K.M. Barlow University of Calgary, Calgary, AB, Canada

R.D. Bhardwaj Sanford Children's Hospital, Sioux Falls, SD, United States

J.B. Carmel Burke-Cornell Medical Research Institute, White Plains, NY, United States; Weill Cornell Medical College, New York, NY, United States

P. Ciechanski University of Calgary, Calgary, AB, Canada

P. Croarkin Mayo Clinic College of Medicine, Rochester, MN, United States

K.M. Friel Burke-Cornell Medical Research Institute, White Plains, NY, United States; Weill Cornell Medical College, New York, NY, United States

R. Gersner Harvard Medical School, Boston, MA, United States

D.L. Gilbert Cincinnati Children's Hospital Medical Center, Cincinnati, OH, United States

B.T. Gillick University of Minnesota, Minneapolis, MN, United States

A.M. Gordon Teachers College of Columbia University, New York, NY, United States; Columbia University Medical Center, New York, NY, United States

W.J. Hader University of Calgary, Calgary, AB, Canada

M.Q. Hameed Harvard Medical School, Boston, MA, United States

N.H. Jung Technische Universität München, Kinderzentrum München gemeinnützige GmbH, Munich, Germany

S.K. Kessler Children's Hospital of Philadelphia, Philadelphia, PA, United States

A. Kirton University of Calgary, Calgary, AB, Canada

K. Limburg Technische Universität München, Kinderzentrum München gemeinnützige GmbH, Munich, Germany

J.-P. Lin Guy's and St. Thomas' NHS Foundation Trust, London, United Kingdom

K. Ma University of Calgary, Calgary, AB, Canada

F.P. MacMaster University of Calgary, Calgary, AB, Canada

V. Mall Technische Universität München, Kinderzentrum München gemeinnützige GmbH, Munich, Germany

J. Menk University of Minnesota, Minneapolis, MN, United States

L.M. Oberman Brown University, Providence, RI, United States

A. Pascual-Leone Harvard Medical School, Boston, MA, United States

E.V. Pedapati Cincinnati Children's Hospital Medical Center, Cincinnati, OH, United States

J.B. Pitcher The University of Adelaide, Adelaide, SA, Australia

T. Rajapakse Alberta Children's Hospital, Calgary, AB, Canada

M. Ranjan University of Calgary, Calgary, AB, Canada

A. Rotenberg Harvard Medical School, Boston, MA, United States

K. Rudser University of Minnesota, Minneapolis, MN, United States

M.J. Sanchez Harvard Medical School, Boston, MA, United States

T.A. Seeger University of Calgary, Calgary, AB, Canada

M. Sembo University of Calgary, Calgary, AB, Canada

M. Staudt Schön Klinik, Vogtareuth, Germany; University Children's Hospital, Tübingen, Germany

S.W. Wu Cincinnati Children's Hospital Medical Center, Cincinnati, OH, United States

E. Zewdie University of Calgary, Calgary, AB, Canada

Foreword

Brain stimulation in various forms has been practiced for centuries, and it has gone in and out of favor, but recently with advances in methodologies, it has achieved a resurgence of interest. Electroconvulsive therapy (ECT) has been used over many decades for treatment of depression, but its side effects are concerning. It is the efficacy of deep brain stimulation (DBS) for various indications such as Parkinson's disease that has pointed the modern direction for therapeutic brain stimulation and increased excitement. This advance was recently recognized with the Lasker prize. In the past two decades a large variety of non-invasive brain stimulation methods have been developed. First was high-voltage pulsed electrical stimulation. The demonstration that such stimulation could activate the motor cortex was met with astonishment. This was quickly followed by transcranial magnetic stimulation (TMS) which was welcomed since the electrical stimulation was painful. TMS has proven to be a superb tool for studying brain physiology, has some clinical utility for diagnosis, and is being explored for therapeutic benefit. Already TMS has achieved US FDA approval for the treatment of depression. Continuous low-voltage electrical stimulation, either direct current or alternating current (tDCS and tACS), is now also being used. It is growing fast in popularity since is it easy to do, cheap to manufacture, and seems also to have clear efficacy in altering brain function.

It should be noted in relation to therapy that some types of brain stimulation, like DBS, can be used continuously and work by online modulation of brain networks. In general, non-invasive brain stimulation is used only intermittently and any prolonged therapeutic effect would depend on brain plasticity to alter brain function.

All of this has applicability in children but has been developing more slowly. There is, of course, always less research in children than in adults. Some of this has to do with safety concerns. Children might well be more vulnerable in many ways, including the possibility of negatively interfering with brain development. Deep brain stimulation for the youngest children would be confounded by growth of the head. With TMS, which creates a loud noise with each pulse, there has been particular concern about possible hearing damage, in part because of the different anatomy of the external auditory canal in children. Recent studies of this latter issue have, fortunately, shown no problem as long as the study is conducted properly.

Nevertheless, the literature on brain stimulation in children has accumulated and it is timely to summarize the field. Dr. Kirton and Dr. Gilbert have put together a fine group of experts to cover the field broadly, from basic mechanisms, to physiology to therapy, including all types of brain stimulation. An area that was explored early on with TMS was the development of the corticospinal system in children, and this has allowed physiological studies of cerebral palsy with brain injury at various times. More recently there has been exploration of other childhood disorders. It is clear that there will be many applications, and a book like this should help firm up the foundation of the field and propel it forward.

Mark Hallett, MD
National Institute of Neurological Disorders
and Stroke, Bethesda, MD, United States

Preface

Children are not little adults and disorders of the developing brain are different. Non-invasive brain stimulation technologies are revolutionizing our understanding of human neurophysiology in health and disease. That the vast majority of progress has occurred in the mature brain is entirely reasonable, as initial uses of experimental devices require substantial safety testing before use in children.

Not long after these early accomplishments were realized the exciting potential to explore the fascinating elements of brain development was harnessed. During the 1990s, several forward-thinking pioneers began applying TMS methodologies to children, including preschoolers and even newborns. The results were nothing short of astounding, providing novel in vivo evidence of fundamental elements of brain development in health and disease previously only suggested by animal models, limited imaging, or simple theory.

Despite this early progress, neurophysiological applications in children continue to lag behind those in adults. A search of transcranial magnetic stimulation (TMS) in PubMed in 2015 demonstrates continued exponential scientific growth, yet <3% of studies are dedicated to children. This trend is even more pronounced when searching therapeutic brain stimulation applications and clinical trials.

The need for novel treatment interventions in childhood neurological and neuropsychiatric disease is pressing. Disorders are increasingly understood but treatment options often lag behind due to complexities of the developing brain, failure of drug companies to develop pediatric-specific pharmaceuticals, and other factors. The appeal of non-pharmacological, individualized, and targeted interventions is particularly appealing to both young patients and their parents.

The time to harness the full potential of non-invasive brain stimulation applications in children is now.

To facilitate this goal, we have attempted to pull together the global leaders in brain stimulation focused on the developing brain. To be clear, the majority of pioneers in virtually all aspects of non-invasive brain stimulation methodology and applications have come from the adult world. Many of the topics reviewed here carry a wealth of experience and data thanks to these many visionary scientists. That none of them are primary contributors to chapters in this volume in no way reflects their many seminal contributions. However, it is also intentional.

Our aim instead was to garner the knowledge and experience of those whose research, clinical care, and experience is focused on children. Validating and supporting this philosophy was the remarkably successful recruitment of authors we encountered. Almost without exception, each contributor responded immediately with keen interest, enthusiasm, and willingness to participate. As editors who also work in the world of pediatric neuroscience, this was not entirely surprising. In our experience, pediatricians and others focused on advancing the wellbeing of children tend to be of a highly generous and collaborative nature. The response we received certainly reflected this. This strong endorsement validated our initiating the project while further inspiring us to generate the best possible product.

The result is a volume that begins by attempting to orient less experienced readers in the fundamentals of currently available non-invasive brain stimulation technologies. Pediatric aspects are emphasized while providing the essential knowledge required to interpret and perform basic brain stimulation research. This is followed by a categorical approach to brain stimulation applications across the main neurological, psychiatric, and developmental disorders most often affecting children. These are intended to inspire those with an interest in these patient populations to pose new questions and conceive potential approaches to advancing understandings and treatment possibilities.

As summarized in the final chapter, the future holds many additional promising directions. We hope our readers will share our enthusiasm and contribute to what will certainly be exciting times to come with real potential to improve the lives of affected children and their families.

Dr. Adam Kirton
Dr. Donald L. Gilbert

FUNDAMENTALS OF NIBS IN CHILDREN

1

TMS Basics: Single and Paired Pulse Neurophysiology

E. Zewdie
University of Calgary, Calgary, AB, Canada

A. Kirton
University of Calgary, Calgary, AB, Canada

O U T L I N E

INTRODUCTION

Transcranial magnetic stimulation (TMS) is a non-invasive technique that painlessly delivers magnetic fields across the scalp and the skull to induce regional activation of cortical neuronal populations in the brain. Since its introduction by Barker in the 1980s, TMS has greatly advanced our ability to explore and understand neural circuitry and physiology in vivo. Using multiple patterns of stimulations (single, paired, or repetitive), TMS is capable of characterizing pathways, excitability of inhibitory and excitatory circuits and other neurophysiological elements, as well as measuring and modulating plasticity. Due to its safe nature and favorable tolerability, TMS has become an invaluable tool across virtually all areas of clinical neuroscience, though its applications in the developing brains of children are only just being realized.

HOW DOES TMS WORK?

Principle of TMS

Merton and Morton[1] were the first to build a high-voltage transcranial electrical stimulator (TES) that could be used to activate the cerebrum via surface electrodes placed over the scalp by passing an electric current in a perpendicular direction to the stimulating electrodes. TES works by transferring a short-lasting electric current from its source to the excitable neurons of the brain. Using this technique a very high voltage is required in order to stimulate the neuronal tissue of the brain as the skull and the scalp have high resistance. Although electrical stimulation has greatly advanced since (see Chapter 5: Transcranial Direct-Current Stimulation (tDCS): Principles and Emerging Applications in Children), magnetic stimulation, a less painful alternative brain stimulation technique, has emerged as an appealing alternative.

TMS was developed in the 1980s as a method to non-invasively stimulate the brain. The main principle of TMS is the induction of a short-lasting electric current in the cerebral cortex. Unlike TES, which involves direct transfer of current through electrodes, TMS uses electromagnetic induction to deliver current to the brain. A very short-lasting (150–300 μs) electric current is applied to a stimulating coil to produce a rapidly changing magnetic field which, in turn, induces a flow of electric current in nearby conductors – including the human brain. The current that is applied to the brain from TMS is governed by the Faraday–Henry law, which states that if an electric conductor is linked by a time-varying magnetic flux, a current is observed in the circuit. The magnetic flux is created due to the current that is passing through the stimulating inductor coil, and its magnitude depends on the material property of the coil and the amount of current that is passing through the coil.

The magnetic stimulator typically consists of a high-current pulse generator (capacitor), which produces a current of 5000 A when charged up to 2.8 kV, and a stimulating coil (inductor coil) producing magnetic pulses with a field strength of 1 T or more and a pulse duration of ~1 ms^2 (Fig. 1.1).

When the capacitor discharges, apart from the electrical energy that is lost in the wiring, about 500 J of energy is transferred to the coil and then returned to the instrument to help reduce coil heating in about 100 μs, which means that, during the discharge, energy initially stored in the capacitor in the form of electrostatic charge is converted into magnetic energy in the stimulating coil in this time frame.[2] This leads to a rapid rise in the magnetic field (around 30 kT/s) which in turn induces current in the brain in the order of 1–20 mA/cm^2.

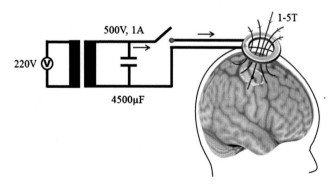

FIGURE 1.1 **The principle of the magnetic stimulator.** The current generated, when the charge stored in the capacitors is discharged through the inductor coils, creates a magnetic field that can penetrate through the skull and induce secondary current in the opposite direction.

The effectiveness of the stimulation in activating the cortical neurons is determined by the waveform of the pulses and the induced current direction. Two types of waveforms are currently used – monophasic and biphasic. Biphasic stimulators are more effective than monophasic stimulators because a biphasic waveform produces currents with longer durations and can therefore produce more current to activate neurons.[3] At the same time, coil orientation is an important determinant of which cortical neuronal populations are affected during TMS. The current that is induced by TMS in the brain tissue flows in a direction that is parallel to the stimulating coil but perpendicular to the magnetic field created by the coil. If the induced current has the same direction as the nerve signal in the axons it results in the most effective stimulation. Therefore, axons that are orientated parallel (or horizontal) to the stimulating coil placed on the surface over the motor cortex are most effectively activated by TMS.

Types of TMS Coil

TMS coils are designed in different shapes (see Fig. 1.2) to focus current in different parts of the cortex by varying the spatial field and depth of the induced electric field. The simplest design is a circular coil which typically has a diameter of 8–15cm (Fig. 1.2A), with a maximum magnetic field directly under the center of the circle and a maximum induced current at the outer edge of the circle. Although this type of coil creates good penetration of the cerebral cortex, its spatial resolution is relatively low. The circular coil activates the motor cortex asymmetrically, with greater activation on the side where the coil current flows from posterior to anterior across the central sulcus.[4] Hence, with the coil placed centrally on the vertex, the induced current predominantly stimulates the left motor cortex as the current flow will be posterior to anterior. Neurons located at a depth of 10.5mm from the surface of the skull will be stimulated by a circular coil.[4]

More focal stimulation can be achieved by putting two smaller circular coils side-by-side in a coil type called a figure-of-eight coil (Fig. 1.2B). The magnetic fields created by currents flowing in opposite directions summate at the junction of the two circular coils and result in a more focal stimulation directly under the junction. Due to the smaller diameter of the coils, the penetration level is more limited at the center of each coil. Neurons located at a depth of 11.5mm from the surface of the skull will be stimulated by a figure-of-eight coil.[5] Typically, side-by-side coils range from very small flat coils for brain mapping work, such as figure-of-eight coils (Fig. 1.2B), to large contoured versions, such as bat-wing coils (Fig. 1.2C) or double cone coils (Fig. 1.2D), which may be used to stimulate deeper neural structures in the brain. Additional coils, such as those with cooling air circulation (Fig. 1.2E) have been designed to further advance the depth or other TMS delivery parameters.

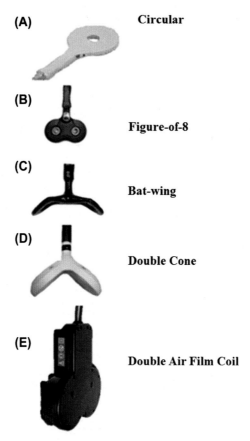

(A) Circular

(B) Figure-of-8

(C) Bat-wing

(D) Double Cone

(E) Double Air Film Coil

FIGURE 1.2 **TMS coil types.** (A) Circular coil has only a single winding with superficial current penetration. (B) Figure-of-eight coil has a flat shape with two adjacent windings and is used to activate more superficial hand areas of the cerebral cortex. (C) Bat-wing coil has a flat center with bent wings and is used for activating leg areas of the cerebral cortex. (D) Double-cone coil has two large cup-shaped windings positioned side-by-side in an angle and can activate deeper areas of the cortex more powerfully than the bat-wing coil. (E) Double air film coil has cold air circulating in order to cool the inductor coil. It is mainly used to deliver repetitive TMS.

NEURONAL STRUCTURES ACTIVATED BY TMS

Stimulation of the Motor Cortex

The cell bodies and axons of neurons in the cerebral cortex that respond to TMS[6] depend on their size, location, orientation, and function. TMS can be delivered in a lateral–medial or anterior–posterior configuration to stimulate specific elements in the cerebral cortex. Compared to TES, TMS produces the most localized electric field. However, a localized field from

TMS also decreases the ability to stimulate neurons buried deep in the brain because a magnetically induced field drops off faster than an electrically produced one.[5] Electrical fields must have a component parallel to the neuron in order to stimulate it.[7] In the motor cortex, specifically in Area 4, there are neurons on the surface that are oriented more vertically and neurons deeper within the central sulcus that are oriented more horizontally (and all configurations in between).

Given that magnetic stimulation produces a current flow that is parallel (in a horizontal direction) to the surface of the cortex, the neurons that are going to be stimulated directly are those that are parallel to the surface of the cortex. These neurons with horizontally aligned axons are typically located in deeper layers of the cortex. Since the strength of an electric field created by magnetic stimulation falls off with depth, these parallel neurons may be less stimulated if they are located very deep. In contrast, the electric field produced during electric stimulation has components both parallel and perpendicular (vertical) to the surface of the head. Therefore, it activated the axons or axon hillocks of the corticospinal tract (CST) neurons.

Descending Volleys

Evidence that multiple descending volleys in the CST can be activated by electrical stimulation was first proposed in the mid-1950s by Patton and Amassian.[8] When a single, square-wave electrical shock was applied to the motor cortex of anesthetized cats and monkeys, multiple CST neurons were activated. The first wave was produced by direct excitation of CST neurons, while later waves originated from indirect activation via cortical interneurons. Therefore, the terms D- and I-waves are used to describe the first (Direct) and the later (Indirect) responses, respectively. The latency of D-waves agreed with the conduction time of large corticospinal fibers. On the other hand, I-waves were found to be more susceptible to cortical injuries than D-waves. In addition, when microstimulating deep white matter, the size of D-waves increased and I-waves only appeared when the stimulating electrode was applied to the gray matter. This was an important finding since it provided initial evidence that multiple descending volleys from I-waves were generated trans-synaptically through excitatory corticocortical fibers when stimulating the surface of the motor cortex with a single pulse.

Similarly, descending corticospinal volleys in response to transcranial stimulation were studied by Day and co-workers.[9,10] They found that the latency of the descending volleys evoked by TMS varied markedly according to the type of stimulation and the stimulation intensity. An increase in stimulation intensity results in an increase in the probability of eliciting shorter-latency motor responses. In particular, at high stimulation intensities, the amount of short-latency volleys was typically increased.

Descending volleys from TMS were typically slower and more complex than the volleys from TES. The fact that TES typically induced firing a few milliseconds before TMS suggests that descending volleys induced by both techniques are likely triggered using different mechanisms. As corticospinal volleys from TMS are typically delayed, it is believed that these volleys are likely I-waves and caused by trans-synaptic activation. In agreement with direct microstimulation of the motor cortex in monkeys, it is predicted that TES predominantly triggers axonal depolarization of pyramidal neurons and leads to a descending volley in the form of a D-wave. As the directions of current flow from TES and TMS are different, it is predicted that the orientation of the current in the cortex is likely responsible for eliciting these different responses. It is believed that the horizontal currents from TMS likely activate interneurons, whereas more vertically orientated currents from TES are more likely to activate the axons of pyramidal neurons directly. Repetitive trans-synaptic activation of pyramidal neurons is likely responsible for the delayed and more numerous volleys observed using TMS.

Descending Pathways From the Cerebral Cortex

The corticospinal tract (CST) is the main descending motor pathway from the cerebral cortex to the spinal cord that can be activated by TMS. The CST originates from large pyramidal cells predominantly in the fifth layer of the cerebral cortex. Over 60% of CST fibers originate from the primary motor cortex, supplementary motor area, and premotor cortex.[11] A substantial proportion of the fibers (about 40%) originate from the primary somatosensory area and the parietal cortex.[12] The corticospinal tract is the only descending motor tract that is known to make monosynaptic connections with spinal motoneurons in humans.[13,14] Unlike in cats where there are no functional corticomotoneuronal (CM) connections,[15] in humans the presence of CM connections on upper limb[16] and lower limb[17] motoneurons has been demonstrated. Functionally, muscles that require greater precision, such as the index finger and thumb, have greater CM connections.[18] In human lower limbs, the monosynaptic projection to the tibialis anterior muscle is particularly strong, comparable in magnitude to the finger muscles.[19] This may be related to the precision required to clear the toes above the ground during the swing phase of human walking.[20]

In addition to activating the cerebral cortex, TMS over the motor cortex can activate projections to other cortical[21] and subcortical regions.[22] For instance, TMS over the primary motor cortex (M1) can trans-synaptically activate corticoreticular[23] and corticovestibular[24] pathways, which can produce or influence spinal motor output via reticulospinal and vestibulospinal pathways, respectively. Single unit recordings from ponto-medullary reticular formation in the brainstem of anesthetized monkeys revealed multiple latency responses evoked by TMS performed over M1.[23] The multiple latency responses may indicate the existence of multiple pathways from M1 to such subcortical structures.

Motor-Evoked Potential

As described above, TMS can generate descending volleys, which in turn activate motoneurons at the spinal cord level. When the action potential of spinal motoneurons arrives at the muscle, it results in muscle depolarization and contraction. Using a pair of surface electrodes over the muscle, it is possible to non-invasively detect the potential difference as muscle action potentials are activated. An electromyography (EMG) can detect the sum of this activation. When the EMG is changed due to the TMS activation, the resulting deflection (increase) of EMG is called the motor-evoked potential (MEP). MEP latency and amplitude reflect the functional integrity of the upper motoneurons, the CST, spinal interneurons, and muscle. MEP amplitudes are measured by subtracting the largest negative peak from the largest positive peak of EMG within the MEP window (Fig. 1.3). On the other hand, the latency of the MEP is measured as the time from TMS trigger to the onset of the MEP. It is possible to calculate the central conduction time by combining TMS latency with peripheral motor latencies. Loss of large myelinated motor axons can cause prolongation of the central conduction time.

A simple mapping procedure is typically used to determine the optimal cortical location for TMS application to generate MEP in the muscle of interest. The center of the coil can be approximated to the area of the motor strip expected to represent the muscle of interest. For hand muscles,

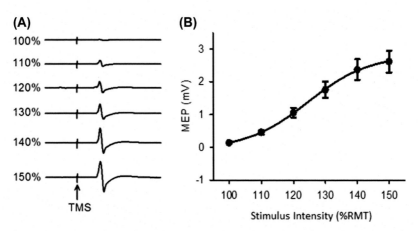

FIGURE 1.3   **Stimulus response curves.** (A) Average of 10 first dorsal interosseous MEPs evoked from incrementing intensities. (B) Mean peak-to-peak amplitude of MEP from the data shown in (A) to produce a TMS response curve. Error bars represent ± standard error (SE).

this is typically close to the central reference point in the 10–20 EEG system (eg, C3 for the muscles of the right hand) (Fig. 1.4). Single pulse applications at modest stimulator output (eg, 30%) can be applied and slowly escalated until MEPs are seen. Using an output slightly above threshold, the coil can then be systematically mapped across the region to determine the location that gives the largest, most consistent MEP. This region can then be marked on the scalp as the "hotspot" and referenced for all subsequent experiments. Modern image-guided neuronavigation systems have further enhanced this process, providing individualized, real-time co-registration of individual anatomy with coil positioning, facilitating very accurate placement within and across experimental sessions.

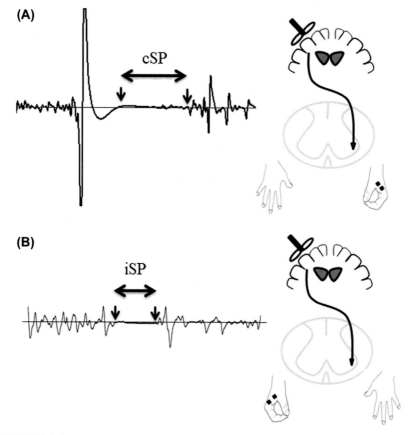

FIGURE 1.4 **Silent period.** (A) An example motor-evoked potential from the FDI muscle of a 15-year-old participant at maximum stimulation intensity (150% RMT) followed by cortical silent period (cSP). (B) An example of ipsilateral EMG suppression called the ipsilateral silent period (iSP). The vertical arrow indicates the onset and offset of the silent period. The horizontal arrow indicates the duration of the silent period. The average EMG before the TMS is shown by the red line.

Motor Thresholds

In their earliest contribution, Merton and Morton[1] noticed that the threshold for evoking a motor response is very low when the muscle under test is contracting (active) compared to when the same muscle is relaxed (rest). This finding was vital for experiments that collect EMG responses to control the contraction level. However, it was necessary to characterize the mechanism and level (spinal or cortical) where it occurs. Voluntary contraction may increase the excitability of cortical neurons so that a given stimulus recruits a larger descending corticospinal volley. On the other hand, voluntary activation may also increase the excitability of spinal motoneurons, making it easier to discharge them with a given descending volley. In their earlier study of motor cortex stimulation, Rothwell and co-workers[25] investigated changes in muscle response to anodal cortical stimulation when a subject exerted a background voluntary contraction. The latency has significantly decreased and the size of peak-to-peak EMG response increased when the muscle was contracting compare to when the muscle was relaxing. The same effect was also seen when using TMS. Mazzochio and co-workers[26] showed that tonic voluntary contraction increased cortical excitability to magnetic stimulation, so that a given stimulus evoked a larger descending corticospinal volley.

Most MEP measurements are normalized across individuals to the resting or active threshold values. Resting motor threshold (RMT) is defined as the minimum TMS intensity (expressed as percentage of maximum stimulator output or MSO) that elicits reproducible MEP responses of at least $50\,\mu V$ in 50% of 5–10 consecutive trials.[27] Similarly, active motor threshold (AMT) is defined as the minimum TMS intensity that elicits reproducible MEP responses of at least $200\,\mu V$ in 50% of 5–10 consecutive trials while the muscle is actively contracting from 10% to 20% of the maximum voluntary contraction (MVC).[28] Visual feedback of EMG activity is often used to obtain consistent levels of contraction. Motor thresholds vary widely across individuals, are higher in young children, and may be influenced by numerous additional factors including the muscle being tested, medications, and level of alertness. As shown below, most additional TMS neurophysiology experiments are referenced to these individualized thresholds.

TMS NEUROPHYSIOLOGY

Single-Pulse TMS

Stimulus Response Curve

Single-pulse TMS can be used to estimate the strength and excitability of the motor cortex and its descending pathways. To produce a TMS response curve, TMS intensities can be incrementally increased either in terms of the maximum stimulator output (usually from 30% to 100% of MSO in steps of 10%) or in proportion to RMT/AMT (usually from 100% to 150% of RMT or AMT). Fig. 1.3A demonstrates an example from a healthy 14-year-old. Each trace is an average of 10 traces given at each of the indicated TMS intensities in random order. By measuring the peak-to-peak MEP amplitude, it is possible to then produce the stimulus response curve (SRC) shown in Fig. 1.3B. SRC from healthy participants typically generates a sigmoidal curve with MEP size plateauing at higher stimulation intensities. The parameters of the SRC may correlate with specific neurophysiological elements. For instance, the maximum MEP can be measured as the largest response on the recruitment curve, and may reflect connectivity or excitability of the CST or changes in the cortical motor map. In addition, SRCs that have shifted leftwards or rightwards without a vertical change could be indicative of alterations in recruitment thresholds or cortical excitability.[29]

Silent Periods

During voluntary background contraction, a temporary suppression in EMG activity has been reported following an MEP induced by TMS to the contralateral motor cortex.[30] This is called the cortical silent period (cSP). This change in excitation is observed through the duration of the cSP and is believed to reflect a lack of cortical drive related to recruitment of intracortical inhibitory circuits. To determine the amount of inhibition induced at spinal and supraspinal levels, Fuhr and co-workers[30] performed H reflex testing at different periods in the silent period. As the H reflex was dramatically depressed at the start of the silent period and strongly recovered by the end, these findings implied that reductions in excitability due to the motoneurons were only involved in inhibition during the start of the silent period. For this reason, the lack of myoelectric activity at the end of the silent period was likely associated with a lack of cortical drive following magnetic stimulation.

Although the cortical silent period is easily demonstrated using TMS in able-bodied subjects, various neurological disorders have been reported to significantly alter and sometimes even abolish the silent period.[31] Changes in cortical silent period may be related to cortical reorganization following neurological injury. For this reason, reorganization and disinhibition in the motor cortex appear to be involved in preserving and enhancing motor function after neurological impairment.

An ipsilateral silent period (iSP) can also be measured when TMS is delivered to the motor cortex on the same side as the contracting hand muscle. The experimental paradigm is essentially the same as that described above for the cortical silent period. The ipsilateral silent period is thought to reflect activation of transcallosal pathways affecting inhibitory interneurons in the contralateral, active motor cortex.[32]

Paired-pulse TMS

All of the above TMS paradigms involve the administration of a single pulse. The addition of one or more earlier pulses of defined intensity and timing can alter these responses. Such paired-pulse TMS can provide additional insight toward neurophysiological mechanisms in health and alterations in disease. The baseline single pulse is referred to as the test stimulus (TS), while the additional modifying pulse is the conditioning stimulus (CS). Conditioning stimuli strength may vary from less than (subthreshold) to greater than (suprathreshold) the RMT. The interstimulus interval (ISI) is the time in ms between the CS and TS and is the key determinant of the response observed.

Short-Interval Intercortical Inhibition (SICI)

Principal types of local intracortical inhibition can be studied using paired-pulse TMS. Kujirai and co-workers[33] discovered that the response evoked by suprathreshold test stimulus (TS) given 1–6 ms after subthreshold conditioning stimulus (CS) is inhibited, compared to the response evoked by the test stimulus alone. This inhibitory phenomenon is called short-interval intracortical inhibition (SICI). As shown in the EMG data from a single 12-year-old participant (Fig. 1.5A), the test stimulus alone elicits an EMG response of about 1-mV peak-to-peak amplitude. The superimposed responses by TS alone (gray line) and paired pulses (solid line) are given at 2 ms after a conditioning stimulus. The paired-pulse

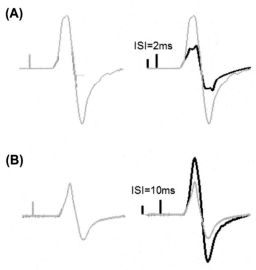

FIGURE 1.5 **Short-interval intercortical inhibition and facilitation in the FDI muscle.** Raw traces showing test MEP (gray traces) and (A) SICI (ISI = 2 ms) and (B) ICF (ISI = 10 ms) shown by black traces. Each of the traces is an average of 10 EMG traces.

response was significantly reduced. In addition, this inhibition is present for ISI of 1–6 ms. This range of ISIs, where the percentage control size (paired-pulse response divided by test alone response) is below 100%, describes the SICI phenomenon.

Intracortical Facilitation (ICF)

Using paired-pulse TMS, two categories of facilitations within M1 have been identified. Using subthreshold conditioning stimulus (CS) to condition suprathreshold test stimulus (TS) at interstimulus interval (ISI) of 6–25 ms reveals TS MEP facilitation, which is known as intracortical facilitation (ICF). In the same experiment, where Kujrai and co-workers[33] described SICI, ICF was also exhibited for ISIs above 6 ms. As shown in Fig. 1.5B, conditioning a test response by a prior conditioning stimulus at 10 ms ISI, can result in an increase in MEP amplitude. Since these ISIs are longer than where SICI happens, it is not called short intracortical facilitation. Subthreshold facilitation of spinal motoneurons was suggested to be the cause of the facilitation of test responses and hence the cause of ICF. Though the exact mechanism has not yet been clarified, the possible mechanism mediating ICF has been suggested by Herwig and co-workers[34] to be the induction of slow excitatory postsynaptic potentials and/or induced intracellular signaling cascades enhancing excitability of pyramidal tract neurons by trans-synaptic activation of metabotrope receptors. However, this suggestion was based only on physiological theory and was not supported by any data.

Short-Interval Intracortical Facilitation (SICF)

Over relatively short intervals, the second type of facilitatory interaction can be demonstrated within M1, called short-interval intracortical facilitation (SICF). This occurs when a suprathreshold stimulus is followed by a subthreshold stimulus.[35] This phenomenon can also be exhibited when two stimuli near motor threshold are given consecutively. In relaxed subjects, when the interval between the stimuli was around 1.0–1.5 ms, 2.5–3.0 ms, or 4.5 ms or later, the size of the response to the pair of stimuli was much greater than the algebraic sum of the response to each stimulus alone. The first, second, and third peaks of facilitations were observed when the second stimulus is fixed to 70% RMT and the first stimulus was 70%, 90%, and 100% RMT, respectively.[36] SICF is also called I-wave facilitation as the three peaks of facilitations observed may correspond to the generation of I-waves. Di Lazzaro and co-workers tested the cortical involvement in SICF and the relation with I-waves by recording descending motor volleys directly from the cervical epidural space of five conscious patients, who had a stimulator implanted in the cervical cord for the treatment of intractable pain.[37] Test stimulus was set at an intensity of 2% (of stimulator output) above active motor threshold (AMT) and condition stimulus was set at AMT. At ISI of 1, 1.2, and 4 ms the amplitude of the total volleys (the sum of individual I-waves minus the responses to test stimulus alone) and of the EMG response is larger after paired stimulation. More importantly, it is clearly shown that the I1-wave is virtually unaffected, while there is a significant increase in the I2- and I3-waves. The ISIs where SICF was observed using EMG recordings might be slightly longer (~0.5 ms) than ISIs where SICF was observed using epidural recordings. This increase could be due to the distance difference of where the recordings are taken, since epidural recordings are taken closer to the stimulation than EMG recordings.

Long-Interval Intracortical Inhibition (LICI)

The second type of local intracortical inhibitory phenomenon that occurs at longer ISI is known as long-interval intracortical inhibition (LICI). LICI is shown by suppression of the MEP to a test stimulus when it is preceded 50–200 ms by a conditioning stimulus that is above motor threshold.[38] Wassermann and co-workers delivered paired stimuli to motor cortex with a circular coil at 1.1 AMT, with various ISIs ranging between 20–200 ms. The active muscle experiments were performed while holding 10% maximum voluntary contraction of wrist extensors of the right arm. The ratio conditioned MEP to the MEP evoked by test stimulus alone was below 100% at all ISIs between 20 and 200 ms, indicating the activation of LICI. The mechanism underlying LICI is similar to that of the cortical silent period (CSP), since the inhibition in both measurements is induced by a suprathreshold stimulus. CSP may represent the duration of inhibition induced by suprathreshold CS, while LICI may reflect its magnitude. Since the spinal inhibitory mechanisms are exerted mainly during the early part (up to 50 ms) of CSP, LICI at early ISIs is presumably mediated by the supraspinal mechanisms, but partial spinal influences cannot be entirely excluded.[38]

Interhemispheric Inhibition (IHI)

In this stimulation paradigm, TMS over the primary motor cortex (M1) of one hemisphere affects the response of the opposite hemisphere to the TMS. Accordingly, this methodology is unique in that it requires the simultaneous placement of two separate TMS coils, one over each motor cortex. Ferbert and colleagues[39] stimulated the left hemisphere M1 (condition stimulus) followed by simulation of the right hemisphere M1 (test stimulus). The MEP in the distal hand muscles by test TMS was inhibited at all ISIs between 6 and 30 ms. They proposed this inhibition is produced at the cortical level via a transcallosal route. The claim was later supported by epidural recordings where I2- and later I-wave are suppressed by conditioning stimuli on the other side of hemisphere, but not D-wave or I1-wave.[37] More detailed studies of IHI have defined both short- (approximately 8–10 ms ISI) and long- (40–50 ms ISI) interval versions of IHI.[40] Mechanisms of IHI may be similar to those of the ipsilateral silent period described above, though differences have also been reported. Our studies of typically developing children suggest similar IHI mechanisms are present by school-age though relative symmetry between the dominant and non-dominant directions may only be established in adolescence (unpublished observation).

TMS Paired With Sensory Activation

Sensory inputs are integrated with motor control commands at spinal, subcortical, and cortical levels and as a result directly or indirectly influence changes in neuronal excitability and reorganization as assessed using TMS. For instance, single-pulse TMS conditioned by sensory afferent inputs allows measurements of corticospinal excitability changes related to specific sensory inputs that are activated artificially through electrical or mechanical stimulation, or naturally through various motor tasks.[41]

Short-Latency Afferent Inhibition (SAI)

SAI is a TMS paradigm that can be used to investigate sensory–motor integration. In this paradigm, a peripheral nerve that is stimulated prior to the activation of the motor cortex reduces the size of the motor-evoked potentials (MEPs) elicited by TMS.[42] SAI requires a minimum ISI that is close to the latency of the N20 component of a somatosensory-evoked potential, and lasts for about 7–8 ms.[42] The pathway mediating SAI is considered to be of cortical origin based on evidence of peripheral nerve electrical stimulation 19 ms prior to TMS of the motor cortex suppressed responses evoked by TMS but not by TES. Similarly, epidural recordings from the cervical epidural space of five patients during TMS over the motor cortex confirmed that the most prominent effect of SAI was on the I2- and I3-waves, whilst the D- and I1-waves were not affected by SAI.[42]

Sensory input can also facilitate on-going EMG activity through transcortical reflex pathways.[43] Using TMS, it was possible to show that MEPs evoked in the flexor pollicis longus muscle were facilitated if they were evoked during the period of the long-latency stretch reflex and not if they were evoked during the short-latency stretch reflex, even when the sizes of the two reflex components were approximately equal.[44] Furthermore, TMS at an intensity that was below threshold for evoking MEPs during the short-latency reflex period could produce a response if given within the long-latency period. Similar effects were not present when low-intensity TES was used instead of TMS.[44] Overall the above findings suggest that the long-latency component of the stretch reflex is mediated, in part, through transcortical pathways and can increase the excitability of the motor cortex.

Paired Associative Stimulation (PAS)

An experimental paradigm called paired associative stimulation (PAS), which involves repetitive activation of sensory inputs to the motor cortex paired with TMS, can produce long-term changes in the excitability of the motor cortex[45] that can last for several hours. The TMS pulse is given at the same time that the sensory afferent inputs that are activated by the peripheral nerve stimulation arrive at the motor cortex. The effect of PAS on MEP size was found to be dependent on the timing of the TMS pulse with respect to the afferent stimulation, which is consistent with the spike timing-dependent plasticity (STDP) paradigm observed in reduced animal experiments (see Chapter 3: Neuroplasticity Protocols: Inducing and Measuring Change, for more detailed discussion).

In short, if the presynaptic neuron is activated during the depolarizing phase of the action potential in the postsynaptic neuron, long-term potentiation (LTP) between the pre- and postsynaptic neuron is induced due to the high calcium level in the postsynaptic neuron.[46] On the other hand, if a presynaptic neuron is activated during the afterhyperpolarization of the postsynaptic neuron (ie, during low calcium levels), then long-term depression (LTD) is induced. Similarly, during PAS with an ISI of −10 ms (PAS10), which ensures the afferent volley arrives at primary motor cortex (M1) after the TMS and likely during the afterhyperpolarization of the CST neuron, LTD-type effects were induced in the motor cortex as reflected in reduced MEPs.[45] On the other hand, when an ISI of 25 ms (PAS 25) was used, which ensures that afferent inputs arrive at the motor cortex when the CST neurons are depolarized, long-term potentiation (LTP-type) effects were induced as evidenced by increases in MEP responses.

When using an excitatory PAS intervention (25 ms ISI) for 30 min (90 pairings at 0.05 Hz), MEPs can be enhanced for 60–120 min after the intervention.[45] This effect can be blocked with the administration of dextromethorphan, an NMDA receptor agonist. The site of PAS-induced plasticity is cortical given that MEPs evoked by electrical cervicomedullary stimulation remained unchanged after PAS. Furthermore, epidural recording of CST activity evoked by TMS showed that PAS changed the later I-waves.[47]

The only study that investigated PAS in children of age 6–18[48] demonstrated that PAS effects are present and reproducible in children. While the effect lasted for more than an hour, the paradigm appeared to be safe and tolerable in the pediatric population.

Repetitive TMS

TMS was originally conceived as a brain-mapping tool to complement information gained from functional magnetic resonance imaging (fMRI) and electroencephalography (EEG). One reason why TMS has been important in neuroscience research is that magnetic stimulation provides a non-invasive approach for activating different regions of the brain. In comparison to fMRI or EEG, which measure changes in activity during particular tasks, TMS can be used to study the direct causal association between activated neurons in the brain and the task. For instance, TMS can be used to deactivate particular regions in the brain to investigate the role of each region involved in the task. For example, subjects asked to memorize and repeat a list of words would likely show increased activity in the prefrontal cortex using fMRI. This increased activity would provide an indirect association between the prefrontal cortex and the task. However, if stimuli from TMS over the prefrontal cortex were found to obstruct the ability to learn and recall the list, then researchers would have more convincing evidence to support the involvement of the prefrontal cortex in short-term memory. Here we can see that by combining TMS with other measurement techniques we can greatly amplify the power of TMS in neuroscience research.

The use of TMS has also been found to induce reorganization in the brain. When nerve cells are conditioned by neighboring neurons, they can form functional circuits in the brain. In particular, it has been found that TMS can be used to alter the circuitry in the brain following repetitive trains of magnetic stimuli. Using low-frequency TMS, the efficiency of intracellular links can be diminished in a process called long-term depression (LTD). High frequency stimulations from rTMS produce the opposite effect known as long-term potentiation (LTP). LTP mechanisms have been found to improve a cell's ability to communicate with other cells, such that changes in the plasticity of the brain are likely involved in memory formation and learning. For this reason, the prospect of using magnetic brain stimulation to alter neural circuitry offers great promise for improving learning, treating neurological disorders, or even just maintaining cortical function as people age. TMS is currently under study as a treatment for psychological disorders including depression, auditory hallucinations, and migraine headaches. The use of magnetic brain stimulation is particularly interesting as it may provide a viable treatment for certain aspects of drug-resistant mental illnesses and may serve as an alternative to electroshock therapy.

TMS RELIABILITY

TMS has appreciable intersubject reliability. Wolf and colleagues[49] evaluated the reliability of several TMS-related parameters contributing to the motor cortical maps of hand muscles of nine healthy participants across three sessions. Their results indicated TMS-related parameters are reliable within participants across the three sessions. Similarly, SRCs obtained using TMS were also tested and were found to be quite reliable[50] with an intraclass coefficient of 0.81. The amplitude and latency of MEPs were also found to be reliable with ICC greater than 0.5.[51] The reliability was the highest when MEPs were recorded during static 40% MVC contraction (ICC=0.8). As compared with other parameters, the silent period showed low reliability.[51]

TMS SAFETY

With thousands of stimulators currently in use, the general understanding of single-pulse stimulators is that they are safe. These devices have yet to induce seizures in healthy subjects, but a few have been reported in patients when used in repetitive paradigms. The high frequency of rTMS provides a much stronger effect on the brain and is known to induce seizures, nausea, arm jerking, pain, and transient loss of vision. To help alleviate these problems, safe intensity limits using rTMS are suggested to help reduce the risk of seizures and any other discomfort. Mild headaches are occasionally reported using both techniques and are likely caused by activation of scalp and neck muscles. Again, the general consensus of TMS is that it is safe; however, operators and subjects should remain mindful of the safe limits of TMS to further minimize the risks involved in stimulating the brain.

All paradigms of TMS described above have been tested to be safe in children of ages 6–18 years old. Single- and double-pulse TMS have also been tested and proved to be safe in children as young as 1-year old (unpublished observation).

References

1. Merton PA, Morton HB. Stimulation of the cerebral cortex in the intact human subject. *Nature.* 1980;285(5762):227. http://dx.doi.org/10.1038/285227a0.
2. Hovey C, Jalinous R. *The Guide to Magnetic Stimulation.* 2008.
3. Sommer M, Alfaro A, Rummel M, et al. Half sine, monophasic and biphasic transcranial magnetic stimulation of the human motor cortex. *Clin Neurophysiol Off J Int Fed Clin Neurophysiol.* 2006;117(4):838–844. http://dx.doi.org/10.1016/j.clinph.2005.10.029.
4. Schmid UD, Walker G, Hess CW, Schmid J. Magnetic and electrical stimulation of cervical motor roots: technique, site and mechanisms of excitation. *J Neurol Neurosurg Psychiatry.* 1990;53(9):770–777.
5. Epstein CM. TMS stimulation coils. In: *Oxford Handbook of transcranial stimulation].* Oxford University Press; 2008.
6. Di Lazzaro V, Ziemann U, Lemon RN. State of the art: physiology of transcranial motor cortex stimulation. *Brain Stimul.* 2008;1(4):345–362. http://dx.doi.org/10.1016/j.brs.2008.07.004.
7. Rushton WA. The effect upon the threshold for nervous excitation of the length of nerve exposed, and the angle between current and nerve. *J Physiol.* 1927;63(4):357–377.
8. Patton HD, Amassian VE. Single and multiple-unit analysis of cortical stage of pyramidal tract activation. *J Neurophysiol.* 1954;17(4):345–363.
9. Day BL, Thompson PD, Dick JP, Nakashima K, Marsden CD. Different sites of action of electrical and magnetic stimulation of the human brain. *Neurosci Lett.* 1987;75(1):101–106.
10. Day BL, Dressler D, Maertens de Noordhout A, et al. Electric and magnetic stimulation of human motor cortex: surface EMG and single motor unit responses. *J Physiol.* 1989;412:449–473.

11. Ropper AH, Samuels MA, Klein JP. *Adams and Victor's Principles of Neurology*. 20th ed. New York: McGraw-Hill Education; 2014.
12. Dum RP, Strick PL. The origin of corticospinal projections from the premotor areas in the frontal lobe. *J Neurosci Off J Soc Neurosci*. 1991;11(3):667–689.
13. Lemon RN. Descending pathways in motor control. *Annu Rev Neurosci*. 2008;31:195–218. http://dx.doi.org/10.1146/annurev.neuro.31.060407.125547.
14. Weber M, Eisen AA. Magnetic stimulation of the central and peripheral nervous systems. *Muscle Nerve*. 2002;25(2):160–175.
15. Illert M, Lundberg A, Tanaka R. Integration in descending motor pathways controlling the forelimb in the cat. 1. Pyramidal effects on motoneurones. *Exp Brain Res*. 1976;26(5):509–519.
16. Colebatch JG, Rothwell JC, Day BL, Thompson PD, Marsden CD. Cortical outflow to proximal arm muscles in man. *Brain J Neurol*. 1990;113(Pt 6):1843–1856.
17. Nielsen J, Petersen N. Changes in the effect of magnetic brain stimulation accompanying voluntary dynamic contraction in man. *J Physiol*. 1995;484(Pt 3):777–789.
18. Courtine G, Bunge MB, Fawcett JW, et al. Can experiments in nonhuman primates expedite the translation of treatments for spinal cord injury in humans? *Nat Med*. 2007;13(5):561–566. http://dx.doi.org/10.1038/nm1595.
19. Nielsen J, Kagamihara Y. The regulation of presynaptic inhibition during co-contraction of antagonistic muscles in man. *J Physiol*. 1993;464:575–593.
20. Capaday C, Lavoie BA, Barbeau H, Schneider C, Bonnard M. Studies on the corticospinal control of human walking. I. Responses to focal transcranial magnetic stimulation of the motor cortex. *J Neurophysiol*. 1999;81(1):129–139.
21. Siebner HR, Peller M, Takano B, Conrad B. New insights into brain function by combination of transcranial magnetic stimulation and functional brain mapping. *Nervenarzt*. 2001;72(4):320–326.
22. Strafella AP, Paus T. Cerebral blood-flow changes induced by paired-pulse transcranial magnetic stimulation of the primary motor cortex. *J Neurophysiol*. 2001;85(6):2624–2629.
23. Fisher KM, Zaaimi B, Baker SN. Reticular formation responses to magnetic brain stimulation of primary motor cortex. *J Physiol*. 2012;590(Pt 16):4045–4060. http://dx.doi.org/10.1113/jphysiol.2011.226209.
24. Kawai N, Nagao S. Origins and conducting pathways of motor evoked potentials elicited by transcranial magnetic stimulation in cats. *Neurosurgery*. 1992;31(3):520–526. discussion 526–527.
25. Rothwell JC, Thompson PD, Day BL, et al. Motor cortex stimulation in intact man. 1. General characteristics of EMG responses in different muscles. *Brain J Neurol*. 1987;110(Pt 5): 1173–1190.
26. Mazzocchio R, Rothwell JC, Day BL, Thompson PD. Effect of tonic voluntary activity on the excitability of human motor cortex. *J Physiol*. 1994;474(2):261–267.
27. Groppa S, Oliviero A, Eisen A, et al. A practical guide to diagnostic transcranial magnetic stimulation: report of an IFCN committee. *Clin Neurophysiol*. 2012;123(1872–8952 (Electronic)):858–882.
28. Rossini PM, Barker AT, Berardelli A, et al. Non-invasive electrical and magnetic stimulation of the brain, spinal cord and roots: basic principles and procedures for routine clinical application. Report of an IFCN committee. *Electroencephalogr Clin Neurophysiol*. 1994;91(2):79–92.
29. Ridding MC, Rothwell JC. Is there a future for therapeutic use of transcranial magnetic stimulation? *Nat Rev Neurosci*. 2007;8(7):559–567. http://dx.doi.org/10.1038/nrn2169.
30. Fuhr P, Agostino R, Hallett M. Spinal motor neuron excitability during the silent period after cortical stimulation. *Electroencephalogr Clin Neurophysiol*. 1991;81(4):257–262.
31. Shimizu T, Hino T, Komori T, Hirai S. Loss of the muscle silent period evoked by transcranial magnetic stimulation of the motor cortex in patients with cervical cord lesions. *Neurosci Lett*. 2000;286(3):199–202.

32. Chen R, Yung D, Li JY. Organization of ipsilateral excitatory and inhibitory pathways in the human motor cortex. *J Neurophysiol.* 2003;89(0022–3077 (Print)):1256–1264.
33. Kujirai T, Caramia MD, Rothwell JC, et al. Corticocortical inhibition in human motor cortex. *JPhysiol.* 1993;471:501–519.
34. Herwig U, Bräuer K, Connemann B, Spitzer M, Schönfeldt-Lecuona C. Intracortical excitability is modulated by a norepinephrine-reuptake inhibitor as measured with paired-pulse transcranial magnetic stimulation. *Psychopharmacol Berl.* 2002;164(2):228–232. http://dx.doi.org/10.1007/s00213-002-1206-z.
35. Ziemann U, Tergau F, Wassermann EM, Wischer S, Hildebrandt J, Paulus W. Demonstration of facilitatory I wave interaction in the human motor cortex by paired transcranial magnetic stimulation. *J Physiol.* 1998;511(Pt 1):181–190.
36. Reis J, Swayne OB, Vandermeeren Y, et al. Contribution of transcranial magnetic stimulation to the understanding of cortical mechanisms involved in motor control. *J Physiol.* 2008;586(2):325–351. http://dx.doi.org/10.1113/jphysiol.2007.144824.
37. Di Lazzaro V, Oliviero A, Profice P, et al. Direct recordings of descending volleys after transcranial magnetic and electric motor cortex stimulation in conscious humans. *Electroencephalogr Clin Neurophysiol Suppl.* 1999;51:120–126.
38. Wassermann EM, Samii A, Mercuri B, et al. Responses to paired transcranial magnetic stimuli in resting, active, and recently activated muscles. *Exp Brain Res.* 1996;109(1):158–163. http://dx.doi.org/10.1007/BF00228638.
39. Ferbert A, Priori A, Rothwell JC, Day BL, Colebatch JG, Marsden CD. Interhemispheric inhibition of the human motor cortex. *J Physiol.* 1992;453(0022–3751 (Print)):525–546.
40. Morishita T, Kubota S, Hirano M, Funase K. Different modulation of short- and long-latency interhemispheric inhibition from active to resting primary motor cortex during a fine-motor manipulation task. *Physiol Rep.* 2014;2(10). http://dx.doi.org/10.14814/phy2.12170.
41. Baker SN, Olivier E, Lemon RN. Task-related variation in corticospinal output evoked by transcranial magnetic stimulation in the macaque monkey. *J Physiol.* 1995;488(Pt 3):795–801.
42. Tokimura H, Di Lazzaro V, Tokimura Y, et al. Short latency inhibition of human hand motor cortex by somatosensory input from the hand. *J Physiol.* 2000;523(Pt 2):503–513. http://dx.doi.org/10.1111/j.1469-7793.2000.t01-1-00503.x.
43. Christensen LO, Petersen N, Andersen JB, Sinkjaer T, Nielsen JB. Evidence for transcortical reflex pathways in the lower limb of man. *Prog Neurobiol.* 2000;62(3):251–272.
44. Day BL, Riescher H, Struppler A, Rothwell JC, Marsden CD. Changes in the response to magnetic and electrical stimulation of the motor cortex following muscle stretch in man. *J Physiol.* 1991;433:41–57.
45. Stefan K, Kunesch E, Benecke R, Cohen LG, Classen J. Mechanisms of enhancement of human motor cortex excitability induced by interventional paired associative stimulation. *J Physiol.* 2002;543(0022–3751 (Print)):699–708.
46. Markram H, Gerstner W, Sjöström PJ. A History of spike-timing-dependent plasticity. *Front Synaptic Neurosci.* 2011;3. http://dx.doi.org/10.3389/fnsyn.2011.00004.
47. Di Lazzaro V, Dileone M, Profice P, et al. LTD-like plasticity induced by paired associative stimulation: direct evidence in humans. *Exp Brain Res.* 2009;194(4):661–664. http://dx.doi.org/10.1007/s00221-009-1774-9.
48. Damji O, Keess J, Kirton A. Evaluating developmental motor plasticity with paired afferent stimulation. *Dev Med Child Neurol.* January 2015. http://dx.doi.org/10.1111/dmcn.12704.
49. Corneal SF, Butler AJ, Wolf SL. Intra- and intersubject reliability of abductor pollicis brevis muscle motor map characteristics with transcranial magnetic stimulation. *Arch Phys Med Rehabil.* 2005;86(8):1670–1675. http://dx.doi.org/10.1016/j.apmr.2004.12.039.
50. Carroll TJ, Riek S, Carson RG. Reliability of the input-output properties of the corticospinal pathway obtained from transcranial magnetic and electrical stimulation. *J Neurosci Methods.* 2001;112(2):193–202.
51. van Hedel HJA, Murer C, Dietz V, Curt A. The amplitude of lower leg motor evoked potentials is a reliable measure when controlled for torque and motor task. *J Neurol.* 2007;254(8):1089–1098. http://dx.doi.org/10.1007/s00415-006-0493-4.

2

Assessing Normal Developmental Neurobiology With Brain Stimulation

K. Limburg, N.H. Jung, V. Mall

Technische Universität München, Kinderzentrum München gemeinnützige GmbH, Munich, Germany

OUTLINE

INTRODUCTION

Although brain stimulation is mostly established in adults, it is an emerging tool in pediatric neurology that is considered safe.[1] Despite its current research application in the context of various disorders with childhood onset, like attention deficit hyperactivity disorder (ADHD), autism spectrum disorder, and cerebral palsy, it could potentially be useful to assess normal developmental neurobiology and thus differentiate it from pathological development. Brain development in children is currently assessed in the context of a childhood preventive medical examination by observing behavior and motor skills.[2] Augmenting this procedure by the use of transcranial magnetic stimulation (TMS) would be extremely useful to also gain neurophysiologic parameters as indicators of normal or pathological neurobiology.

This chapter aims to first give an overview of maturational changes to the brain and central nervous system as well as the course of neuromotor development. Second, it discusses published data on TMS measures which quantify aspects of normal development. Third, it discusses these neurophysiological measures in three heterogeneous conditions affecting development (cerebral palsy, ADHD, and autism spectrum disorders) and in one group of disorders with a common molecular pathway – the RASopathies. Finally, based on these descriptions, an outlook toward future applications is discussed.

NORMAL DEVELOPMENTAL NEUROBIOLOGY

Theoretical Background of Excessive Production of Synapses/ Pruning of Synaptic Connectivity

Childhood is a phase in life characterized by an enormous amount of new and previously unknown experiences.[3] Newborn infants explore a whole new world with unfamiliar situations, requirements, and experiences; thus, they have to adapt to a great amount of new and interesting situations, which are demanding on the developing mind. This requires the acquisition of new skills and thus constant learning. Next to learning vital skills it is commonly known that children are faster at learning new languages or how to play musical instruments; in contrast, it is often harder for adults to learn these skills.[4] This difference between children and adults in the ability to learn is explainable on a neurological level. During early childhood, there is an ongoing production of new synapses, synaptic connections are pruned and, dependent on activity, redefined later during childhood and adolescence.[4–6] It has been found out early that postnatal synapse production occurs most rapidly at the age between 2 and 4 months and begins in the visual cortex, followed by an elimination of nearly half of the synapses beginning at approximately 8 months.[4,7] Synaptogenesis in other brain regions appears at different points in time, for example, synaptic density in frontal areas of the brain has been shown to reach its maximum after the age of 15 months.[8] After this overproduction of synapses, activity-dependent pruning of synaptic connections occurs.[4,9–12] During this process, the previously produced excess of synapses is useful to allow an experience-based selection of important connections and programmed cell death of neurons that are no longer used.[4] In line with clinical findings, which show that cognitive abilities like judgment develop later, the number of synapses in frontal cortical areas remains higher until early adult age.[4] A more recent study replicated findings of extended production of synapses and pruning of synaptic connections in early childhood. In addition, the authors extended the already-existing knowledge by explaining that the production and refining of synaptic spines continues until about 30 years of age.[11]

Synaptic plasticity, the activity-dependent change in the strength of synaptic signal transmission rate, is a mechanism of brain plasticity that plays an important role during pruning of synaptic connections. It can lead either to an increase or decrease in the strength of synaptic connectivity, the former referred to as long-term potentiation (LTP), the latter as long-term depression (LTD).[13] Both LTP and LTD can be experimentally induced by various forms of transcranial magnetic stimulation.[14] There is evidence that plasticity is enhanced during development,[4,15,16] possibly due to reduced intracortical GABAergic inhibition in children compared to adults.[17] For example, Eyre et al.[18] studied a sample of neonates and found evidence that corticospinal axonal connections are withdrawn depending on activity during development, with the most noticeable changes occurring during the first 18 months of life.

Structural Changes to the Developing Brain

Thanks to the development of new imaging techniques, like functional magnetic resonance imaging (fMRI) or diffusion tensor imaging (DTI), it is possible to evaluate structural changes to the brain. For example, it has been possible to investigate developmental changes to the corpus callosum, a structure that connects the two hemispheres of the brain and thus allows exchange and integration of information between the hemispheres as well as inhibition of cortical functions.[19] Evidence exists that this structure first grows in size in the early postnatal period by addition of fibers. From about 6 months of age, myelination of fibers becomes increasingly important.[19] Myelination of axons in the white matter of the brain mainly serves to enhance conductivity and isolation of axons, thus improving the quality and velocity of information transmission. However, important tracts of the central nervous system are already operating before myelination is completed. Besides faster transmission, another aim of myelination is to allow more precise processing of information and integration of inputs from other cortical areas.[20] The ongoing integrity of the white matter connections within the brain is important for cognitive development. For example, Qiu et al.[21] reviewed and summarized findings from DTI studies indicating that language acquisition seems to progress along with ongoing myelination; also, an increasing working memory capacity may be associated with myelination progression and individual differences in cognitive performance may indicate the stage of overall brain development. In addition to cognitive skills, fine motor skills also develop in the same stage as structural and functional maturation of the motor cortex, the corpus callosum, and the myelination of the white matter. Accordingly, the observably progressing acquisition of motor skills seems to reflect neural correlates of motor development.[22] Taken together, these findings indicate that myelination is the most important precondition to learning in the developing brain as synaptic plasticity can only occur in myelinated axons.

Different Aspects of Neuromotor Development

TMS is a technique that is most commonly used to examine corticomotor pathways.[23] Hence, in order to use TMS to probe motor cortex function and acquisition of motor skills during development, it is important to gain a general understanding of the developmental course of motor skills during maturation. Along with the aforementioned structural changes to the developing brain, motor skill acquisition happens rapidly and extensively.[22] Thus, different developmental states need to be considered when applying TMS in children at various ages and interpreting the results.

Due to their rapid development, motor skills are constantly assessed, starting in the newborn or even prenatally.[2] General movements, one category of spontaneous movements occurring during fetal and early postnatal life, involve the whole body in variable sequences and appear to be complex and variable. If those movements appear less fluent and smooth it can be an early indicator of impairment to the nervous system.[24] Rather basic gross motor skills like grasping and holding are developing during the first year of life and further skills like, for example, throwing a ball, climbing steps, or using cutlery are added constantly during the following months.[2] Fine motor skills start to improve later and develop most rapidly during early school years with the acquisition of skills like writing and drawing. Hand dexterity continuously improves until mid to late adolescence in a rather subtle way, with movements becoming finer, faster, and more coordinated. Associated movements, like mirror movements that appear with complex motor tasks, are most prominent in children under the age of 10 years and continuously decrease in their intensity and frequency during puberty, although they can still be observed during adult life in relation to very complex tasks.[22]

Handedness or hand preference, an indicator of cortical asymmetry, is first observable as early as about 9 months of age with right-handers showing their preference earlier and more distinctly. The dominant hand is determined in right- and left-handers at about 3 years of age, in some cases later, at about 5 or 6 years.[2]

TMS-EVOKED PARAMETERS THAT REFLECT NEUROMOTOR MATURATION

After discussing structural changes to the central nervous system and clinical changes in motor skills, the following paragraphs will summarize central neurophysiological parameters that can be measured with TMS and that reflect the above-described changes. In this context, the procedures of assessment, as well as typical developmental changes in the TMS-evoked parameters, will be explained.

Age Specificity of Resting Motor Threshold

The resting motor threshold (RMT) refers to the TMS stimulus intensity at which motor-evoked potentials (MEPs) of a defined minimum peak-to-peak amplitude can be elicited when the target muscle is relaxed. Although different methods to estimate the motor threshold exist, the guidelines of the International Federation of Clinical Neurophysiology recommend the "adaptive stair-case procedure" that uses a maximum-likelihood procedure to predict the TMS intensity that yields a 50% probability of eliciting an MEP.[23] MEP thresholds can be interpreted as a measure of normal electrophysiological maturation as they reflect the stage of myelination as well as synaptic efficacy.[25] It has been demonstrated that the resting motor threshold decreases with age[25] and reaches adult-like levels in early adolescence.[18] This finding can be interpreted as reflecting ongoing myelination during maturation, leading to increased cortical excitability. Interestingly, the threshold remains higher in the hemisphere of the non-dominant hand, demonstrating a tendency for laterality in hand performance on a cortical level.[25]

Maturational Aspects of MEP Latencies and CMCT

The MEP latency refers to the time period between the application of a magnetic stimulus with TMS and the onset of the MEP[26] and is commonly used to estimate the central motor conduction time (CMCT) as a marker for corticospinal integrity.[27] CMCT can be assessed after measuring two estimates of motor latency: First, the shortest peripheral motor latency (PML) is estimated after magnetic stimulation of the proximal spinal nerve at the level of the intervertebral foramen; second, the shortest corticomotor latency (CML) is assessed after motor cortex magnetic stimulation. Finally, the CMCT is calculated as the difference between PML and CML.[23]

Several studies have demonstrated that CMCT decreases with age, reaching adult-like levels at the beginning of early adolescence.[26,28,29] Possible reasons for shorter CMCTs with increasing age may include growth processes of the corticospinal axons, such as myelination, and developing central motoneuronal recruitment.[22,26,30] Interestingly, the facilitated CMCT (measured when the target muscle is contracted) has been shown to mature earlier than the relaxed CMCT (measured when the target muscle is relaxed), the former becoming stable at about 5 years of age, the latter at about 10 years of age. This so-called "latency jump" between relaxed and facilitated CMCT, that is known to be four times greater in preschool children than in adults,[31] may indicate an earlier maturation of excitability at the spinal level versus a later maturation of excitability at the cortical level (see Fig. 2.1).[26] The findings should be interpreted while keeping in mind that TMS initiates an indirect activation of the target muscle via interneurons connected to the pyramidal tract. At higher levels of muscle activation, faster neurons of the pyramidal tract are activated which may lead to a shorter facilitated CMCT.[22,26,32]

Ipsilateral Versus Contralateral Projections – MEP Amplitudes (Contra-vs. Ipsilateral)

Amplitudes of MEPs can be interpreted as another marker for corticospinal integrity.[27] The size of an MEP amplitude is measured by calculating the voltage difference between the maximal negative and the maximal positive deflection (peak-to-peak amplitude). Scientific TMS studies usually record various MEP traces and calculate the average MEP size. For diagnostic purposes however, it is sufficient to record five to six

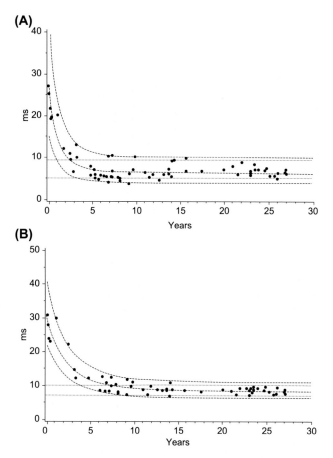

FIGURE 2.1   Latency jump – the difference between the CMCT facilitated (A) and relaxed (B) decreases with age. *Figure taken from: Fietzek U, Heinen F, Berweck S. et al. Development of the corticospinal system and hand motor function: central conduction times and motor performance tests.* Dev Med Child Neurol. 2000;42(4):220–227.

consecutive MEP traces and choose the MEP with the largest amplitude for analysis, in order to reflect optimal corticomotor conduction.[23] The amplitude of an MEP increases with a higher stimulus intensity. Noticeably, the increase in MEP amplitudes with increasing stimulus intensities has been shown to be steeper in adolescents and adults than in children.[31] This finding is in line with the decreasing MEP threshold which occurs during maturation, reflecting increased cortical excitability with age.

Although MEPs are usually recorded from the contralateral muscle due to the contralateral organization of the motor cortex, ipsilateral motor-evoked potentials (iMEPs) of upper extremity muscles have been recorded in children and adults after TMS. This indicates the presence of an ipsilateral corticofugal motor projection that controls ipsilateral movements in healthy subjects although the occurrence of ipsilateral responses is more rare in adults.[18] Ipsilateral connections are especially prominent in children with brain lesions, demonstrating the ability of the developing brain for reorganization or adaptation to changed conditions.[33]

Because ipsilateral connections have been found to be more prominent in younger children[29] it appears valid to assume that these connections may be the reason for developmental mirror movements that commonly occur in young children and decrease during development.[34] The term refers to unintended movements of the opposite side of the body during voluntary activity of one side. However, there is evidence that suggests simultaneous activation of crossed corticospinal pathways descending from both motor cortices and the inability of the developing active motor cortex to suppress activity of the contralateral motor cortex during unimanual movements. Consequently, the decrease in mirror movements with age may reflect an increasing transcallosal inhibitory influence during development rather than the decrease in ipsilateral connections.[31,35]

Intracortical Inhibition and Connectivity Between Brain Areas – Silent Period, Intracortical Inhibition Paradigms, Interhemispheric Inhibition

As mentioned above, there is evidence for increasing intracortical inhibition during development, explaining the need for a method to study the inhibitory functions of the cortex. There are two methods that have been used in children for this purpose, the paired-pulse paradigm and the investigation of silent periods.

Using the short-interval intracortical inhibition (SICI) paradigm, one form of the paired-pulse paradigm, it has been found that cortical inhibition increases with age, being approximately four times greater in adults compared to children less than 10 years of age.[17] Utilizing an interstimulus interval (ISI) of 2 ms, that has previously been found to approximate the GABA-mediated ICI that leads to a decrease in MEP amplitude,[36] the authors explained their findings with the maturation process of GABAergic interneurons. GABA is known to be an important inhibitory neurotransmitter in human motor cortex,[37] thus reduced intracortical GABAergic inhibition during early stages of life may serve to enhance plasticity and the capacity of motor learning, which is especially important for development.[36]

The second paradigm to study intracortical inhibition involves investigation of the ipsilateral silent period (iSP), a term that refers to the reduction of voluntarily produced electromyographic activity (EMG) in the muscles of the ipsilateral hand after suprathreshold stimulation of the motor cortex.[38] This phenomenon is known as interhemispheric or transcallosal inhibition and

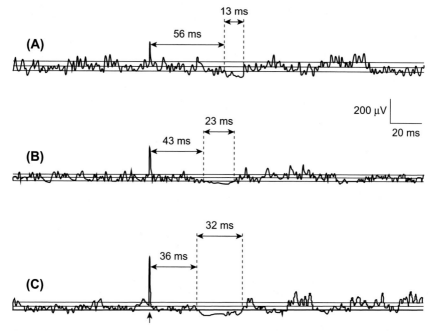

FIGURE 2.2 Development of the ipsilateral silent period (iSP). With increasing age, the duration of the iSP increases while its onset-latency decreases. (A) displays EMG tracings of a 7-year-old boy; (B) of a 12-year-old girl; and (C) of a 22-year-old man. *Taken from Garvey M, Ziemann U, Bartko J, Denckla M, Barker C, Wassermann E. Cortical correlates of neuromotor development in healthy children.* Clin Neurophysiol. 2003;114(9):1662–1670.

could not be detected in preschool children up to the age of 5 years.[39] In a study of children, adolescents, and adults, aged 6–26 years, an iSP was detected in all age groups. With increasing age, the duration of the iSP increased while its onset-latency decreased; there were fewer iSPs in the hand which showed more mirror movements (see Fig. 2.2).[25] Taken together, these findings point at an increasing interhemispheric inhibition being responsible for decreasing presence of mirror movements during development, although the iSP reflects only one of many aspects of interhemispheric inhibition.[25]

Finally, there is the cortical silent period, evoked over active motor cortex with a supra-active-threshold stimulus. Similar to the ipsilateral silent period, this is a postexcitatory silent period in the TMS-activated muscle. The duration is linearly correlated to the stimulus intensity, reflecting a suppression of muscle activity after a TMS-induced MEP and thus constituting another marker of cortical inhibitory interneuron activity.[27,40]

Synaptic Plasticity

Synaptic plasticity is the underlying neurophysiological correlate of learning and thus constitutes a central ability of the brain, its ability for reorganization and adaptation in response to changing environmental demands. TMS can be used to measure and understand synaptic plasticity by inducing it via different protocols, e.g., theta burst stimulation[41] or quadripulse stimulation protocols.[42] These protocols are subsumed under the term repetitive TMS (rTMS) and have in common that a train of TMS stimuli is applied to one area of the primary motor cortex; after a set period of time following this plasticity-inducing train, the cortical or corticospinal response to single pulses of TMS is measured by investigating the MEPs elicited by the target muscle. An average increase in MEP amplitudes is interpreted as long-term potentiation (LTP)-like plasticity, a decrease hints at the induction of long-term depression (LTD)-like plasticity.[41,43] Another TMS protocol that can be used to induce synaptic plasticity is paired associative stimulation[43] (PAS). This protocol is somewhat different from rTMS protocols because it consists of pairs of stimuli that combine peripheral electric stimulation, usually of the left median nerve at the wrist, with TMS of the contralateral hemisphere.[44]

Using paired-associative stimulation (PAS), an increase in MEP amplitude after the stimulation protocol has been observed in healthy children,[43] other studies using this protocol have been conducted in adults and also resulted in an increased MEP amplitude up to about 30 min after stimulation.[44,45] As mentioned previously, synaptic plasticity is enhanced in typically developing children compared to adults.[5] Indeed, using continuous theta burst stimulation (cTBS) protocols it has been shown that the extent of modulation of corticospinal excitability is inversely correlated with age, hinting at a declining capacity for corticomotor plasticity across the lifespan.[46,47] While the safety of the use of TBS in children and adolescents has been demonstrated,[1] quadripulse stimulation protocols have not yet been used in children but do offer a promising tool to induce stable and long-lasting plasticity.[48]

ASSESSING NEUROPHYSIOLOGIC FUNCTIONING WITH BRAIN STIMULATION IN DEVELOPMENT AND DEVELOPMENTAL DISORDERS

In the following paragraphs we are integrate the findings of TMS-evoked parameters that change during developmental neurobiology stated above with current knowledge about parameters that are altered in specific neurodevelopmental disorders. Consequently, it will become clear how to use TMS as a tool to differentiate between normal and pathological neurodevelopment. However, it is important to note that TMS is at present most frequently used for research purposes rather than as a diagnostic tool.

Cerebral Palsy

Cerebral palsy (CP) refers to a developmental disorder that occurs after a lesion to the developing fetal or infant brain. Its most prominent symptoms are permanent disorders of movement and posture as well as disordered sensation, perception, cognition, communication, and behavior.[49] Vry et al.[27] conducted a study of children with cerebral palsy after bilateral periventricular leukomalacia (PVL) (damage to the white matter of the brain caused by low oxygen levels) compared to healthy children. The authors found that corticospinal integrity did not appear to be impaired in bilateral cerebral palsy as CMCT and MEP amplitudes did not significantly differ between healthy children and patients with CP. However, there was a trend towards a reduction of MEP amplitudes in the CP patient group. The authors interpret their findings of a normal CMCT and a trend towards reduced MEP amplitudes as an indicator that the pathogenetic mechanism of CP may rather be a loss of axons instead of demyelination of the corticospinal tract. As the main finding of their study, the authors revealed a shorter postexcitatory silent period in the children with bilateral CP, pointing at impaired intercortical inhibition. This may be due to cortical dysfunction caused by inflammatory mechanisms during the occurrence of PVL in the immature brain which leads to an altered function of GABAergic interneurons.[27]

In patients with congenital unilateral CP after three different types of prenatal brain lesions, Staudt et al.[33] assessed sensorimotor reorganization and found both patients with preserved crossed corticospinal projections from the lesioned hemisphere and patients with ipsilateral corticospinal projections from the contralesional hemisphere that controlled the paretic hand. This finding suggests that ipsilateral connections can be maintained after lesions of the developing brain so that the paretic side of the body can still be controlled by the ipsilateral corticospinal projection, proving the ability of the infant brain for reorganization in response to lesions.[18,33] Nevertheless, sensorimotor organization in children with retained ipsilateral connections appears to differ from those with preserved crossed projections, as the two groups have been shown to respond differently to constraint-induced movement therapy (CIMT), a therapeutic approach to train the paretic hand. Although the quality of hand function improved in both groups, the children with ipsilateral connections needed more time for the given tasks, whereas the group with contralateral projections showed a time reduction, thus pointing out that children with ipsilateral projections respond differently to CIMT.[50]

In a more recent study, Juenger et al.[51] examined the same sample as above and measured hand function as well as neuroplastic effects via single-pulse TMS, fMRI, and magnetoencephalography before and after CIMT. They also found a difference between the two groups: the patients with ipsilateral connections showed reduced trans-synaptic excitability in the primary motor cortex detected via TMS and decreased synaptic activity during movements of the paretic hands detected via fMRI. In contrast, both parameters increased in patients with contralateral projections.

Attention Deficit Hyperactivity Disorder (ADHD)

ADHD is a neurodevelopmental disorder characterized by inattention as well as hyperactive, restless behavior and the inability to suppress impulses.[52] Corresponding to the impaired ability to inhibit off-task behavior, the iSP latency in boys with ADHD did not decrease with age as it did in healthy controls, pointing out an abnormality in interhemispheric interactions in ADHD.[53] Under medication with methylphenidate, a drug that is currently used to improve hyperkinetic behavior, the iSP duration has been shown to increase while the iSP latency decreased and clinical symptoms improved, further supporting a disturbed interhemispheric inhibition as an underlying neural correlate of ADHD.[54] Furthermore, in a study of healthy adults it has been shown that methylphenidate influenced intracortical excitability by leading to an increased intracortical inhibition and intracortical facilitation. This finding was unexpected as it had been assumed that a drug can either enhance intracortical inhibition while reducing intracortical facilitation, or vice versa. The authors explained their findings by stating that the drug may act on the motor systems through other neurotransmitter systems than only the dopaminergic one; for example, the enhanced intracortical facilitation may be explainable by noradrenergic effects.[55] Of note, the quality of the motor anomalies in ADHD is similar to associated movements in normally developing children but the movements are more prominent and persist up to a higher age.[22] Further evidence for the relation between reduced cortical inhibition and ADHD symptom severity comes from a paired-pulse study in children with Tourette's syndrome and ADHD symptoms. Cortical disinhibition was more strongly related to severity of ADHD symptoms than to tic severity. Thus, there may be distinct neuronal abnormalities between the two disorders which cannot be identified using TMS alone.[56]

Autism Spectrum Disorders (ASDs)

ASDs are severe neurodevelopmental disorders characterized by deficits in social communication and interaction, as well as restricted and often repetitive patterns of behavior, frequently accompanied by stereotyped movements. ASD can be diagnosed in patients with or without accompanying intellectual impairment.[52] A study in patients ages 15–29 years with ASDs in comparison to healthy adult subjects demonstrated that whereas in typically developing children MEP amplitudes after application of a

PAS protocol increased, the patients with ASD did not show a significant increase in their corticospinal responses. This finding hints at an impaired capacity for excitatory synaptic plasticity and connectivity and reduced sensory-motor integration in autistic patients.[43] In contrast, Oberman et al.[41] used cTBS in 20 adults with ASD and found enhanced, longer-lasting cortical depression, suggesting increased inhibitory synaptic plasticity. Using iTBS, a stimulation pattern that has been shown to facilitate the MEP, in an LTP-like manner, the authors found enhanced LTP-like changes.[41,57] These LTP- and LTD-like protocols were repeated in a crossover design on the second day. In the ASD adults, the reaction was completely blocked on the second day of the experiment.[41] Jung et al.[43] explained these contradictory findings with differences between the stimulation principles. In PAS, a stimulation protocol where a peripheral electrical stimulus is followed by TMS of the primary motor cortex, the integrity of the sensorimotor system may play an important role. This may be different for TBS, a form of direct motor cortex stimulation. Taken together, both studies demonstrate an altered synaptic plasticity in patients with autism that can be investigated by the use of TMS. In addition, Oberman et al.[58] recently applied cTBS in a sample of high-functioning boys with ASD aged 9–18. They found that the duration of modulation of cTBS after-effects became longer with increasing age. Because cTBS has been found to model LTD-like plasticity, and this is thought to be mediated by GABA, the results may be indicative of an increased LTD-like plasticity or GABAergic inhibition during adolescence,[58] also going in line with previously introduced findings of a decrease in synaptic plasticity from child to adulthood.

Both Oberman et al.[41] and Jung et al.[43] also measured short-interval intracortical inhibition (SICI) using the paired-pulse protocol. While Oberman et al.[41] demonstrated a heterogeneity with one of five individuals with autism showing a reduced SICI, Jung et al.[43] did not find differences between the samples regarding SICI. This heterogeneity is currently not explainable on a neurophysiological level but the observation points out that patients with ASD do show interindividual differences and similarities.

RASopathies

The term "RASopathies" describes a group of nine developmental syndromes (capillary malformation–AV malformation syndrome, autoimmune lymphoproliferative syndrome, cardiofaciocutaneous syndrome (CFC syndrome), hereditary gingival fibromatosis type 1, neurofibromatosis type 1, Noonan syndrome, Costello syndrome, Legius syndrome, and LEOPARD syndrome) that are caused by germline mutations in genes that alter the rat sarcoma (RAS)/mitogen-activated protein kinase (MAPK) pathway, a pathway that controls various intracellular forms of signal transduction, e.g. those responsible for axon growth. Most of these mutations lead to an increased signal transduction along the RAS/MAPK pathway.[59] Neurofibromatosis type 1 (NF1) is one of these syndromes characterized by an RAS pathway hyperactivity caused by a mutation of the neurofibromin gene. Clinical characteristics of NF1 include cutaneous manifestations (café-au-lait spots), fibromas (benign tumors of the peripheral nervous system), and secondary bone abnormalities like scoliosis, long bone dysplasia, and osteoporosis. In addition, cognitive impairments like learning disabilities or ADHD occur in about 50–60% of patients with NF1.[60,61]

Using a mouse model, it has been shown that the NF1 mutation leads to increased GABA release in the hippocampus, and thus to decreased LTP-like synaptic plasticity.[62] This observation was replicated and extended in a placebo-controlled, double-blind, randomized trial of adult patients with NF1 and healthy controls by Mainberger et al.[63] Using paired associative stimulation, the authors revealed a diminished LTP-like synaptic plasticity in NF1 patients compared to controls; accordingly, they found increased intracortical inhibition in patients, demonstrated by an increased SICI in the paired-pulse paradigm. Attentional performance was lower in patients compared to healthy controls. After recording these baseline data, the authors examined the effect of 200-mg lovastatin for 4 days on these parameters. Lovastatin is a potent inhibitor of RAS/MAPK activity that has been shown to improve LTP deficits, as well as spatial learning and attention impairments in mice.[64] Similarly, in the human sample the drug lead to an improvement in all these parameters, demonstrated by decreased SICI and thus increased LTP-like plasticity after PAS (see Fig. 2.3), and better attentional performance in patients that received the drug compared to the placebo patient group.[63]

Another syndrome that is caused by an alteration in the RAS/MASKP pathway is the Noonan syndrome, a disorder caused by a mutation in the gene *PTPN11* and characterized by a short stature, facial anomalies, a range of heart defects, neck and chest deformity, usually mild cognitive deficits, developmental delays, and learning disabilities.[65] Next to patients with NF1, Mainberger et al.[66] also studied patients with Noonan syndrome using paired associative stimulation and found an impaired synaptic plasticity in these patients compared to healthy controls, even during maximal specific attention control.

OUTLOOK

After introducing the changes that occur in TMS-evoked parameters in patients with neurodevelopmental disorders, the final section will address important emerging and future applications of TMS. For instance, the

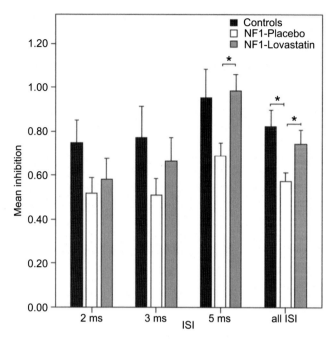

FIGURE 2.3 Lovastatin leads to an improvement in TMS-evoked parameters (SICI and LTP-like plasticity). *Asterisks* display significant effects between groups ($p < 0.05$) and *error bars* represent standard errors of the mean. *Taken from Mainberger F, Jung NH, Zenker M, et al. Lovastatin improves impaired synaptic plasticity and phasic alertness in patients with neurofibromatosis type 1. BMC Neurol. 2013;13(1):131.*

method of TMS is starting to be used in fields like neurorehabilitation or even brain surgery. The following paragraphs will briefly introduce some of these new applications of TMS. Although most research concerning the various applications of TMS has been conducted in adult samples, the method can potentially be equally useful for pediatric patients.

Cortical Maps/Presurgical Mapping/Virtual Lesion Technique

Despite the fact that different methods of navigation of TMS exist,[67,68] the most accurate approach that takes into account individual anatomical differences is neuronavigated or image-guided TMS. This method implies monitoring the position of the freely movable head in relation to the equally movable coil while visualizing the stimulated brain area on a computer screen. Therefore, the brain coordinates as investigated in MRI scans need to be merged in a common reference system together with the head and coil coordinates recorded by a camera system. This allows determination and visualization of the ideal coil position in real time, thus improving the spatial accuracy of the method.[68,69]

An emerging and very useful clinical application of image-guided TMS is presurgical mapping, e.g., motor or expressive language mapping. This approach allows identifying areas engaged in important cortical functions like speech production or motor function in preparation for brain surgery with the aim of avoiding damaging eloquent cortical areas.[69–71] For example, expressive language mapping is conducted preoperatively by delivering bursts of rTMS while a patient performs a task like reading, object naming, or counting; meanwhile, the stimulus, the response, and the cortical location of TMS are monitored. The map is then compiled by comparing the patient's response during TMS with their baseline performance and checking for errors or disrupted speech. The brain areas where TMS resulted in errors are marked on the patient's MRI. These areas are considered to be crucial for expressive language and thus resecting these regions will be avoided during brain surgery.[69] Thus, preoperative TMS mapping is considered to be a useful and technically feasible tool[72] before brain surgery, in addition to intraoperative monitoring through direct electrical stimulation. In a case study on three patients with recurring left-side gliomas that underwent repeated resections, preoperative language mapping by rTMS was compared to intraoperative mapping via direct cortical stimulation (DCS). DCS is performed after craniotomy by directly stimulating cortical sites while the patient carries out an object-naming task; sites where errors occur during the process are marked. While the results of initial TMS and DCS during the first resection showed good correlations, the authors revealed that TMS mapping is less sensitive to plastic changes in language function after the first surgery and prior to the second DCS. Thus, although it is a useful tool for presurgical mapping, TMS is not yet accurate enough to be able to replace intraoperative monitoring via direct electrical stimulation. The authors explain the need for more reliable and accurate rTMS protocols in order to be able to assess stability or plasticity in brain function before surgery.[73]

Next to presurgical language mapping, there is evidence that preoperative motor mapping can result in a lower rate of residual tumor and better postoperative motor function in patients that receive preoperative TMS.[70] Thus, it may also depend on the type of lesion, brain function, as well as number of surgeries and therefore the extent of possibly ongoing plastic changes, whether TMS should be regarded as a reliable source of information before brain surgery. In addition, it is important to know that, to date, TMS can be used pre- and postoperatively while only direct electrical stimulation can be applied intraoperatively.[74]

TMS and Other Neuroimaging Methods

As it becomes clear through the image-guided use, TMS is especially useful in combination with other functional neuroimaging technologies.

Next to the combination of TMS with fMRI, studies have applied TMS together with EEG. For example, Premoli et al.[75] used the two methods to study the modulatory influence of inhibitory GABAergic neurotransmission on cortical excitability as reflected by TMS-evoked EEG potentials in a drug trial. They revealed a differential influence of two different GABA receptors on EEG potentials, thus offering more detailed insights into the role of CNS-active drugs in the human motor cortex then what can be gained with TMS or EEG alone.[75]

Another TMS-EEG study investigated patients with drug-resistant frontal-lobe epilepsy and revealed that TMS allows interrupting epileptiform discharges as shown by a change in EEG records. Although TMS cannot yet be applied as a replacement for deep brain stimulation in epilepsy for a number of reasons (e.g. effects are not long-lasting enough, TMS can only be applied superficially, etc.) it may at least help to predict responsiveness to invasive stimulation techniques.[76] Furthermore, the combination of various methods has been proven to be useful in children with congenital hemiparesis as shown recently by Juenger et al.[51] and introduced above in the context of cerebral palsy.

Neuromodulation/Therapeutic Applications of TMS

Besides its use in diagnostics and cortical mapping, TMS is also an emerging tool in therapy in the context of various disorders. For example, Sokhadze et al.[77] applied neuromodulation based on low-frequency rTMS in pediatric patients with autism and found better executive functioning after the intervention, demonstrated by decreased hyperactivity and less stereotypic behaviors, as well as an improvement on TMS-evoked parameters, a higher accuracy of responses, and a slower post-error reaction time. Furthermore, in a case study of four children with low-functioning autism, Panerai et al.[78] applied high-frequency rTMS and an eye–hand integration training. The children showed improvements in eye–hand performance after TMS over the left premotor cortex that lasted up to 1 h after the stimulation. Outcomes were even better when a combination of TMS and eye–hand integration training was delivered.

Next to studies in pediatric autism samples, several authors have investigated the effects of neuromodulation in adults. For example, Butts et al.[79] found that a combination of intermittent theta burst stimulation (iTBS), transcranial direct current stimulation (tDCS), and motor training lead to better hand function that lasted more than 7 days after the intervention in the TMS group compared to the group that solely received motor training. Similar results were observed by Neva et al.[80] after comparing an intervention of inhibitory continuous theta burst stimulation (cTBS) to the healthy motor cortex combined with bimanual movement training to the training alone. Moreover, the authors even found an increased volume of the cortical map of the affected muscle. Taken together, these results suggest that TMS may serve as a promising tool to accompany motor training in the context of various disorders that involve an impairment of motor functions.

References

1. Hong YH, Wu SW, Pedapati EV, et al. Safety and tolerability of theta burst stimulation versus single and paired pulse transcranial magnetic stimulation: a comparative study of 165 pediatric subjects. *Front Hum Neurosci*. 2015;9:29.
2. Baumann T. *Atlas der Entwicklungsdiagnostik: Vorsorgeuntersuchungen von U1 bis U10/J1*. vol. 3. Georg Thieme Verlag; 2013.
3. Thompson RA. Development in the first years of life. *Future Child*. 2001:21–33.
4. Johnston MV. Plasticity in the developing brain: implications for rehabilitation. *Dev Disabil Res Rev*. 2009;15(2):94–101.
5. Johnston MV. Clinical disorders of brain plasticity. *Brain Dev*. 2004;26(2):73–80.
6. Kolb B, Mychasiuk R, Gibb R. Brain development, experience, and behavior. *Pediatr Blood Cancer*. 2014;61(10):1720–1723.
7. Huttenlocher PR, De Courten C. The development of synapses in striate cortex of man. *Hum Neurobiol*. 1986;6(1):1–9.
8. Huttenlocher PR, Dabholkar AS. Regional differences in synaptogenesis in human cerebral cortex. *J Comp Neurol*. 1997;387(2):167–178.
9. Huttenlocher PR. Synaptic density in human frontal cortex-developmental changes and effects of aging. *Brain Res*. 1979;163(2):195–205.
10. Changeux JP, Danchin A. Selective stabilisation of developing synapses as a mechanism for the specification of neuronal networks. *Nature*. 1976;264(5588):705–712.
11. Petanjek Z, Judaš M, Šimić G, et al. Extraordinary neoteny of synaptic spines in the human prefrontal cortex. *Proc Natl Acad Sci*. 2011;108(32):13281–13286.
12. Lüthi A, Schwyzer L, Mateos JM, Gähwiler BH, McKinney RA. NMDA receptor activation limits the number of synaptic connections during hippocampal development. *Nat Neurosci*. 2001;4(11):1102–1107.
13. Abraham WC. Metaplasticity: tuning synapses and networks for plasticity. *Nat Rev Neurosci*. 2008;9(5):387.
14. Lang N, Siebner HR. Repetitive transkranielle magnetstimulation. In: *Das TMS-Buch*. Springer; 2007:499–511.
15. Eyre J. Development and plasticity of the corticospinal system in man. *Neural Plast*. 2003;10(1–2):93.
16. Crair MC, Malenka RC. A critical period for long-term potentiation at thalamocortical synapses. *Nature*. 1995;375(6529):325–328.
17. Mall V, Berweck S, Fietzek U, et al. Low level of intracortical inhibition in children shown by transcranial magnetic stimulation. *Neuropediatrics*. 2004;35(2):120–125.
18. Eyre J, Taylor J, Villagra F, Smith M, Miller S. Evidence of activity-dependent withdrawal of corticospinal projections during human development. *Neurology*. 2001;57(9):1543–1554.
19. Fabri M, Pierpaoli C, Barbaresi P, Polonara G. Functional topography of the corpus callosum investigated by DTI and fMRI. *World J Radiol*. 2014;6(12):895.
20. Michaelis R, Niemann GW. *Entwicklungsneurologie und Neuropädiatrie: Grundlagen und diagnostische Strategien*. Georg Thieme Verlag; 2010.
21. Qiu A, Mori S, Miller MI. Diffusion tensor imaging for understanding brain development in early life. *Annu Rev Psychol*. 2015;66:853–876.
22. Garvey MA, Mall V. Transcranial magnetic stimulation in children. *Clin Neurophysiol*. 2008;119(5):973–984.

23. Groppa S, Oliviero A, Eisen A, et al. A practical guide to diagnostic transcranial magnetic stimulation: report of an IFCN committee. *Clin Neurophysiol.* 2012;123(5):858–882.
24. Einspieler C, Prechtl HF. Prechtl's assessment of general movements: a diagnostic tool for the functional assessment of the young nervous system. *Ment Retard Dev Disabil Res Rev.* 2005;11(1):61–67.
25. Garvey M, Ziemann U, Bartko J, Denckla M, Barker C, Wassermann E. Cortical correlates of neuromotor development in healthy children. *Clin Neurophysiol.* 2003;114(9):1662–1670.
26. Fietzek U, Heinen F, Berweck S, et al. Development of the corticospinal system and hand motor function: central conduction times and motor performance tests. *Dev Med Child Neurol.* 2000;42(4):220–227.
27. Vry J, Linder-Lucht M, Berweck S, et al. Altered cortical inhibitory function in children with spastic diplegia: a TMS study. *Exp Brain Res.* 2008;186(4):611–618.
28. Koh T, Eyre J. Maturation of corticospinal tracts assessed by electromagnetic stimulation of the motor cortex. *Arch Dis Child.* 1988;63(11):1347–1352.
29. Müller K, Kass–Iliyya F, Reitz M. Ontogeny of ipsilateral corticospinal projections: a developmental study with transcranial magnetic stimulation. *Ann Neurol.* 1997;42(5):705–711.
30. Caramia M, Desiato M, Cicinelli P, Iani C, Rossini P. Latency jump of "relaxed" versus "contracted" motor evoked potentials as a marker of cortico-spinal maturation. *Electroencephalogr Clin Neurophysiol/Evoked Potentials Sect.* 1993;89(1):61–66.
31. Garvey MA, Gilbert DL. Transcranial magnetic stimulation in children. *Eur J Paediatr Neurol.* 2004;8(1):7–19.
32. Abbruzzese G, Trompetto C. Clinical and research methods for evaluating cortical excitability. *J Clin Neurophysiol.* 2002;19(4):307–321.
33. Staudt M, Gerloff C, Grodd W, Holthausen H, Niemann G, Krägeloh–Mann I. Reorganization in congenital hemiparesis acquired at different gestational ages. *Ann Neurol.* 2004;56(6):854–863.
34. Koerte I, Eftimov L, Laubender RP, et al. Mirror movements in healthy humans across the lifespan: effects of development and ageing. *Dev Med Child Neurol.* 2010;52(12): 1106–1112.
35. Mayston MJ, Harrison LM, Stephens JA. A neurophysiological study of mirror movements in adults and children. *Ann Neurol.* 1999;45:583–594.
36. Ziemann U, Muellbacher W, Hallett M, Cohen LG. Modulation of practice-dependent plasticity in human motor cortex. *Brain.* 2001;124(6):1171–1181.
37. McCormick DA. GABA as an inhibitory neurotransmitter in human cerebral cortex. *J Neurophysiol.* 1989;62(5):1018–1027.
38. Heinen F, Brodbeck V. Besonderheiten im Kindes-und Jugendalter. In: *Das TMS-Buch.* Springer; 2007:139–146.
39. Heinen F, Glocker FX, Fietzek U, Meyer BU, Lücking CH, Korinthenberg R. Absence of transcallosal inhibition following focal mangnetic stimulation in preschool children. *Ann Neurol.* 1998;43(5):608–612.
40. Terao Y, Ugawa Y. Basic mechanisms of TMS. *J Clin Neurophysiol.* 2002;19(4):322–343.
41. Oberman L, Ifert-Miller F, Najib U, et al. Transcranial magnetic stimulation provides means to assess cortical plasticity and excitability in humans with fragile X syndrome and autism spectrum disorder. *Front Synaptic Neurosci.* 2010;2.
42. Hamada M, Terao Y, Hanajima R, et al. Bidirectional long–term motor cortical plasticity and metaplasticity induced by quadripulse transcranial magnetic stimulation. *J Physiol.* 2008;586(16):3927–3947.
43. Jung NH, Janzarik WG, Delvendahl I, et al. Impaired induction of long–term potentiation–like plasticity in patients with high–functioning autism and Asperger syndrome. *Dev Med Child Neurol.* 2013;55(1):83–89.
44. Stefan K, Kunesch E, Cohen LG, Benecke R, Classen J. Induction of plasticity in the human motor cortex by paired associative stimulation. *Brain.* 2000;123(3):572–584.

45. Battaglia F, Quartarone A, Rizzo V, et al. Early impairment of synaptic plasticity in patients with Down's syndrome. *Neurobiol Aging*. 2008;29(8):1272–1275.
46. Freitas C, Perez J, Knobel M, et al. Changes in cortical plasticity across the lifespan. *Front Aging Neurosci*. 2011;3.
47. Pascual-Leone A, Freitas C, Oberman L, et al. Characterizing brain cortical plasticity and network dynamics across the age-span in health and disease with TMS-EEG and TMS-fMRI. *Brain Topogr*. 2011;24(3–4):302–315.
48. Nakatani-Enomoto S, Hanajima R, Hamada M, et al. Bidirectional modulation of sensory cortical excitability by quadripulse transcranial magnetic stimulation (QPS) in humans. *Clin Neurophysiol*. 2012;123(7):1415–1421.
49. Rosenbaum P, Paneth N, Leviton A, et al. A report: the definition and classification of cerebral palsy April 2006. *Dev Med Child Neurol Suppl*. 2007;109(suppl 109):8–14.
50. Kuhnke N, Juenger H, Walther M, Berweck S, Mall V, Staudt M. Do patients with congenital hemiparesis and ipsilateral corticospinal projections respond differently to constraint–induced movement therapy? *Dev Med Child Neurol*. 2008;50(12):898–903.
51. Juenger H, Kuhnke N, Braun C, et al. Two types of exercise–induced neuroplasticity in congenital hemiparesis: a transcranial magnetic stimulation, functional MRI, and magnetoencephalography study. *Dev Med Child Neurol*. 2013;55(10):941–951.
52. Association AP. *Diagnostic and statistical manual of mental disorders, (DSM-5®)*. American Psychiatric Pub; 2013.
53. Garvey MA, Barker CA, Bartko JJ, et al. The ipsilateral silent period in boys with attention-deficit/hyperactivity disorder. *Clin Neurophysiol*. 2005;116(8):1889–1896.
54. Buchmann J, Gierow W, Weber S, et al. Modulation of transcallosally mediated motor inhibition in children with attention deficit hyperactivity disorder (ADHD) by medication with methylphenidate (MPH). *Neurosci Lett*. 2006;405(1):14–18.
55. Kirschner J, Moll G, Fietzek U, et al. Methylphenidate enhances both intracortical inhibition and facilitation in healthy adults. *Pharmacopsychiatry*. 2003;36(2):79–82.
56. Gilbert DL, Bansal AS, Sethuraman G, et al. Association of cortical disinhibition with tic, ADHD, and OCD severity in Tourette syndrome. *Mov Disord*. 2004;19(4):416–425.
57. Oberman L, Eldaief M, Fecteau S, Ifert-Miller F, Tormos JM, Pascual-Leone A. Abnormal modulation of corticospinal excitability in adults with Asperger's syndrome. *Eur J Neurosci*. 2012;36(6):2782–2788.
58. Oberman LM, Pascual-Leone A, Rotenberg A. Modulation of corticospinal excitability by transcranial magnetic stimulation in children and adolescents with autism spectrum disorder. *Front Hum Neurosci*. 2014;8.
59. Tidyman WE, Rauen KA. The RASopathies: developmental syndromes of Ras/MAPK pathway dysregulation. *Curr Opin Genet Dev*. 2009;19(3):230–236.
60. Boyd KP, Korf BR, Theos A. Neurofibromatosis type 1. *J Am Acad Dermatol*. 2009;61(1):1–14.
61. Muntau A. *Intensivkurs Pädiatrie*. Elsevier, Urban & Fischer; 2011.
62. Cui Y, Costa RM, Murphy GG, et al. Neurofibromin regulation of ERK signaling modulates GABA release and learning. *Cell*. 2008;135(3):549–560.
63. Mainberger F, Jung NH, Zenker M, et al. Lovastatin improves impaired synaptic plasticity and phasic alertness in patients with neurofibromatosis type 1. *BMC Neurol*. 2013;13(1):131.
64. Li W, Cui Y, Kushner SA, et al. The HMG-CoA reductase inhibitor lovastatin reverses the learning and attention deficits in a mouse model of neurofibromatosis type 1. *Curr Biol*. 2005;15(21):1961–1967.
65. Allanson JE. Noonan syndrome. In: *Paper Presented at: American Journal of medical Genetics Part C: Seminars in Medical Genetics*; 2007.
66. Mainberger F, Zenker M, Jung NH, et al. Impaired motor cortex plasticity in patients with Noonan syndrome. *Clin Neurophysiol*. 2013;124(12):2439–2444.

67. Herwig U, Satrapi P, Schönfeldt-Lecuona C. Using the international 10-20 EEG system for positioning of transcranial magnetic stimulation. *Brain Topogr.* 2003;16(2):95–99.
68. Herwig U, Schönfeldt-Lecuona C, Wunderlich AP, et al. The navigation of transcranial magnetic stimulation. *Psychiatry Res Neuroimaging.* 2001;108(2):123–131.
69. Narayana S, Papanicolaou AC, McGregor A, Boop FA, Wheless JW. Clinical applications of transcranial magnetic stimulation in pediatric neurology. *J Child Neurol.* 2014;30(9):1111–1124.
70. Krieg SM, Sabih J, Bulubasova L, et al. Preoperative motor mapping by navigated transcranial magnetic brain stimulation improves outcome for motor eloquent lesions. *Neuro-Oncology.* 2014;16(9):1274–1282.
71. Ottenhausen M, Krieg SM, Meyer B, Ringel F. Functional preoperative and intraoperative mapping and monitoring: increasing safety and efficacy in glioma surgery. *Neurosurg Focus.* 2015;38(1):E3.
72. Choudhri AF, Narayana S, Rezaie R, et al. Same day tri-modality functional brain mapping prior to resection of a lesion involving eloquent cortex: technical feasibility. *Neuroradiol J.* 2013;26(5):548.
73. Krieg SM, Sollmann N, Hauck T, Ille S, Meyer B, Ringel F. Repeated mapping of cortical language sites by preoperative navigated transcranial magnetic stimulation compared to repeated intraoperative DCS mapping in awake craniotomy. *BMC Neurosci.* 2014;15(1):20.
74. Takahashi S, Picht T. Comparison of navigated transcranial magnetic stimulation to direct electrical stimulation for mapping the motor cortex prior to brain tumor resection. *Tumors of the Central nervous system.* vol. 12. Springer; 2014:261–276.
75. Premoli I, Castellanos N, Rivolta D, et al. TMS-EEG signatures of GABAergic neurotransmission in the human cortex. *J Neurosci.* 2014;34(16):5603–5612.
76. Kimiskidis VK, Kugiumtzis D, Papagiannopoulos S, Vlaikidis N. Transcranial magnetic stimulation (TMS) modulates epileptiform discharges in patients with frontal lobe epilepsy: a preliminary EEG-TMS study. *Int J Neural Syst.* 2013;23(01).
77. Sokhadze EM, El-Baz AS, Sears LL, Opris I, Casanova MF. rTMS neuromodulation improves electrocortical functional measures of information processing and behavioral responses in autism. *Front Syst Neurosci.* 2014;8.
78. Panerai S, Tasca D, Lanuzza B, et al. Effects of repetitive transcranial magnetic stimulation in performing eye-hand integration tasks: four preliminary studies with children showing low-functioning autism. *Autism.* August 2014;18(6):638–650.
79. Butts RJ, Kolar MB, Newman-Norlund RD. Enhanced motor skill acquisition in the nondominant upper extremity using intermittent theta burst stimulation and transcranial direct current stimulation. *Front Hum Neurosci.* 2014;8.
80. Neva JL, Singh AM, Vesia M, Staines WR. Selective modulation of left primary motor cortex excitability after continuous theta burst stimulation to right primary motor cortex and bimanual training. *Behav Brain Res.* 2014;269:138–146.

Neuroplasticity Protocols: Inducing and Measuring Change

S.W. Wu, E.V. Pedapati

Cincinnati Children's Hospital Medical Center, Cincinnati, OH, United States

OUTLINE

BACKGROUND (IN VITRO AND ANIMAL WORK)

Over the past decade, the term neuroplasticity has been increasingly used both inside and outside the scientific community, even as far as to be applied as a marketing synonym for "brain enhancement." Even within the community of neuroscientists, the term neuroplasticity can be ambiguous with widely different meanings depending on the academic subspecialty.[1] Though there is no mutually agreed upon definition, the term neuroplasticity can be broadly defined as the ability of the nervous system to reorganize in response to intrinsic or environmental demands and underlies the conceptual framework of learning, memory, and development.[2,3]

As far back as the 19th century, there is evidence in the scientific literature that neuroscientists were discussing the possibility of a dynamic and adaptive nervous system. In his 1898 paper entitled "Regeneration of Nerve Fibers in the Central Nervous System," the physician and pathologist W.L. Worcester described the unexpected recovery of facial movement and sensation in an institutionalized woman who had previously suffered two massive strokes.[4] Worcester presented postmortem histopathology of the patient's brain at the 1898 American Neurological Association where he displayed "outgrowths from neurons" within the heavily infarcted tissue which he termed a "curiosity." Ernest Lindley, a psychology fellow at Clark University, wrote in a 1897 manuscript detailing how children solve puzzles: "the whole nervous system of organisms is a differentiation of tissues with the supreme function of preserving the results of former adjustments and the effecting of new ones (pp. 474)"[5] – a statement which closely reflects a contemporary perspective on learning and memory.

The technological advancements of the 20th century have allowed for an unprecedented investigation into the molecular and physiological underpinnings of neuroplastic mechanisms. The preponderance of evidence today suggests that, although genetic and early environmental factors dictate the potential scope of brain development, neuroplastic processes play a crucial role throughout life in configuring and optimizing neural circuits, including the maturation of complex sensory, cognitive, and regulatory functions.[6] Since direct human studies of inducible plasticity remain limited and preliminary, the majority of the evidence has been drawn from molecular, animal, and in vitro studies.[7]

Hebbian/Konorski Theories of Plasticity

The Canadian neurophysiologist Donald Hebb theorized that the optimization and pruning of connections during development and learning was a fundamental principle of the operation of neuronal networks.[8] His well-known phrase "neurons that fire together wire together" would become known as Hebb's Postulate. Hebb extrapolated even further, suggesting that non-genetic information was encoded within the number and strength of synaptic connections within specific configurations of neurons and supporting cells. Along with Hebb, a contemporary physiologist, Jerzy Konorski proposed that pre-existing connections between neurons could transform into new patterns of functional connectivity with repetitive associations which would lead to morphological changes in neuronal structures.[3] Though the Hebb/Konorski view is now widely accepted in neuroscience, at the time, both scientists had little empirical evidence to support their views.

Despite a great interest in uncovering the fundamental mechanisms of learning and memory, progress was limited in humans due to the immense complexity of the brain as well as due to ethical concerns. Thus, it is not surprising that the earliest evidence for Hebb's theories came from relatively simple animal models. Eric Kandel of Columbia University has spent nearly a half century developing a research program around the nervous system of the sea slug *Aplysia*.[9] Whereas the mammalian brain has approximately a billion neurons, *Aplysia* contains 20,000 neurons, many of which are visible by the naked eye and well-suited for experimental manipulation. Kandel and his colleagues found that only approximately 100 of these neurons were involved in learning and memory. Kandel demonstrated Hebbian learning, tied to reflexive behaviors, within simple neuronal circuits that could be easily quantified. The significance of this work was recognized in 2000, when Kandel and his collaborators were awarded the Nobel Prize in Medicine for key discoveries of the physiological and molecular substrates of short- and long-term memory.[10]

Mammalian Brain and Hippocampus

Work on the mammalian brain was focused on the hippocampus which would (and continues to) take a prominent role in experimental neuroscience as clinical and observational studies uncovered the importance of this region in memory, learning, and navigation.[11,12] The hippocampus is a bilateral cashew-shaped structure that can be readily dissected and often considered an anatomical elaboration of the edge of the medial temporal lobes in vertebrates. Early anatomists used the term "cornu Ammonis (CA; horn of Amun)" to described the horn-shaped appearance of the hippocampus proper when viewed in three-dimensional space. The acronym, CA, continues to be used to describe the four major subgroups of cells (CA1–CA4) within the hippocampus. The term "hippocampal formation" has been loosely used to encompass the dentate gyrus (including CA4), fields CA1–CA3 (hippocampus proper), and the subiculum (also known as the parahippocampal gyrus).

When the hippocampal formation is freshly sectioned in the coronal plane, a three-synapse neuronal circuit can be prepared for electrophysiological recording. Neocortical axonal projections of the entorhinal cortex (so-called "perforant path (PP)") synapse among the dendrites of the granule cells of the dentate gyrus (synapse 1). Axons from the granule cells, known as "mossy fibers (MF)" for their characteristic elaborate appearance, form major inputs to the pyramidal cells of CA3 (synapse 2). Information from the CA3 region leaves via Schaffer collaterals forming excitatory glutamatergic synapses with pyramidal cells of CA1 (synapse 3).

At the 1966 Scandinavian Physiological Society, Terje Lomo, a physician and doctoral student at the University of Oslo, presented recordings from rabbit hippocampus that demonstrated a marked increase in the amplitude and number of population spikes within the dentate gyrus following high-frequency stimulation to the perforant path.[13] Lomo and his mentor, Per Anderson, hypothesized that the changes they observed were "too short" to account for learning processes, but "if frequency potentiation takes place in a set of neurons constituting a poly-synaptic chain, the individual effects may be greatly enhanced." Indeed, in a landmark publication in 1973, Lomo and his colleague, Tim Bliss, presented evidence of increased synaptic efficacy of granule cells that lasted up to 10h after stimulation of the perforant path. They termed this effect long-term potentiation (LTP).[14] These data would serve as a key confirmation of several theoretical neuroplastic mechanisms. Though outside the scope of this chapter, the details of this work have been reviewed by the primary research team.[15]

The phenomenon LTP has been extensively confirmed in other mammalian hippocampi including hippocampal slices from humans undergoing temporal lobe surgery.[16,17] From a bird's eye perspective, experimental LTP can be divided into an induction phase and an expression phase. The induction phase consists of a transient event, such as a brief train of stimulation to a local neuronal circuit (encompassing both the pre- and postsynaptic responses). The expression phase refers to the subsequent changes following the original stimulation period, including specific and durable enhancement of synaptic transmission. Such changes may further persist by downstream cellular events which employ mechanisms for continued maintenance of these enhancements.

Theta Burst Stimulation (TBS): A Physiological Approach to LTP Induction

Despite the success of LTP induction in vitro, it was well known through electrical recordings of the brains of live animals that patterns of brain activity, including synchronicity between regions, were much more sophisticated than the experimental stimuli (ie, high-frequency train of stimulation).[18,19] Regions of neurons arranged in parallel fashion, such as in cortical areas, can give rise to large-amplitude extracellular potentials that can be measured by electroencephalography (EEG) as oscillatory patterns. Neuronal oscillations consistent with large-amplitude, slow-wave theta frequency are characteristically present in the mammalian hippocampus during voluntary movements, exploratory behaviors, and states of arousal, but absent during more sedentary behaviors.[18,20]

In an attempt to mimic this complexity, Larson et al. demonstrated that the application of patterned high-frequency bursts (four pulses at 100 Hz) with a duration as short as 200 ms to the Schaffer projections of the CA1 field of rat hippocampal slices resulted in reliable, robust LTP.[21] This so-called "theta burst" stimulation train was developed based primarily on observations of physiological firing patterns naturally occurring within the hippocampus and in contrast with the longer (1–2 s, 100 Hz trains) stimulation trains commonly used experimentally for LTP induction at the time.

Non-human mammalian studies have demonstrated theta oscillations most regular in frequency and largest in amplitude reside within the hippocampal CA1 region, but are also found in the dentate gyrus and CA3 regions of the hippocampal formation.[22] Of interest to learning processes, subpopulations of neurons within these regions fire short bursts of activity in phase with these rhythms. Such "phase lock" neuronal discharges to theta frequency have been identified in other brain structures including the cingulate cortex, amygdala, entorhinal cortex, and perirhinal cortex.[22] Such activity was hypothesized to represent natural induction of LTP through incremental spontaneous activity.

By administering four 100-Hz pulses in a burst, with bursts repeating at 5 Hz, and repeating this sequence 10 times, Larson et al. attempted to replicate the phase locked firing patterns measured in behaving animals. In line with this prediction, burst repetition at frequencies outside of theta frequency, such as 1 or 10 Hz, were less effective for LTP induction.[23] More recent studies have added further complexities, including the observation that robust LTP can be induced by similar burst patterns in the delta frequency range, mimicking low-frequency hippocampal activity present during slow-wave sleep and putatively involved in memory processes.[24] Rather than cast doubt, the ramifications of such findings further reinforce the complex network dynamics involved in plastic processes which can be influenced by other brain activity and with different behavioral states.

Physiological Significance of LTP

The experimental induction of LTP within the hippocampus can be considered a highly significant accomplishment from a technical perspective, but questions of how LTP mechanisms contribute to naturally occurring brain function and development remain unanswered. Despite an incomplete understanding of the underlying electrochemical mechanisms, behavioral and histopathological studies of neonate and juvenile animals have begun to fill the gaps in knowledge of the complicated interplay of neuronal activity, genetic predisposition, and environmental cues to create an array of specialized sensory, motor, and cognitive networks.[6]

Though investigations of LTP at the cellular level in children have obviously been limited, electrophysiological studies of neonatal and juvenile animals have shed light on the purpose and mechanisms of LTP during development.[25,26] During the intrauterine period, neurons complete their migration, resulting in their axons and dendrites spanning local and distant regions of the brain. The assembly of these early circuits and synapses is supported by scaffolding cells and molecular gradients. Although the resulting structures demonstrate a complex anatomical and histological organization, this alone is insufficient for the development of an organism's higher functional capacities. Several additional processes are necessary. There is pruning and remodeling of existing synapses, mediated by activity-dependent strengthening and weakening of connections in response to an organism's interactions with the surrounding environment. In addition, there are certain epochs during maturation, so-called "critical periods," in which experiential input is obligatory for normal development of specialized neural circuits, and after which development is no longer possible.[27]

LTP mechanisms were suspected to play a critical role in this patterning of neuronal connections during early development, but little experimental evidence initially supported this link.[28] To better understand the role of LTP, Kirkwood et al.[29] studied the visual system of rats in which normal development requires simultaneous input from both eyes during a postnatal critical period. They examined the effect of LTP induction on hippocampal slices of light- and dark-reared rats and made a striking observation that susceptibility to LTP coincided with the critical period. Furthermore, by rearing animals in complete darkness, they were able to prolong this susceptibility period. Such findings support the hypothesis that LTP plays a key role in the modification of experience-dependent synapses in the mammalian brain. Later reports have confirmed this finding and further extended the results such that the developmental age in rodents has been associated with varying susceptibility and efficacy of induced LTP in hippocampal slices.[30–33] Although there continues to be a large gap between understanding LTP of hippocampal synapses and understanding behavioral plasticity, this mechanism of synaptic plasticity provides a plausible neural mechanism for long-lasting changes in the brain.

OVERVIEW OF REPETITIVE TRANSCRANIAL MAGNETIC STIMULATION (rTMS)

Not long after the initial development of TMS,[34] rTMS experiments were conducted to study cortical physiology for potential clinical applications.[35–37] The initial reports to systematically document the safety[38] and neurophysiologic effects of rTMS[39] in healthy volunteers were published in the 1990s. This "conventional" rTMS is delivered using specific frequencies and stimulation intensities to produce either facilitatory or inhibitory cortical changes. Due to the potential for seizure induction,[40] two international consensus statements on rTMS use and safety parameters were eventually established.[41,42] The conventional rTMS paradigm subsequently paved the way for other forms of rTMS with resulting cortical changes that outlast the stimulation duration. These include paired associative stimulation (PAS),[43] theta burst stimulation (TBS),[44] repetitive paired-pulse (PPS), and quadripulse stimulation (QPS).[45,46] PAS is unique in that it utilizes the principle of spike-timing-dependent plasticity by repeatedly pairing a cortical TMS pulse with a preceding peripheral electrical stimulation. In contrast, TBS and PPS/QPS are patterned rTMS protocols where all pulses are delivered centrally, to the cerebral or cerebellar cortex.

The following sections will focus on the discussion of neuroplasticity-inducing protocols on the primary motor cortex (M1) of healthy volunteers as most studies have focused on M1. Based on available data, the neurophysiologic effects of rTMS on motor threshold, motor-evoked potential (MEP), cortical inhibition, and facilitation will be discussed. A review of basic single- and paired-pulse TMS measures and underlying mechanism was discussed in Chapter 1, "TMS Basics: Single and Paired Pulse Neurophysiology." Available pediatric rTMS studies will be discussed in this chapter. Disease-specific applications of rTMS are discussed in their respective chapters.

TYPES OF rTMS

Conventional rTMS

Conventional rTMS can been further categorized as low- vs. high-frequency rTMS. Historically this delineation is based on the inhibitory and facilitating effects of low-frequency (≤1 Hz) and high-frequency rTMS (>1 Hz) over M1, respectively. However, this classification is likely an oversimplification as the response to rTMS can be quite variable.[47]

Very Low-Frequency rTMS

Most very low-frequency rTMS (0.1–0.3 Hz) studies have not been shown to have significant effects on MEP over M1[48-51] although a 0.2 Hz study showed a decrease in MEP amplitude.[52] Changes in cortical inhibition have been reported as 0.1 Hz rTMS delivered at 80% of RMT resulted in increases of short- and long-interval cortical inhibition (SICI, LICI)[48] and suprathreshold 0.3 Hz rTMS was reported to prolong the cortical silent period (CSP).[51] Most rTMS studies over M1 use multiple single TMS pulses in this frequency range to characterize the baseline M1 excitability. Therefore, if the purpose of the study is to examine the effects of plasticity protocol(s) on cortical inhibition, then it may be important to assess baseline M1 excitability with <0.1 Hz single TMS pulses.

~One Hz rTMS

There are numerous studies that have examined the effect of ~1 Hz rTMS on M1 physiology. While many studies showed decreases in cortical excitability, these effects are variable. The variability depends on factors such as the total number of pulses and the stimulation intensity. In general, suprathreshold 1 Hz rTMS is more likely to result in cortical excitability changes.[53] Low-frequency rTMS (0.9–1 Hz) has been shown to reduce MEP amplitudes.[49,53-55] One Hz rTMS can also increase M1 motor threshold.[53,54,56] The effects of 1 Hz rTMS on M1 SICI and intracortical facilitation (ICF) are mixed, with studies showing a poststimulation decrease in SICI[57] and ICF.[58] However, ICF has been shown to be increased or decreased after 1 Hz rTMS, depending on stimulation intensity.[59] Likewise the effect on the cortical silent period is also mixed with no change,[60] decrease,[61] or increase[59,62] in cSP duration. The number of pulses used in these studies differed significantly (20–900 pulses) and therefore might explain the variable results. Studies have also examined the effect of 1 Hz rTMS on interhemispheric inhibition (IHI). IHI can be quantified by measuring the ipsilateral silent period (iSP) or paired-pulse (one coil on each side of the brain) IHI (ppIHI). No significant changes on iSP have been found after 1 Hz rTMS.[62,63] However, healthy adult ppIHI studies showed that the interhemispheric inhibition decreased predominantly from the stimulated M1 to non-stimulated M1 following 1 Hz rTMS.[63,64]

Attempts to use 1 Hz rTMS as a therapeutic intervention for children occurred as early as the late 1990s[65,66] but overall remains quite rare. To date, there are two randomized, sham-controlled trials using 1 Hz rTMS. Both studies focused on motor rehabilitation in children and adolescents with perinatal or childhood stroke.[67,68] The significant clinical findings in these studies are discussed further in Chapter 10, "The Right Stimulation of the Right Circuits: Merging Understanding of Brain Stimulation Mechanisms and Systems Neuroscience for Effective Neuromodulation in Children." One of these studies provided detailed neurophysiologic outcomes in 10 pediatric (13.9 ± 4.4 years) stroke patients.[69] In this study, patients received 1200 pulses of 1 Hz rTMS at an intensity of 100% of RMT per day for 8 days. All participants completed the rTMS procedure without seizure induction, although two teens experienced syncope.[70] Motor threshold did not change in either the stroke or contralesional M1 after 8 days of 1 Hz rTMS. After repeated rTMS, there was an increase in ppIHI (interstimulus intervals of 10, 40 ms) from the stroke to the contralesional hemisphere. This change did not reach statistical significance, possibly explained by the small sample size.

There are also a few open label studies using 1 Hz rTMS to reduce tics in Tourette syndrome, suggesting significant benefit.[71–73] Repetitive stimulation (1200 pulses per day for 5–20 days) was delivered over the supplementary motor area (SMA) at intensities of either 100% or 110% of RMT. SMA was targeted because it is an important site thought to generate tics.[74] All pediatric subjects completed these studies without any serious adverse events. M1 motor thresholds were increased after repeated sessions of 1 Hz rTMS over the SMA.

>One Hz rTMS ("High"-Frequency rTMS)

As with 1 Hz rTMS, the poststimulation effects are also variable for higher-frequency rTMS. Although few studies show an increase in cortical excitability with 2 and 3 Hz rTMS,[39,75] most high-frequency protocols employ ≥5 Hz rTMS. At 5 Hz, rTMS generally increases cortical excitability.[39,76–78] The effect on MEP amplitude is also dependent on stimulation intensity[39,78] and number of pulses.[76,78] Motor threshold is not changed after 5 Hz rTMS.[62,78,79] Several studies have shown that 5 Hz rTMS reduces SICI.[77,78,80] The cortical silent period has been prolonged by 5 Hz rTMS.[60,81] ICF has not been reported to change afterwards.[62] Although iSP has not been shown to change after 5 Hz rTMS,[82] one study did show a significant interhemispheric effect after repetitive stimulation at 5 Hz.[83] Here they probed the left M1 with single- and paired-pulse TMS after 5 Hz rTMS to the right M1. The result showed increased MEP in the right first dorsal interosseous (FDI) muscle, but no significant changes to SICI/ICF.

Conventional rTMS has also been delivered at frequencies higher than 5 Hz. Most of these studies used either 10 or 20 Hz with very few at 25 Hz and one study at 50 Hz.[84] Based on safety parameters,[41,42] increasing the frequency is counterbalanced with lower stimulation intensity and shorter duration to avoid serious adverse events such as seizures. In one high-frequency (10 and 20 Hz) rTMS study, motor threshold was not changed,[85] although numerous studies have shown increased MEP amplitudes.[39,47,86,87] However, there were significant variations and some post-rTMS tracing showed inhibition interspersed with facilitated MEPs.[39,75] Again, rTMS effects were dependent on the number of pulses,[47] stimulation intensity,[86] and stimulation train duration.[87] When stimulation frequency is increased to 25 Hz, more consistent MEP facilitation can be seen.[39]

Few studies have addressed the effects of 10 or 20 Hz rTMS on intra-cortical inhibition and facilitation. After 10 Hz rTMS, SICI changes have been variable in opposite directions.[87,88] The differences in stimulation intensity, stimulation train and total pulses may contribute to the difference in the SICI changes. SICI changes after 20 Hz rTMS were variable and depended on baseline measures.[85] CSP duration is increased by both 10 and 20 Hz rTMS.[60,85] ICF was shown to decrease in one 10 Hz rTMS study.[87] Interhemispheric physiology can also be modulated by 10 Hz rTMS as iSP duration has been reported to increase.[82]

There is one published study using 25 Hz rTMS in healthy adult subjects.[89] Sixteen adults received 1500 pulses of rTMS using 12 5-s trains with an intertrain interval of 10 s. All participants received rTMS at 100% of RMT with the exception of one subject with a higher resting threshold. This person received rTMS at 90% of RMT for safety reasons. No adverse events were reported. After rTMS, motor threshold and cSP duration were decreased. There were no significant changes to SICI and ICF.

Clinically, 10 Hz rTMS over the left dorsolateral prefrontal cortex has received FDA clearance to treat adult patients with depression.[90] To date, there are two published pediatric studies using this protocol to treat depression in an open-label design.[91,92] Croarkin et al.[91] recruited eight teen participants (16.1 ± 1.2 years old). After one dropout due to scalp discomfort, seven received 30 sessions of 10 Hz rTMS over the left dorsolateral prefrontal cortex (DLPFC) at a stimulation intensity of 120% of RMT. Stimulation train was 4 s each with an intertrain interval of 26 s. Each teen received 3000 TMS pulses per treatment session. No serious adverse events were reported in this small study. The primary aim of the study was to examine cortical excitability over 30 sessions of rTMS. To this end, the mixed model repeated measures analysis did not find a significant change in left M1 motor threshold. However, a post hoc non-parametric comparison revealed a decrease in RMT at week 5 compared to baseline. Yang et al.[92] recruited six patients (four females; 15–21 years of age) who received 10 Hz rTMS (40 pulses per train, intertrain interval of 26 s; total of 3000 pulses/day; 120% of RMT as stimulation intensity) over left DLPFC for 15 consecutive weekdays. In addition to assessment of clinical response, the investigators also measured left DLPFC glutamate level using MR spectroscopy. After rTMS, four patients had improvements in depressive symptoms (average of 68% reduction in Hamilton Depression Rating Scale score). Corresponding to this symptom reduction was an 11% increase in the left DLPFC glutamate level. No adverse events were reported.

Paired Associative Stimulation (PAS)

The PAS protocol delivers repetitive pairs of peripheral electrical and cortical magnetic stimuli to induce a form of spike-timing-dependent plasticity.[43] This technique involves an initial suprathreshold electrical shock of the median nerve followed by suprathreshold cortical magnetic stimulation of contralateral M1. The variables in PAS include the stimulation intensity, interstimulus interval (ISI), number of stimulation pairs, and frequency of stimulation.[93] One important factor that determines poststimulation facilitation versus inhibition is the duration of ISI.[94] In the initial PAS report,[43] 90 pairs of stimuli were delivered over 30 min with ISI of 25 ms (PAS25), such that the afferent input from the peripheral stimulation and the cortical magnetic pulse arrive simultaneously at sensorimotor cortex. MEP amplitudes subsequently increased on average by 55%, with this effect lasting at least 30 min. RMT did not change after PAS in three subjects. If the ISI is decreased to 10 ms (PAS10), such that the magnetic stimulation over M1 precedes the afferent input from the peripheral stimulation, the post-PAS MEP amplitudes decrease $25 \pm 10\%$, with the effect lasting about 90 min.[95] RMT was also not changed after PAS with ISI of 10 ms. As with other forms of rTMS, the response to excitatory and inhibitory PAS can be variable.[96]

The effects of PAS on intracortical excitability have also been investigated. Several studies found that PAS25 does not alter SICI,[97–100] whereas two other studies showed that PAS10 decreases SICI.[62,100] Two studies showed that PAS25 reduces LICI[100,101] but the effect of PAS10 was inconsistent.[100,102] Duration of cSP has been prolonged by PAS25[43,98,102] but not altered by PAS10.[62,102] PAS25 has not been shown to alter ICF.[62,103,104]

There is currently one published pediatric PAS study.[105] Twenty-eight typically developing children (12.1 ± 3.9 years) participated. The protocol involved 90 pairs of peripheral and cortical stimulations delivered at 0.2 Hz. The ISI was 25 ms (PAS25) and the TMS intensity was percentage of maximal stimulator output required to produce on average a ~1 mV MEP. Some participants reported adverse events – neck pain, headache, nausea, and discomfort from peripheral nerve stimulation. However, no serious adverse event was reported. Post-PAS25 cortical excitability was assessed at four time points (15, 30, 45, and 75 min). The authors stratified PAS25 response based on how many time points exhibited significant MEP increase (>33%) above baseline. Those who had this increase in at least two time points were categorized as "definite" responders ($n = 11$). Probable responders ($n = 7$) had significant increases at one time point while those with less than 33% increase at all time points were defined as nonresponders ($n = 10$). Of these participants, nine had pre- and post-PAS25 stimulus–response curve (SRC) determined and quantified as area under curve (AUC). Seven of these nine children were classified as "definite" responders but the average SRC-AUC for all nine subjects increased. Test/retest reliability was assessed in nine children who received PAS25 2–4 weeks after the first session. The post-PAS25 cortical excitability changes were statistically similar between the first and second sessions. This is an important study demonstrating that a plasticity-inducing protocol can be safely delivered in healthy children with reliable results over time. The variability of PAS25 response in this pediatric population is somewhat expected given significant variability in numerous adult rTMS studies.

Theta Burst Stimulation (TBS)

Non-invasive human TBS was first introduced in 2005 and was based on animal work showing plasticity changes after theta burst stimulation.[106] One major advantage of TBS is that the stimulation duration is relatively short (40s for continuous and 190s for intermittent TBS) and the stimulation intensity is subthreshold.[44] Both of these features make TBS ideal for pediatric research from a standpoint of comfort and tolerability. The original stimulation paradigm consists of three 50-Hz pulses in a burst, with bursts repeating at 5 Hz. For intermittent TBS (iTBS), each 2-s train consisting of 10 bursts (30 pulses) is repeated every 10s for a total of 600 pulses (20 trains). For continuous TBS (cTBS), the trains are repeated 20 times for a total of 600 pulses without any intertrain interval. Stimulation intensity was originally set at 80% of active motor threshold (AMT). The different variables for the stimulation paradigm are stimulation intensity, pulse frequency within the burst, burst frequency, and total pulses. In the original TBS report,[44] iTBS led to LTP-like facilitation that lasted about 20 min, while cTBS produced long-term depression (LTD)-like inhibition lasting ~1 h. A subsequent study from the same laboratory with a much large sample ($n = 56$) showed that the expected facilitation and depression from iTBS and cTBS were highly variable.[107] TBS has also been studied in rat neocortex to better understand the differential effects of iTBS and cTBS, especially at cellular and histochemical levels.[108]

Multiple factors can affect neurophysiologic outcome after TBS. TBS protocol variations such as differences in pulse/burst frequencies,[109] stimulation intensity,[110–112] number of total pulses,[112–115] and coil direction[107,116] can result in changes in post-TBS LTP- and LTD-like plasticity. Hereditary and physiologic factors can also contribute to variations in post-TBS cortical excitability. These include genetic polymorphism,[117,118] hormone,[119] fine motor movement,[114,120,121] and exercise.[122]

Similar to other rTMS protocols, neurophysiologic changes after theta burst rTMS are not always consistent. This may be due to slight differences in TMS protocol (eg, conditioning stimulus in paired pulse stimulation, TBS). RMTs have been reported as increased or unchanged after cTBS.[44,123] The effect of cTBS on SICI has been reported by many groups with some reporting no change[62,124–126] or a reduction.[44,110,111,120,123,127,128] SICI has been increased by iTBS[44,111,127,128] but others reported no change.[110,124,125] Other inhibitory measures such as LICI[123,128] and cSP[62] did not change after i- or cTBS. ICF has been shown to be reduced[44,126] or unchanged after cTBS,[110,128] while it remains unaffected by iTBS.[44,82,125]

There are very few published studies that used TBS in children. The original iTBS protocol[129] was used to study one teen female with fragile X syndrome (FXS) while a modified iTBS protocol[130] was used to study M1 physiology of 14 typically developing children. None of the participants reported any serious adverse reaction. The FXS group ($n = 2$, one teen, one adult) showed a prolonged M1 cortical facilitation after iTBS. The other iTBS study used a variant of TBS with three 30-Hz pulses per burst, with bursts repeating at 5 Hz for total of 30 pulses per train. Each train is repeated every 10 s for a total of 300 pulses at a stimulation intensity of 70% of RMT.[130] Although the post-iTBS response was variable, most children had, as expected, facilitation after iTBS with the 3-min time point being statistically significant in post hoc analysis after correcting for multiple comparisons.

cTBS was also conducted and described in three published studies – one FXS teen,[129] 19 children/adolescents with ASD[131] and five with Tourette syndrome (TS).[132] The Oberman group used the original cTBS settings[44] with one subject reporting a mild headache and two reporting mild fatigue.[131] The cTBS protocol in the Tourette syndrome study[132] used 30 Hz as the pulse frequency within each burst. The bursts were repeated at 5 Hz for a total of 600 pulses. The resting motor thresholds were measured with Magstim 200 stimulator (monophasic pulse). cTBS was delivered at 90% of RMT on the Magstim SuperRapid2 machine (biphasic pulse). TS patients received four sessions of cTBS for two consecutive days. For each day, cTBS sessions 2, 3, and 4 were delivered 15, 60, and 75 min after the first train. This study was a sham-controlled study so only five children/adolescents received active cTBS stimulation, of these one reported having abdominal pain while another reported headaches and dry eyes after cTBS. Despite the only case of TBS-induced seizure happening after cTBS in an adult,[133] no serious adverse events were reported by any of these three pediatric cTBS studies. In the FXS study, the patient not showed significant M1 changes after one session of cTBS. cTBS also did not result in excitability changes when it was delivered 1 day after iTBS.[129] The ASD study showed that post-cTBS response varied and some participants showed paradoxical facilitation afterwards. They also find that age correlated with post-cTBS time period for neurophysiologic changes to return to baseline.[131] In the TS study, the primary aim was to determine a difference in tic severity after eight sessions of sham-controlled cTBS over 2 days. Unfortunately, there was no statistically significant difference between the sham and active cTBS groups. However, a secondary analysis was performed to compare function MRI BOLD signals in the motor network after cTBS. Relative to sham stimulation, the active cTBS group resulted in a significant reduction in bilateral M1 and supplementary motor area BOLD activation.[132]

Repetitive Paired-Pulse (PPS) and Quadripulse Stimulation (QPS)

PPS and QPS paradigms are rTMS that have been used in healthy adults to evoke plastic changes in M1. These protocols have been used in limited fashion by only a few groups. No pediatric studies have been published. For PPS, suprathreshold pulse pairs with ISI of 1.5 ms were repeated over 30 min at 0.2 Hz. Although motor threshold was not changed afterwards, MEP size facilitation was observed.[46] Instead of pulses delivered in pairs, QPS delivers a series of four suprathreshold pulses at a time with variable ISI. This series of four pulses is repeated to result in LTP- or LTD-like changes in M1 depending on the ISI. In the original report, the ISI was set at 1.5 ms and each train of quadripulses was repeated at 0.2 Hz.[45] The motor threshold did not change after QPS-1.5 ms but the M1 stimulus–response curve slope became steeper. LTP-like effects were detected with the duration of stimulation as an important factor affecting outcome. The same group did a separate QPS study showing that MEP facilitation is seen with ISI 1.5–10 ms, whereas MEPs were suppressed with ISIs between 30 and 100 ms. They also showed that SICI was not changed by QPS but ICF was enhanced by QPS-5 ms and decreased by QPS-50 ms. This group also published another study showing that QPS-5 ms can increase IHI but not after QPS-50 ms.[134]

Metaplasticity

Metaplasticity is the alteration of synaptic plasticity that is dependent on the prior activity of the same synapse.[135] It is likely an important factor that contributes to the significant variability to rTMS response. Different forms of non-invasive brain stimulation have been used to study metaplasticity.[136] Some discussion is warranted here as one pediatric study utilized this type of rTMS stimulation.

The first human metaplasticity study design used a theta frequency primed 1 Hz rTMS to enhance LTD-like M1 changes.[137] The protocol design was based on data showing that theta-primed stimulation of rat dentate gyrus resulted in prolonged LTD.[138] The rTMS protocol involved 10 min of subthreshold (90% of RMT) priming with 6 Hz conventional rTMS followed by 10 min of 1 Hz rTMS at 115% of RMT as stimulation intensity. This resulted in enhanced LTD-like changes in M1.[137] Recently, one group used this primed rTMS protocol for a randomized sham-controlled rTMS clinical trial in hemiparetic children.[67] Nineteen children were randomized to five sessions of either sham or active primed rTMS over the healthy cerebral hemisphere M1. The stimulations were alternated daily with constraint-induced movement therapy. The priming protocol consisted of 10 min of 6 Hz rTMS at 90% of RMT, delivered in two trains per minute with 5 s per train and an intertrain interval of 25 s. The priming procedure was followed immediately by 10 min of 1 Hz rTMS at 90% of RMT. Mild headaches were reported during the study but no serious adverse events were reported. The primary outcome based on bimanual hand functions was improved more in the active rTMS group than the sham group.

STUDY DESIGN CONSIDERATIONS

Type of Studies

Different rTMS paradigms have been used in healthy or case–control studies to study normal and disease-specific cortical physiology. Results characterizing reduced or aberrant cortical plasticity in neuropsychiatric diseases have been mixed due to various factors – variability of rTMS-induced changes, different severity/symptom duration within disease populations and potential confounders (eg, concurrent CNS medication use). A good example is the review by Udapa and Chen on Parkinson's disease that summarizes the various cortical physiological findings in the context of multiple potential confounders.[139] Given that multiple factors can contribute to variability in case–control studies, care must be taken in interpreting differences in cortical plasticity in this type of study design.

Rather than using case–control studies to study cortical plasticity in a disease population, another approach is to use rTMS to detect how neuro-plasticity changes in response to a medication or an intervention. In this design, the comparison of plasticity change is within each subject and therefore eliminating the noise produced by high interindividual variability. A few studies addressed intraindividual test–retest variability and showed that MEP measures in M1 immediately following TBS have reasonable reproducibility.[120,140] As an example of this approach, one group showed that deficient cortical LTP-like plasticity in Parkinson's disease was partially restored after dosing with levodopa.[141]

Another potential and promising use of rTMS is for clinical trials. As with pharmacotherapy trials, it is likely that rTMS trials will result in variable clinical outcomes, especially since many studies have shown variability in rTMS-induced neurophysiologic changes in both adult and pediatric participants. Therefore, trials should be designed to include non-clinical measures (eg, neurophysiologic, neuroimaging) such that if clinical/behavioral differences are not apparent, these measures will provide some insight into disease pathophysiology.[69,91,132]

Patient Selection and Safety

Few laboratories have extensive experience in pediatric rTMS research, therefore safety literature is limited. However, pediatric rTMS safety data have recently started to emerge,[142–144] with two reports specifically monitoring for cognitive adverse events.[145,146] Since rTMS-induced serious adverse events such as seizure and syncope have been reported in pediatric subjects,[70,147] it is vital to ensure the safety of the research participants.

One major safety consideration is rTMS-induced seizure, which is more prone in certain patient populations (eg, epilepsy, cerebral palsy, patients taking medications that reduce seizure threshold). A TMS screening questionnaire has been published based on international consensus.[42,148] The investigator(s) must be able to recognize and handle medical emergencies such as seizure and syncope. The TMS lab should be equipped with appropriate resuscitation equipment. The research setting should provide comfort to minimize excessive motion as this can potentially affect the rTMS effects. If possible, the most up-to-date TMS devices should be used as these machines generate power more efficiently so that lower stimulation intensity can be used to produce similar effects.

In addition to lowering seizure thresholds, concurrent CNS medications can also affect neurophysiology studies and clinical trials. Sodium channel blockers can raise motor threshold,[149] and therefore require rTMS to be delivered at a higher intensity. Medications that affect rTMS-induced LTP- and LTD-like plasticity include NMDA-R agonists/antagonists, voltage-gated sodium/calcium channel antagonists, GABA-A and GABA-B receptor agonists, dopamine receptor agonists/antagonists, acetylcholine receptor agonists/antagonists, and alpha adrenergic receptor antagonists. A comprehensive review has been published on this topic.[150]

rTMS Procedure

The rTMS target site will likely determine the complexity of the study procedure. If primary motor or visual cortex is the target, TMS can be used to identify the stimulation site. For other cortical regions, localization may either be based on electroencephalography (EEG) lead placement convention or individualized neuronavigation. Anatomical and functional brain MRIs are the primary methods for identifying the stimulation target. Studies have found better clinical and neurophysiologic results with neuronavigation.[151,152]

Depending on the disease of interest and stimulation site, the outcome measure(s) will vary significantly. For all regions, relevant clinical and behavioral measures can be obtained. For any stimulation site that is a critical part of the motor network, TMS-evoked neurophysiologic measures over M1 can be useful. For regions that have a less direct effect on the motor system, other technologies such as fMRI, diffusion MRI, MR spectroscopy, positron emission tomography, near infrared spectroscopy (NIRS), EEG, and magnetoencephalography may be necessary.[153–155] For some of these technologies (eg, fMRI, EEG, NIRS), it may be possible to deliver rTMS and simultaneously record data if compatible hardware is available. If offline data are to be obtained after rTMS, it is also important to account for the time required to travel from the TMS lab to another location (eg, MRI suite) as the time delay may affect the data.

Choosing a particular rTMS technique depends on the disease population and the pathophysiologic mechanism. For example, neurofibromatosis (NF) animal models have deficient theta-burst-induced LTP. Therefore one possible approach to study NF patients is to use iTBS to characterize LTP-like cortical plasticity. Another example is the use of inhibitory rTMS over the non-lesional hemisphere for stroke/hemiparetic CP patients. This approach theoretically enhances recovery of motor function via reducing the interhemispheric inhibition exerted by the healthy hemisphere on the affected hemisphere. Others have used excitatory rTMS over the lesioned hemisphere to promote clinical improvement.[156] For clinical trials, additional questions may arise regarding how many sessions of rTMS and whether rTMS should be combined with therapies.

Statistical Analysis

In situations when the dependent variable is measured only once after rTMS, the appropriate parametric verus non-parametric test (ie, t-test vs. Wilcoxon–Mann–Whitney test) can be chosen for comparison against baseline values. However in most rTMS studies, multiple measurements are usually obtained after rTMS. This type of data acquisition applies both to clinical trials and studies that strictly focus on cortical neurophysiology. For these datasets, repeated measures analysis of variance (ANOVA) is often used. The assumptions for repeated measures ANOVA are that (1) the dependent variable is normally distributed in the population being sampled, (2) the variance of the dependent variable in each group (if applicable) is equal, and (3) the variances of the differences between all combinations of related groups are equal (sphericity). If sphericity is violated, then correction may be applied. If any baseline characteristic is different between groups, it is also important to include whichever independent variable in the analysis as a covariate.

For MEP analysis, most groups have traditionally used ratio (ie, post-rTMS MEP/baseline MEP) for repeated measures ANOVA. MEP sizes are highly variable at baseline or after rTMS. By using this approach to create a ratio, the baseline amplitude is artificially transformed to 1.0 with a variance of 0. Therefore, the variance of the baseline raw MEP size is lost in the final analysis. TMS researchers have attempted to minimize the baseline variance by using stimulation intensities such that the baseline average peak-to-peak MEP amplitudes for all participants are 1 mV. By doing so, the baseline MEP size variance is likely reduced but still not eliminated. A recent pediatric iTBS study used both repeated measures ANOVA and a repeated measures linear mixed model (LMM) and found similar LTP-like results.[130] The LMM approach has several advantages: (1) it maintains the variability of the baseline measure, (2) it allows the incorporation of intraindividual correlations, and (3) this model can account for missing observations.

CONCLUSION

Based on in vitro and animal work, there are now multiple rTMS paradigms that can be used non-invasively to induce neuroplastic changes in humans. These techniques have been used to further the understanding of pathophysiology for various neuropsychiatric disorders and show promise as potential therapeutic options. Although most studies have been conducted in adults, there is tremendous interest in pediatric studies. As a child's brain undergoes many developmental changes with the unfortunate possibility of developing neuropsychiatric conditions in childhood, a technique that can probe neuroplasticity in this age range can potentially advance the understanding of CNS pathophysiology. For the same reason, there is also reasonable safety concern that plasticity-altering rTMS protocols may interfere with brain maturation during this developmental stage. Therefore, care should be taken to ensure careful monitoring for adverse events in all future pediatric rTMS studies. Studies should also be designed to minimize the various potential confounding effects, and to collect and analyze data using statistically rigorous methods.

Disclosures

Dr. Wu receives support from NIMH and as a clinical trial site investigator for Psyadon Pharmaceuticals.

Dr. Pedapati receives research support from the Cincinnati Children's Hospital Research Foundation and the American Academy Child and Adolescent Psychiatry.

References

1. Shaw CA, McEachern JC. *Toward a Theory of Neuroplasticity*. Psychology Press; 2001.
2. Pascual-Leone A, Amedi A, Fregni F, Merabet LB. The plastic human brain cortex. *Annu Rev Neurosci*. 2005;28:377–401.
3. Lamprecht R, LeDoux J. Structural plasticity and memory. *Nat Rev Neurosci*. 2004;5(1):45–54.
4. Worcester WL. Regeneration of nerve fibres in the central nervous system. *J Exp Med*. 1898;3(6):579–583.
5. Lindley EH. A study of puzzles with special reference to the psychology of mental adaptation. *Am J Psychol*. 1897;8(4):431–493.
6. Tau GZ, Peterson BS. Normal development of brain circuits. *Neuropsychopharmacology*. 2009;35(1):147–168.

7. Fuchs E, Flugge G. Adult neuroplasticity: more than 40 years of research. *Neural Plast.* 2014;2014:10.
8. Hebb DO. *The Organization of Behavior: A Neuropsychological Approach.* John Wiley & Sons; 1949.
9. Hunter P. Ancient rules of memory. The molecules and mechanisms of memory evolved long before their 'modern' use in the brain. *EMBO Rep.* 2008;9(2):124–126.
10. Bailey CH, Kandel ER. Synaptic remodeling, synaptic growth and the storage of long-term memory in Aplysia. *Prog Brain Res.* 2008;169:179–198.
11. Scoville WB, Milner B. Loss of recent memory after bilateral hippocampal lesions. *J Neurol, Neurosurg Psychiatry.* 1957;20(1):11.
12. Moser MB, Moser EI. Functional differentiation in the hippocampus. *Hippocampus.* 1998;8(6):608–619.
13. Lomo T. The discovery of long-term potentiation. *Phil Trans R Soc Lond Ser B, Biol Sci.* 2003;358(1432):617–620.
14. Bliss TV, Lomo T. Long-lasting potentiation of synaptic transmission in the dentate area of the anaesthetized rabbit following stimulation of the perforant path. *J Physiol.* 1973;232(2):331–356.
15. Bliss TV, Collingridge GL. A synaptic model of memory: long-term potentiation in the hippocampus. *Nature.* 1993;361(6407):31–39.
16. Brown TH, Chapman PF, Kairiss EW, Keenan CL. Long-term synaptic potentiation. *Science.* 1988;242(4879):724–728.
17. Beck H, Goussakov IV, Lie A, Helmstaedter C, Elger CE. Synaptic plasticity in the human dentate gyrus. *J Neurosci.* 2000;20(18):7080–7086.
18. Vanderwolf CH. Hippocampal electrical activity and voluntary movement in the rat. *Electroencephalogr Clin Neurophysiol.* 1969;26(4):407–418.
19. Bland BH, Andersen P, Ganes T, Sveen O. Automated analysis of rhythmicity of physiologically identified hippocampal formation neurons. *Exp Brain Res.* 1980;38(2):205–219.
20. Rudell AP, Fox SE, Ranck Jr JB. Hippocampal excitability phase-locked to the theta rhythm in walking rats. *Exp Neurol.* 1980;68(1):87–96.
21. Larson J, Wong D, Lynch G. Patterned stimulation at the theta frequency is optimal for the induction of hippocampal long-term potentiation. *Brain Res.* 1986;368(2):347–350.
22. Buzsáki G. Theta oscillations in the Hippocampus. *Neuron.* 2002;33(3):325–340.
23. Larson J, Munkácsy E. Theta-burst LTP. *Brain Res.* 2015;1621:38–50.
24. Grover LM, Kim E, Cooke JD, Holmes WR. LTP in hippocampal area CA1 is induced by burst stimulation over a broad frequency range centered around delta. *Learn Mem.* 2009;16(1):69–81.
25. Katz LC, Shatz CJ. Synaptic activity and the construction of cortical circuits. *Science.* 1996;274(5290):1133–1138.
26. Hensch TK. Critical period regulation. *Annu Rev Neurosci.* 2004;27(1):549–579.
27. Morishita H, Hensch TK. Critical period revisited: impact on vision. *Curr Opin Neurobiol.* 2008;18(1):101–107.
28. Stryker MP. Neuroscience. Growth through learning. *Nature.* 1995;375(6529):277–278.
29. Kirkwood A, Lee H-K, Bear MF. Co-regulation of long-term potentiation and experience-dependent synaptic plasticity in visual cortex by age and experience. *Nature.* 1995;375(6529):328–331.
30. Cao G, Harris KM. Developmental regulation of the late phase of long-term potentiation (L-LTP) and metaplasticity in hippocampal area CA1 of the rat. *J Neurophysiol.* 2012;107(3):902–912.
31. Leinekugel X, Khazipov R, Cannon R, Hirase H, Ben-Ari Y, Buzsaki G. Correlated bursts of activity in the neonatal hippocampus in vivo. *Science.* 2002;296(5575):2049–2052.

32. Harris KM, Jensen FE, Tsao B. Three-dimensional structure of dendritic spines and synapses in rat hippocampus (CA1) at postnatal day 15 and adult ages: implications for the maturation of synaptic physiology and long-term potentiation. *J Neurosci.* 1992;12(7):2685–2705.

33. Swartzwelder HS, Wilson W, Tayyeb M. Age-dependent inhibition of long-term potentiation by ethanol in immature versus mature Hippocampus. *Alcohol Clin Exp Res.* 1995;19(6):1480–1485.

34. Barker AT, Jalinous R, Freeston IL. Non-invasive magnetic stimulation of human motor cortex. *Lancet.* 1985;1(8437):1106–1107.

35. Dhuna A, Gates J, Pascual-Leone A. Transcranial magnetic stimulation in patients with epilepsy. *Neurology.* 1991;41(7):1067–1071.

36. Pascual-Leone A, Gates JR, Dhuna A. Induction of speech arrest and counting errors with rapid-rate transcranial magnetic stimulation. *Neurology.* 1991;41(5):697–702.

37. Amassian VE, Cracco RQ, Mcaccabee PJ, Cracco JB, Rudell A, Eberle L. Repetitive magnetic coil stimulation of human occipital cortex prolongs suppression of visual perception. *Neurology.* 1990;40(Suppl 1):311.

38. Pascual-Leone A, Houser CM, Reese K, et al. Safety of rapid-rate transcranial magnetic stimulation in normal volunteers. *Electroencephalogr Clin Neurophysiol.* 1993;89(2):120–130.

39. Pascual-Leone A, Valls-Sole J, Wassermann EM, Hallett M. Responses to rapid-rate transcranial magnetic stimulation of the human motor cortex. *Brain.* 1994;117(Pt 4):847–858.

40. Pascual-Leone A, Valls-Sole J, Brasil-Neto JP, Cohen LG, Hallett M. Seizure induction and transcranial magnetic stimulation. *Lancet.* 1992;339(8799):997.

41. Wassermann EM. Risk and safety of repetitive transcranial magnetic stimulation: report and suggested guidelines from the International Workshop on the Safety of Repetitive Transcranial Magnetic Stimulation, June 5–7, 1996. *Electroencephalogr Clin Neurophysiol.* 1998;108(1):1–16.

42. Rossi S, Hallett M, Rossini PM, Pascual-Leone A. Safety, ethical considerations, and application guidelines for the use of transcranial magnetic stimulation in clinical practice and research. *Clin Neurophysiol.* 2009;120(12):2008–2039.

43. Stefan K, Kunesch E, Cohen LG, Benecke R, Classen J. Induction of plasticity in the human motor cortex by paired associative stimulation. *Brain.* 2000;123:572–584.

44. Huang YZ, Edwards MJ, Rounis E, Bhatia KP, Rothwell JC. Theta burst stimulation of the human motor cortex. *Neuron.* 2005;45(2):201–206.

45. Hamada M, Hanajima R, Terao Y, et al. Quadro-pulse stimulation is more effective than paired-pulse stimulation for plasticity induction of the human motor cortex. *Clin Neurophysiol.* 2007;118(12):2672–2682.

46. Thickbroom GW, Byrnes ML, Edwards DJ, Mastaglia FL. Repetitive paired-pulse TMS at I-wave periodicity markedly increases corticospinal excitability: a new technique for modulating synaptic plasticity. *Clin Neurophysiol.* 2006;117(1):61–66.

47. Maeda F, Keenan JP, Tormos JM, Topka H, Pascual-Leone A. Interindividual variability of the modulatory effects of repetitive transcranial magnetic stimulation on cortical excitability. *Exp Brain Res.* 2000;133(4):425–430.

48. Delvendahl I, Jung NH, Mainberger F, Kuhnke NG, Cronjaeger M, Mall V. Occlusion of bidirectional plasticity by preceding low-frequency stimulation in the human motor cortex. *Clin Neurophysiol.* 2010;121(4):594–602.

49. Chen R, Classen J, Gerloff C, et al. Depression of motor cortex excitability by low-frequency transcranial magnetic stimulation. *Neurology.* 1997;48(5):1398–1403.

50. Furukawa T, Toyokura M, Masakado Y. Suprathreshold 0.2 Hz repetitive transcranial magnetic stimulation (rTMS) over the prefrontal area. *Tokai J Exp Clin Med.* 2010;35(1):29–33.

51. Cincotta M, Borgheresi A, Gambetti C, et al. Suprathreshold 0.3 Hz repetitive TMS prolongs the cortical silent period: potential implications for therapeutic trials in epilepsy. *Clin Neurophysiol.* 2003;114(10):1827–1833.

52. Ikeguchi M, Touge T, Kaji R, et al. Durable effect of very low-frequency repetitive transcranial magnetic stimulation for modulating cortico-spinal neuron excitability. *Int Congr Ser.* 2005;1278:272–275.

53. Fitzgerald PB, Brown TL, Daskalakis ZJ, Chen R, Kulkarni J. Intensity-dependent effects of 1 Hz rTMS on human corticospinal excitability. *Clin Neurophysiol.* 2002;113(7):1136–1141.

54. Muellbacher W, Ziemann U, Boroojerdi B, Hallett M. Effects of low-frequency transcranial magnetic stimulation on motor excitability and basic motor behavior. *Clin Neurophysiol.* 2000;111(6):1002–1007.

55. Wassermann EM, Grafman J, Berry C, et al. Use and safety of a new repetitive transcranial magnetic stimulator. *Electroencephalogr Clin Neurophysiol.* 1996;101(5):412–417.

56. Fitzgerald PB, Brown TL, Marston NA, et al. Reduced plastic brain responses in schizophrenia: a transcranial magnetic stimulation study. *Schizophr Res.* 2004;71(1):17–26.

57. Modugno N, Curra A, Conte A, et al. Depressed intracortical inhibition after long trains of subthreshold repetitive magnetic stimuli at low frequency. *Clin Neurophysiol.* 2003;114(12):2416–2422.

58. Romero JR, Anschel D, Sparing R, Gangitano M, Pascual-Leone A. Subthreshold low frequency repetitive transcranial magnetic stimulation selectively decreases facilitation in the motor cortex. *Clin Neurophysiol.* 2002;113(1):101–107.

59. Lang N, Harms J, Weyh T, et al. Stimulus intensity and coil characteristics influence the efficacy of rTMS to suppress cortical excitability. *Clin Neurophysiol.* 2006;117(10):2292–2301.

60. Romeo S, Gilio F, Pedace F, et al. Changes in the cortical silent period after repetitive magnetic stimulation of cortical motor areas. *Exp Brain Res.* 2000;135(4):504–510.

61. Fierro B, Piazza A, Brighina F, La Bua V, Buffa D, Oliveri M. Modulation of intracortical inhibition induced by low- and high-frequency repetitive transcranial magnetic stimulation. *Exp Brain Res.* 2001;138(4):452–457.

62. Di Lazzaro V, Dileone M, Pilato F, et al. Modulation of motor cortex neuronal networks by rTMS: comparison of local and remote effects of six different protocols of stimulation. *J Neurophysiol.* 2011;105(5):2150–2156.

63. Gilio F, Rizzo V, Siebner HR, Rothwell JC. Effects on the right motor hand-area excitability produced by low-frequency rTMS over human contralateral homologous cortex. *J Physiol.* 2003;551(Pt 2):563–573.

64. Pal PK, Hanajima R, Gunraj CA, et al. Effect of low-frequency repetitive transcranial magnetic stimulation on interhemispheric inhibition. *J Neurophysiol.* 2005;94(3):1668–1675.

65. Walter G, Tormos JM, Israel JA, Pascual-Leone A. Transcranial magnetic stimulation in young persons: a review of known cases. *J Child Adolesc Psychopharmacol.* 2001;11(1):69–75.

66. Wedegaertner FR, Garvey MA, Cohen LG, Hallett M, Wassermann EM. Low frequency repetitive transcrnial magnetic stimulation can reduce action myoclonus. *Neurology.* 1997;48(3 suppl 2):A119.

67. Gillick BT, Krach LE, Feyma T, et al. Primed low-frequency repetitive transcranial magnetic stimulation and constraint-induced movement therapy in pediatric hemiparesis: a randomized controlled trial. *Dev Med Child Neurol.* 2014;56(1):44–52.

68. Kirton A, Chen R, Friefeld S, Gunraj C, Pontigon AM, Deveber G. Contralesional repetitive transcranial magnetic stimulation for chronic hemiparesis in subcortical paediatric stroke: a randomised trial. *Lancet Neurol.* 2008;7(6):507–513.

69. Kirton A, Deveber G, Gunraj C, Chen R. Cortical excitability and interhemispheric inhibition after subcortical pediatric stroke: plastic organization and effects of rTMS. *Clin Neurophysiol.* 2010;121(11):1922–1929.

70. Kirton A, Deveber G, Gunraj C, Chen R. Neurocardiogenic syncope complicating pediatric transcranial magnetic stimulation. *Pediatr Neurol.* 2008;39(3):196–197.

71. Kwon HJ, Lim WS, Lim MH, et al. 1-Hz low frequency repetitive transcranial magnetic stimulation in children with Tourette's syndrome. *Neurosci Lett.* 2011;492(1):1–4.

72. Le K, Liu L, Sun M, Hu L, Xiao N. Transcranial magnetic stimulation at 1 hertz improves clinical symptoms in children with Tourette syndrome for at least 6 months. *J Clin Neurosci Official J Neurosurg Soc Australasia.* 2013;20(2):257–262.

73. Mantovani A, Leckman JF, Grantz H, King RA, Sporn AL, Lisanby SH. Repetitive transcranial magnetic stimulation of the supplementary motor area in the treatment of Tourette Syndrome: report of two cases. *Clin Neurophysiol.* 2007;118(10):2314–2315.

74. Bohlhalter S, Goldfine A, Matteson S, et al. Neural correlates of tic generation in Tourette syndrome: an event-related functional MRI study. *Brain.* 2006;129(8):2029–2037.

75. Jennum P, Winkel H, Fuglsang-Frederiksen A. Repetitive magnetic stimulation and motor evoked potentials. *Electroencephalogr Clin Neurophysiol.* 1995;97(2):96–101.

76. Peinemann A, Reimer B, Loer C, et al. Long-lasting increase in corticospinal excitability after 1800 pulses of subthreshold 5 Hz repetitive TMS to the primary motor cortex. *Clin Neurophysiol.* 2004;115(7):1519–1526.

77. Takano B, Drzezga A, Peller M, et al. Short-term modulation of regional excitability and blood flow in human motor cortex following rapid-rate transcranial magnetic stimulation. *NeuroImage.* 2004;23(3):849–859.

78. Quartarone A, Bagnato S, Rizzo V, et al. Distinct changes in cortical and spinal excitability following high-frequency repetitive TMS to the human motor cortex. *Exp Brain Res.* 2005;161(1):114–124.

79. Siebner HR, Mentschel C, Auer C, Lehner C, Conrad B. Repetitive transcranial magnetic stimulation causes a short-term increase in the duration of the cortical silent period in patients with Parkinson's disease. *Neurosci Lett.* 2000;284(3):147–150.

80. Di Lazzaro V, Oliviero A, Mazzone P, et al. Short-term reduction of intracortical inhibition in the human motor cortex induced by repetitive transcranial magnetic stimulation. *Exp Brain Res.* 2002;147(1):108–113.

81. Berardelli A, Inghilleri M, Gilio F, et al. Effects of repetitive cortical stimulation on the silent period evoked by magnetic stimulation. *Exp Brain Res.* 1999;125(1):82–86.

82. Cincotta M, Giovannelli F, Borgheresi A, et al. Modulatory effects of high-frequency repetitive transcranial magnetic stimulation on the ipsilateral silent period. *Exp Brain Res.* 2006;171(4):490–496.

83. Gorsler A, Baumer T, Weiller C, Munchau A, Liepert J. Interhemispheric effects of high and low frequency rTMS in healthy humans. *Clin Neurophysiol.* 2003;114(10):1800–1807.

84. Benninger DH, Lomarev M, Wassermann EM, et al. Safety study of 50 Hz repetitive transcranial magnetic stimulation in patients with Parkinson's disease. *Clin Neurophysiol.* 2009;120(4):809–815.

85. Daskalakis ZJ, Moller B, Christensen BK, Fitzgerald PB, Gunraj C, Chen R. The effects of repetitive transcranial magnetic stimulation on cortical inhibition in healthy human subjects. *Exp Brain Res.* 2006;174(3):403–412.

86. Arai N, Okabe S, Furubayashi T, et al. Differences in after-effect between monophasic and biphasic high-frequency rTMS of the human motor cortex. *Clin Neurophysiol.* 2007;118(10):2227–2233.

87. Jung SH, Shin JE, Jeong YS, Shin HI. Changes in motor cortical excitability induced by high-frequency repetitive transcranial magnetic stimulation of different stimulation durations. *Clin Neurophysiol.* 2008;119(1):71–79.

88. Pascual-Leone A, Tormos JM, Keenan J, Tarazona F, Canete C, Catala MD. Study and modulation of human cortical excitability with transcranial magnetic stimulation. *J Clin Neurophysiol.* 1998;15(4):333–343.

89. Khedr EM, Rothwell JC, Ahmed MA, Shawky OA, Farouk M. Modulation of motor cortical excitability following rapid-rate transcranial magnetic stimulation. *Clin Neurophysiol.* 2007;118(1):140–145.

90. O'Reardon JP, Solvason HB, Janicak PG, et al. Efficacy and safety of transcranial magnetic stimulation in the acute treatment of major depression: a multisite randomized controlled trial. *Biol Psychiatry.* 2007;62(11):1208–1216.

91. Croarkin PE, Wall CA, Nakonezny PA, et al. Increased cortical excitability with prefrontal high-frequency repetitive transcranial magnetic stimulation in adolescents with treatment-resistant major depressive disorder. *J Child Adolesc Psychopharmacol.* 2012;22(1):56–64.

92. Yang XR, Kirton A, Wilkes TC, et al. Glutamate alterations associated with transcranial magnetic stimulation in youth depression: a case series. *J ECT.* 2014;30(3): 242–247.

93. Carson RG, Kennedy NC. Modulation of human corticospinal excitability by paired associative stimulation. *Front Hum Neurosci.* 2013;7:823.

94. Paulsen O, Sejnowski TJ. Natural patterns of activity and long-term synaptic plasticity. *Curr Opin Neurobiol.* 2000;10(2):172–179.

95. Wolters A, Sandbrink F, Schlottmann A, et al. A temporally asymmetric Hebbian rule governing plasticity in the human motor cortex. *J Neurophysiol.* 2003;89(5): 2339–2345.

96. Huber R, Maatta S, Esser SK, et al. Measures of cortical plasticity after transcranial paired associative stimulation predict changes in electroencephalogram slow-wave activity during subsequent sleep. *J Neurosci.* 2008;28(31):7911–7918.

97. Stefan K, Kunesch E, Benecke R, Cohen LG, Classen J. Mechanisms of enhancement of human motor cortex excitability induced by interventional paired associative stimulation. *J Physiol.* 2002;543(Pt 2):699–708.

98. Quartarone A, Bagnato S, Rizzo V, et al. Abnormal associative plasticity of the human motor cortex in writer's cramp. *Brain.* 2003;126(Pt 12):2586–2596.

99. Rosenkranz K, Rothwell JC. Differences between the effects of three plasticity inducing protocols on the organization of the human motor cortex. *Eur J Neurosci.* 2006;23(3): 822–829.

100. Russmann H, Lamy JC, Shamim EA, Meunier S, Hallett M. Associative plasticity in intracortical inhibitory circuits in human motor cortex. *Clin Neurophysiol.* 2009;120(6): 1204–1212.

101. Meunier S, Russmann H, Shamim E, Lamy JC, Hallett M. Plasticity of cortical inhibition in dystonia is impaired after motor learning and paired-associative stimulation. *Eur J Neurosci.* 2012;35(6):975–986.

102. De Beaumont L, Tremblay S, Poirier J, Lassonde M, Theoret H. Altered bidirectional plasticity and reduced implicit motor learning in concussed athletes. *Cereb Cortex.* 2012;22(1):112–121.

103. Elahi B, Gunraj C, Chen R. Short-interval intracortical inhibition blocks long-term potentiation induced by paired associative stimulation. *J Neurophysiol.* 2012;107(7): 1935–1941.

104. Sale MV, Ridding MC, Nordstrom MA. Factors influencing the magnitude and reproducibility of corticomotor excitability changes induced by paired associative stimulation. *Exp Brain Res.* 2007;181(4):615–626.

105. Damji O, Keess J, Kirton A. Evaluating developmental motor plasticity with paired afferent stimulation. *Dev Med Child Neurol.* 2015;57(6):548–555.
106. Hess G, Aizenman CD, Donoghue JP. Conditions for the induction of long-term potentiation in layer II/III horizontal connections of the rat motor cortex. *J Neurophysiol.* 1996;75(5):1765–1778.
107. Hamada M, Murase N, Hasan A, Balaratnam M, Rothwell JC. The role of interneuron networks in driving human motor cortical plasticity. *Cereb Cortex.* 2013;23(7): 1593–1605.
108. Funke K, Benali A. Modulation of cortical inhibition by rTMS – findings obtained from animal models. *J Physiol.* 2011;589(Pt 18):4423–4435.
109. Goldsworthy MR, Pitcher JB, Ridding MC. A comparison of two different continuous theta burst stimulation paradigms applied to the human primary motor cortex. *Clin Neurophysiol.* 2012;123(11):2256–2263.
110. McAllister SM, Rothwell JC, Ridding MC. Selective modulation of intracortical inhibition by low-intensity theta burst stimulation. *Clin Neurophysiol.* 2009;120(4):820–826.
111. Huang Y, Chen R, Rothwell J, Wen H. The after-effect of human theta burst stimulation is NMDA receptor dependent. *Clin Neurophysiol.* 2007;118(5):1028–1032.
112. Doeltgen SH, Ridding MC. Low-intensity, short-interval theta burst stimulation modulates excitatory but not inhibitory motor networks. *Clin Neurophysiol.* 2011;122(7):1411–1416.
113. Gamboa OL, Antal A, Moliadze V, Paulus W. Simply longer is not better: reversal of theta burst after-effect with prolonged stimulation. *Exp Brain Res.* 2010;204(2): 181–187.
114. Gentner R, Wankerl K, Reinsberger C, Zeller D, Classen J. Depression of human corticospinal excitability induced by magnetic theta-burst stimulation: evidence of rapid polarity-reversing metaplasticity. *Cereb Cortex.* 2008;18(9):2046–2053.
115. Huang YZ, Rothwell JC, Lu CS, Chuang WL, Lin WY, Chen RS. Reversal of plasticity-like effects in the human motor cortex. *J Physiol.* 2010;588(Pt 19):3683–3693.
116. Talelli P, Cheeran BJ, Teo JT, Rothwell JC. Pattern-specific role of the current orientation used to deliver theta burst stimulation. *Clin Neurophysiol.* 2007;118(8):1815–1823.
117. Lee M, Kim SE, Kim WS, et al. Interaction of motor training and intermittent theta burst stimulation in modulating motor cortical plasticity: influence of BDNF Val66Met polymorphism. *PloS One.* 2013;8(2):e57690.
118. Lee NJ, Ahn HJ, Jung KI, et al. Reduction of continuous theta burst stimulation-induced motor plasticity in healthy elderly with COMT Val158Met polymorphism. *Ann Rehabil Med.* 2014;38(5):658–664.
119. Clow A, Law R, Evans P, et al. Day differences in the cortisol awakening response predict day differences in synaptic plasticity in the brain. *Stress.* 2014;17(3):219–223.
120. Huang YZ, Rothwell JC, Edwards MJ, Chen RS. Effect of physiological activity on an NMDA-dependent form of cortical plasticity in human. *Cereb Cortex.* 2008;18(3): 563–570.
121. Iezzi E, Conte A, Suppa A, et al. Phasic voluntary movements reverse the aftereffects of subsequent theta-burst stimulation in humans. *J Neurophysiol.* 2008;100(4): 2070–2076.
122. McDonnell MN, Buckley JD, Opie GM, Ridding MC, Semmler JG. A single bout of aerobic exercise promotes motor cortical neuroplasticity. *J Appl Physiol.* 2013;114(9): 1174–1182.
123. Goldsworthy MR, Pitcher JB, Ridding MC. Neuroplastic modulation of inhibitory motor cortical networks by spaced theta burst stimulation protocols. *Brain Stimul.* 2013;6(3):340–345.
124. Doeltgen SH, Ridding MC. Modulation of cortical motor networks following primed theta burst transcranial magnetic stimulation. *Exp Brain Res.* 2011;215(3–4): 199–206.

125. Murakami T, Muller-Dahlhaus F, Lu MK, Ziemann U. Homeostatic metaplasticity of corticospinal excitatory and intracortical inhibitory neural circuits in human motor cortex. *J Physiol*. 2012;590(Pt 22):5765–5781.

126. Jacobs MF, Tsang P, Lee KG, Asmussen MJ, Zapallow CM, Nelson AJ. 30 Hz theta-burst stimulation over primary somatosensory cortex modulates corticospinal output to the hand. *Brain Stimul*. 2014;7(2):269–274.

127. Murakami T, Sakuma K, Nomura T, Nakashima K, Hashimoto I. High-frequency oscillations change in parallel with short-interval intracortical inhibition after theta burst magnetic stimulation. *Clin Neurophysiol*. 2008;119(2):301–308.

128. Suppa A, Ortu E, Zafar N, et al. Theta burst stimulation induces after-effects on contralateral primary motor cortex excitability in humans. *J Physiol*. 2008;586(Pt 18): 4489–4500.

129. Oberman L, Ifert-Miller F, Najib U, et al. Transcranial magnetic stimulation provides means to assess cortical plasticity and excitability in humans with fragile x syndrome and autism spectrum disorder. *Front Synaptic Neurosci*. 2010;2:26.

130. Pedapati EV, Gilbert DL, Horn PS, et al. Effect of 30 Hz theta burst transcranial magnetic stimulation on the primary motor cortex in children and adolescents. *Front Hum Neurosci*. 2015;9:91.

131. Oberman LM, Pascual-Leone A, Rotenberg A. Modulation of corticospinal excitability by transcranial magnetic stimulation in children and adolescents with autism spectrum disorder. *Front Hum Neurosci*. 2014;8:627.

132. Wu SW, Maloney T, Gilbert DL, et al. Functional MRI-navigated repetitive transcranial magnetic stimulation over supplementary motor area in chronic tic disorders. *Brain Stimul*. 2014;7(2):212–218.

133. Oberman LM, Pascual-Leone A. Report of seizure induced by continuous theta burst stimulation. *Brain Stimul*. 2009;2(4):246–247.

134. Tsutsumi R, Hanajima R, Terao Y, et al. Effects of the motor cortical quadripulse transcranial magnetic stimulation (QPS) on the contralateral motor cortex and interhemispheric interactions. *J Neurophysiol*. 2014;111(1):26–35.

135. Abraham WC, Bear MF. Metaplasticity: the plasticity of synaptic plasticity. *Trends Neurosci*. 1996;19(4):126–130.

136. Muller-Dahlhaus F, Ziemann U. Metaplasticity in human cortex. *Neuroscientist (A Rev J bringing Neurobiol Neurol Psychiatry*. 2015;21(2):185–202.

137. Iyer MB, Schleper N, Wassermann EM. Priming stimulation enhances the depressant effect of low-frequency repetitive transcranial magnetic stimulation. *J Neurosci*. 2003;23(34):10867–10872.

138. Christie BR, Abraham WC. Priming of associative long-term depression in the dentate gyrus by theta frequency synaptic activity. *Neuron*. 1992;9(1):79–84.

139. Udupa K, Chen R. Motor cortical plasticity in Parkinson's disease. *Front Neurol*. 2013;4:128.

140. Vernet M, Bashir S, Yoo WK, et al. Reproducibility of the effects of theta burst stimulation on motor cortical plasticity in healthy participants. *Clin Neurophysiol*. 2014; 125(2):320–326.

141. Ueki Y, Mima T, Kotb MA, et al. Altered plasticity of the human motor cortex in Parkinson's disease. *Ann Neurol*. 2006;59(1):60–71.

142. Hong YH, Wu SW, Pedapati EV, et al. Safety and tolerability of theta burst stimulation vs. single and paired pulse transcranial magnetic stimulation: a comparative study of 165 pediatric subjects. *Front Hum Neurosci*. 2015;9:29.

143. Wu SW, Shahana N, Huddleston DA, Lewis AN, Gilbert DL. Safety and tolerability of theta-burst transcranial magnetic stimulation in children. *Dev Med Child Neurol*. 2012;54(7):636–639.

144. Krishnan C, Santos L, Peterson MD, Ehinger M. Safety of noninvasive brain stimulation in children and adolescents. *Brain Stimul*. 2014;8(1):76–87.

145. Wall CA, Croarkin PE, McClintock SM, et al. Neurocognitive effects of repetitive transcranial magnetic stimulation in adolescents with major depressive disorder. *Front Psychiatry*. 2013;4:165.
146. Mayer G, Aviram S, Walter G, Levkovitz Y, Bloch Y. Long-term follow-up of adolescents with resistant depression treated with repetitive transcranial magnetic stimulation. *J ECT*. 2012;28(2):84–86.
147. Chiramberro M, Lindberg N, Isometsa E, Kahkonen S, Appelberg B. Repetitive transcranial magnetic stimulation induced seizures in an adolescent patient with major depression: a case report. *Brain Stimul*. 2013;6(5):830–831.
148. Rossi S, Hallett M, Rossini PM, Pascual-Leone A. Screening questionnaire before TMS: an update. *Clin Neurophysiol*. 2011;122(8):1686.
149. Ziemann U. TMS and drugs. *Clin Neurophysiol*. 2004;115(8):1717–1729.
150. Ziemann U, Reis J, Schwenkreis P, et al. TMS and drugs revisited 2014. *Clin Neurophysiol*. 2014;126(10):1847–1868.
151. Fitzgerald PB, Hoy K, McQueen S, et al. A randomized trial of rTMS targeted with MRI based neuro-navigation in treatment-resistant depression. *Neuropsychopharmacology*. 2009;34(5):1255–1262.
152. Julkunen P, Saisanen L, Danner N, et al. Comparison of navigated and non-navigated transcranial magnetic stimulation for motor cortex mapping, motor threshold and motor evoked potentials. *NeuroImage*. 2009;44(3):790–795.
153. Siebner HR, Bergmann TO, Bestmann S, et al. Consensus paper: combining transcranial stimulation with neuroimaging. *Brain Stimul*. 2009;2(2):58–80.
154. Stagg CJ, Wylezinska M, Matthews PM, et al. Neurochemical effects of theta burst stimulation as assessed by magnetic resonance spectroscopy. *J Neurophysiol*. 2009; 101(6):2872–2877.
155. Allendorfer JB, Storrs JM, Szaflarski JP. Changes in white matter integrity follow excitatory rTMS treatment of post-stroke aphasia. *Restor Neurol Neurosci*. 2012;30(2):103–113.
156. Szaflarski JP, Vannest J, Wu SW, DiFrancesco MW, Banks C, Gilbert DL. Excitatory repetitive transcranial magnetic stimulation induces improvements in chronic post-stroke aphasia. *Med Sci Monit Int Med J Exp Clin Res*. 2011;17(3):CR132–CR139.

Therapeutic rTMS in Children

D.L. Gilbert

Cincinnati Children's Hospital Medical Center, Cincinnati, OH,
United States

INTRODUCTION: FROM BIOMARKER OF PLASTICITY AND EXCITABILITY TO INDUCER OF THERAPEUTIC CHANGES?

Transcranial magnetic stimulation (TMS) inhabits a fairly uncommon location within non-invasive brain technologies: it can be used to probe and measure brain physiology and it can also modulate that physiology to improve brain function.[1] Current devices are based on prototypes developed and tested in the 1980s.[2] Since that time a number of companies have developed and tested TMS devices for a wide range of purposes, in laboratories in many countries.

The basic principles of single- and paired-pulse TMS and repetitive TMS (rTMS) emerged from multiple laboratories worldwide in the 1990s.[3] In the United States, this occurred during the so-called "Decade of the Brain," from 1990 to 1999, a joint project between the US Library of Congress and the National Institutes of Health, to enhance awareness of the benefits of brain research across government agencies and employees as well as the general public. More recently, in 2014, US President Barack Obama announced the Brain Research through Advancing Innovative Neurotechnologies (BRAIN) initiative to support development and application of innovative technologies to understanding brain function (http://braininitiative.nih.gov/) (https://www.whitehouse.gov/share/brain-initiative). This has become an international effort spanning many research organizations and industries.

While the majority of TMS research to date involves probing human physiology, there is an imperative where possible to move from understanding to treatment. Parents visiting our TMS laboratory for functional studies, even after advance explanations of the purpose of the research, not infrequently ask, "can you use this to help my kid's brain?" Of course this is important to parents, and doctors, and it's important to keep this in mind in our research plans.

Unfortunately, despite many high-profile efforts and investments, our ability to "help kids' brains" remains hampered by the enormous gap between our observations and measurements of central nervous system function and our understanding at the molecular, cellular, and network level of diseases of the central nervous system. In diseases at the end of life, as well as at the end-stage of many diseases, knowledge has been guided in part by postmortem pathological science. However, for normal childhood development and the myriad genetic and acquired conditions affecting childhood brain function, many noninvasive, complementary techniques must be applied, painstakingly, to gain partial insights. Synthesizing these complementary datasets and analyses comes next. And finally, better understanding of the biology will be critical to designing childhood treatments that arrest the progression of degenerative disease, prevent complications, and restore or improve function.

TMS is one non-invasive, innovative neurotechnology which can provide insights into normal and diseased brain function.[4] The majority of this book and of pediatric TMS research to date involves the use of TMS for understanding normal development, function, and diseases and disorders. But can TMS be used effectively and safely to treat pediatric neurological or psychiatric illness? Treatment protocols involve repetitively stimulating the brain in regular patterns – "driving" the brain's cortex externally in inhibitory or excitatory paradigms – to try to achieve a longer-lasting improvement in brain activity and benefit patients. In single-visit protocols, many of these repetitive stimulation techniques can generate short-term changes, indicative in some cases of potential for adaptive neuroplasticity. The emphasis of this chapter will be on clinical trials, usually involving multiple sessions or visits, to improve brain function. While most such studies have been conducted in adults, with important lessons learned, the discussion of adult studies in this chapter is aimed at guidance for pediatric studies. So, after a brief review of adult research, this chapter will describe current studies in children, results, limitations, and possible future directions.

METHODS: TYPES OF REPETITIVE TRANSCRANIAL MAGNETIC STIMULATION (rTMS)

Repetitive TMS protocols are described in detail in Chapter 3, Neuroplasticity Protocols: Inducing and Measuring Change and will be briefly discussed here. The basic idea is that certain stimulation protocols tend to transiently (up to 60 min) excite or inhibit cortex in the area within the coils' magnetic fields. Demonstrating transient excitatory or inhibitory effects resulting from TMS protocols can be carried out in several ways. In motor cortex (M1), this demonstration typically involves comparing TMS motor-evoked potential (MEP) amplitudes pre- and post-rTMS. Excitatory rTMS protocols should make the average MEP amplitudes larger compared to pre-RMTS baseline, and inhibitory protocols make them smaller compared to pre-rTMS baseline. In M1 and other areas, other physiological techniques such as EEG and metabolic or blood flow imaging studies can also be utilized as a "readout" of rTMS effects.

Once transient excitatory or inhibitory effects of an rTMS technique have been demonstrated in multiple single-session studies using several "readout" modalities, then investigators may modify those protocols for treatment purposes. Usually this involves single-session rTMS daily visits for 1 week or more, although some protocols use multiple daily sessions.[5,6] The idea is that the mechanisms underlying the transient cortical effects, when engaged repeatedly (10 or more times), will establish a new, more functional cortical "baseline" that counteracts those involved in the symptoms of disease. For example, if one has evidence in a particular disease that a focal cortical region is underactivated, compared to that same region in healthy controls, then one could bring affected individuals to the lab for 10 or more individual rTMS excitatory sessions and attempt to generate longer-term normalization of the physiology of that region. Some biomarkers such as fMRI blood oxygen level dependent (BOLD) signal may be used as a readout at a later time point, for example 1–3 months later, to show that a brain effect has lasted. The more important point would be to both verify the long-term neural changes and improvement in disease-related symptoms or function. As is common practice in drug studies, if there is a promising result from "open-label" rTMS, then randomized controlled trials are performed comparing rTMS to sham rTMS. Sham TMS takes various forms, but the basic idea is that the operator places a coil on the participant's head that mimics the sound and feel of active treatment but does not activate the brain in the possibly therapeutic manner. In the past a variety of methods have been used including tilting and shielding an active coil. This has largely been replaced through the use of sham coils, available from multiple manufacturers.

MOST COMMON REPETITIVE TMS METHODS

Conventional rTMS involves administering pulses at a regular rate in an attempt to inhibit or excite the cortical area beneath the coil. "Downstream" effects on connected cortical or subcortical structures may also be of clinical or scientific interest. Based on multiple complementary techniques, it appears in most cases that high-frequency rTMS (5 Hz or faster) has an excitatory effect on local cortex.[7] In contrast, low-frequency rTMS (1 Hz or slower) has inhibitory effects.[8] At higher frequencies of 10–20 Hz, one typically administers pulses for a short interval, like 2 s, with interleaved pauses.

Paired associative stimulation (PAS) requires an electrical stimulator for peripheral sensory nerves as well as a TMS device, and the ability to pair these at biologically meaningful interstimulus intervals, ie, indexed to the peripheral to central signal latencies.[9] A typical protocol involves administering a suprathreshold intensity electrical pulse to a peripheral nerve, eg, the median nerve, followed approximately 25 ms later by a TMS pulse over sensorimotor cortex (S1/M1). These paired sensory electrical/cortical magnetic stimuli are repeated, for example every 5 s (0.2 Hz) for a period of approximately 15 min. Depending on the paired intervals and other parameters, transient excitation or inhibition may occur. Administration of PAS at 25-ms interstimulus intervals tends to be excitatory.[10] A recent study carefully quantified the extent to which excitatory PAS25 effects occur and are reproducible in school-aged children.[11]

Theta burst stimulation (TBS) involves administering low-intensity pulses in particular, clustered patterns in an attempt to inhibit or excite the cortical area beneath the coil. The patterns used are based on neuroplasticity-inducing firing patterns observed in invasive animal studies. They generally involve iterative patterns of 30–50 Hz rTMS at submotor threshold intensities. In the earliest human protocol, a three-pulse burst at 50 Hz burst was administered five times per second.[12,13] So this means $3 \times 5 = 15$ pulses (in five bursts) per second. So-called "continuous TBS (cTBS)" for 600 pulses would require 40 s. In TBS the stimulation intensity is usually set below rest motor threshold (RMT) and even below active motor threshold (AMT). Based on intensity and duration, cTBS would probably be the most tolerable form of rTMS for a sensitive or restless child. A typical protocol of intermittent TBS (iTBS) interleaves regular pauses. The pattern of 15 pulses per second, in three-pulse high-frequency bursts, is administered for 2 s (30 pulses), followed by an 8-s pause. Based on 30 high-frequency, low-intensity pulses administered every 10 s, 600 pulses would require just over 3 min. cTBS sequences generally inhibit cortical activity, whereas iTBS sequences generally excite cortical activity.[12] These are often referred to in the literature as analogous to long-term depression (LTD)-like and long-term potentiation (LTP)-like changes.

BRIEF OVERVIEW OF CLINICAL TRIALS AND INDICATIONS IN YOUNG ADULTS

rTMS interventional research is much more extensive in adult than in pediatric populations. At present, a MEDLINE literature search for rTMS treatment studies or clinical trials in children yields just a handful of studies. While adult research is not the focus of this chapter, there are several benefits to understanding the adult literature which warrant a brief review in this section.

First, it's worth noting why adult research is more prevalent. Under standard research regulations, children constitute a vulnerable population for whom special protections are ethically mandated. In vulnerable populations, ethics and risk minimization outweigh science. So, while biological research holds the key for future better treatments, there has to be an indication of potential benefit before non-significant risk intervention studies can be performed in children. Adults can provide informed consent for studies where risks are unknown, and these studies may ultimately provide a foundation for reasonable risk analyses for pediatric research. Adults also cooperate more readily with study protocols.

Adult studies can also provide indications of potential benefit in many circumstances where diseases with similar symptoms or altered pathways can occur in children. Here are some examples where adult rTMS studies should possibly inform pediatric ones:

1. Significant therapeutic benefit of rTMS in young adults with illnesses that adolescents also have. For example, positive efficacy results in young adults with major, medication-refractory depression could support extending that research to adolescents.
2. Safety (lack of adverse events in study of reasonable size), with or without strong statistical evidence of benefit, of rTMS in conditions where early intervention could be superior. Adults may enroll in interventional studies with refractory disease present for many years. At that point, a disease process may be less amenable to intervention than the same disease would be in a child or adolescent, shortly after onset.

3. Safety with evidence of enhanced or augmented function in healthy adults. For example, rTMS studies showing improved strength, motor learning, memory, or pain tolerance might point toward a role for rTMS in a variety of pediatric diseases affecting those functions.
4. Studies which alter stimulation protocols using methods which may be more effective. An example would be frequency priming prior to rTMS,[14,15] or TBS studies with different intensity or frequency parameters.[16]
5. Studies involving a location that may have particular pediatric relevance, for example, cerebellum.[17–19]

In order to discern, empirically, promising areas for pediatric rTMS treatment research based on lessons from adult rTMS studies, it's helpful to illustrate using an OVID (or other MEDLINE) literature search for clinical trials and treatment reports in young adults. In OVID, searching MEDLINE from 1996 to September 2015, a "Transcranial Magnetic Stimulation" keyword search, limited to English Language, Humans, "Therapy (maximizes sensitivity) and young adult (19–24 years)" generated 491 papers, which was reduced to 250 by changing to Therapy to "(maximizes specificity)". Abstract review of the 250 papers yielded a total of 159 papers of which 126 had some relevance to clinical therapeutics in children and adolescents. These included many pilot and proof-of-concept studies, some reviews and follow-up biomarker studies after therapeutic interventions, and a few randomized trials. The breakdown of therapeutic studies by condition is in Table 4.1. As can be

TABLE 4.1   Published TMS Therapeutic Studies in Young Adults

Disease Area	Number of rTMS Treatment Publications
Depression	27
Pain (migraine, fibromyalgia, mechanisms of pain tolerance)	15
Schizophrenia	10
Cognition (mechanisms of attention, memory, specific domains)	9
Stroke (recovery, motor control, spasticity)	8
Epilepsy	3
Addiction	3

MEDLINE search results, see text. Additional studies, one each, involved ADHD, autism, bipolar, general anxiety, tinnitus, Tourette syndrome.

seen, disabling psychiatric conditions and pain are the most common, but there are studies involving rehabilitation of and modulation of motor function and of epilepsy as well. Bearing in mind both similarities and differences between pathologies in the central nervous system in children and adults, it would seem reasonable to conduct research in similar areas in younger children.

BRIEF OVERVIEW OF REGISTERED CLINICAL TRIALS INVOLVING CHILDREN

The extent to which rTMS interventional research in children has been conducted remains quite limited. Registries of clinical trials are important to consider. This is because studies in progress may not yet be published and, importantly, negative or potentially harmful studies may never be published. For this chapter, an additional source of information on the use of rTMS in children was utilized, the ClinicalTrials.gov registry, at https://clinicaltrials.gov/. This registry has a search interface allowing topics to be narrowed by a variety of factors of interest, including geography, so that the public can readily find studies nearby, and enrollment status (open or closed). It is established and maintained by the United States National Library of Medicine and contains clinical trial registrations for tens of thousands of clinical trials from more than 170 countries. As many Institutional Review Boards and many journals require ClinicalTrials.gov registration, this is a helpful comprehensive source of information on ongoing and recently completed studies. While not a comprehensive resource for study results, this registry does provide results in studies were investigators either decide to post results or at least indicate that results are available.

Searching ClinicalTrials.gov in September of 2015 for "transcranial magnetic stimulation" yielded 735 studies. Restricting this to interventional studies reduces the number to 587. Limiting this further to "child (ages 0–18)" reduces the number to 52 studies of known status, of which 21 are currently enrolling. Not all of these actually use TMS for treatment, for example, TMS measures in motor cortex may be used as a biomarker of results of another form of treatment. As is the case for MEDLINE searches, searching the Clinical Trials Registry for children's studies can result in misclassification. For example, a study with no clear minimum age listing, eg, enrolling participants "up to age 85 years," may appear under a pediatric search, even though no children are enrolled. Table 4.2 shows studies which, after reviewing information available on the Website, appear to be both therapeutic and reasonably pediatric-focused.

TABLE 4.2 Registered Clinical Trials Using Repetitive Transcranial Magnetic Stimulation (rTMS) and Enrolling Persons Under 18 Years of Age

Study Registration Identifier	Diagnosis Treated	Recruitment Status in 2015	Use of Sham	rTMS Stimulation Parameters	Stimulation Location	Duration of Treatment	Ages Eligible (years)
NCT02311751	Autism spectrum disorder	Recruiting	Yes	20Hz high frequency	Sequential to both DLPFC	4 weeks	16–35
NCT01388179	Autism spectrum disorder	Recruiting	Yes	1Hz low frequency standard plus high frequency deep coil[a]	L-DLPFC (low); left fusiform area through the superior temporal sulcus (high, deep)	Not specified	10–18
NCT01502033	Depression	Completed	No	10Hz high frequency	L-DLPFC	6 weeks	13–18
NCT01731678	Depression	Recruiting	No	10Hz high frequency	L-DLPFC	3 weeks	12–21
NCT01804296	Depression	Recruiting	No	10Hz high frequency	L-DLPFC	Extension	12–21
NCT01804270	Depression	Active	Yes	10Hz high frequency	L-DLPFC	6 weeks	12–21
NCT00587639	Depression	Completed	No	10Hz high frequency	L-DLPFC	6–8 weeks	13–18
NCT02472470	Depression	Recruiting	No	iTBS L-DLPFC; cTBS R-DLPFC[a]	B-DLPFC	10 days	16–24
NCT00878787	Dystonia	Completed	No	cTBS	Motor cortex	1 session	2–29
NCT00048490	Epilepsy	Completed	Yes	1Hz low frequency	Epileptic focus	5 days	5–65
NCT01745952	Epilepsy, refractory focal	Active	Yes	1Hz low frequency	Epileptic focus	2 weeks	16–75
NCT01352910	Functional paralysis (psychogenic)	Recruiting	Yes	2Hz low frequency	Motor cortex	2 days	14 and Older

NCT	Condition	Status		Stimulation	Target	Duration	Age
NCT02518867	Motor control: Cerebral palsy	Recruiting	Yes	Intermittent theta burst stimulation (iTBS)	SMA, leg motor cortex	3 days	7–20
NCT01189058	Motor control: Cerebral palsy, stroke	Active	Yes	1 Hz low frequency	Contralesional hemisphere motor cortex	10 days	6–18
NCT01637129	Motor control: Stroke	Completed	Yes	1 Hz low frequency	Contralesional hemisphere motor cortex	2 days (8 sessions)	6–18
NCT01104064	Motor control: Stroke, hemiplegia	Completed	Yes	6 Hz priming +1 Hz low frequency	Contralesional hemisphere motor cortex	10 days, 5 sessions alternating with CIMT	8–16
NCT01022489	Schizophrenia	Completed	Yes	20 Hz high frequency	L superior temporal sulcus	2 days (4 sessions)	16–65
NCT01373866	Schizophrenia: Hallucinations	Recruiting	No	1 Hz low frequency	fMRI hallucination focus, vs T3/P3	5 days	13–60
NCT01258790	Tourette syndrome	Completed	Yes	Continuous theta burst stimulation (CTBS)	SMA	2 days (8 sessions)	10 and Older
NCT02356003	Tourette syndrome	Recruiting	No	1 Hz low frequency	SMA	3 weeks	7–12

Hz, Hertz (pulses per second); *DLPFC*, dorsolateral prefrontal cortex; *SMA*, supplementary motor area; *R/L/B*, right/left/bilateral; *CIMT*, constraint-induced movement therapy; *fMRI*, functional magnetic resonance imaging for coil guidance (note – a number of studies use "navigated TMS" – see study registration for details); T3/P3 EEG standard lead location guidance.

ᵃStudies involving multiple stimulation parameters.

BRIEF OVERVIEW OF PUBLISHED CLINICAL TRIALS INVOLVING CHILDREN

Details on published rTMS therapeutic studies in children are presented in the topic-based chapters in this book. The role of this section is to provide an overview of the indications and stimulation parameters used. Table 4.3 contains information on the types of stimulation used for therapeutic studies in children. It should be noted that a number of groups have several publications for the same clinical trial, looking at biomarkers of response, for example. In addition, some studies included both children and adults. In those, the table includes only children.

It is notable that the studies for depression and motor control are similar, likely based on adult data. Both high- and low-frequency rTMS protocols are published in children. Most protocols stimulate at or just below resting motor threshold, but the details in terms of total pulses, number of days or weeks of treatment vary widely. Given the higher thresholds in children than adults, this means the intensity and volume of the pulses is high. However, overall based on the published studies the procedures seem well tolerated. At least one study of dystonia, whose results are described but unpublished, did acknowledge challenges related to high motor thresholds and difficulties with mapping motor cortex accurately.[28]

WHERE TO NEXT WITH rTMS TREATMENT IN CHILDREN?

A primary consideration in pediatric research is safety. It is reassuring that, to date, after over 15 years of repetitive TMS research in humans, safety concerns for seizures[29] and other adverse events continue to be low. Safety recommendations have been published.[30] It is to be anticipated that trials using rTMS will continue, with close monitoring and transparent reporting of adverse events.

Based on currently registered clinical trials it appears that rTMS clinical trials in children may continue to focus on common, relatively disabling conditions. This is driven in part by promising data in adults, and in part by practical considerations. TMS treatment protocols for the most part are incredibly time-consuming. Although an individual treatment session may require less than 30 min, the large number of treatments, often daily, administered after travel to the hospital setting pose a major time and economic burden to busy families. However, in children with asymmetric or unilateral cortical spinal tract injuries as part of cerebral palsy or sequelae of stroke, parents often visit various therapists regularly. So, in such circumstances, adding a TMS treatment before or after may not have much higher opportunity costs than the frequent therapy already does.

TABLE 4.3 Published Studies of Clinical Trials Using rTMS for Treatment in Children

Diagnosis Treated	n*	Location	Intensity (% RMT)	Frequency	Number of Pulses	Details	Duration	Design	Effects	References
Autism spectrum disorder	18	R And L DLPFC	90%	0.5 Hz	160	40s Trains, 20s intertrain, 8 trains	18 weekly Sessions	Open, wait list control	Improved	Casanova*[20]
Depression	2	L-DLPFC	110%	10 Hz	2000	5s Trains, 25s intertrain, 40 trains	4 weeks	Open	Improved	Loo[21]
Depression	8	L DLPFC	120%	10 Hz	3000	4s Trains, 26s intertrain, 75 trains	6–8 weeks	Open	Improved	Wall[22]
Motor control: Hemiparetic cerebral palsy	10	Motor cortex	90%	1 Hz	600	After 6 Hz priming	5 rTMS treatments alternating with 5 days CIMT	Double blind sham controlled	Improved	Gillick[23]
Motor control: Spasticity	10	Motor cortex	90%	5 Hz, 1 Hz	1500	60s trains, 120s intertrain, 5 trains	1 week	Double blind sham controlled	5 Hz improved	Valle[24]
Motor control: Stroke	5	Contralesional motor cortex	100%	1 Hz	1200	20 min	2 weeks	Double blind sham controlled	Improved	Kirton[25]
Tourette's syndrome	10	SMA	100%	1 Hz	1200	20 min	2 weeks	Open	Improved	Kwon[26]
Tourette's syndrome	5	SMA	90%	30 Hz CTBS	600	3 pulse Burst at 30 Hz, repeated 5× per second	8 sessions over 2 days	Double blind sham controlled	No different from sham	Wu[27]

Hz, Hertz (pulses per second); DLPFC, dorsolateral prefrontal cortex; SMA, supplementary motor area; R/L/B, right/left/bilateral; CIMT, constraint-induced movement therapy; (note – a number of studies use "navigated TMS" – see citation for details). n* indicates number of participants in the study who received active (not sham) TMS.

Similarly, if an adolescent has severe, treatment refractory depressive illness, she may be in a day program, residential program, or inpatient unit where again adding TMS sessions daily is straightforward.

For other, less-disabling conditions like high-functioning autism spectrum disorder, epilepsy, or Tourette's syndrome, it may not be feasible to enroll mildly to moderately affected patients in studies that require 15 or more visits on separate days. Severely affected patients may participate, but this represents a small fraction of children with these conditions and therefore poses challenges for adequate enrollment. If protocols combining multiple visits over several days can be effective, or if changing stimulation parameters or introducing priming protocols is effective, then shorter therapeutic courses of TMS could become more broadly feasible for less disabled children.

References

1. Hallett M. Transcranial magnetic stimulation and the human brain. *Nature.* 2000;406(6792): 147–150.
2. Barker AT, Jalinous R, Freeston IL. Non-invasive magnetic stimulation of human motor cortex. *Lancet.* 1985;1(8437):1106–1107.
3. Chen R. Studies of human motor physiology with transcranial magnetic stimulation. *Muscle Nerve.* 2000; suppl(9):S26–S32.
4. Dayan E, Censor N, Buch ER, Sandrini M, Cohen LG. Noninvasive brain stimulation: from physiology to network dynamics and back. *Nat Neurosci.* 2013;16(7):838–844.
5. George MS, Wassermann EM, Williams WA, et al. Daily repetitive transcranial magnetic stimulation (rTMS) improves mood in depression. *Neuroreport.* 1995;6(14):1853–1856.
6. Wassermann EM. Risk and safety of repetitive transcranial magnetic stimulation: report and suggested guidelines from the International Workshop on the Safety of Repetitive Transcranial Magnetic Stimulation, June 5–7, 1996. *Electroencephalogr Clin Neurophysiol.* 1998;108(1):1–16.
7. Pascual-Leone A, Valls-Sole J, Wassermann EM, Hallett M. Responses to rapid-rate transcranial magnetic stimulation of the human motor cortex. *Brain.* 1994;117(Pt 4):847–858.
8. Wassermann EM, Wedegaertner FR, Ziemann U, George MS, Chen R. Crossed reduction of human motor cortex excitability by 1-Hz transcranial magnetic stimulation. *Neurosci Lett.* 1998;250(3):141–144.
9. Stefan K, Kunesch E, Cohen LG, Benecke R, Classen J. Induction of plasticity in the human motor cortex by paired associative stimulation. *Brain.* 2000;123(Pt 3):572–584.
10. Litvak V, Zeller D, Oostenveld R, et al. LTP-like changes induced by paired associative stimulation of the primary somatosensory cortex in humans: source analysis and associated changes in behaviour. *Eur J Neurosci.* 2007;25(9):2862–2874.
11. Damji O, Keess J, Kirton A. Evaluating developmental motor plasticity with paired afferent stimulation. *Dev Med Child Neurol.* 2015;57(6):548–555.
12. Huang YZ, Edwards MJ, Rounis E, Bhatia KP, Rothwell JC. Theta burst stimulation of the human motor cortex. *Neuron.* 2005;45(2):201–206.
13. Huang YZ, Rothwell JC. The effect of short-duration bursts of high-frequency, low-intensity transcranial magnetic stimulation on the human motor cortex. *Clin Neurophysiol.* 2004;115(5):1069–1075.
14. Iyer MB, Schleper N, Wassermann EM. Priming stimulation enhances the depressant effect of low-frequency repetitive transcranial magnetic stimulation. *J Neurosci.* 2003;23(34):10867–10872.

15. Nongpiur A, Sinha VK, Praharaj SK, Goyal N. Theta-patterned, frequency-modulated priming stimulation enhances low-frequency, right prefrontal cortex repetitive transcranial magnetic stimulation (rTMS) in depression: a randomized, sham-controlled study. [Erratum appears in J Neuropsychiatry Clin Neurosci. 2012 Dec 1;24(1):118]. *J Neuropsychiatry Clin Neurosci.* 2011;23(3):348–357.

16. Wu SW, Shahana N, Huddleston DA, Gilbert DL. Effects of 30Hz theta burst transcranial magnetic stimulation on the primary motor cortex. *J Neurosci Methods.* 2012;208(2):161–164.

17. Desmond JE, Chen SH, Shieh PB. Cerebellar transcranial magnetic stimulation impairs verbal working memory. *Ann Neurol.* 2005;58(4):553–560.

18. Torriero S, Oliveri M, Koch G, Caltagirone C, Petrosini L. Interference of left and right cerebellar rTMS with procedural learning. *J Cogn Neurosci.* 2004;16(9):1605–1611.

19. Schutter DJ, Enter D, Hoppenbrouwers SS. High-frequency repetitive transcranial magnetic stimulation to the cerebellum and implicit processing of happy facial expressions. *J Psychiatry Neurosci.* 2009;34(1):60–65.

20. Casanova MF, Hensley MK, Sokhadze EM, et al. Effects of weekly low-frequency rTMS on autonomic measures in children with autism spectrum disorder. *Front Hum Neurosci.* 2014;8:851.

21. Loo C, McFarquhar T, Walter G. Transcranial magnetic stimulation in adolescent depression. *Australas.* 2006;14(1):81–85.

22. Wall CA, Croarkin PE, Sim LA, et al. Adjunctive use of repetitive transcranial magnetic stimulation in depressed adolescents: a prospective, open pilot study. *J Clin Psychiatry.* 2011;72(9):1263–1269.

23. Gillick BT, Krach LE, Feyma T, et al. Primed low-frequency repetitive transcranial magnetic stimulation and constraint-induced movement therapy in pediatric hemiparesis: a randomized controlled trial. *Dev Med Child Neurol.* 2014;56(1):44–52.

24. Valle AC, Dionisio K, Pitskel NB, et al. Low and high frequency repetitive transcranial magnetic stimulation for the treatment of spasticity. *Dev Med Child Neurol.* 2007;49(7):534–538.

25. Kirton A, Chen R, Friefeld S, Gunraj C, Pontigon AM, Deveber G. Contralesional repetitive transcranial magnetic stimulation for chronic hemiparesis in subcortical paediatric stroke: a randomised trial. *Lancet neurol.* 2008;7(6):507–513.

26. Kwon HJ, Lim WS, Lim MH, et al. 1-Hz low frequency repetitive transcranial magnetic stimulation in children with Tourette's syndrome. *Neurosci Lett.* 2011;492(1):1–4.

27. Wu SW, Maloney T, Gilbert DL, et al. Functional MRI-navigated repetitive transcranial magnetic stimulation over supplementary motor area in chronic tic disorders. *Brain Stimul.* 2014;7(2):212–218.

28. Bertucco M, Sanger TD. Current and emerging strategies for treatment of childhood dystonia. *J Hand Ther.* 2015;28(2):185–193. quiz 194.

29. Bae EH, Schrader LM, Machii K, et al. Safety and tolerability of repetitive transcranial magnetic stimulation in patients with epilepsy: a review of the literature. *Epilepsy Behav.* 2007;10(4):521–528.

30. Rossi S, Hallett M, Rossini PM, Pascual-Leone A. Safety, ethical considerations, and application guidelines for the use of transcranial magnetic stimulation in clinical practice and research. *Clin Neurophysiol.* 2009;120(12):2008–2039.

Transcranial Direct-Current Stimulation (tDCS): Principles and Emerging Applications in Children

P. Ciechanski, A. Kirton

University of Calgary, Calgary, AB, Canada

INTRODUCTION

Application of electricity to alter neuronal firing patterns predates the times of Galen. Using torpedo electric fish, Galen, and others, delivered strong transient currents to relieve conditions such as severe headaches, and later suggested applying this method to treat epilepsy. As early as 1804 physicist Giovanni Aldini applied galvanic (direct) currents to treat individuals suffering from melancholia. Over the next 130 years physicists, physicians, and scientists applied galvanic currents towards a range of mental disorders. Variable results were produced due to inconsistent procedures and poorly documented clinical descriptions of the treatment. With the discovery of electroconvulsive therapy in the 1930s, the use of galvanic current brain stimulation was abandoned. However, throughout the 1960s, it re-emerged with investigations focused on treating schizophrenia and depression, as well as altering mood and alertness. This re-emergence was short-lived, as pharmaceutical advances in psychiatry placed other research in its shadow. Direct-current brain stimulation research dwindled, yet again. In 1998, Priori and colleagues[1] were among the first to apply controlled polarized direct current across the scalp, producing small changes in cortical excitability. Two years following Priori's experiments, Michael Nitsche and Walter Paulus[2] published their seminal paper that established much of the modern approach to transcranial direct-current stimulation (tDCS). Since the turn of the century, tDCS research has exponentially increased, permeating into the fields of psychiatry, rehabilitation, and cognitive and motor performance enhancement. Since its most recent re-emergence, over 1500 papers have been published on the topic of tDCS, with less than 2% of those involving pediatric populations.

The aim of this chapter is to describe modern tDCS approaches towards application in children and adolescents. We will describe stimulation principles and standard procedures to ensure safe application of direct current in the pediatric population. Borrowing from both advanced computational current modeling studies and neurophysiology evidence, we will discuss differences in the electric fields stimulating the cortex of adults and children. Motor performance enhancement represents one of the most widely studied topics in the field of adult tDCS. We will discuss pairing direct-current stimulation with motor learning in children. In addition, we will describe adverse events, side effects, and tolerability of tDCS in children. Finally, we will discuss future clinical applications of direct-current stimulation in various neurological and psychiatric conditions, basing these applications on the best-available adult evidence.

PRINCIPLES AND CORTICAL EFFECTS OF TRANSCRANIAL DIRECT-CURRENT STIMULATION

Transcranial direct-current stimulation is an emerging form of non-invasive brain stimulation. The application of current through regions of the brain via two electrodes, an anode and a cathode, may shift regional cortical excitability to a state of excitation or inhibition.[2-4] A subthreshold current passes from the anode to the site of the cathode, generating weak electric fields in surface neuronal populations, which in turn change neuronal excitability and spontaneous firing patterns.[2,5,6] These changes in neuronal properties are dependent on both polarity (direction of the current) and intensity (amount of current).[4,7] The polarity of the tDCS application is conventionally referred to as anodal or cathodal stimulation. Traditionally, anodal stimulation involves placing the anode on the scalp over the target location, such as the primary motor cortex (M1), and the cathode at a distant inert location, such as the contralateral supraorbital area or shoulder. The electrode locations are reversed in cathodal stimulation, with the cathode being placed over a target location and the anode over an inert location. As the stimulation begins, changes primarily occur in the membrane potential of layer V/VI pyramidal neurons, leading to depolarization or hyperpolarization of neurons at the target region.[8] Anodal tDCS depolarizes target neurons, whereas cathodal tDCS hyperpolarizes those same neurons. The underlying principle for these effects stems from neuronal–electric field interactions, which suggest that electric fields polarized from the dendrites towards the axon will depolarize the axon and soma, but hyperpolarize the dendritic tree; the axon and soma will be hyperpolarized if electric fields point from the axon to the dendrites.[9] These changes in cortical excitability can be measured using transcranial magnetic stimulation (TMS), whereby a focused magnetic field depolarizes neurons, while responses are measured from respective muscles in the form of a motor-evoked potential (MEP). TMS is described in greater detail in Chapter 1, "TMS Basics: Single and Paired Pulse Neurophysiology."

Pharmacological studies have advanced our understanding of the immediate effects of tDCS on neurons. Application of the calcium channel-blocker flunarizine reduces the increases in neuronal excitability seen with anodal tDCS.[10,11] Similarly, blockage of sodium channels with the application of carbamazepine completely abolishes these effects.[10,11] Neither application of dextromethorphan, an NMDA-receptor antagonist, nor application of the $GABA_A$ receptor agonist lorazepam alters these changes in cortical excitability, suggesting that glutamatergic and GABAergic interneurons are less affected by anodal tDCS.[10,11] In contrast, the hyperpolarizing effects of cathodal stimulation remain unaffected by the application of flunarizine or carbamazepine, which follows evidence that cathodal stimulation inactivates calcium and sodium voltage-gated channels.[6] These changes also appear to be independent of NMDA activity.

The effects of both anodal and cathodal tDCS outlast the duration of the stimulation. Nitsche and Paulus[2,5] demonstrated that as little as 5 min of tDCS produces changes in cortical excitability that outlast the stimulation, often lasting 90 min after termination of the stimulation. These lasting changes may be dependent on synaptic modulation of interneurons. Application of D-cycloserine, a partial NDMA agonist, extends the length of time that increased cortical excitability is present following anodal tDCS,[12] whereas the application of an NMDA antagonist abolished this increase.[11] In addition, a reduction in GABAergic tone is associated with anodal stimulation, as demonstrated through TMS[13] and magnetic resonance spectroscopy (MRS) studies.[14] The after-effects of cathodal tDCS are less clear. Although cathodal stimulation produces a lasting decrease in cortical excitability, the role of glutamatergic interneurons in this decrease remains unclear due to the observation that NMDA antagonists,[6] but not agonists,[12] alter the after-effects.

A variety of neurotransmitters have been implicated in cortical excitability changes generated by tDCS. Amphetamines, catecholamine reuptake blockers, enhance and prolong hyperexcitability following anodal tDCS, producing changes that may last 24h after stimulation.[15] Similarly, dopamine appears to consolidate cortical excitability increases, particularly via the D2 receptor, and to a lesser extent through the D1 receptor.[16] GABA, the principal inhibitory neurotransmitter of the central nervous system (CNS), strengthens and prolongs cortical excitability increases following anodal tDCS, however this enhancement is delayed.[17] In contrast to catecholamines, dopamine, and GABA, which prolong cortical excitation following anodal stimulation, increasing acetylcholine appears to hinder hyperexcitability.[18] Nicotinic receptor agonists influence the effects of anodal and cathodal tDCS, however effects are dose-dependent.[19] Serotonin exhibits strong facilitatory effects on both anodal and cathodal tDCS[20]; application of citalopram, a serotonin reuptake blocker, enhances cortical hyperexcitability following anodal tDCS, and shifts the inhibitory effects of cathodal tDCS towards excitation. The application of tDCS towards a clinical population employing pharmacological management of symptoms, which alter neurotransmitter concentrations in the CNS, requires a thorough understanding of the influence of these neurotransmitters on the effects of tDCS.

Recent advances in MRS have presented the opportunity to quantify changes in neurotransmitter and metabolite concentrations within the brain. Largely, changes in cortical neurochemistry occur under the target electrode, primarily affecting GABA, glutamine, and glutamate. Decreases in GABA have been observed following anodal stimulation,[14,21,22] with increases in combined glutamine and glutamate concentrations[23,24] and myoinositol.[25] Cathodal tDCS has been studied less extensively in regard to neurochemical changes, however early studies suggest that cathodal tDCS decreases the concentration of combined glutamate and glutamine, and, due to its location in the metabolic pathway, GABA.[14] Resting-state imaging studies suggest that both anodal and cathodal tDCS may influence default mode networks, however additional studies are required.[24,26–29]

Over the past 15 years, the basic mechanisms of tDCS have been increasingly elucidated. By convention, anodal tDCS increases cortical excitability, and cathodal stimulation inhibits excitability, although exceptions to these rules are increasingly recognized. Pharmacological studies and advanced neuroimaging have demonstrated that baseline neurotransmitter concentrations influence the cortical responses to tDCS. Subsequently, applying subthreshold current to the cortex modifies metabolite concentrations under the site of the target electrode. How well these neuropharmacological and imaging studies apply to understanding the mechanisms of tDCS in the developing brain remains to be determined.

tDCS METHODOLOGY: ELECTRODES, CURRENT, AND PROTOCOLS

By adjusting various parameters, tDCS can be tailored to achieve a multitude of behavioral and neurophysiological effects. In this section we will discuss the importance and influence of electrode positioning and size, as well as current intensity.

Over the past 15 years, two general types of tDCS have been employed: conventional tDCS and, more recently, high-definition (HD) tDCS. Conventional tDCS (also referred to as 1 × 1 tDCS) is the most commonly applied and best-studied form of tDCS. This form of stimulation involves placing two relatively large electrodes, an anode and a cathode, on the head. Electrodes are encased in saline-soaked sponges. In contrast, HD tDCS (4 × 1 HD tDCS) involves positioning five small gel-based disk electrodes in the casing of an electroencephalogram recording cap. A central anode is surrounded by four cathodal electrodes in a ring-like orientation. Stimulator units should be current-controlled, due to changing electrode impedance. The majority of tDCS application targets the M1, where TMS can be easily used to monitor cortical excitability changes.

Conventional tDCS can further be divided into three montages: anodal, cathodal, or bihemispheric (Fig. 5.1A–C). Of these subdivisions, anodal tDCS is the most commonly applied form of tDCS. Anodal tDCS involves placing an anode over a target location (such as the M1), with the cathode being placed either at the contralateral supraorbital area (SO) or at an extracephalic location, such as the shoulder. The placement of the cathode on the head or upper body appears to influence cortical excitability

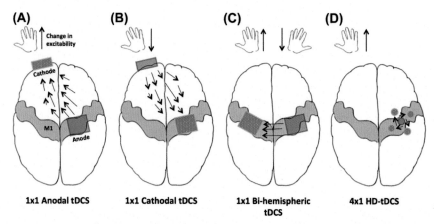

| 1x1 Anodal tDCS | 1x1 Cathodal tDCS | 1x1 Bi-hemispheric tDCS | 4x1 HD-tDCS |

FIGURE 5.1   Commonly applied tDCS montages, including: (A) anodal, (B) cathodal, (C) bihemispheric, and (D) HD tDCS. The anode is displayed as the solid red color, and the cathode as the dotted blue. The direction of current flow, from the anode to cathode, is displayed through arrows embedded on the brain. The change in cortical excitability, which would be assessed through motor-evoked potentials using TMS measures, and the target hand, is represented by the *hand* and a directional *arrow*, where "up" is an increase, and "down" is a decrease in MEP size.

changes with stimulation effects.[30] The target location is typically determined using the 10–20 international EEG system. More accurate electrode placement can be accomplished using neuronavigation technologies (such as BrainSight®), which allow users to determine the location of the M1 (or other structures) based on an individual's anatomical or functional neuroimaging. For additional precision, TMS can be performed to determine muscle-specific "hotspots" within M1.

Rubber electrodes are encased in square or rectangular saline-soaked sponge electrodes, typically 20–35 cm^2 in size. Sponges should be well-soaked in a standard 154 mM (0.9%) saline solution. Stronger saline solutions can be used to reduce voltage levels, however this may result in less tolerable stimulation.[31] Alternatively, deionized water can be used, however this is at the cost of applying greater voltage, which may trigger the built-in voltage limit restrictions set by stimulator units, automatically terminating stimulation. The sponge electrodes can be held in place by a non-conducting head strap, allowing uniform electrode contact with the scalp.

Once the electrodes are set up accordingly, the user may begin the stimulation. Typically, 0.5–2 mA currents are applied for 5–30 min. Published studies typically report the current applied, however the current properties are partially a function of the electrode size. Therefore, when presenting data reporting the current density (mA/cm^2) should be commonplace; a standard 1 mA current applied through a 25 cm^2 electrode produces a current density of 0.040 mA/cm^2. Reporting current density allows users to apply consistent currents when various electrode sizes are used. Current should be ramped up from 0 mA to the target strength over a period of time, especially in tDCS-naïve children, so the participant has a chance to inform the researcher if the stimulation is painful. Ramp-up times of 10–60 s are commonly used. Throughout the ramp-up phase the researcher should monitor voltage and impedance, as stimulator units will terminate the stimulation if voltage and impedance exceed predefined limits. As the stimulation begins, subjects typically report tingling and itching under the anode, however these sensations usually fade within 1–2 min. Users should ensure that sponges remain damp throughout the stimulation, as drying of the sponges may result in inconsistent current density. New sponges should be used routinely, as electrode fraying may lead to non-uniform current density under the electrode.

The application of cathodal or bihemispheric tDCS is comparable to that of anodal tDCS. With cathodal tDCS, the electrode locations are reversed from those of anodal tDCS; the cathode is positioned over the target M1, and the anode is placed on the contralateral SO. Bihemispheric tDCS involves direct modulation of both hemispheres concurrently. The anode is placed over the target M1, with the cathode centered on the contralateral M1.[32]

Conventional tDCS (specifically the M1/SO montage) does not produce focal stimulation. Advanced computational current modeling studies have demonstrated that a large portion of the cortex may be stimulated with this tDCS approach[32–34]; stimulation may occur under the anode, cathode, and parenchyma between these two regions. Various protocols have been tested to reduce this vast extent of stimulation. One possible modification is increasing the size of the cathode and reducing the size of the anode,[35] although the current strength should be reduced to maintain a constant current density under the anode. Furthermore, the interelectrode distance may influence the regions receiving stimulation. The effects of interelectrode distance require further investigation, as current modeling studies suggest that increasing the interelectrode distance does not significantly affect the focality,[36] whereas neurophysiology studies suggest that electrode separation does influence the degree of change in excitability.[30] Current shunting (current following a different path, such as moving across the scalp rather than passing through cortex) may increase when two large electrodes are relatively close to each other,[36] which is relevant in children who have a small head circumference. Importantly, peak electric fields are not localized directly under the anode; to achieve maximum electric fields at the M1 the anode should be positioned slightly posterior to the M1.[36,37]

When more focused current distribution is desired, HD tDCS methods can be employed.[38] HD tDCS involves positioning five electrodes, one center electrode and four return electrodes, into an EEG recording cap (Fig. 5.1D). In the case of M1 stimulation the center anode is positioned directly over the M1, surrounded in a ring-like fashion by the return electrodes. The return electrodes are typically positioned within a radius of approximately 7.5 cm. These small ring electrodes are Ag/AgCl-based electrodes embedded in an electrically conductive gel. The EEG cap should sit snugly on the participant's head, and hair should be separated at the electrode locations to ensure the scalp is exposed to the electrodes. Additional information for using the Soterix Medical HD-tDCS system is provided elsewhere.[39] Current modeling (described in the following section) suggests that current is restricted to the location of the return electrodes, and therefore is more focal than conventional tDCS.[40,41] Adult studies suggest that the neurophysiological effects of HD tDCS are comparable but not equivalent to those of conventional tDCS. Peak cortical excitability changes with HD tDCS occur 30 min later and last 2 h longer than those from conventional tDCS.[42] The effects of HD tDCS have not been evaluated in a pediatric population, and therefore warrant extra caution.

Sham tDCS conditions can be generated using commercially available tDCS units. Typical sham conditions involve completing all steps performed in active tDCS, with the lone difference being a greatly reduced stimulation period with early ramping-off of stimulation. Recently, Ambrus et al.[43] demonstrated that when sham tDCS was ramped-up to 1 mA over 20 s, held for 30 s, then ramped-down over 10 s, neither tDCS-naïve nor experienced subjects were able to reliably predict whether or not they received real or sham stimulation. Blinding participants may be more difficult when stronger currents are applied[44]; a longer ramp-down period may be necessary to accurately mimic the fading of stimulation sensation. Alternatively, a ramp-up ramp-down phase can be added towards the end of the sham stimulation phase to further mimic active stimulation.

Prior to stimulation the researcher should screen the participant for any of the following tDCS contraindications: implanted medical devices (ie, pacemakers), metallic objects in the head (braces and other metallic dental accessories excluded), a history of epilepsy, currently pregnant, head injuries resulting in wounds to the stimulation location, or previous adverse reactions to tDCS. Recent prescription and recreational drug-use should be noted; if these drugs act on the central nervous system then special considerations should be made (ie, exclude, or interpret results cautiously), as certain drugs can shape neurophysiological effects (see above). In addition, traits such as handedness may be controlled for, as hand-dominance may influence the effects of tDCS.[45]

tDCS DOSAGE: COMPUTATIONAL MODELING EVIDENCE

The effects of tDCS, as we will see in the next section, are largely dependent on the bioavailable dose. Although each stimulator is able to produce consistent current intensity, neuroanatomical factors, as well as electrode position, can largely influence the strength of the electric fields experienced at the neuronal level. Direct monitoring of electric fields in humans has not been performed to date, as these procedures are invasive. Although we cannot directly investigate electric field strengths, advanced computational current models have been developed to assist in our understanding of this concept.[40,41,46–48] High-resolution magnetic resonance imaging (MRI) of the head combined with finite element modeling displays the impact of tissue architecture on current flow. The first step in this process involves segmenting MRI scans into individual elements (typically soft tissues, bone, air, cerebrospinal fluid, eyes, white matter, and cortical and deep gray matter). Next, digitally rendered electrodes are positioned over the target location and a current is applied. The model is then solved using computational methods, where each segmented layer is given an isotropic conductivity value. This method generates electric field plots, which estimate the electric field strength throughout the brain given the electrode position and current strength.

To date two studies have employed current modeling to rationalize tDCS dosing in children.[33,49] Current modeling studies demonstrate that even in adults of similar age, peak electric fields reaching the cortex can vary by as much as 1.5 times,[32] therefore this variation is likely aggravated in developing children where skull thickness,[33] myelination,[50] CSF volume,[51] and brain–scalp distance[52] vary greatly with age. When comparing current distributions between an 8-year-old, a 12-year-old, and three adults, Kessler et al.[33] concluded that when applying a fixed current, peak electric fields are higher in children compared to adults, and also occupy a wider range of intensities. These observations may be attributable to multiple neurodevelopmental factors. First and foremost, bone is a poor conductor and therefore the skull largely blocks the transmission of current from the scalp to cortex. Adult modeling studies suggest that the skull may decrease current transmission by up to 50%[46]; the thinner skull of a developing child allows for greater current transmission, compared to adults. Furthermore, children have lower cerebrospinal fluid volumes compared to adults. Cerebrospinal fluid is highly conductive and fluid, therefore current is likely to be shunted (reducing current density) to a greater extent in adults. In addition, head circumference is age- and, to a lesser extent, gender-dependent.[53] As in adult studies, tDCS studies in children typically apply 25–35 cm² electrodes. Given the smaller head circumference and similar electrode sizes, the electrodes occupy a larger proportion of the scalp, leading to less focal stimulation in children.[33] HD tDCS may be used to produce more focal stimulation in both children[33] and adults,[40] although this is at the expense of more varied peak electric fields and experience in the pediatric population is limited.

Pediatric tDCS dosing is further complicated when applying stimulation in children with lesions or skull defects. For example, following an ischemic or hemorrhagic stroke, the lesioned area accumulates cerebrospinal fluid and surrounding brain structure may be deformed. Given the high conductivity of cerebrospinal fluid, applying tDCS over a lesioned area may result in extensive current shunting. In a case of a 10-year-old boy who experienced an arterial perinatal ischemic stroke, Gillick et al.[49] applied a computational current modeling method to determine appropriate doses required to produce electric fields comparable to adults receiving 1 mA cathodal tDCS. Using a standard M1-contralateral supraorbital area montage, they demonstrated that reducing the current by 30% resulted in comparable electric fields. Overall, these results suggest that lower current strengths can be applied in children to achieve electric field strengths comparable to those in adults.

tDCS DOSAGE: NEUROPHYSIOLOGY EVIDENCE

The effects of prolonged electrical fields on the cortex can be assessed using TMS measures. As described, anodal tDCS typically shifts cortical excitability towards a state of increased excitation, as seen through larger MEP amplitudes, whereas cathodal tDCS generates an inhibitory shift. More complex TMS measures can be performed to further assess the effects of tDCS on cortical neurophysiology in vivo.

Paired-pulse TMS protocols typically combine a submotor-threshold TMS conditioning pulse followed by a suprathreshold test pulse, allowing for the investigation of excitatory and inhibitory intracortical systems (see Chapter 1: TMS Basics: Single and Paired Pulse Neurophysiology).[54] When two such magnetic stimuli are separated by an interstimulus interval of 1–5 ms, inhibitory systems are activated, specifically $GABA_A$-mediated "short interval cortical inhibition" (SICI).[55–57] In contrast, interstimulus intervals of 10–15 ms activate "intracortical facilitation" (ICF), excitatory systems mediated by the NMDA receptor. In adults, anodal tDCS generally shifts SICI and ICF towards a state of excitation, such that the SICI response is less pronounced, and ICF is greater.[4,58] The effects of cathodal tDCS on SICI and ICF remain unclear. Early research suggested that 1 mA cathodal tDCS changes ICF, where less facilitation was present following stimulation,[6] however subsequent studies demonstrated no changes in SICI or ICF.[59] Importantly, the effects of cathodal tDCS are nonlinear, with 2 and 1 mA stimulation producing different, and possibly opposite effects. Recently, Batsikadze et al.[4] applied either 2 mA anodal, 2 mA cathodal, or 1 mA cathodal tDCS to a group of young adults. Two mA cathodal stimulation appeared to mimic the effects of 2 mA anodal tDCS, where cortical excitability was increased, as seen by larger MEP amplitudes, less pronounced SICI, and greater ICF responses. In contrast, 1 mA cathodal tDCS decreased MEP amplitudes, increased SICI, and decreased ICF. These findings suggest that the traditional anodal-excitatory/cathodal-inhibitory model may be more complex and requires additional study, particularly in children (see below).

In the previous section we described evidence suggesting that bioavailability of current (the electric field strength experienced by the cortex) may greatly differ between adults and children. Depending on multiple factors, the cortex of children may experience twofold stronger electric field strengths compared to adults receiving the same current. When combined with findings that 2 mA cathodal tDCS reverses neurophysiological outcomes, compared to 1 mA, we must consider the possibility that when we apply a 1 mA cathodal current to a child, their cortex may actually experience similar currents seen at 2 mA in adults. By extrapolation, the 2 mA currents commonly applied in adult studies may produce current densities as high as 4 mA in children.

To date, only a single study has been published examining the effects of tDCS on cortical excitability in healthy children.[60] This study indirectly confirmed that current densities experienced by children are higher than those in adults. Application of 1 mA cathodal tDCS over the M1 of 19 children (mean age 13.9 years) shifted cortical excitability to a state of excitation as measured by change in MEP amplitude. The magnitude of effect appeared comparable to the Batsikadze study[4] effects of 2 mA cathodal tDCS in adults. Adjusting this cathodal current stimulation in children to 0.5 mA produced the expected shift towards reduced excitability. These results are in contrast to the finding that 0.5 mA anodal tDCS is not sufficient to increase MEP amplitudes in children, whereas in adult studies current greater than 0.4 mA typically produces such changes in excitability.[2] The neurophysiological effects of tDCS in children have not been investigated in great detail, with an increasing need to describe effects of tDCS on cortical neurophysiology.

tDCS IN HEALTHY CHILDREN TO ENHANCE MOTOR LEARNING

The effects of tDCS on motor learning have been studied extensively in adults, where tDCS is efficacious in improving skill acquisition among a wide range of motor tasks. The general principle behind facilitating motor learning is seeded in long-term potentiation-like mechanisms in the cortical motor network, however recent studies have also implied a role for GABA.[61,62] Functional-MRI studies have shown that such motor learning is associated with changes in cortical networks and structures.[63] Three distinct tDCS montages have been applied to alter cortical function to modulate motor learning: anodal, cathodal, and bihemispheric tDCS.

Anodal tDCS over the M1 contralateral to the trained hand is the most commonly applied stimulation. Possibly by modulating NMDA receptors, which may increase postsynaptic Ca^{2+} levels, the effects of motor learning may be additive with anodal tDCS, allowing for increased synaptic modification.[64] In accordance with Hebbian plasticity models, synaptic plasticity may occur in the M1 when a motor task is repeated, leading to a lasting increase in that skill. The majority of motor learning training in past studies has been completed using the non-dominant hand; training the dominant hand may lead to a learning "ceiling," as participants already likely have high motor function in that hand they may be less able to further improve function. Indeed, when examining the effects of anodal tDCS on simple daily tasks, assessed using the Jebsen–Taylor hand function test (JTHFT), improvements were noted in the non-dominant hand but not the dominant hand.[65] Furthermore, fine motor tasks, which require greater dexterity, may be more sensitive to change than gross motor tasks.[66] Stimulating the ipsilateral (dominant) M1 with anodal tDCS is insufficient to improve motor skill.[67]

Increased motor skill and learning is not limited to simple tasks, with improvements noted in more difficult skills requiring speed and accuracy. Using the Purdue pegboard test (PPT) Kidgell et al.[68] demonstrated improvements in function with anodal tDCS. Although the majority of studies employ a single application of tDCS, multiple days of tDCS may facilitate further improvements. Using novel motor learning tasks that involved a high-speed–accuracy trade-off function, Reis et al.[69] and Fritsch et al.[70] demonstrated large improvements with anodal tDCS versus sham over 5 days of stimulation in adults, with skill remaining higher up to 3 months following stimulation. Interestingly these improvements came predominantly in the form of consolidation (so-called "offline") rather than daily practice ("online") learning effects. The timing of tDCS application relative to motor task execution has also been investigated in a series of experiments. Concurrent application of anodal tDCS during training or immediately following training consistently leads to improved motor performance, whereas application prior to training may not be effective.[61]

Neurophysiological correlates of tDCS-facilitated motor learning are increasingly defined. Advances in TMS and MRS have improved understanding regarding the role of interhemispheric inhibition and GABA on motor learning modulation. Improvements in motor function, assessed on the non-dominant hand using the JTHFT, suggest that decreases in inhibition from the dominant to non-dominant hemisphere (allowing for greater activation of the non-dominant hemisphere) are correlated with improvements in motor function.[71] In addition, changes in GABA have been implicated in motor learning modulation; the extent of decrease in GABA following anodal tDCS correlates with the degree of motor learning in sequence-learning tasks.[21]

The effects of cathodal or bihemispheric tDCS on motor learning have been studied less extensively with varying results. The principle behind these stimulation paradigms stems from modulating interhemispheric inhibition; cathodal tDCS aiming to decrease the influence of the M1 ipsilateral to the trained hand, whereas bihemispheric approaches may have the added effects of simultaneously increasing excitability in the contralateral M1. The effects of cathodal tDCS on sequence learning or finger-tapping tasks have produced varying results, where cathodal tDCS either disrupts or does not affect performance,[72–74] however bihemispheric tDCS can improve task performance.[75] The effects of cathodal tDCS in particular are difficult to interpret in regards to mechanisms, as excitability increases associated with training may counteract the decreases in cortical excitability seen with stimulation.

The effects of tDCS on motor learning have been investigated in children. Preliminary experiments by our group (unpublished) suggest that tDCS can facilitate motor learning. Over a 3-day period, 19 school-aged children (aged 8–18 years of age) performed the PPT with their left hand. They were randomized to receive either: contralateral 1 mA anodal tDCS, ipsilateral 1 mA cathodal tDCS, ipsilateral 2 mA cathodal tDCS, or sham tDCS during the first 20 min of training each day. All groups improved their PPT scores; however the active stimulation groups had significantly greater improvements (>50% gains in performance) at the end of day 3, compared to those receiving sham stimulation (Fig. 5.2). Further investigations are required to elucidate the effects of tDCS on motor learning in children.

SAFETY, ADVERSE EVENTS, AND TOLERABILITY

tDCS is widely regarded as a safe form of non-invasive brain stimulation. A recent review by Bikson et al. (unpublished) demonstrated the safety of tDCS, taking into consideration over 26,000 tDCS sessions. This analysis concluded that tDCS is not associated with any tissue damage, irreversible behavioral changes, or serious adverse events. Medwatch, the FDA-based adverse-event reporting system, lists no cases of serious adverse events, and likewise, the FDA acknowledges that tDCS trials pose no significant risks to health and safety.

Animal studies demonstrated that current densities of $142.9–285.7\,A/m^2$ applied for 10 min resulted in parenchymal lesions.[76] Lesion size was

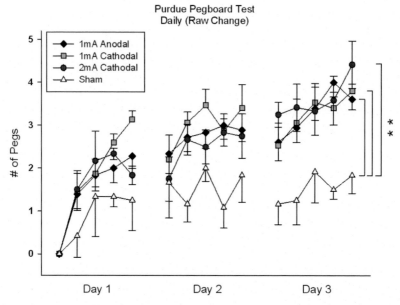

FIGURE 5.2   Effects of tDCS on motor learning in healthy children. The Y-axis depicts improvements in left (non-dominant) hand Purdue pegboard test (PPT) scores, and time-course is along the X-axis. Participants receive 20 min of tDCS daily over 3 days. Stimulation conditions included: right hemisphere 1 mA anodal tDCS (black diamond), left hemisphere 1 mA cathodal tDCS (light gray square), left hemisphere 2 mA cathodal tDCS (dark gray square), or sham tDCS. All groups improved their PPT scores, however all active tDCS conditions facilitated improvements. ** p-value < 0.01.

correlated with charge density, with the minimal charge necessary to generate a lesion being $52 \, kC/m^2$. Standard tDCS procedures in human involve the application of up to $0.480 \, kC/m^2$, approximately 100 times lower than the lesion-producing charge density. Moreover, charge density experienced by the parenchyma is further reduced due to current shunting by the scalp, skull, and cerebrospinal fluid. To prevent tissue damage, current densities exceeding $14.29 \, mA/cm^2$ should not be applied in adults.[77] Substantially lower current density limits in children should be considered as peak electric fields may be higher in children as discussed above.

Evidence from human experiments demonstrates that minimal side-effects are experienced with tDCS. A review of the safety of tDCS in children and adolescents by Krishnan et al.[78] reported the adverse events from 16 studies, including 191 participants (3–18 years of age) (see Table 5.1). The most commonly reported events were tingling (22/191), itching (9/191), and redness (under the site of the anode) (9/191). Tingling and itching sensations were transient, whereas redness would persist as much as 2 h following stimulation. Of these 191 participants, only

TABLE 5.1 Description of all Pediatric tDCS Studies Completed to Date, Including Sample Size, Stimulation Parameters, and Reports of Adverse Events. No Adverse Events Were Classified as Serious. Event Count was not Listed if not Explicitly Stated in the Article

Author, Year	Indication	# of Subjects	Current Strength (mA)	Current Density (mA/cm²)	Montage Anode–Cathode	Time (min)	# of Sessions per Subject	Adverse Events (Event Count, if Reported)
Bogdanov et al. (1994)	Infantile cerebral palsy (ICP)	21	0.2–0.8	0.04–0.16	F1 – C3 F1 – mastoid process	10–50	7–15	Slight heating or burning
Alon et al. (1998)	ICP	7	0.5	0.044	Left – right temporal areas	10	112 (2/day)	No adverse events reported
Shelyakin et al. (2001)	ICP CNS lesions	18	0.3–0.7	0.06–0.14	Posterior temporal area – parietal area	20–40	15	No adverse events reported
Ilyukhina et al. (2005)	Delayed neuropsych. development	30	0.03–0.08	Unknown	F1 – Not reported Parietal area – not reported	15–20	6	No adverse events reported
Mattai et al. (2011)	Schizophrenia	13	2.0	0.080	F3 – forearm STG - forearm	20	10	Redness (4), tingling, itching, fatigue (3)
San-juan et al. (2011)	Rasmussen's encephalitis	1	1.0–2.0	6.6–13.2	Contralateral SOA – C3 F2 – F8	60	4	No adverse events reported
Schneider et al. (2011)	Autism	10	2.0	0.080	F3 – contralateral SOA	30	1	No adverse events reported
Varga et al. (2011)	Epilepsy	5	1.0	0.040	-ve Epileptogenic focus – +ve epileptogenic focus	20	1	No adverse events reported

Continued

TABLE 5.1 Description of all Pediatric tDCS Studies Completed to Date, Including Sample Size, Stimulation Parameters, and Reports of Adverse Events. No Adverse Events Were Classified as Serious. Event Count was not Listed if not Explicitly Stated in the Article —cont'd

Author, Year	Indication	# of Subjects	Current Strength (mA)	Current Density (mA/cm²)	Montage Anode–Cathode	Time (min)	# of Sessions per Subject	Adverse Events (Event Count, if Reported)
Yook et al. (2011)	Seizure	1	2.0	0.080	Contralateral SOA – right tempo-parietal area	20	10	No adverse events reported
Andrade et al. (2013)	Language disorders	14	2	0.057	Broca area – contralateral SOA	30	5	Tingling (4), itching (4), slight burning (2), scalp pain, redness, headache (2), sleepiness (2), difficulty concentrating (2), acute mood change (6), irritability (5)
Auvichayapat et al. (2013)	Epilepsy	36	1.0	0.029	Contralateral shoulder – epileptogenic focus	20	1	Redness (1)
Young et al. (2013)	Dystonia	11	1.0	0.029	Contralateral SOA – M1	2 × 9 min	1	Skin discomfort (2)
Grecco et al. (2014)	Delayed neuropsych. Development	1	1.0	0.040	M1 – contralateral SOA	20	10	Unknown (reported as "mild")

Study	Group	N			Montage	Duration	N	Adverse events
Greco et al. (2014)	CP	20	1.0	Unknown	M1 – contralateral SOA	20	1	Unknown
Aree-uea et al. (2014)	CP	46	1.0	Unknown	M1 – contralateral SOA	20	5	Acute mood change, irritability, tingling, itching, headache, slight burning, sleepiness, difficulty concentrating
Greco et al. (2014)	CP	24	1.0	0.040	M1 – contralateral SOA	20	10	No adverse events reported
Duarte et al. (2014)	CP	24	1.0	0.040	M1 – contralateral SOA	20	10	Tingling (18), redness (3)
Prehn-Kristensen et al. (2014)	ADHD	12	0.25	0.497	F3 or F4 – M1 or M2	5 × 5 min	1	No adverse events
Young et al. (2014)	Dystonia	14	1.0	0.029	Contralateral SOA – M1	2 × 9 min	1	Skin discomfort (3)
Collange Grecco et al. (2015)	CP	20	1.0	0.040	M1 – contralateral SOA	20	10	Tingling (4)
Moliadze et al. (2015)	Healthy	19	0.5–1.0	0.014–0.029	M1 – contralateral SOA Contralateral SOA – M1	10	1	Itching

Continued

TABLE 5.1 Description of all Pediatric tDCS Studies Completed to Date, Including Sample Size, Stimulation Parameters, and Reports of Adverse Events. No Adverse Events Were Classified as Serious. Event Count was not Listed if not Explicitly Stated in the Article —cont'd

Author, Year	Indication	# of Subjects	Current Strength (mA)	Current Density (mA/cm^2)	Montage Anode–Cathode	Time (min)	# of Sessions per Subject	Adverse Events (Event Count, if Reported)
Gillick et al. (2015)	CP	13	0.7	0.020	Perilesional M1 – contralesional M1	10	1	Itching, sleepiness, difficulty concentrating, discomfort (1)
Bhanpuri et al. (2015)	Dystonia	9	1.5–2.0	0.054–0.071	C3/C4 – contralateral SOA Contralateral SOA – C3/C4	9	5	Tingling, discomfort (2), headache (1)
Lazzari et al. (2015)	CP	10	1.0	0.040	M1 – contralateral SOA	20	1	No adverse events reported
Amatachaya et al. (2015)	Autism	20	1.0	0.029	F3 – contralateral shoulder	20	1	Redness (3)
Total # of studies		Total # of subjects	Current strength range (mA)	Current density range (mA/cm^2)		Time range (min)	# of sessions range	
25		498	0.03–2.0 Average=0.95	0.014–13.2		5–60 Average=20.2	1–112	

Krishnan C, Santos L, Peterson MD, Ehinger M. Safety of noninvasive brain stimulation in children and adolescents. Brain Stimul. 2015;8(1):76–87. http://dx.doi.org/10.1016/j.brs.2014.10.012.

a single drop-out was reported, suggesting high tolerability. Adverse effects appear to occur at a lower frequency in children and adolescents as compared to adults. A systematic review of 117 adult tDCS trials, which reported adverse effects, revealed adverse events including itching (39.3% of studies), tingling (22.2% of studies), headache (14.8% of studies), a burning sensation (8.7% of studies), and discomfort (10.4% of studies).[79] Trials that included sham stimulation reported adverse events at a similar frequency. Recently, Kessler et al.[80] pooled adverse effects data from eight trials, consisting of 131 adults receiving 227 sessions of sham or active tDCS. As tDCS was applied, adverse events noted included (in order of decreasing frequency): tingling, itching, burning, difficulty concentrating, pain, fatigue, changes in visual perception, headache, nervousness, and acute mood changes. The frequency of these adverse events ranged from 77% to 7%. Of these adverse events, only tingling, itching, burning, pain, and difficulty concentrating were significantly different between active and sham tDCS. With the exception of mood changes, all adverse events were reported at a lower frequency after the stimulation was terminated. Of note, adult and child adverse effect data were pooled from multiple studies that employed varying electrode placement, current intensities, current ramping, and saline strengths. Reporting of adverse events is variable and incomplete in many tDCS trials. Published trials should include: the type of adverse event, severity, as reported by the participant (ie, mild, moderate, or severe), duration, and proportion of participants that reported the event.

tDCS is well-tolerated in adults and children. Evidence from adult studies suggests that the 50% detection threshold of tDCS is at approximately 0.4 mA.[81] Current intensities lower than 3 mA have been described as being tolerable,[82] however sensations involved with application of a 3 mA current in children may produce a more painful sensation (due to higher peak electric fields in children). Researchers should be cautious when considering applying these strong currents in children. Detection thresholds of children have not been explored to date.

Preliminary investigations by our group (unpublished) of 57 sessions of 1–2 mA anodal, cathodal, or sham tDCS in 19 children aged 8–18 years suggest that tDCS is tolerable in this population. Following 20 min of stimulation (1 or 2 mA), children completed a tolerability questionnaire (adapted from Garvey et al.[83]) in which they rank tDCS in comparison to seven common childhood experiences. Overall, tDCS ranked ~4/8, comparably tolerable to a long car ride (Fig. 5.3). No significant differences in scores were noted between 1 mA anodal, 1 mA cathodal, 2 mA cathodal, and sham. tDCS tolerability and comfort may be increased by adjusting the sponge-saline concentration; high concentration saline (220 mM) is associated with less comfortable sensation, whereas lower concentrations (15 mM) are favorable.[31] It is important to keep in mind that lower ionic strengths result in higher-voltage outputs, which may be limited in certain tDCS units.

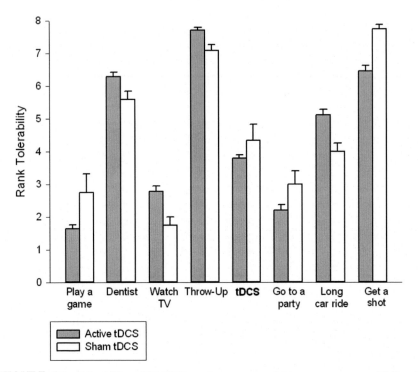

FIGURE 5.3   Tolerability of 57 tDCS sessions compared to seven common childhood experiences. Healthy children received either: 1 mA anodal tDCS, 1 mA cathodal tDCS, 2 mA cathodal tDCS, and sham tDCS. No differences were noted between the three active stimulation groups, and therefore data were grouped. Overall tDCS was tolerable, ranking 4/8, where eight is least tolerable, and one is most tolerable.

CLINICAL APPLICATION OF tDCS

In this section, we will briefly review the clinical use of tDCS in rehabilitation, and neurological and psychiatric conditions. To date there have not been any large clinical trials documenting the effects of tDCS in children and adolescents, therefore we will draw our conclusions from the best available adult evidence, while suggesting possible parallel applications in the developing brain. It should be noted that such extrapolation is not a sufficient replacement for clinical trials investigating the effects of tDCS in children. Throughout this section we will report effect sizes in the form of "number of participants (number of studies); effect size (where greater

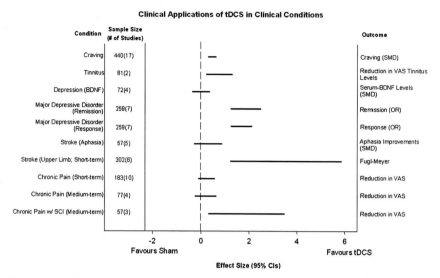

FIGURE 5.4   Clinical application of tDCS in neurological and psychiatric conditions. Conditions investigated are listed along the far left axis, followed by sample sizes and the number of studies included in the meta-analysis. Confidence intervals of 95% are reported. Positive effect sizes favor active over sham tDCS, and negative effect sizes favor sham tDCS. When confidence intervals cross the null-line (zero) no effect is noted. Outcome measures, from which confidence intervals are drawn, are listed along the right axis. BDNF, Brain-derived neurotrophic factor; OR, odds ratio; SCI, spinal cord injury; SMD, standardized mean difference; VAS, visual analog scale.

than zero favors active tDCS over sham, and less than zero favors sham); 95% confidence intervals (CIs)". When 95% CIs cross zero, the effect is considered as null. Fig. 5.4 summarizes the findings we will discuss here. There is a pressing need to explore the effects of tDCS in children in various clinical settings.

Stroke Rehabilitation

Perinatal stroke is the leading cause of hemiparetic cerebral palsy (CP).[84] CP has classically been regarded as a permanent and non-modifiable condition.[85] Survivors of perinatal stroke will be afflicted by this disability for decades; therefore, there is an increasing need to investigate technologies that might partially restore motor function or enhance rehabilitation strategies. Combining animal and human evidence, models of motor system reorganization following such early focal brain injury have identified cortical targets for neuromodulation including the non-lesioned M1.[86] Another example of such altered neurophysiology in childhood and adult stroke is an imbalance of interhemispheric inhibition, where the unaffected hemisphere exerts a relative excess of inhibition over the lesioned hemisphere.[87,88] While the poststroke reorganization models differ based on age at injury, a general concept of enhancing lesioned M1 or inhibiting contralesional M1 are shared targets for non-invasive neuromodulation. As with applying tDCS to facilitate motor learning, three strategies have been proposed in the application of tDCS across these stroke populations: (1) increasing the excitability of the lesioned cortex using anodal tDCS, (2) decreasing the excitability of the contralesional hemisphere with cathodal tDCS, or (3) a combination of (1) and (2) using bihemispheric tDCS. The effects of tDCS on perinatal and childhood stroke rehabilitation have not been published to date, however Gillick et al.[89] have demonstrated the feasibility and safety of these trials (see Chapter 10: The Right Stimulation of the Right Circuits: Merging Understanding of Brain Stimulation Mechanisms and Systems Neuroscience for Effective Neuromodulation in Children).

Multiple controlled tDCS trials have been completed in adult hemiparesis secondary to stroke. A recent meta-analysis by The Cochrane Collaboration revealed that active tDCS produces favorable outcomes compared to sham stimulation.[90] Outcomes for these studies included measures of activities of daily living (ADL; Barthel Index) and upper extremity function (Fugl–Meyer). When assessed at the end of the intervention, tDCS had no effect on ADL measures ($n = 172[5]$; 5.31 standardized mean difference [SMD]; −0.52, 11.14), however at follow-up months later, tDCS was favorable to sham ($n = 63[3]$; 11.16 SMD; 2.89, 19.43). Intervention with active tDCS is favorable in the late subacute to chronic phase of recovery ($n = 70[2]$; 10.78 SMD; 2.53, 19.02) compared to acute and subacute timeframes ($n = 102[3]$; 5.23 SMD; −3.74, 14.21). Furthermore, the application of cathodal tDCS over the non-lesioned hemisphere ($n = 50[3]$; 8.31 SMD; 1.11, 15.50) may be more effective as compared to anodal tDCS of the lesioned hemisphere ($n = 52[3]$; 4.92 SMD; −6.54, 16.39) in regards to ADL. Application of tDCS over the lesioned hemisphere is also less desirable with less defined targets and possible shunting of the current by the large volume of cerebrospinal fluid that may occupy the lesioned area.

In addition to motor disabilities, approximately 20–30% of stroke survivors suffer from aphasia.[91,92] The most common approach to combating aphasia is speech and language therapy (SLT). Although evidence for SLT is modest, early studies are promising.[93] Pairing tDCS with SLT may improve the efficacy of therapy, although ideal stimulation locations and montages are not well defined. A recent Cochrane meta-analysis evaluated the effects of tDCS paired with SLT in adults suffering from aphasia following stroke.[94] Moderate-quality evidence suggests that tDCS does not improve aphasia symptoms in adults ($n = 54[5]$; 0.31 SMD; −0.26, 0.87). Due to the limited number of studies and some evidence of success with rTMS, further investigations are needed. Recently we demonstrated that pairing contralesional inferior frontal gyrus rTMS with SLT was well tolerated with large functional improvement in a single case of childhood stroke aphasia.[94a] The effects of pairing tDCS with SLT in perinatal and childhood stroke have not yet been investigated in clinical trials.

Major Depression

Major depressive disorder (MDD) is a chronic psychiatric illness that afflicts millions of people worldwide. With an estimated prevalence of up to 17%,[95] MDD is projected to be the second most common condition by the year 2020.[96] There is an increasing need to investigate therapies that combat MDD. Currently, pharmacological management is a first-line treatment but response rates are poor and nearly one third of patients will not achieve remission after two or more antidepressants.[97] Limitations are particularly evident in adolescents and young adults where suicide rates are highest. Non-invasive brain stimulation has emerged as a novel treatment for MDD (see Chapter 15: Brain Stimulation in Pediatric Depression: Biological Mechanisms and Chapter 16: Brain Stimulation in Childhood Mental Health: Therapeutic Applications). Modern imaging in MDD has demonstrated altered metabolic activity and structural differences in multiple brain areas, particularly the dorsolateral prefrontal cortex (DLPFC).[98,99] These findings have identified the DLPFC as a target for neurostimulation. It is now well established that excitatory repetitive TMS over the left DLPFC improves MDD symptoms in adults[100] and we have recently demonstrated similar preliminary evidence in pediatric populations.[101]

As an excitatory neuromodulator, tDCS has been examined as a potential non-pharmacological intervention for individuals suffering from MDD. A 2014 meta-analysis by Shiozawa et al.[102] including 259 patients in seven randomized control trials demonstrated that, similar to rTMS, tDCS may also be beneficial in treating MDD with odds ratios for response to treatment (1.73; 1.26, 2.12) and remission (2.50; 1.26, 2.49) favoring active tDCS over sham. The meta-analysis did not identify any response predictors, however Yang et al.[101] reported that high baseline glutamate levels may decrease response rates in a pediatric MDD population receiving rTMS. Other neurotrophic factors, such as brain-derived neurotrophic factor (BDNF) have been implicated in MDD[103,104] and treatment may increase serum BDNF levels.[105] Although it was suggested that tDCS promotes BDNF-dependent plasticity at the synaptic level,[70] a meta-analysis by Brunoni et al.[106] demonstrated that BDNF serum levels are unchanged following tDCS ($n = 72[4]$; 0.00 SMD; $-0.35, 0.36$). Early investigations suggest that tDCS may be efficacious in the treatment of MDD, although precise mechanisms are yet to be established.

Additional fields of therapeutic tDCS investigation include chronic pain, management of craving symptoms, tinnitus, and epilepsy. Early investigations into reducing chronic pain have produced only low-grade evidence with no clear suggestion that tDCS is effective. Pain levels, assessed using the visual analog scale (VAS; pain scale rating from 0 to 10) have not favored active over sham tDCS at short-term ($n = 183[10]$; 0.18 mean reduction in VAS; −0.09, 0.56) or medium-term follow-up ($n = 77[4]$; mean 0.20 reduction in VAS; −0.24, 0.63).[107] Although tDCS may be ineffective in general chronic pain, stimulation holds potential in reducing chronic pain following spinal cord injury ($n = 57[3]$; 1.87 mean reduction in VAS; 0.45, 3.30).[108] Neuromodulation of the DLPFC may influence the craving of illicit substances and alcohol dependence, as well as craving for food. A meta-analysis of 440 subjects enrolled over 17 trials demonstrated the positive effect of active tDCS over sham tDCS in reducing craving (0.476 SMD; 0.316, 0.636).[109] It is important to note that this meta-analysis combined repetitive TMS (nine studies) and tDCS (eight studies) studies, although no differences were found between the two stimulation types. Tinnitus is a sound perception that occurs in the absence of any auditory stimuli. As of 2012, 17 studies had been performed, however only two randomized control trials of tDCS were included in a meta-analysis performed by Song et al.[110] This analysis suggested that tDCS may reduce auditory perception in tinnitus ($n = 81[2]$; 0.77 reduction in tinnitus VAS; 0.23, 1.31). Finally, an emerging field in which tDCS is being tested is epilepsy (see Chapter 14: Noninvasive Brain Stimulation in Pediatric Epilepsy: Diagnostic and Therapeutic Uses). Using the inhibitory properties of cathodal tDCS, researchers are attempting to suppress epileptiform discharges and clinical seizures. A recent systematic review by San-juan et al.[111] identified six studies that enrolled 65 patients with epilepsy. Although a meta-analysis was not performed, the systematic review provided evidence that tDCS may reduce seizure activity (four of six studies) and epileptiform discharges (five of six studies). Although these studies have shown the safety of applying tDCS in patients with epilepsy, further investigations and larger randomized controlled trials are needed to determine the efficacy of tDCS in epilepsy.

SUMMARY

The relative simplicity and portability of tDCS, significant safety data, and a substantial evidence base for efficacy in motor learning modulation and multiple clinical conditions, makes it an appealing method for non-invasive neuromodulation in children. However, experience in children remains limited with genuine concern of significant developmental differences and mechanisms of action are poorly understood. A cautious, systematic progression of carefully designed studies and clinical trials promises to see increased tDCS applications in the developing brain in the near future.

References

1. Priori A, Berardelli A, Rona S, Accornero N, Manfredi M. Polarization of the human motor cortex through the scalp. *Neuroreport*. 1998;9(10):2257–2260.
2. Nitsche MA, Paulus W. Excitability changes induced in the human motor cortex by weak transcranial direct current stimulation. *J Physiol*. 2000;527(3):633–639.
3. Lang N, Nitsche MA, Paulus W, Rothwell JC, Lemon RN. Effects of transcranial direct current stimulation over the human motor cortex on corticospinal and transcallosal excitability. *Exp Brain Res*. 2004;156(4):439–443. http://dx.doi.org/10.1007/s00221-003-1800-2.
4. Batsikadze G, Moliadze V, Paulus W, Kuo M-F, Nitsche MA. Partially non-linear stimulation intensity-dependent effects of direct current stimulation on motor cortex excitability in humans: effect of tDCS on cortical excitability. *J Physiol*. 2013;591(7):1987–2000. http://dx.doi.org/10.1113/jphysiol.2012.249730.
5. Nitsche MA, Paulus W. Sustained excitability elevations induced by transcranial DC motor cortex stimulation in humans. *Neurology*. 2001;57(0028–3878 (Print)):1899–1901.
6. Nitsche MA, Nitsche MS, Klein CC, Tergau F, Rothwell JC, Paulus W. Level of action of cathodal DC polarisation induced inhibition of the human motor cortex. *Clin Neurophysiol*. 2003;114(1388–2457 (Print)):600–604.
7. Bastani A, Jaberzadeh S. Differential modulation of corticospinal excitability by different current densities of anodal transcranial direct current stimulation. Wenderoth N, ed. *PLoS One*. 2013;8(8):e72254. http://dx.doi.org/10.1371/journal.pone.0072254.
8. Radman T, Datta A, Ramos RL, Brumberg JC, Bikson M. One-dimensional representation of a neuron in a uniform electric field. *Conf Proc Annu Int Conf IEEE Eng Med Biol Soc IEEE Eng Med Biol Soc Annu Conf*. 2009;2009:6481–6484. http://dx.doi.org/10.1109/IEMBS.2009.5333586.
9. Ruffini G, Wendling F, Merlet I, et al. Transcranial current brain stimulation (tCS): models and technologies. *IEEE Trans Neural Syst Rehabil Eng*. 2013;21(3):333–345. http://dx.doi.org/10.1109/TNSRE.2012.2200046.
10. Nitsche MA, Fricke K, Henschke U, et al. Pharmacological modulation of cortical excitability shifts induced by transcranial direct current stimulation in humans. *J Physiol*. 2003;553(1):293–301. http://dx.doi.org/10.1113/jphysiol.2003.049916.
11. Liebetanz D, Nitsche MA, Tergau F, Paulus W. Pharmacological approach to the mechanisms of transcranial DC-stimulation-induced after-effects of human motor cortex excitability. *Brain*. 2002;125(10):2238–2247.
12. Nitsche MA, Jaussi W, Liebetanz D, Lang N, Tergau F, Paulus W. Consolidation of human motor cortical neuroplasticity by D-Cycloserine. *Neuropsychopharmacology*. 2004;29(8):1573–1578. http://dx.doi.org/10.1038/sj.npp.1300517.
13. Nitsche MA, Seeber A, Frommann K, et al. Modulating parameters of excitability during and after transcranial direct current stimulation of the human motor cortex: cortical excitability and tDCS. *J Physiol*. 2005;568(1):291–303. http://dx.doi.org/10.1113/jphysiol.2005.092429.
14. Stagg CJ, Best JG, Stephenson MC, et al. Polarity-sensitive modulation of cortical neurotransmitters by transcranial stimulation. *J Neurosci*. 2009;29(16):5202–5206. http://dx.doi.org/10.1523/JNEUROSCI.4432-08.2009.
15. Nitsche MA. Catecholaminergic consolidation of motor cortical neuroplasticity in humans. *Cereb Cortex*. 2004;14(11):1240–1245. http://dx.doi.org/10.1093/cercor/bhh085.
16. Nitsche MA, Lampe C, Antal A, et al. Dopaminergic modulation of long-lasting direct current-induced cortical excitability changes in the human motor cortex: dopamine in human neuroplasticity. *Eur J Neurosci*. 2006;23(6):1651–1657. http://dx.doi.org/10.1111/j.1460-9568.2006.04676.x.
17. Nitsche MA, Liebetanz D, Schlitterlau A, et al. GABAergic modulation of DC stimulation-induced motor cortex excitability shifts in humans. *Eur J Neurosci*. 2004;19(10):2720–2726. http://dx.doi.org/10.1111/j.0953-816X.2004.03398.x.

18. Kuo M-F, Grosch J, Fregni F, Paulus W, Nitsche MA. Focusing effect of acetylcholine on neuroplasticity in the human motor cortex. *J Neurosci*. 2007;27(52):14442–14447. http://dx.doi.org/10.1523/JNEUROSCI.4104-07.2007.

19. Batsikadze G, Paulus W, Grundey J, Kuo M-F, Nitsche MA. Effect of the nicotinic 4 2-receptor partial agonist varenicline on non-invasive brain stimulation-induced neuroplasticity in the human motor cortex. *Cereb Cortex*. June 2014. http://dx.doi.org/10.1093/cercor/bhu126.

20. Nitsche MA, Kuo M-F, Karrasch R, Wächter B, Liebetanz D, Paulus W. Serotonin affects transcranial direct current–induced neuroplasticity in humans. *Biol Psychiatry*. 2009;66(5):503–508. http://dx.doi.org/10.1016/j.biopsych.2009.03.022.

21. Stagg CJ, Bachtiar V, Johansen-Berg H. The role of GABA in human motor learning. *Curr Biol*. 2011;21(6):480–484. http://dx.doi.org/10.1016/j.cub.2011.01.069.

22. Kim S, Stephenson MC, Morris PG, Jackson SR. tDCS-induced alterations in GABA concentration within primary motor cortex predict motor learning and motor memory: a 7T magnetic resonance spectroscopy study. *NeuroImage*. 2014;99:237–243. http://dx.doi.org/10.1016/j.neuroimage.2014.05.070.

23. Clark VP, Coffman BA, Trumbo MC, Gasparovic C. Transcranial direct current stimulation (tDCS) produces localized and specific alterations in neurochemistry: a ^{1}H magnetic resonance spectroscopy study. *Neurosci Lett*. 2011;500(1):67–71. http://dx.doi.org/10.1016/j.neulet.2011.05.244.

24. Hunter MA, Coffman BA, Gasparovic C, Calhoun VD, Trumbo MC, Clark VP. Baseline effects of transcranial direct current stimulation on glutamatergic neurotransmission and large-scale network connectivity. *Brain Res*. 2015;1594:92–107. http://dx.doi.org/10.1016/j.brainres.2014.09.066.

25. Rango M, Cogiamanian F, Marceglia S, et al. Myoinositol content in the human brain is modified by transcranial direct current stimulation in a matter of minutes: a ^{1}H-MRS study. *Magn Reson Med*. 2008;60(4):782–789. http://dx.doi.org/10.1002/mrm.21709.

26. Stagg CJ, Bachtiar V, Amadi U, et al. Local GABA concentration is related to network-level resting functional connectivity. *Elife*. 2014;3:e01465.

27. Peña-Gómez C, Sala-Lonch R, Junqué C, et al. Modulation of large-scale brain networks by transcranial direct current stimulation evidenced by resting-state functional MRI. *Brain Stimul*. 2012;5(3):252–263. http://dx.doi.org/10.1016/j.brs.2011.08.006.

28. Park C, Chang WH, Park J-Y, Shin Y-I, Kim ST, Kim Y-H. Transcranial direct current stimulation increases resting state interhemispheric connectivity. *Neurosci Lett*. 2013;539:7–10. http://dx.doi.org/10.1016/j.neulet.2013.01.047.

29. Amadi U, Ilie A, Johansen-Berg H, Stagg CJ. Polarity-specific effects of motor transcranial direct current stimulation on fMRI resting state networks. *NeuroImage*. 2014;88:155–161. http://dx.doi.org/10.1016/j.neuroimage.2013.11.037.

30. Moliadze V, Antal A, Paulus W. Electrode-distance dependent after-effects of transcranial direct and random noise stimulation with extracephalic reference electrodes. *Clin Neurophysiol*. 2010;121(12):2165–2171. http://dx.doi.org/10.1016/j.clinph.2010.04.033.

31. Dundas JE, Thickbroom GW, Mastaglia FL. Perception of comfort during transcranial DC stimulation: effect of NaCl solution concentration applied to sponge electrodes. *Clin Neurophysiol Off J Int Fed Clin Neurophysiol*. 2007;118(5):1166–1170. http://dx.doi.org/10.1016/j.clinph.2007.01.010.

32. Datta A, Truong D, Minhas P, Parra LC, Bikson M. Inter-individual variation during transcranial direct current stimulation and normalization of dose using MRI-derived computational models. *Front Psychiatry*. 2012;3. http://dx.doi.org/10.3389/fpsyt.2012.00091.

33. Kessler SK, Minhas P, Woods AJ, Rosen A, Gorman C, Bikson M. Dosage considerations for transcranial direct current stimulation in children: a computational modeling study. Chambers C, ed. *PLoS One*. 2013;8(9):e76112. http://dx.doi.org/10.1371/journal.pone.0076112.

34. Bikson M, Rahman A, Datta A. Computational models of transcranial direct current stimulation. *Clin EEG Neurosci.* 2012;43(3):176–183. http://dx.doi.org/10.1177/1550059412445138.

35. Nitsche MA, Doemkes S, Karakose T, et al. Shaping the effects of transcranial direct current stimulation of the human motor cortex. *J Neurophysiol.* 2007;97(4):3109–3117. http://dx.doi.org/10.1152/jn.01312.2006.

36. Faria P, Hallett M, Miranda PC. A finite element analysis of the effect of electrode area and inter-electrode distance on the spatial distribution of the current density in tDCS. *J Neural Eng.* 2011;8(6):066017. http://dx.doi.org/10.1088/1741-2560/8/6/066017.

37. Lee M, Kim Y-H, Im C-H, et al. What is the optimal anodal electrode position for inducing corticomotor excitability changes in transcranial direct current stimulation? *Neurosci Lett.* 2015;584:347–350. http://dx.doi.org/10.1016/j.neulet.2014.10.052.

38. Minhas P, Bikson M, Woods AJ, Rosen AR, Kessler SK. Transcranial direct current stimulation in pediatric brain: a computational modeling study. In: *Engineering in Medicine and Biology Society (EMBC). Annual International Conference of the IEEE.* vol. IEEE. 2012:859–862. http://ieeexplore.ieee.org/xpls/abs_all.jsp?arnumber=6346067; Accessed 11.03.15.

39. Villamar MF, Volz MS, Bikson M, Datta A, Dasilva AF, Fregni F. Technique and considerations in the use of 4x1 ring high-definition transcranial direct current stimulation (HD-tDCS). *J Vis Exp JoVE.* 2013;(77):e50309. http://dx.doi.org/10.3791/50309.

40. Datta A, Elwassif M, Battaglia F, Bikson M. Transcranial current stimulation focality using disc and ring electrode configurations: FEM analysis. *J Neural Eng.* 2008;5(2):163–174. http://dx.doi.org/10.1088/1741-2560/5/2/007.

41. Datta A, Bansal V, Diaz J, Patel J, Reato D, Bikson M. Gyri-precise head model of transcranial direct current stimulation: improved spatial focality using a ring electrode versus conventional rectangular pad. *Brain Stimul.* 2009;2(4):201–207, 207.e1. http://dx.doi.org/10.1016/j.brs.2009.03.005.

42. Kuo H-I, Bikson M, Datta A, et al. Comparing cortical plasticity induced by conventional and high-definition 4 × 1 ring tDCS: a neurophysiological study. *Brain Stimul.* 2013;6(4):644–648. http://dx.doi.org/10.1016/j.brs.2012.09.010.

43. Ambrus GG, Al-Moyed H, Chaieb L, Sarp L, Antal A, Paulus W. The fade-in – short stimulation – fade out approach to sham tDCS – reliable at 1 mA for naïve and experienced subjects, but not investigators. *Brain Stimul.* 2012;5(4):499–504. http://dx.doi.org/10.1016/j.brs.2011.12.001.

44. O'Connell NE, Cossar J, Marston L, et al. Rethinking clinical trials of transcranial direct current stimulation: participant and assessor blinding is inadequate at intensities of 2 mA. Eldabe S, ed. PLoS One. 2012;7(10):e47514. http://dx.doi.org/10.1371/journal.pone.0047514.

45. Schade S, Moliadze V, Paulus W, Antal A. Modulating neuronal excitability in the motor cortex with tDCS shows moderate hemispheric asymmetry due to subjects' handedness: a pilot study. *Restor Neurol Neurosci.* 2012;30(3):191–198.

46. Miranda PC, Lomarev M, Hallett M. Modeling the current distribution during transcranial direct current stimulation. *Clin Neurophysiol.* 2006;117(7):1623–1629. http://dx.doi.org/10.1016/j.clinph.2006.04.009.

47. Wagner T, Fregni F, Fecteau S, Grodzinsky A, Zahn M, Pascual-Leone A. Transcranial direct current stimulation: a computer-based human model study. *NeuroImage.* 2007;35(3):1113–1124. http://dx.doi.org/10.1016/j.neuroimage.2007.01.027.

48. Oostendorp TF, Hengeveld YA, Wolters CH, Stinstra J, van Elswijk G, Stegeman DF. Modeling transcranial DC stimulation. *Conf Proc Annu Int Conf IEEE Eng Med Biol Soc IEEE Eng Med Biol Soc Annu Conf.* 2008;2008:4226–4229. http://dx.doi.org/10.1109/IEMBS.2008.4650142.

49. Gillick BT, Kirton A, Carmel JB, Minhas P, Bikson M. Pediatric stroke and transcranial direct current stimulation: methods for rational individualized dose optimization. *Front Hum Neurosci.* 2014;8:739. http://dx.doi.org/10.3389/fnhum.2014.00739.

50. Pfefferbaum A, Mathalon DH, Sullivan EV, Rawles JM, Zipursky RB, Lim KO. A quantitative magnetic resonance imaging study of changes in brain morphology from infancy to late adulthood. *Arch Neurol.* 1994;51(9):874–887.
51. Brain Development Cooperative Group. Total and regional brain volumes in a population-based normative sample from 4 to 18 years: the NIH MRI study of normal brain development. *Cereb Cortex N Y N 1991.* 2012;22(1):1–12. http://dx.doi.org/10.1093/cercor/bhr018.
52. Beauchamp MS, Beurlot MR, Fava E, et al. The developmental trajectory of brain-scalp distance from birth through childhood: implications for functional neuroimaging. *PloS One.* 2011;6(9):e24981. http://dx.doi.org/10.1371/journal.pone.0024981.
53. Nellhaus G. Head circumference from birth to eighteen years. Practical composite international and interracial graphs. *Pediatrics.* 1968;41(1):106–114.
54. Kujirai T, Caramia MD, Rothwell JC, et al. Corticocortical inhibition in human motor cortex. *J Physiol.* 1993;471:501–519.
55. Di Lazzaro V, Pilato F, Dileone M, et al. GABA$_A$ receptor subtype specific enhancement of inhibition in human motor cortex. *J Physiol.* 2006;575(Pt 3):721–726. http://dx.doi.org/10.1113/jphysiol.2006.114694.
56. Ziemann U, Lönnecker S, Steinhoff BJ, Paulus W. The effect of lorazepam on the motor cortical excitability in man. *Exp Brain Res.* 1996;109(1):127–135.
57. Di Lazzaro V, Oliviero A, Meglio M, et al. Direct demonstration of the effect of lorazepam on the excitability of the human motor cortex. *Clin Neurophysiol Off J Int Fed Clin Neurophysiol.* 2000;111(5):794–799.
58. Cengiz B, Murase N, Rothwell JC. Opposite effects of weak transcranial direct current stimulation on different phases of short interval intracortical inhibition (SICI). *Exp Brain Res.* 2013;225(3):321–331. http://dx.doi.org/10.1007/s00221-012-3369-0.
59. Di Lazzaro V, Manganelli F, Dileone M, et al. The effects of prolonged cathodal direct current stimulation on the excitatory and inhibitory circuits of the ipsilateral and contralateral motor cortex. *J Neural Transm.* 2012;119(12):1499–1506. http://dx.doi.org/10.1007/s00702-012-0845-4.
60. Moliadze V, Schmanke T, Andreas S, Lyzhko E, Freitag CM, Siniatchkin M. Stimulation intensities of transcranial direct current stimulation have to be adjusted in children and adolescents. *Clin Neurophysiol Off J Int Fed Clin Neurophysiol.* October 2014. http://dx.doi.org/10.1016/j.clinph.2014.10.142.
61. Reis J, Fritsch B. Modulation of motor performance and motor learning by transcranial direct current stimulation. *Curr Opin Neurol.* 2011;24(6):590–596. http://dx.doi.org/10.1097/WCO.0b013e32834c3db0.
62. Madhavan S, Shah B. Enhancing motor skill learning with transcranial direct current stimulation – a concise review with applications to stroke. *Front Psychiatry.* 2012;3. http://dx.doi.org/10.3389/fpsyt.2012.00066.
63. Ungerleider LG, Doyon J, Karni A. Imaging brain plasticity during motor skill learning. *Neurobiol Learn Mem.* 2002;78(3):553–564.
64. Stagg CJ, Nitsche MA. Physiological basis of transcranial direct current stimulation. *Neurosci.* 2011;17(1):37–53. http://dx.doi.org/10.1177/1073858410386614.
65. Boggio PS, Castro LO, Savagim EA, et al. Enhancement of non-dominant hand motor function by anodal transcranial direct current stimulation. *Neurosci Lett.* 2006;404 (1–2):232–236. http://dx.doi.org/10.1016/j.neulet.2006.05.051.
66. Hummel FC, Heise K, Celnik P, Floel A, Gerloff C, Cohen LG. Facilitating skilled right hand motor function in older subjects by anodal polarization over the left primary motor cortex. *Neurobiol Aging.* 2010;31(12):2160–2168. http://dx.doi.org/10.1016/j.neurobiolaging.2008.12.008.
67. Sohn MK, Kim BO, Song HT. Effect of stimulation polarity of transcranial direct current stimulation on non-dominant hand function. *Ann Rehabil Med.* 2012;36(1):1. http://dx.doi.org/10.5535/arm.2012.36.1.1.

68. Kidgell DJ, Goodwill AM, Frazer AK, Daly RM. Induction of cortical plasticity and improved motor performance following unilateral and bilateral transcranial direct current stimulation of the primary motor cortex. *BMC Neurosci.* 2013;14(1):64.

69. Reis J, Schambra HM, Cohen LG, et al. Noninvasive cortical stimulation enhances motor skill acquisition over multiple days through an effect on consolidation. *Proc Natl Acad Sci.* 2009;106(5):1590–1595.

70. Fritsch B, Reis J, Martinowich K, et al. Direct current stimulation promotes BDNF-dependent synaptic plasticity: potential implications for motor learning. *Neuron.* 2010;66(2):198–204. http://dx.doi.org/10.1016/j.neuron.2010.03.035.

71. Williams JA, Pascual-Leone A, Fregni F. Interhemispheric modulation induced by cortical stimulation and motor training. *Phys Ther.* 2010;90(3):398–410. http://dx.doi.org/10.2522/ptj.20090075.

72. Zimerman M, Heise K-F, Gerloff C, Cohen LG, Hummel FC. Disrupting the ipsilateral motor cortex interferes with training of a complex motor task in older adults. *Cereb Cortex.* 2014;24(4):1030–1036. http://dx.doi.org/10.1093/cercor/bhs385.

73. Stagg CJ, Jayaram G, Pastor D, Kincses ZT, Matthews PM, Johansen-Berg H. Polarity and timing-dependent effects of transcranial direct current stimulation in explicit motor learning. *Neuropsychologia.* 2011;49(5):800–804. http://dx.doi.org/10.1016/j.neuropsychologia.2011.02.009.

74. Nitsche MA, Schauenburg A, Lang N, et al. Facilitation of implicit motor learning by weak transcranial direct current stimulation of the primary motor cortex in the human. *J Cogn Neurosci.* 2003;15(0898–929X (Print)):619–626.

75. Vines BW, Cerruti C, Schlaug G. Dual-hemisphere tDCS facilitates greater improvements for healthy subjects' non-dominant hand compared to uni-hemisphere stimulation. *BMC Neurosci.* 2008;9(1):103. http://dx.doi.org/10.1186/1471-2202-9-103.

76. Liebetanz D, Koch R, Mayenfels S, König F, Paulus W, Nitsche MA. Safety limits of cathodal transcranial direct current stimulation in rats. *Clin Neurophysiol.* 2009;120(6):1161–1167. http://dx.doi.org/10.1016/j.clinph.2009.01.022.

77. Nitsche MA, Liebetanz D, Lang N, Antal A, Tergau F, Paulus W. Safety criteria for transcranial direct current stimulation (tDCS) in humans. *Clin Neurophysiol.* 2003;114(11):2220–2222.

78. Krishnan C, Santos L, Peterson MD, Ehinger M. Safety of noninvasive brain stimulation in children and adolescents. *Brain Stimul.* 2015;8(1):76–87. http://dx.doi.org/10.1016/j.brs.2014.10.012.

79. Brunoni AR, Vanderhasselt M-A, Boggio PS, et al. Polarity- and valence-dependent effects of prefrontal transcranial direct current stimulation on heart rate variability and salivary cortisol. *Psychoneuroendocrinology.* 2013;38(1):58–66. http://dx.doi.org/10.1016/j.psyneuen.2012.04.020.

80. Kessler SK, Turkeltaub PE, Benson JG, Hamilton RH. Differences in the experience of active and sham transcranial direct current stimulation. *Brain Stimul.* 2012;5(2):155–162. http://dx.doi.org/10.1016/j.brs.2011.02.007.

81. Ambrus GG, Paulus W, Antal A. Cutaneous perception thresholds of electrical stimulation methods: comparison of tDCS and tRNS. *Clin Neurophysiol.* 2010;121(11):1908–1914. http://dx.doi.org/10.1016/j.clinph.2010.04.020.

82. Furubayashi T, Terao Y, Arai N, et al. Short and long duration transcranial direct current stimulation (tDCS) over the human hand motor area. *Exp Brain Res.* 2008;185(2):279–286. http://dx.doi.org/10.1007/s00221-007-1149-z.

83. Garvey MA, Kaczynski KJ, Becker DA, Bartko JJ. Subjective reactions of children to single-pulse transcranial magnetic stimulation. *J Child Neurol.* 2001;16(0883–0738 (Print)):891–894.

84. Kirton A, Deveber G. Life after perinatal stroke. *Stroke J Cereb Circ.* 2013;44(11):3265–3271. http://dx.doi.org/10.1161/STROKEAHA.113.000739.

85. Bax M, Goldstein M, Rosenbaum P, et al. Proposed definition and classification of cerebral palsy, April 2005. *Dev Med Child Neurol*. 2005;47(8):571–576.
86. Kirton A. Modeling developmental plasticity after perinatal stroke: defining central therapeutic targets in cerebral palsy. *Pediatr Neurol*. 2013;48(2):81–94. http://dx.doi.org/10.1016/j.pediatrneurol.2012.08.001.
87. Kirton A, deVeber G, Gunraj C, Chen R. Cortical excitability and interhemispheric inhibition after subcortical pediatric stroke: plastic organization and effects of rTMS. *Clin Neurophysiol*. 2010;121(1872–8952 (Electronic)):1922–1929.
88. Murase N, Duque J, Mazzocchio R, Cohen LG. Influence of interhemispheric interactions on motor function in chronic stroke. *AnnNeurol*. 2004;55(0364–5134 (Print)):400–409.
89. Gillick BT, Feyma T, Menk J, et al. Safety and feasibility of transcranial direct current stimulation in pediatric hemiparesis: a randomized controlled pilot study. *Phys Ther*. November 2014. http://dx.doi.org/10.2522/ptj.20130565.
90. Elsner B, Kugler J, Pohl M, Mehrholz J. Transcranial direct current stimulation (tDCS) for improving function and activities of daily living in patients after stroke. *Cochrane Database Syst Rev*. 2013;11:CD009645. http://dx.doi.org/10.1002/14651858.CD009645.pub2.
91. Dickey L, Kagan A, Lindsay MP, Fang J, Rowland A, Black S. Incidence and profile of inpatient stroke-induced aphasia in Ontario, Canada. *Arch Phys Med Rehabil*. 2010;91(2):196–202. http://dx.doi.org/10.1016/j.apmr.2009.09.020.
92. Pedersen PM, Jørgensen HS, Nakayama H, Raaschou HO, Olsen TS. Aphasia in acute stroke: incidence, determinants, and recovery. *Ann Neurol*. 1995;38(4):659–666. http://dx.doi.org/10.1002/ana.410380416.
93. Brady MC, Kelly H, Godwin J, Enderby P. Speech and language therapy for aphasia following stroke. *Cochrane Database Syst Rev*. 2012;5:CD000425. http://dx.doi.org/10.1002/14651858.CD000425.pub3.
94. Elsner B, Kugler J, Pohl M, Mehrholz J. Transcranial direct current stimulation (tDCS) for improving aphasia in patients after stroke. In: The Cochrane Collaboration, ed. *Cochrane Database of Systematic Reviews*. vol. 6. Chichester, UK: John Wiley & Sons, Ltd; 2013. http://doi.wiley.com/10.1002/14651858.CD009760.pub2. Accessed 16.03.15.
94a. Carlson HL, Jadavji Z, Mineyko A, et al. Treatment of dysphasia with rTMS and language therapy after childhood stroke: multimodal imaging of plastic change. *Brain Lang*. (Under review).
95. Andrade L, Caraveo-Anduaga JJ, Berglund P, et al. The epidemiology of major depressive episodes: results from the International Consortium of Psychiatric Epidemiology (ICPE) Surveys. *Int J Methods Psychiatr Res*. 2003;12(1):3–21.
96. Murray CJ, Lopez AD. Alternative projections of mortality and disability by cause 1990–2020: Global Burden of Disease Study. *Lancet*. 1997;349(9064):1498–1504. http://dx.doi.org/10.1016/S0140-6736(96)07492-2.
97. Berlim MT, Turecki G. Definition, assessment, and staging of treatment-resistant refractory major depression: a review of current concepts and methods. *Can J Psychiatry Rev Can Psychiatr*. 2007;52(1):46–54.
98. Baxter LR, Schwartz JM, Phelps ME, et al. Reduction of prefrontal cortex glucose metabolism common to three types of depression. *Arch Gen Psychiatry*. 1989;46(3):243–250.
99. Caetano SC, Fonseca M, Olvera RL, et al. Proton spectroscopy study of the left dorsolateral prefrontal cortex in pediatric depressed patients. *Neurosci Lett*. 2005;384(3):321–326. http://dx.doi.org/10.1016/j.neulet.2005.04.099.
100. Lefaucheur J-P, André-Obadia N, Antal A, et al. Evidence-based guidelines on the therapeutic use of repetitive transcranial magnetic stimulation (rTMS). *Clin Neurophysiol*. 2014;125(11):2150–2206. http://dx.doi.org/10.1016/j.clinph.2014.05.021.
101. Yang X-R, Kirton A, Wilkes TC, et al. Glutamate alterations associated with transcranial magnetic stimulation in youth depression: a case series. *J ECT*. 2014;30(3):242–247. http://dx.doi.org/10.1097/YCT.0000000000000094.

102. Shiozawa P, Fregni F, Benseñor IM, et al. Transcranial direct current stimulation for major depression: an updated systematic review and meta-analysis. *Int J Neuropsychopharmacol Off Sci J Coll Int Neuropsychopharmacol CINP*. 2014;17(9):1443–1452. http://dx.doi.org/10.1017/S1461145714000418.

103. Hashimoto K, Shimizu E, Iyo M. Critical role of brain-derived neurotrophic factor in mood disorders. *Brain Res Brain Res Rev*. 2004;45(2):104–114. http://dx.doi.org/10.1016/j.brainresrev.2004.02.003.

104. Levinson DF. The genetics of depression: a review. *Biol Psychiatry*. 2006;60(2):84–92. http://dx.doi.org/10.1016/j.biopsych.2005.08.024.

105. Teche SP, Nuernberg GL, Sordi AO, et al. Measurement methods of BDNF levels in major depression: a qualitative systematic review of clinical trials. *Psychiatr Q*. 2013;84(4):485–497. http://dx.doi.org/10.1007/s11126-013-9261-7.

106. Brunoni AR, Baeken C, Machado-Vieira R, Gattaz WF, Vanderhasselt M-A. BDNF blood levels after non-invasive brain stimulation interventions in major depressive disorder: a systematic review and meta-analysis. *World J Biol Psychiatry Off J World Fed Soc Biol Psychiatry*. 2015;16(2):114–122. http://dx.doi.org/10.3109/15622975.2014.958101.

107. O'Connell NE, Wand BM, Marston L, Spencer S, Desouza LH. Non-invasive brain stimulation techniques for chronic pain. *Cochrane Database Syst Rev*. 2014;4:CD008208. http://dx.doi.org/10.1002/14651858.CD008208.pub3.

108. Boldt I, Eriks-Hoogland I, Brinkhof MW, de Bie R, Joggi D, von Elm E. *Nonpharmacological Interventions for Chronic Pain in People with Spinal Cord Injury*. Cochrane Libr; 2011. http://onlinelibrary.wiley.com/doi/10.1002/14651858.CD009177.pub2/full. Accessed 16.03.15.

109. Jansen JM, Daams JG, Koeter MWJ, Veltman DJ, van den Brink W, Goudriaan AE. Effects of non-invasive neurostimulation on craving: a meta-analysis. *Neurosci Biobehav Rev*. 2013;37(10 Pt 2):2472–2480. http://dx.doi.org/10.1016/j.neubiorev.2013.07.009.

110. Song J-J, Vanneste S, Van de Heyning P, De Ridder D. Transcranial direct current stimulation in tinnitus patients: a systemic review and meta-analysis. *Scientific World Journal*. 2012;2012:427941. http://dx.doi.org/10.1100/2012/427941.

111. San-juan D, Morales-Quezada L, Orozco Garduño AJ, et al. Transcranial direct current stimulation in epilepsy. *Brain Stimul*. January 2015. http://dx.doi.org/10.1016/j.brs.2015.01.001.

6

Insights Into Pediatric Brain Stimulation Protocols From Preclinical Research

M.Q. Hameed, M.J. Sanchez, R. Gersner, A. Rotenberg

Harvard Medical School, Boston, MA, United States

STIMULATING THE IMMATURE BRAIN

The therapeutic effects of brain stimulation, particularly cortical stimulation, are reliant on mechanisms of neuronal excitability and neuronal plasticity; and insight into such mechanisms, at the synaptic and molecular levels, has come largely from preclinical research in animal models. The scientific literature contains relatively few published in vivo brain stimulation experiments in immature animals,[1,2] however important aspects of developmental neurobiology, such as the maturational patterns of γ-aminobutyric acid (GABA)-mediated inhibition, glutamate-mediated excitation, and use-dependent synaptic plasticity are well described.[1-7] These are largely derived from in vitro and ex vivo experiments, rather than from clinical brain stimulation protocols that have been adapted to animal subjects. However, the results of extensive basic science studies in the immature brain may nevertheless inform the design of pediatric brain stimulation protocols.

THE IMMATURE BRAIN IS HYPEREXCITABLE

As often underscored by child neurologists, the developing brain is a unique physiological entity, and not merely a small adult brain. Therefore, adapting brain stimulation to pediatric populations requires special considerations of safety and efficacy in the context of the neurodevelopmental sequences and unique characteristics of the immature brain. Especially relevant may be the child's vulnerability to seizure, which has been studied extensively in immature rodent epilepsy models.[5] Given that seizures are a realistic potential side effect of cortical stimulation, as by repetitive transcranial magnetic stimulation (rTMS),[8] insights into the developmental biology of the cortical excitation:inhibition (E:I) ratio may translate to adjustments of adult brain stimulation protocols for pediatric populations.

Normal brain and synaptic development are critically dependent on neuronal activity, and so excitatory processes outweigh inhibition in cortical and subcortical circuits during the first few years after birth.[6] Glutamate – the main excitatory neurotransmitter – acts on both ionotropic and metabotropic G-protein-coupled receptors to trigger intracellular signaling cascades. The first category includes N-methyl-D-aspartate (NMDA) and alpha-amino-3-hydroxy-5-methylisoxazole-4-propionic acid (AMPA) receptors, both consisting of distinct subunits, and the relative abundance of individual subunits within the receptor, which varies with age, has important functional consequences.[4]

NMDA receptors (NMDARs) are heterotetrameric complexes consisting of an NR1 subunit along with different combinations of isoforms of the NR2 and NR3 subunits. When activated, NMDARs are a major source of calcium influx into the intracellular space at the synapse. NMDAR-mediated calcium signaling is essential for normal synaptic plasticity, and is required for the long-term effects of brain stimulation.[9–11] However, NMDAR overactivation, and with it excessive calcium influx, can also contribute to excitotoxicity and to pathophysiologic processes such as epileptogenesis.[5]

NR2B is the predominant NR2-type NMDAR subunit in the immature rodent brain, and NR2B-containing NMDARs, once activated by the glutamate ligand, have a longer current decay time compared to the NR2A-containing NMDARs found on mature neurons.[7,12] Longer current decay leads to prolonged excitatory transmission and calcium entry, which may affect brain stimulation by lowering effective thresholds for injury and plasticity.

NR2C, NR2D, and NR3A are the other important subunits highly expressed in the first 2 weeks after birth in rodents – roughly equivalent to the first 2 years after birth in humans.[13] The cation (largely calcium) channels of NMDARs in the adult rodent brain are blocked by a magnesium ion at resting membrane potentials via a binding site within the channel, preventing further ion movements,[14] and the abundance of the NR2C, NR2D, and NR3A subunits in NMDARs in developing rodent brains reduces the receptors' sensitivity to magnesium, thus increasing activation.[7]

In sum, enhanced NMDAR activation in immature rodent brains results in increased calcium influx as compared to adult brains, which – while critical for developmental plasticity and synaptogenesis – results in lower seizure and excitotoxic injury thresholds.[4,5,7] In addition, as discussed below, the adult mechanisms for glutamate clearance from the synapse that also control NMDA receptor activation are not established in the immature rat neocortex until the fourth week of life (human age 4–11 years), permitting extrasynaptic receptor activation and synaptic crosstalk.[15]

AMPA receptors (AMPARs) are also heterotetramers of distinct GluR1, GluR2, GluR3, and GluR4 subunits. Most AMPARs in adult rodent brains are impermeable to calcium because they contain a GluR2 subunit. However, GluR2 expression is low in the immature rodent brain up to 21 days after birth (human age 2–3 years) leading to increased calcium influx through the AMPARs.[7] Thus, like NMDARs, AMPAR composition in the developing rat brain renders it hyperexcitable, and predisposes to an enhanced vulnerability to injury and epileptogenesis,[16] and may therefore be a consideration during brain stimulation.

Much less is known about the maturational trajectory of metabotropic glutamate receptors – a family of G-protein-coupled glutamate receptors that modulate excitability and synaptic transmission via intracellular second messenger systems. mGluR1α expression peaks by the ninth postnatal day in rats (term human infant), particularly in CA1 stratum oriens-alveus interneurons, with mGluR2, mGluR3, and mGluR5 maturing by 15 days after birth (human age 0–24 months).[17] Glutamate-mediated activation of these mGluR1α receptors may facilitate presynaptic GABA release, which may explain the resistance of the developing brain to seizure-induced damage in spite of increased susceptibility to seizures.[17] Plausibly, the abundance of presynaptic mGluR1α also contributes to a relatively high threshold for activation in young brains, although this may be better explained by incomplete myelination in early life.

In addition to the increased sensitivity of glutamate receptors, the major glutamate clearance mechanism in the rodent brain, glutamate uptake into astrocytes via glutamate transporter 1 (GLT-1), is also impaired in the immature brain, resulting in slower removal of glutamate from the synaptic cleft. GLT-1 (called excitatory amino acid transporter 2, EAAT-2 in humans) normally provides 95% of the total glutamate clearance capacity in the mammalian brain,[18] and is highly expressed – accounting for as much as 1% of total brain protein.[19] Astrocytic GLT-1 expression and glutamate uptake are extremely low in the rat hippocampus for the first postnatal week (23–36 weeks gestation in humans), and increase to adult levels by the 30th postnatal day (human age 4–11 years).[20] Neocortical GLT-1 expression is even lower than hippocampal expression in rat brain on the seventh postnatal day (term human infant), and the two levels become similar by 28 days after birth (4–11 years in humans).[15]

In summary, immature neurons are more sensitive to, and stay activated for longer after, glutamate exposure. Given that glutamate removal from the synapse is also slower in the developing brain, it follows that increased glutamate signaling leads to considerably more excitable cortical and subcortical circuits which are more prone to seizures during both postnatal and juvenile periods (Table 6.1). This vulnerability is consistent with clinical observations that seizures are more likely in children than in adults, and can inform the risk–benefit assessments in design of clinical brain stimulation protocols for children.

INHIBITORY CONTROL DEVELOPS WITH AGE

The effectiveness of brain stimulation protocols in modulating brain excitability in adults depends not only on the excitatory glutamate-dependent transmission mentioned above, but also on modulation of inhibitory GABAergic signaling.[21,22] The strong GABAergic inhibitory mechanisms normally found in the adult brain are compromised in the developing brain. For instance, the expression of glutamic acid decarboxylase (GAD – the enzyme responsible for GABA synthesis), and GABA receptors reaches adult levels in the fourth postnatal week in rats (human age 4–11 years).[5,7,23]

TABLE 6.1 Distinctions in Excitatory and Inhibitory Mechanisms Between Immature and Mature Brains

	Developing Brain	Mature Adult Brain	Relevance to Brain Stimulation
Overall Excitability	⇑	⇓	
Glutamate and Excitation			
NMDA Receptor	⇓ NR2A, ⇑ NR2B (⇑ excitatory postsynaptic current)	⇑ NR2A, ⇓ NR2B (⇓ excitatory postsynaptic current)	
	⇑ NR2C, NR2D, NR3A (⇓ Mg^{2+} block)	⇓ NR2C, NR2D, NR3A (⇑ Mg^{2+} block)	Increased seizure susceptibility
AMPA Receptor	⇓ GluR2: (⇑ Ca^{2+} permeability)	⇑ GluR2: (⇓ Ca^{2+} permeability)	Unexpected behavioral or functional consequences
Glutamate Transport	⇓ GLT-1 (⇓ glutamate clearance)	⇑ GLT-1 (⇑ glutamate clearance)	
GABA and Inhibition			
Intracellular Chloride	⇑ NKCC1:KCC2 ratio (⇑ intracellular Cl⁻) (⇓ $GABA_A$ inhibition)	⇑ NKCC1:KCC2 ratio (⇓ intracellular Cl⁻) (⇑ $GABA_A$ inhibition)	
$GABA_A$ Receptor	⇓ α_1 (⇓ benzodiazepine sensitivity)	⇑ α_1 (⇑ benzodiazepine sensitivity)	

GABA receptor function also undergoes major developmental change in childhood. The $GABA_A$ receptor ($GABA_A R$) is a transmembrane protein consisting of five subunits arranged around a central pore, and its functional properties are determined by subunit composition, as with the NMDA and AMPA receptors. GABA's inhibitory effect in the mature brain is a result of hyperpolarization caused by a net influx of chloride into the neuron that is driven by a high extracellular to intracellular chloride concentration gradient. However, while $GABA_A Rs$ are embryonically expressed, GABA release in the first week after birth in rodents (36–40 weeks gestation in humans) paradoxically results in excitation due to reversed chloride gradients in the immature brain, causing a net efflux of chloride from the neuron through the receptor.[5–7,24] This reversal is due to overexpression of the chloride importer NKCC1 and decreased expression of the chloride exporter KCC2 relative to the adult rat brain,[25–27] which results in higher intracellular chloride concentration in early development. Thus, brain stimulation protocols that rely on promoting GABAergic inhibition may not work well, if at all, in the very immature brain. GABAergic inhibition gradually strengthens over the first 3 postnatal weeks in rodents (first 3–4 years after birth in humans), and correlates with the gradually increasing expression of neuronal KCC2. $GABA_A Rs$ in the immature brain also have slower kinetics and decreased sensitivity to benzodiazepines due to lower quantities of the α_1 subunit, which also increases gradually to adult levels with age (Table 6.1).[6,7]

Which form of brain stimulation will be most affected by such maturation of the GABA inhibitory systems is difficult to predict. However, given that some protocols such as theta burst TMS likely affect cortical GABAergic inhibition,[28,29] this $GABA_A R$ biology is important to consider in the context of stimulating the young central nervous system. Recent data, for example, indicate exciting new directions in increasing the efficacy of benzodiazepines in immature brains (discussed above), as cathodal transcranial direct current stimulation (tDCS) is shown to augment the efficacy of subtherapeutic doses of lorazepam in achieving suppression of seizures in the rat pentylenetetrazol (PTZ) status epilepticus model.[30]

Pre- and post-synaptic G-protein-coupled $GABA_B$ receptors ($GABA_B R$) also play a prominent role in neuronal development. Activation of postsynaptic receptors results in a potassium efflux, hyperpolarizing the membrane, while presynaptic $GABA_B R$ activation inhibits neurotransmitter release by closing calcium channels.[31] $GABA_B R$ binding increases in a region-specific temporal profile during the first 3 weeks of life in rodents (human age 2–3 years), peaking at postnatal day 14 (human age 0–24 months) in the thalamus and the neocortex and in the first postnatal week (term infant in humans) in the hippocampus.[31] In the hippocampus, presynaptic $GABA_B R$ effects are seen earlier than postsynaptic effects, possibly increasing excitability.[6,31] $GABA_B Rs$ modulate neuronal migration and differentiation, and $GABA_B R$ activation by ambient GABA promotes GABAergic synaptogenesis and inhibitory plasticity in rats, presumably via the induction of BDNF and TrkB cascades.[32] The relevance of $GABA_B R$ maturation to pediatric brain stimulation protocols is less clear, but, as with $GABA_A Rs$, the preclinical data may indicate a vulnerability to seizure in the developing brain.

Notably, not all preclinical research into the maturation of the E:I ratio indicates increased risk of brain stimulation to the immature brain. Developing rodent brains, while more prone to seizures, are less susceptible to seizure-induced neuronal damage compared to adult brains. For instance, kainate-induced seizures in adult rodents are followed by neuronal death in the hippocampus and limbic system[33] and compensatory mossy fiber sprouting, creating novel recurrent excitatory circuits between neurons.[34] In contrast, there is no cell death after kainate-induced seizures in developing rodent brains during the first 2 postnatal weeks[35,36] regardless of kainate dose. These findings have been confirmed in other rodent epilepsy models as well, and may be due to the absence of well-developed hippocampal mossy fiber terminals which would otherwise amplify excitatory signaling in the hippocampus and help to sustain epileptic seizures.[37] Yet, despite the lack of cell death, recurrent seizures in the immature brain may lead to long-lasting behavioral changes and decreased seizure thresholds in adulthood, presumably owing to the disruption of vital neurodevelopmental milestones that result in deranged neuronal networks.[37] This may be a consideration in pediatric brain stimulation, as seizures, a plausible side effect,[8] may have distinct effects on the immature brain.

GABAERGIC DYSMATURITY IN NEURODEVELOPMENTAL DISEASE

Neuropsychiatric treatments, and brain stimulation protocols in particular, should also include considerations that immature brain biology in a human may be for one of two reasons: the brain is chronologically immature or, often for genetic reasons, appropriate maturation did not occur, and the brain is pathologically dysmature into adulthood. Such dysmature brain biology is identified in a range of prevalent neuropsychiatric syndromes, and is particularly relevant to maturation of the GABAergic inhibitory system. For instance, animal models and limited data from human subjects indicate that GABAergic dysfunction is of major importance in the pathogenesis of some epilepsies, autism spectrum disorders, and associated conditions such as fragile X, tuberous sclerosis, and Rett syndrome.

Specific mutations in $GABA_AR$ subunits are also associated with epileptic and behavioral disorders – α_1 subunit deficiencies in older patients, reminiscent of immature α_1-deficient $GABA_ARs$ with slow kinetics and reduced benzodiazepine sensitivity, are seen in early infantile epileptic encephalopathy and Dravet syndrome.[38] Expression of the α_1 subunit is also reduced in the cerebellum in schizophrenia and major depression, and in the superior frontal and parietal cortex and cerebellum in autism.[39] As stimulation protocols are developed across ranges of disease states, including those that affect primarily young patients, such region-specific immature physiology may also be considered in the treatment of adults with these conditions. Two examples are now described.

Fragile X syndrome (FXS) is an inherited disease characterized by mutations in the fragile X mental retardation (*FMR1*) gene on the X chromosome – resulting in decreased or absent expression of the fragile X mental retardation protein (FMRP) required for normal neurological development. Affected patients suffer from intellectual disabilities and neuropsychiatric disorders including autism, and are at a higher risk for developing seizures. Juvenile *Fmr1* knockout mice – a model of fragile X syndrome, which is commonly associated with autism – exhibit a delayed switch from depolarizing to hyperpolarizing GABA signaling in the somatosensory cortex. Specifically, GABAergic excitatory depolarization of the neuronal membrane (rather than inhibitory hyperpolarization) is seen in knockout but not wild-type mice on postnatal day 10 (term human infant), and was associated with an increase in NKCC1 expression in the somatosensory cortex of the mutants at the same time point, while KCC2 expression was not affected. However, KCC2 expression is decreased in the hippocampus in Fmr1 knockout mice on postnatal days 15 (human age 0–24 months) and 30 (human age 4–11 years). Pretreatment of pregnant mice with the NKCC1 blocker bumetanide restored the developmental excitatory-to-inhibitory GABA switch in neonatal knockout mice,[40,41] and confirms the immature KCC2:NKCC1 ratio. Similarly, decreased KCC2 levels and a lower KCC2:NKCC1 ratio are seen in the CSF of human patients with Rett syndrome.[42] As above, novel therapeutic protocols aimed at treating patients with fragile X and Rett syndromes may take into account the aberrant excitatory GABA signaling.

Tuberous sclerosis (TSC) is a neurodevelopmental disease in which mutations of either the *TSC1* or *TSC2* genes – which code for inhibitors of the central cell growth control the mechanistic target of rapamycin (mTOR) pathway – often result in early-life refractory epilepsy and autism spectrum disorders. GABAergic dysmaturity and E:I imbalance are a hallmark of TSC. Surgically removed tubers from human patients with TSC demonstrate decreased expression of the $GABA_AR$ α_1 subunit (which explains benzodiazepine insensitivity in these patients), as well as increased NKCC1 and decreased KCC2 expression, resulting in immature excitatory $GABA_AR$ responses seen on patch clamp recordings which were significantly decreased after administration of the NKCC1 blocker, bumetanide.[43] These findings suggest that TSC is another syndrome with dysmature GABAergic transmission, and these defects may be considered when designing novel brain stimulation protocols for the treatment of this condition.

SYNAPTIC PLASTICITY CONSIDERATIONS IN THE DEVELOPING BRAIN

In the design of brain stimulation protocols, investigators and clinicians make a safe assumption that similar mechanisms underlie both the desired brain stimulation effect, and the innate capacity of the brain to register experience and store it as memories or changes in specific neuronal functions. In predominant neuroscience views, such brain plasticity is thought to rely on mechanisms of long-term potentiation (LTP) and long-term depression (LTD) of synaptic strength. Here also, as with brain excitability, preclinical animal research indicates a developmental regulation of synaptic plasticity that the brain stimulation field should consider in the design of child brain stimulation protocols.

LTP, a use-dependent enhancement of excitatory synaptic strength was first identified in 1973 in the hippocampus of an anesthetized rabbit.[44] Subsequently, LTP has been reliably reproduced in vitro in isolated rodent brain slices.[45,46] Homosynaptic LTD was identified in 1992 in rat hippocampal slices, is reliably reproducible in vitro in numerous laboratories, and mechanistically appears to be due to the reversal of the molecular mechanisms seen in LTP.[45,46] More recently, LTP-like potentiation and LTD-like depression of the corticospinal response have been reliably produced by repetitive TMS (rTMS) and tDCS in humans.[47,48] Yet, such LTP- and LTD-like phenomena have not been studied across the pediatric ages in humans, and for future considerations the field can turn to preclinical literature.

Extrapolating relevance to human brain stimulation protocols from animal LTP/LTD studies is more difficult than from studies aimed to identify the developmental trajectory of cortical inhibition and excitability. For instance, synaptic plasticity matures at distinct rates in separate brain regions of the rodent brain. Rat barrel cortex LTP is present at birth, peaks at postnatal days 3–5 (32–36 weeks gestation in humans), and disappears by postnatal day 14 (human age 0–24 months).[1] In contrast, theta burst stimulation first induces LTP in rat hippocampal slices on postnatal day 12 (0–24 months in humans), and LTP improves with age up to postnatal day 35 (humans: 11 years).[3] Similarly, the capacity for LTP is greater in the visual cortex of adult mice more than 6 months old as compared to 4–5-week-old mice (human age 4–11 years).[49] Whether such region-specific differences in maturation of synaptic plasticity exist in humans is unknown, but potentially they could be tested with non-invasive brain stimulation methods such as rTMS.

LTD in the hippocampus, in contrast to hippocampal LTP, is enhanced in immature relative to older rats. Hippocampal LTD is maximal in rats less than 2 weeks old (human age <2 years), and progressively declines to stabilize at adult levels by the fifth postnatal week (human age 4–11 years).[50,51] Additionally, not only is the magnitude of LTD greater in younger rats, the threshold for LTD induction is lower in younger rodents.[50] These data, along with classic descriptions of plasticity critical periods in the visual and auditory cortex,[52,53] underscore that a desired modulation of cortical excitability in an immature brain may require specialized stimulation protocols that are distinct from those that may be applied to adult subjects.

In vivo animal experiments aimed to test distinctions between immature and mature brain physiology are few. However, one published report shows that activity-dependent functional modification of cortical GABAergic parvalbumin-positive (PV+) cells, as evidenced by decreased labeling after intermittent theta burst (iTBS) TMS, occurs only in rats more than 32 days old (human age 4–11 years), reaching a maximal reduction by the 40th postnatal day (12–18 years in humans). As the development and modulation of synaptic input to the PV+ interneurons is mediated by the perineuronal nets surrounding them, it follows that immature PV+ cells with restricted perineuronal nets are not connected enough and therefore do not receive sufficient excitatory input to be affected by iTBS.[2] However, these data only reflect the absolute change in magnitude of the evoked postsynaptic response, not the induction threshold, which may well be lower in the immature brain.

RELEVANCE TO TRANSLATIONAL BRAIN STIMULATION

The authors hope that the basic science data discussed above raise concerns that merit consideration in pediatric brain stimulation. However, whether and to what extent these data translate to meaningful differences in brain stimulation of the young animal or human subject is unknown, since data on the physiologic results of brain stimulation in immature rodents are limited.

For instance, one important aspect to consider when translating results of in vitro synaptic plasticity studies, is that most are derived from microscopic electrical stimulation of an isolated afferent pathway (such as the Schaffer collateral in a hippocampal slice) (Fig. 6.1A), whereas anatomic localization and focality of brain stimulation is unavoidably much coarser with in vivo brain stimulation protocols such as tDCS and TMS (Fig. 6.1B), resulting in activation of entire neuronal networks as well as supporting glial cells. An additional wrinkle is the fact that while the individual neuron is more excitable in the immature brain due to the relative expression and composition of receptors and ion channels, the brain as a whole has less synaptic connectivity in developing brains as compared to adult brains.

Further preclinical studies are therefore required to investigate the effects of these basic age-dependent mechanistic differences on the safety and efficacy of brain stimulation in the pediatric brain, before any meaningful attempts can be made to redesign brain stimulation protocols for pediatric age groups.

CONCLUSION

Important mechanistic differences in the E:I balance and plasticity between mature and immature brains necessitate the reoptimization of brain stimulation protocols in pediatric populations, to account for critical neurodevelopmental processes which directly and indirectly influence the safety and efficacy of modulating excitability in the developing brain. Such optimization is especially important given the vulnerability of the

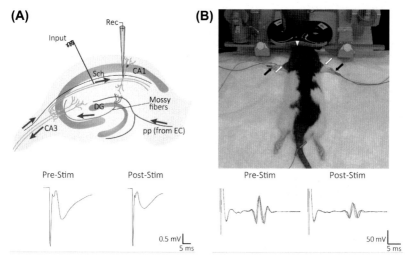

FIGURE 6.1 (A) Diagram of a typical hippocampal electrophysiological plasticity experiment in an isolated hippocampal slice. Most results that pertain to synaptic plasticity are derived from such setups. Note the microscopic electrical stimulation of isolated afferent Schaffer collateral pathway and subsequent recording of a field excitatory postsynaptic potential (fEPSP) from CA1. Repetitive low-frequency stimulation results in decreased fEPSP amplitude, reflecting long-term depression (LTD). *(Adapted with permission from Macmillan Publishers Ltd: Neves G, Cooke SF, Bliss TV. Synaptic plasticity, memory and the hippocampus: a neural network approach to causality. Nature Reviews Neuroscience. 2008;9(1):65–75, copyright (2008).)* (B) Transcranial magnetic stimulation in rats. Compared to in vitro brain slice stimulation, brain stimulation is significantly less focal with in vivo protocols such as TMS. A large electromagnetic coil is discharged in close proximity to the scalp of an anesthetized rat, resulting in the generation of intracranial stimulating currents and activation of entire neuronal networks as well as supporting glial cells. Motor-evoked potentials (MEPs) are then recorded from limb muscles using needle electromyography (EMG), and as in (A), repetitive low-frequency stimulation again results in a decrease in MEP amplitude, reflecting LTD-like cortical plasticity. *(Reprinted from Rotenberg A, Muller PA, Vahabzadeh-Hagh AM, et al. Lateralization of forelimb motor evoked potentials by transcranial magnetic stimulation in rats. Clinical Neurophysiology. January 2010;121(1):104–108, copyright (2010), with permission from Elsevier.)*

child to seizures, as these are a realistic side effect of brain stimulation. Preclinical research in animal models has provided crucial insights on the maturational trajectory of excitatory, inhibitory, and use-dependent synaptic plasticity mechanisms, which will inform the design and application of pediatric brain stimulation protocols to allow for maximal efficacy while avoiding the disruption of key neurodevelopmental milestones that could potentially result in aberrant neuronal circuitry.

Preclinical as well as clinical data also reveal that certain neurodevelopmental diseases exhibit a pathologically dysmature brain regardless of chronological age, due to delayed or absent molecular milestones, and the design of stimulation protocols should include these considerations to allow for the development of safe and effective therapeutic options.

Acknowledgments

This work was supported by the Boston Children's Hospital Translational Research Program and the King Saud University Visiting Professors Program.

References

1. An S, Yang JW, Sun H, Kilb W, Luhmann HJ. Long-term potentiation in the neonatal rat barrel cortex in vivo. *J Neurosci.* 2012;32(28):9511–9516.

2. Mix A, Hoppenrath K, Funke K. Reduction in cortical parvalbumin expression due to intermittent theta-burst stimulation correlates with maturation of the perineuronal nets in young rats. *Dev Neurobiol.* 2015;75(1):1–11.

3. Cao G, Harris KM. Developmental regulation of the late phase of long-term potentiation (L-LTP) and metaplasticity in hippocampal area CA1 of the rat. *J Neurophysiol.* 2012;107(3):902–912.

4. Guerriero RM, Giza CC, Rotenberg A. Glutamate and GABA imbalance following traumatic brain injury. *Curr Neurol Neurosci Rep.* 2015;15(5):27.

5. Rakhade SN, Jensen FE. Epileptogenesis in the immature brain: emerging mechanisms. *Nat Rev Neurol.* 2009;5(7):380–391.

6. Sanchez RM, Jensen FE. Maturational aspects of epilepsy mechanisms and consequences for the immature brain. *Epilepsia.* 2001;42(5):577–585.

7. Silverstein FS, Jensen FE. Neonatal seizures. *Ann Neurol.* 2007;62(2):112–120.

8. Rosa MA, Picarelli H, Teixeira MJ, Rosa MO, Marcolin MA. Accidental seizure with repetitive transcranial magnetic stimulation. *J ECT.* 2006;22(4):265–266.

9. Sui L, Huang S, Peng B, Ren J, Tian F, Wang Y. Deep brain stimulation of the amygdala alleviates fear conditioning-induced alterations in synaptic plasticity in the cortical-amygdala pathway and fear memory. *J Neural Transm.* 2014;121(7):773–782.

10. Tawfik VL, Chang SY, Hitti FL, et al. Deep brain stimulation results in local glutamate and adenosine release: investigation into the role of astrocytes. *Neurosurgery.* 2010;67(2):367–375.

11. Chaieb L, Antal A, Paulus W. Transcranial random noise stimulation-induced plasticity is NMDA-receptor independent but sodium-channel blocker and benzodiazepines sensitive. *Front Neurosci.* 2015;9:125.

12. Monyer H, Burnashev N, Laurie DJ, Sakmann B, Seeburg PH. Developmental and regional expression in the rat brain and functional properties of four NMDA receptors. *Neuron.* 1994;12(3):529–540.

13. Semple BD, Blomgren K, Gimlin K, Ferriero DM, Noble-Haeusslein LJ. Brain development in rodents and humans: identifying benchmarks of maturation and vulnerability to injury across species. *Prog Neurobiol.* 2013;106–107:1–16.

14. Blanke ML, VanDongen AMJ. Activation mechanisms of the NMDA receptor. In: Van Dongen AM, ed. *Biology of the NMDA Receptor.* ; 2009. Boca Raton (FL).

15. Hanson E, Armbruster M, Cantu D, et al. Astrocytic glutamate uptake is slow and does not limit neuronal NMDA receptor activation in the neonatal neocortex. *Glia.* 2015;63(10):1784–1796.

16. Sanchez RM, Jensen FE. Modeling hypoxia-induced seizures and hypoxic encephalopathy in the neonatal period. In: Pitkanen A, Moshe SL, Schwartzkroin PA, eds. *Models of Seizures and Epilepsy.* San Diego: Elsevier; 2006.

17. Avallone J, Gashi E, Magrys B, Friedman LK. Distinct regulation of metabotropic glutamate receptor (mGluR1 alpha) in the developing limbic system following multiple early-life seizures. *Exp Neurol.* 2006;202(1):100–111.

18. Danbolt NC, Storm-Mathisen J, Kanner BI. An [Na$^+$ + K$^+$]coupled L-glutamate transporter purified from rat brain is located in glial cell processes. *Neuroscience.* 1992;51(2):295–310.

19. Lehre KP, Danbolt NC. The number of glutamate transporter subtype molecules at glutamatergic synapses: chemical and stereological quantification in young adult rat brain. *J Neurosci.* 1998;18(21):8751–8757.

20. Ullensvang K, Lehre KP, Storm-Mathisen J, Danbolt NC. Differential developmental expression of the two rat brain glutamate transporter proteins GLAST and GLT. *Eur J Neurosci.* 1997;9(8):1646–1655.

21. Kim S, Stephenson MC, Morris PG, Jackson SR. tDCS-induced alterations in GABA concentration within primary motor cortex predict motor learning and motor memory: a 7 T magnetic resonance spectroscopy study. *Neuroimage.* 2014;99:237–243.

22. Amadi U, Allman C, Johansen-Berg H, Stagg CJ. The homeostatic interaction between anodal transcranial direct current stimulation and motor learning in humans is related to GABAA activity. *Brain Stimul.* 2015;8(5):898–905.

23. Swann JW, Brady RJ, Martin DL. Postnatal development of GABA-mediated synaptic inhibition in rat hippocampus. *Neuroscience.* 1989;28(3):551–561.

24. LoTurco JJ, Owens DF, Heath MJ, Davis MB, Kriegstein AR. GABA and glutamate depolarize cortical progenitor cells and inhibit DNA synthesis. *Neuron.* 1995;15(6):1287–1298.

25. Plotkin MD, Snyder EY, Hebert SC, Delpire E. Expression of the Na-K-2Cl cotransporter is developmentally regulated in postnatal rat brains: a possible mechanism underlying GABA's excitatory role in immature brain. *J Neurobiol.* 1997;33(6):781–795.

26. Rivera C, Voipio J, Payne JA, et al. The K^+/Cl^- co-transporter KCC2 renders GABA hyperpolarizing during neuronal maturation. *Nature.* 1999;397(6716):251–255.

27. Ganguly K, Schinder AF, Wong ST, Poo M. GABA itself promotes the developmental switch of neuronal GABAergic responses from excitation to inhibition. *Cell.* 2001;105(4):521–532.

28. Trippe J, Mix A, Aydin-Abidin S, Funke K, Benali A. theta burst and conventional low-frequency rTMS differentially affect GABAergic neurotransmission in the rat cortex. *Exp Brain Res.* 2009;199(3–4):411–421.

29. Labedi A, Benali A, Mix A, Neubacher U, Funke K. Modulation of inhibitory activity markers by intermittent theta-burst stimulation in rat cortex is NMDA-receptor dependent. *Brain Stimul.* 2014;7(3):394–400.

30. Dhamne SC, Ekstein D, Zhuo Z, et al. Acute seizure suppression by transcranial direct current stimulation in rats. *Ann Clin Transl Neurol.* 2015;2(8):843–856.

31. Gaiarsa JL, McLean H, Congar P, et al. Postnatal maturation of gamma-aminobutyric acid A and B-mediated inhibition in the CA3 hippocampal region of the rat. *J Neurobiol.* 1995;26(3):339–349.

32. Gaiarsa JL, Porcher C. Emerging neurotrophic role of GABAB receptors in neuronal circuit development. *Front Cell Neurosci.* 2013;7:206.

33. Ben-Ari Y. Limbic seizure and brain damage produced by kainic acid: mechanisms and relevance to human temporal lobe epilepsy. *Neuroscience.* 1985;14(2):375–403.

34. Epsztein J, Represa A, Jorquera I, Ben-Ari Y, Crepel V. Recurrent mossy fibers establish aberrant kainate receptor-operated synapses on granule cells from epileptic rats. *J Neurosci.* 2005;25(36):8229–8239.

35. Nitecka L, Tremblay E, Charton G, Bouillot JP, Berger ML, Ben-Ari Y. Maturation of kainic acid seizure-brain damage syndrome in the rat. II. Histopathological sequelae. *Neuroscience.* 1984;13(4):1073–1094.

36. Tremblay E, Nitecka L, Berger ML, Ben-Ari Y. Maturation of kainic acid seizure-brain damage syndrome in the rat. I. Clinical, electrographic and metabolic observations. *Neuroscience.* 1984;13(4):1051–1072.

37. Ben-Ari Y. The developing cortex. *Handb Clin Neurol.* 2013;111:417–426.

38. Braat S, Kooy RF. The GABAA receptor as a therapeutic target for neurodevelopmental disorders. *Neuron.* 2015;86(5):1119–1130.

39. Fatemi SH, Folsom TD. GABA receptor subunit distribution and FMRP-mGluR5 signaling abnormalities in the cerebellum of subjects with schizophrenia, mood disorders, and autism. *Schizophr Res.* 2014;167(1–3):42–56.
40. He Q, Nomura T, Xu J, Contractor A. The developmental switch in GABA polarity is delayed in fragile X mice. *J Neurosci.* 2014;34(2):446–450.
41. Braat S, Kooy RF. Insights into GABAAergic system deficits in fragile X syndrome lead to clinical trials. *Neuropharmacology.* 2015;88:48–54.
42. Duarte ST, Armstrong J, Roche A, et al. Abnormal expression of cerebrospinal fluid cation chloride cotransporters in patients with Rett syndrome. *PLoS One.* 2013;8(7):e68851.
43. Talos DM, Sun H, Kosaras B, et al. Altered inhibition in tuberous sclerosis and type IIb cortical dysplasia. *Ann Neurol.* 2012;71(4):539–551.
44. Lomo T. The discovery of long-term potentiation. *Philos Trans R Soc Lond B Biol Sci.* 2003;358(1432):617–620.
45. Bear MF, Malenka RC. Synaptic plasticity: LTP and LTD. *Curr Opin Neurobiol.* 1994;4(3):389–399.
46. Malenka RC, Bear MF. LTP and LTD: an embarrassment of riches. *Neuron.* 2004;44(1):5–21.
47. Fitzgerald PB, Fountain S, Daskalakis ZJ. A comprehensive review of the effects of rTMS on motor cortical excitability and inhibition. *Clin Neurophysiol.* 2006;117(12):2584–2596.
48. Nitsche MA, Paulus W. Transcranial direct current stimulation–update 2011. *Restor Neurol Neurosci.* 2011;29(6):463–492.
49. Kirkwood A, Silva A, Bear MF. Age-dependent decrease of synaptic plasticity in the neocortex of alphaCaMKII mutant mice. *Proc Natl Acad Sci USA.* 1997;94(7):3380–3383.
50. Dudek SM, Bear MF. Bidirectional long-term modification of synaptic effectiveness in the adult and immature hippocampus. *J Neurosci.* 1993;13(7):2910–2918.
51. Lante F, Cavalier M, Cohen-Solal C, Guiramand J, Vignes M. Developmental switch from LTD to LTP in low frequency-induced plasticity. *Hippocampus.* 2006;16(11):981–989.
52. Froemke RC. Plasticity of cortical excitatory-inhibitory balance. *Annu Rev Neurosci.* 2015;38:195–219.
53. Takesian AE, Hensch TK. Balancing plasticity/stability across brain development. *Prog Brain Res.* 2013;207:3–34.

Pediatric Issues in Neuromodulation: Safety, Tolerability and Ethical Considerations

K.M. Friel

Burke-Cornell Medical Research Institute, White Plains, NY, United States;
Weill Cornell Medical College, New York, NY, United States

A.M. Gordon

Teachers College of Columbia University, New York, NY, United States;
Columbia University Medical Center, New York, NY, United States

J.B. Carmel

Burke-Cornell Medical Research Institute, White Plains, NY, United States;
Weill Cornell Medical College, New York, NY, United States

A. Kirton

University of Calgary, Calgary, AB, Canada

B.T. Gillick

University of Minnesota, Minneapolis, MN, United States

Neuromodulation encompasses a variety of methods by which brain function can be monitored or influenced using non-invasive brain stimulation (NIBS). Two of the most commonly studied and used modalities include transcranial magnetic stimulation (TMS) and transcranial direct-current stimulation (tDCS). The use of both these technologies is rapidly increasing in pediatric populations. NIBS protocols have shown excellent safety profiles in adults. In children, the use of NIBS has greatly increased in recent years. Safety is paramount when working with children. The safety and ethical ramifications of NIBS use in children must be carefully considered. Very few serious adverse events have been reported across large adult populations and many children, suggesting that pediatric NIBS is probably safe. However, there is some disagreement in the literature about the safety and ethics of NIBS in children. Some[1,2] have concluded that NIBS in children is lower risk, citing the very low incidence of adverse events. Others[3] point out that the low number of pediatric NIBS studied (compared with adult NIBS studies) and lack of information about the long-term effects of NIBS warrant greater caution.

This chapter will discuss safety issues and ethical issues related to the use of NIBS in children. We will consider the direct effects (both known and unknown) of NIBS on the developing brain, particularly as it relates to the safety profile of NIBS in children. We will discuss known risks and side effects, and how these influence the ethics of applying NIBS to children. We will highlight the most common uses of NIBS in children, and parse the important ethical considerations regarding the application of NIBS to these potentially vulnerable populations. We will describe our laboratory practices to optimize safety of NIBS in children and adhere faithfully to ethical principles.

BRAIN STIMULATION TECHNIQUES USED IN CHILDREN

The most widely tested and used NIBS methods used in pediatrics are TMS and tDCS. TMS induces a focal magnetic field to a region of cortex, which results in depolarization of neurons (see Chapter 1: TMS Basics: Single and Paired Pulse Neurophysiology). TMS pulses are delivered through a coil, the shape of which affects the physical properties of the pulse. Most commonly, TMS pulses are delivered via a figure-of-eight shaped coil. Three general types of TMS protocols are used: single-pulse, paired-pulse, and repetitive TMS. In single-pulse TMS, a single TMS pulse is applied over the child's scalp. Single-pulse TMS can elicit a motor-evoked potential in a targeted muscle, allowing understanding of the connectivity between the cortex and the muscle. Single-pulse TMS can be applied to different locations within motor cortex to map the cortical areas involved in movement of particular muscles.[4] Single-pulse TMS mapping can be applied before and after an intervention to measure changes in cortical organization associated with the intervention. In paired-pulse TMS, two TMS pulses are applied in rapid succession (milliseconds) to the same cortical area or to two different areas. Paired-pulse TMS can evaluate the integrity of inhibitory or facilitatory processes in the brain dependent upon interstimulus intervals and stimulus intensities.[5] Gilbert et al.[1] conducted an extensive review of the use of single- and paired-pulse TMS in children, concluding that these protocols could be considered minimal risk.

In repetitive TMS (rTMS, see Chapter 4: Therapeutic rTMS in Children), trains of pulses are applied with the intention of modulating brain excitability. If pulses are delivered at a frequency of less than or equal to 1 Hz, the effects are typically inhibitory, and can be used to create temporary "virtual lesions" to probe function within a region or the relationship between one region and others. In contrast, pulse frequencies greater than 5–10 Hz typically result in facilitation and can be used to "prime" circuits to enhance neuronal activity.[6,7] Repetitive TMS also has a very good safety profile, though greater caution is recommended regarding the use of rTMS in children compared to single-pulse TMS due to its potential modulatory effects.[3] In adults, there is a higher, but still low, incidence of adverse events associated with rTMS compared to single-pulse TMS. Although there have been no reported incidents of serious adverse events in children associated with rTMS, the use of rTMS in children has been limited in comparison to adults.[6,7]

Over the past 10 years, tDCS has emerged as a safe, promising method for neuromodulation (see Chapter 5: Transcranial Direct-Current Stimulation (tDCS): Principles and Emerging Applications in Children).[8] In tDCS, sponge electrodes (usually an anode and a cathode) are secured to make contact with the skull. A low level (0.7–2 mA) of direct, constant current stimulation is then applied to the scalp, which changes the resting membrane potential of cortical neurons within the electrical field, which is usually centered underneath the stimulating electrode but may span broad cortical regions.[9] This results in a slight shift in the resting potential of the neurons, making them slightly more or less likely to fire, depending on the polarity of the stimulation. Similar to rTMS, tDCS is used to modulate the excitability of neural networks. tDCS has an excellent safety profile in adults and may readily be paired with physical rehabilitation due to its convenient portability.[8]

Both TMS and tDCS have been applied to children,[2] with good safety outcomes and promising evidence of efficacy. Specifically, there have been no serious adverse events associated with TMS or rTMS, such as seizure, reported in hundreds of children and only extremely rarely across thousands of adults.[10] Possible adverse effects associated with tDCS in children and adults are minor, such as scalp tingling, transient, and have resolved without medical care.

DIRECT EFFECTS OF BRAIN STIMULATION IN CHILDREN

Children differ from adults in ways that affect the physiological actions of NIBS, including both head anatomy and brain physiology.[3] There are important, often underappreciated differences in head anatomy between children and adults that affect how NIBS acts in the central nervous system. Not only is head size smaller in children than in adults, but there are also clear differences in physiological and physical properties of heads in children versus adults. These differences must be considered carefully when applying NIBS to children. If the same type and intensity of stimulation is applied to an adult versus a child head, the effective stimulus intensity that reaches the brain may be stronger in the child than the adult, as evidenced by modeling of tDCS current flow in pediatric populations.[11,12]

When setting experimental parameters for NIBS in children, the impact of head size on dosing and safety must be carefully considered and additional studies are required. Moreover, brain physiology is different in children versus adults. For example, brains of children and adults have different relative amounts of neurotransmitters and neurotrophins.[13] During development, particularly early in development, inhibitory neurotransmitters can have excitatory actions.[14] Pediatric-onset brain disorders can also result in an alteration of neurochemistry[15] that may alter the brain's responsiveness to stimulation.

Moliadze et al. directly tested the impact of smaller head size on tDCS effects.[16] Ten minutes of 1 mA anodal tDCS in adolescents increased the amplitude of TMS motor-evoked potentials (MEPs), starting 20 min after cessation of tDCS and lasting longer than 1 h. The increases in MEP size were more robust than the MEP changes found using the same tDCS protocol in adults.[17] Importantly, in adults cathodal tDCS usually reduces the size of the MEP and has often been used to reduce excitability of a brain area. However, in adolescents cathodal tDCS was shown in one study to *increase* MEP size. These effects may be dependent on multiple other variables such as current intensity where 1 mA versus 2 mA cathodal tDCS was recently shown to have divergent effects on cortical excitability.[18] Caution, therefore, is advised in using tDCS in children, since the effects of cathodal tDCS are different in children versus adults. More studies are greatly needed in both typically developing children and those affected by neurodevelopmental disease to better characterize the effects of tDCS.

DEVELOPMENTAL NEUROPLASTICITY

During development and into young adulthood, neural circuits in the brain and spinal cord are strengthened and refined.[19–21] Damage or loss of normal activity in a brain region can derail normal development and have permanent effects on function. Motor[22] and cognitive[23,24] networks develop most protractedly in children and are therefore most vulnerable. While development provides a window of opportunity to intervene and redirect a previously existing maladaptive developmental process, it also represents a period of heightened susceptibility. This susceptibility is particularly apparent in children who experience traumatic brain injury (TBI).[25] Many children who experience even mild to moderate TBI show long-term psychosocial consequences, such as poorer academic performance and increased incidence of attention-deficit hyperactivity disorder (ADHD).[26] The long-term consequences of TBI tend to be more severe the younger a child was at the time of injury.[27] The long-term effects of NIBS in adults and children are unknown. Although evidence to date supports the safety of NIBS in children, these unique elements of the developing nervous system must be considered with care taken to appropriately target and dose NIBS as best possible to minimize risk. Since developmental neuroplasticity decreases with age, it would be prudent to test new protocols in older children, then apply the protocols to progressively younger children as safety evidence becomes more established.[28]

SIDE EFFECTS OF STIMULATION

Several side effects have been identified in the use of NIBS in adults. These side effects may be different or more severe in children, given differences in head size, body composition, neurophysiology, and possible unknown factors.

Heating

TMS, particularly rTMS, can cause heating of the scalp over time. This heating has the potential to become painful or cause damage to the skin. Heating may occur more rapidly in children than in adults due to the smaller skull thickness of children. Care must be taken to monitor the temperature of NIBS devices during a protocol and employ appropriate cooling mechanisms. Many systems already have such safeguards built in such as constant temperature detection with predetermined automatic shut-off mechanisms that must be used during studies. Experimenters must also be mindful that some children with neurological impairments may have reduced scalp sensation and thus may not be able to promptly report scalp heating.

Hearing Deficits

Noise from TMS can be bothersome and has the theoretical potential to cause hearing damage. A single TMS pulse at maximum stimulator output produces a brief but very loud clicking noise (200–300-μs duration, up to 120 dB). Overexposure to loud noise can cause damage to the hair cells in the cochlea. Due to the brevity of the pulse duration, the true loudness of the pulse can be difficult to perceive, tempting some to forego protection of hearing. It is very important to protect the hearing of children during TMS administration. Occupational Safety and Health Administration (OSHA) regulations require that workers be exposed to no more than 15 min/day of continuous 120-dB noise.[29] Since TMS pulse duration is very brief, a typical TMS session involves less than 15 min of total 120-dB noise. A higher frequency of rTMS is associated with greater risk of hearing damage.[30] Earplugs should be worn to protect the hearing of those exposed to the noise of a TMS pulse, including study participants, experimenters, and family members who may be near the TMS. Child-sized earplugs should be used, since adult-sized earplugs may not sufficiently protect the hearing of children. Despite these theoretical concerns, no evidence of NIBS-related hearing deficits has been reported in children who have received TMS.[31]

Seizures

Seizure is a potential risk of NIBS, particularly high-frequency rTMS. Of note, there have been no reported cases of seizures produced by single-pulse TMS in children.[10,32] This includes more than 1000 subjects, many of whom have pre-existing conditions associated with lower seizure thresholds. Several studies have deliberately stimulated known focal seizure foci with either no induced seizures or even an improvement of EEG epileptiform discharges with certain paradigms.[46] Despite this, seizures have occasionally been reported in close proximity to brain stimulation procedures. No seizures have been reported during tDCS sessions[8] but there is a case report of a 4-year-old boy who had a seizure 4h after the third in a series of tDCS stimulation sessions.[33] The boy had a diagnosis of left dominant spastic tetraparesis, and was enrolled in an outpatient tDCS study to improve upper limb function and reduce spasticity. The boy had a history of seizures, and his antiepileptic medication had recently been adjusted. As a potential association of tDCS with seizure cannot be excuded in all subjects, caution is warranted when delivering tDCS to patients who have a higher than normal risk. Reports of seizure after single-pulse TMS have been reported in several adults, but this is out of hundreds of thousands of applications across thousands of individuals.[6] In all of these cases, the seizure occurred during the TMS session. Seven cases involved patients with a brain lesion or pre-existing seizure disorder (intractable epilepsy).[1,2,10] In one case, the subject was a healthy 33-year-old male with no known predispositions to seizure.[34] No other seizures associated with single-pulse TMS have been reported. More than 1000 children have received single-pulse or repetitive TMS in the past decade. Only one incident of seizure was reported – in a 15-year-old who was receiving 10Hz repetitive TMS at an intensity of 80% resting motor threshold over the left dorsolateral prefrontal cortex to treat depression.[35] The adolescent had no known predispositions to seizure, and was concurrently taking 100mg/day sertraline in accordance with his treatment protocol. Sertraline is a selective serotonin reuptake inhibitor that may decrease seizure threshold, ie, making one more susceptible to seizures. It is recommended that subjects with known reductions in seizure threshold, such as poorly controlled epilepsy and perhaps those taking some medications, should be carefully counseled about the relative risks and benefits of participating in a specific NIBS intervention or study.

Even though the risk of seizure appears to be extremely low for children who receive TMS or tDCS, we have excluded children from NIBS protocols if they have an active seizure disorder. This is particularly recommended for experimental protocols in which there is no potential benefit of NIBS for an individual. Studies that have used TMS to probe or modulate the motor system in children with cerebral palsy, including our own studies, have excluded children with an active seizure disorder.[36-45] In several of these studies, children were required to be seizure-free for 6 months to 5 years before the application of TMS. No seizures occurred in any child in these protocols.

Other centers experienced in pediatric brain stimulation have only excluded children with "unstable epilepsy" defined as more than one seizure per month or a recent change in seizure frequency or medication (Kirton, personal communication). That center has performed extensive single, paired, and rTMS procedures on >180 children (many with epilepsy and/or brain injuries) totaling >250,000 stimulations without a single seizure.

TMS is increasingly being used to treat patients with epilepsy. Inhibitory rTMS, given over a series of sessions, may decrease the frequency of seizures in children[46-48] without adverse events, though controlled studies are lacking. Though these studies are limited in number, they provide evidence that not only is TMS likely safe in children with active seizure disorders, but may also have therapeutic potential for treating epilepsy (see Chapter 4: Therapeutic rTMS in Children).

Syncope

There have been two reports of neurocardiogenic syncope in adolescents who received rTMS.[49] Both children had a prior history of presyncope, and one child had not eaten breakfast before the TMS session. One child withdrew from the study, while the other child continued enrollment. Syncope is not believed to be directly caused by TMS, rather it is likely to be a combination of anxiety and exposure to a mildly painful or noxious sensation in an individual predisposed to vasovagal syncope. To minimize the risk of syncope, it is recommended that children be screened for these risk factors. If the child previously experienced syncope, the probable triggers (hunger, nervousness) should be identified and minimized. Kirton et al.[49] also recommend assuring recent intake of food and water, disclosure of the risk of syncope, and gradual increase of TMS stimulus intensity at the beginning of a session to optimize tolerability. Gradual increase of stimulus intensity, starting below 30% of maximum stimulator output, may help ease a child's anxiety and help them acclimate to the sensations of TMS stimulation.

Headache and Scalp Sensations

Mild headache is a common side effect of TMS and tDCS. The most common type of headache reported in our studies is pain underneath a headband that is used for frameless neuronavigation (matching the location of TMS stimulation to a participant's MRI) or underneath the band of a swim cap on which the locations of TMS stimulation sites are marked. Of children who report headache, approximately 75% say the headache is associated with the headband or swim cap, not the TMS itself. In the Friel laboratory, we have conducted 136 sessions using single-pulse TMS to map motor cortex in children with CP. After eight (5.8%) sessions, participants reported mild headache. In all reported pediatric NIBS cases, headache was mild when it occurred, resolved without medical treatment, and did not result in the participant withdrawing from the protocol. In the Gillick laboratory, one of 19 participants withdrew from an rTMS protocol due to scalp tenderness.[42] In 133 sessions of rTMS, no serious adverse events occurred. Minor headaches occurred in some participants, and were mainly due to the swim cap used for marking stimulation locations. Headaches resolved within 1 h after rTMS.

A randomized, controlled, clinical trial of 45 children aged 6–18 years with perinatal stroke and hemiparesis combined intensive motor learning therapy with rTMS.[42a] Subjects received daily contralesional inhibitory rTMS (1 Hz) for 20 min (1200 stimuli) immediately prior to the 2-h intensive therapy session within a 2-week daily learning camp for 10 consecutive weekdays. Safety outcomes including function of both the paretic and unaffected upper limb (with separate analysis of those with no stroke side motor-evoked potentials) found no evidence of adverse effect. Tolerability scores were favorable with mild, self-limiting headache in 40–50% and no dropouts or serious adverse events were reported.

An open label trial of high-frequency rTMS to the left DLPFC in adolescents and young adults (aged 15–21 years) with refractory major depression included daily treatment for 40 min for 15 consecutive weekdays.[50] Although some subjects reported discomfort on day 1, procedures were generally well tolerated with no serious adverse events or drop-outs.

The risk of scalp irritation sensations is higher for tDCS than TMS. Scalp itching and tingling is commonly associated with tDCS, though symptoms are typically mild, transient, and resolve without medical intervention. During tDCS application, care should be taken to maintain moisture of sponges; many systems have built-in safety features to alert the operator if impedance becomes high. Further dosing studies are needed to determine optimal intensity and duration of stimulation, as limiting exposure may also reduce the incidence of side effects.

A study of tDCS enhancement of motor learning in 22 typically developing children aged 6–18 years found no serious adverse events (A. Kirton, unpublished observations). Children performed a motor learning task repeatedly over 3 days, randomized to sham, 1 mA contralateral M1 anodal, 1 mA or 2 mA ipsilateral cathodal M1 stimulation for the first 20 min of each training session. Specific safety outcomes included any decrease in either the trained or untrained hand as well as decline in multiple non-trained motor tasks before and after intervention. All functional outcomes improved with tDCS. Mild tingling or itching of the scalp was reported in 55% of subjects but never precluded participation.

A randomized, controlled, clinical trial of 24 children aged 6–18 years with perinatal stroke and hemiparesis combined intensive motor learning therapy with tDCS. Subjects received contralesional M1 cathodal 1 mA tDCS (or sham) for the first 20 min of a 2 h therapy session for 10 consecutive weekdays. Safety outcomes after 12 and 24 subjects including function of both the paretic and unaffected upper limb found no evidence of adverse effect. Aside from scalp tingling/itching (42%), no other adverse events were reported (A. Kirton, unpublished observations).

Cognitive and Psychiatric Changes

Safety guidelines for the use of TMS in adults include cognitive and psychiatric effects as potential side effects. Early studies of brain stimulation in adults often included neuropsychological testing as safety measures.[51] rTMS and tDCS have indeed been attempted to enhance cognitive performance and are proven to reduce psychiatric symptoms.[6] Several studies have examined the effects of NIBS on selective attention, reaction time, working memory, and psychomotor speed (ie, Stroop test). In these studies, there was no measurable diminution of cognitive measures after NIBS. In pediatric NIBS use, there have been no reports of unintended cognitive or psychiatric effects. Indeed, TMS has shown efficacy in improving symptoms of depression[52–56] and possibly ADHD.[52] Novel applications of NIBS in the developing brain should consider these more difficult to measure outcomes with appropriate addition of safety outcome measures.

ETHICAL APPLICATION OF NEUROMODULATION IN CHILDREN

Childress and Beauchamp[57] outlined four basic principles that are now considered pillars of biomedical ethics: autonomy, beneficence, nonmaleficence, and justice. Each of these important issues must be considered regarding the use of NIBS in children.

It is important to note that NIBS can be used to either modulate or to measure brain activity. Malsen et al.[58] discuss the ethical use of NIBS in children. They draw a distinction between methods used to treat disorders versus methods aimed at enhancement in healthy children. We extend these distinctions to include another category: using NIBS to study the brain without therapeutic intent. Different ethical considerations come into play for each of these types of NIBS. For example, single-pulse or paired-pulse TMS is used to probe brain circuitry without modulating the function of those circuits. For the most part, the use of single- or paired-pulse TMS is not therapeutic or modulatory in nature, minimizing the risk of altering brain function. Rather, these studies add to our understanding of brain circuitry associated with healthy function or a particular condition. In these studies, NIBS does not typically benefit the particular child receiving the NIBS. Benefit is conferred at the population level. Specific findings about brain connections that these studies provide could guide the development or prescription of therapies. In situations in which NIBS does not provide a direct benefit to the child receiving the stimulation, risks should be very low. Currently, the known risks of single- and paired-pulse TMS are very low, allowing these types of paradigms to be ethically applied to children who are identified as having no risk factors.

Autonomy

Persons have the right to make informed decisions about their involvement in medical therapies or research protocols. The principle of autonomy can be complex when children are involved. Autonomy assumes that a person is able to make an informed decision about a procedure, carefully weighing potential risks and benefits. While children may have difficulty fully understanding a protocol and associated risks, investigators, medical staff, children, and their families can take steps that maximize a child's autonomy.

Children, regardless of their health status or the reason for receiving NIBS, should be provided with age-appropriate educational materials about a procedure. In our research, we earnestly invest in the education of our study participants. Children are provided documentation that describes brain stimulation in a child-friendly way. We have developed a series of videos in which former study participants explain the procedures to children who would be watching the video. It is also important to remember that education and assent are not only preprocedure issues. Throughout the procedure, engagement of children is important. Children must be aware of the purpose of each step, and must have the freedom to withdraw their consent or assent if they do not wish to continue and understand how to express this wish in a supportive environment. Our experience indicates that continuous engagement of children eases anxiety. When invited to do so, many children are eager to understand the rationale for the use of NIBS and enjoy a feeling of partnership with scientists during study visits.

Beneficence

The benefits of a procedure must outweigh risks and costs. A challenge in the field of NIBS is that nearly all clinical applications of pediatric NIBS are experimental. Benefits and risks are not well understood. Nonetheless, the principle of beneficence can rest on potential benefits versus potential risks. The most promising evidence of efficacy is found in the treatment of depression.[59] TMS is being used more widely to manage symptoms of ADHD and autism, with promising findings.[10] In addition, NIBS has shown efficacy in magnifying the efficacy of motor rehabilitation in children with cerebral palsy.[43,60] Nonetheless, the benefits and risks of NIBS in children are not well characterized.[3] While NIBS is considered safe in children[2,10] given the very low incidence of serious side effects, it is equally important that potential benefits not be overstated, to allow subjects and their caregivers to make an informed decision.

Decisions regarding benefits and risks must consider the purpose of the NIBS and the health of the child. If a child does not have any health-related reason for receiving NIBS, such as for neuro-enhancement (discussed below), risks must be very low to justify the use of NIBS, since benefits are not well understood.

If a child has a neurological or psychiatric condition, the purpose of NIBS impacts the ethics of use. In some protocols, particularly most single-pulse protocols, TMS is used to examine brain physiology of a particular condition. In these protocols, NIBS is not used to improve a person's symptoms, rather it is used to better understand the condition. Since single-pulse TMS is very low risk, the use of single-pulse TMS to probe brain function can usually be easily justified. Benefits and risks, when risks are believed to be quite low, should also consider a population-wide view. Although the benefits of NIBS for a particular child may not be clear, continued safe research has the potential to benefit society with the development of effective therapy protocols. Single-pulse TMS may also have the potential to directly benefit a child. For example, there is emerging evidence that children with unilateral CP may respond differently to particular types of intensive hand therapy based on the connectivity of their corticospinal system.[60a] TMS might therefore be used to identify which therapy might be best for a particular child and such personalized medicine is an important future goal in pediatric neurorehabilitation.

If a NIBS protocol is being used to modulate brain function using a protocol that has shown efficacy, possible benefits to a particular child may become more clear but are still likely hard to estimate accurately. In these cases, it could be ethically appropriate to endure greater risk, since a greater benefit is expected. Repetitive TMS and tDCS have shown preliminary efficacy in treating some disorders as discussed above. Potential benefits of these forms of NIBS can apply to the child receiving the treatment. Therefore, the risk–benefit ratio can be different from the risk–benefit ratio of single- or paired-pulse TMS: slightly higher risks may be ethically tolerable if benefits to individuals are also higher. It should be reiterated that risks of all types of NIBS appear to be quite low, even in children.

For example, one can envision justifying the use of rTMS for seizure treatment in children with uncontrolled seizure disorders. The risk of a TMS-provoked seizure is higher in these children, but the potential therapeutic benefit may justify this risk. Just the opportunity to try a potential new treatment for a condition where existing treatments are limited or have failed may carry meaningful benefit to children and their parents.

Non-maleficence

A procedure should not cause harm that is disproportionate to the benefits. Non-maleficence is closely related to beneficence. Since the benefits of NIBS in children remain uncharacterized for many neurological disorders – although evidence of benefits is mounting – it is ethically important that NIBS not be harmful to children. Further investigation of the short- and long-term effects of NIBS is needed. If future studies uncover harmful effects of NIBS in children, these findings must be promptly communicated to the scientific community such that risk–benefit ratios can be recalibrated. Again, the benefits and risks must be weighed based on the health of the child and the purpose of the NIBS. If benefits are clearer for a particular child, greater risk may be ethically tolerated.

Justice

The benefits and risks of a procedure must be balanced across patients. Many methods of NIBS remain quite expensive. Care must be taken to provide access to promising NIBS methods to children of varied backgrounds and socioeconomic statuses. The issue of justice particularly comes into play as evidence mounts regarding efficacy of NIBS to treat a specific condition. As efficacy trials are done, care must be taken to enroll children of similar backgrounds in each experimental arm (eg, placebo versus active stimulation).

NEURO-ENHANCEMENT IN HEALTHY INDIVIDUALS

As methods of NIBS continue to be studied and refined, the breadth of applications is expanding. An emerging, and controversial, use of NIBS is neuro-enhancement in healthy individuals – that is, the use of NIBS to improve motor, cognitive, and attentional abilities in a population without a diagnosis being targeted. It can be argued that some applications of NIBS for neuro-enhancement confer benefits to individuals and to society, such as the use of TMS to enhance cognitive ability or decrease fatigue[61] in persons who work to protect others. Establishing the principles of an NIBS intervention in typically developing brains in order to apply it systematically to a specific disease population is another reasonable justification of neuromodulation in healthy subjects, For example, testing the effects and optimal paradigms for tDCS enhancement of motor learning is directly applicable to clinical trials in children with motor disabilities such as cerebral palsy. However, other current and potential uses of NIBS, particularly in children, are controversial.

It should also be remembered that many, if not all, NIBS neuromodulation procedures are most likely affecting *endogenous neurophysiological processes that are unique to the individual*. It is generally accepted that this must occur for neuromodulation to have effect, rather than expecting that brain stimulation could simply generate or change brain functions on its own. While this does not preclude the possibility of effects in an adverse direction, it does imply that modulation is only altering the brain processes of the individual themselves, rather than "changing who they are."

As mentioned above, Malsen et al.[58] consider separately the ethical issues surrounding use of NIBS for therapeutic purposes versus use of NIBS for neuro-enhancement in healthy children. As new NIBS protocols become available to enhance cognitive performance, reading speed, attention, and athletic ability, it is important to consider the ethics of their use. Since the efficacy, and therefore the potential benefits, of such protocols are largely unknown, what is the risk–benefit tradeoff? While some risks can be justified in the application of NIBS for treatment of a clinical condition, what level of risk is justified for the use of NIBS in a healthy child? It could be foreseeable that children may feel pressure to take on unknown risks for the possible benefit of faster reading speed, better attention, higher test scores, and better athletic abilities. Pressure may be from peers or even from parents to succeed. The ready availability of "do it yourself" stimulation devices online further increases this risk. As the use of NIBS for neuro-enhancement becomes more widespread, great care must be taken that children and their families fully understand the risks and benefits of any protocol they might consider.

ADVISEMENTS FOR SAFE ADMINISTRATION OF NIBS IN CHILDREN

Although NIBS is associated with a very low occurrence of adverse events and side effects in children, we recognize that there is a need for greater evidence of safety, particularly regarding long-term side effects, for NIBS in children. Based on the practices of our laboratories, we advise the following strategies for optimizing safety when using NIBS with children:

1. Participant screening: participants should be screened for the presence of magnetic implanted devices, which could be affected by NIBS, for history of seizures, and history of syncope. In our laboratories,

we do not enroll children with an active seizure disorder or the use of medications that lower the seizure threshold in NIBS protocols. Exceptions are warranted if the NIBS protocol to be used is given with the intent of treating seizures and if there is evidence that the protocol can do so. If children have a history of syncope, the triggers of the prior syncopal episodes should be identified and minimized. Care should be taken that children have eaten within 2h of NIBS. Snacks and water should be provided during NIBS sessions to prevent fatigue or dehydration.

2. Participant and family education and consent: Participants and their families should be provided with age-appropriate explanations and written materials about the NIBS protocol. Children must be given the opportunity to have questions adequately answered, and must give informed assent to the protocol. An option for a practice trial before final informed consent may be considered. During procedures, children should be kept informed of what is being done and why. Children should be given the opportunity to withdraw assent as desired during the procedure. Caregivers must also provide informed consent. Special care must be given to children who have cognitive deficits or communication challenges, to be sure the children understand and assent to the procedures. We keep a cue within reach of the child during the procedures: a "stop sign" that children can point to if they are unable to articulate their wish to stop the procedures.

3. Safety procedures during NIBS application: Even though the risk of adverse events is low, we perform NIBS procedures in a laboratory with easy access to emergency medical care, if needed. Laboratory or clinic staff are trained in seizure management and first aid, to provide assistive care until medical professionals can be summoned. During procedures, the comfort and well-being of children should be monitored frequently. Again, special consideration must be given to children who have difficulty communicating. Signs or pictures can be employed; children can point to appropriate diagrams to communicate their needs.

4. We encourage all NIBS administrators to have a written, bullet-pointed plan for how to deal with adverse events or other emergencies. The list should be readily accessible by all staff during NIBS administration. If an adverse event occurs, our laboratory protocols are as follows: Immediately stop the NIBS protocol and turn off stimulating devices. Monitor vital signs and maintain participant in position that maximizes safety, such as turning on the side during a seizure. Dispatch emergency medical personnel if needed. Details will depend on the nature of the event. Follow up with the child and family after the event to monitor lingering effects. Report all adverse

events to the governing ethics committee(s) and publish the incident in the scientific literature. It is important not only to report the incidence of adverse events (serious and minor), but also to report the methods concerning how safety outcomes were assessed.

5. Outcomes. Apply standardized measures of safety and tolerability such as the Pediatric TMS Tolerability Measure[62] in all studies and sessions.

CONCLUSIONS

The safety profile of NIBS in children is very good. The use of NIBS in children is expanding rapidly. However, much is unknown about the efficacy, appropriate targeting and dosing, and long- and short-term safety outcomes in children. More research is needed to better characterize the risks and benefits of NIBS for different disorders. NIBS holds great promise as a safe, effective method for promoting recovery in children with neurological disorders.

References

1. Gilbert DL, Garvey MA, Bansal AS, Lipps T, Zhang J, Wassermann EM. Should transcranial magnetic stimulation research in children be considered minimal risk? *Clin Neurophysiol (Official J Int Fed Clin Neurophysiol)*. August 2004;115(8):1730–1739.
2. Rajapakse T, Kirton A. Non-invasive brain stimulation in children: applications and future directions. *Transl Neurosci*. June 2013;4(2).
3. Davis NJ. Transcranial stimulation of the developing brain: a plea for extreme caution. *Front Hum Neurosci*. 2014;8:600.
4. Wittenberg GF. Motor mapping in cerebral palsy. *Dev Med Child Neurol*. October 2009;51(suppl 4):134–139.
5. Reis J, Swayne OB, Vandermeeren Y, et al. Contribution of transcranial magnetic stimulation to the understanding of cortical mechanisms involved in motor control. *J Physiol*. January 15, 2008;586(2):325–351.
6. Rossi S, Hallett M, Rossini PM, Pascual-Leone A. Safety, ethical considerations, and application guidelines for the use of transcranial magnetic stimulation in clinical practice and research. *Clin Neurophysiol (Official J Int Fed Clin Neurophysiol)*. December 2009;120(12):2008–2039.
7. Frye RE, Rotenberg A, Ousley M, Pascual-Leone A. Transcranial magnetic stimulation in child neurology: current and future directions. *J Child Neurol*. January 2008;23(1):79–96.
8. Bikson M, Datta A, Elwassif M. Establishing safety limits for transcranial direct current stimulation. *Clin Neurophysiol (Official J Int Fed Clin Neurophysiol)*. June 2009;120(6):1033–1034.
9. Bikson M, Rahman A, Datta A. Computational models of transcranial direct current stimulation. *Clin EEG Neurosci*. July 2012;43(3):176–183.
10. Krishnan C, Santos L, Peterson MD, Ehinger M. Safety of noninvasive brain stimulation in children and adolescents. *Brain Stimul*. Jan-Feb 2015;8(1):76–87.
11. Gillick BT, Kirton A, Carmel JB, Minhas P, Bikson M. Pediatric stroke and transcranial direct current stimulation: methods for rational individualized dose optimization. *Front Hum Neurosci*. 2014;8:739.

12. Kessler SK, Minhas P, Woods AJ, Rosen A, Gorman C, Bikson M. Dosage considerations for transcranial direct current stimulation in children: a computational modeling study. *PloS One*. 2013;8(9):e76112.
13. Stoneham ET, Sanders EM, Sanyal M, Dumas TC. Rules of engagement: factors that regulate activity-dependent synaptic plasticity during neural network development. *Biol Bull*. October 2010;219(2):81–99.
14. Akerman CJ, Cline HT. Refining the roles of GABAergic signaling during neural circuit formation. *Trends Neurosci*. August 2007;30(8):382–389.
15. Ben-Ari Y. Basic developmental rules and their implications for epilepsy in the immature brain. *Epileptic Disord Int Epilepsy J Videotape*. June 2006;8(2):91–102.
16. Moliadze V, Schmanke T, Andreas S, Lyzhko E, Freitag CM, Siniatchkin M. Stimulation intensities of transcranial direct current stimulation have to be adjusted in children and adolescents. *Clin Neurophysiol (Official J Int Fed Clin Neurophysiol)*. October 28, 2014;126(7):1392–1399.
17. Nitsche MA, Paulus W. Sustained excitability elevations induced by transcranial DC motor cortex stimulation in humans. *Neurology*. November 27, 2001;57(10):1899–1901.
18. Batsikadze G, Moliadze V, Paulus W, Kuo MF, Nitsche MA. Partially non-linear stimulation intensity-dependent effects of direct current stimulation on motor cortex excitability in humans. *J Physiol*. April 1, 2013;591(Pt 7):1987–2000.
19. Friel KM, Williams PT, Serradj N, Chakrabarty S, Martin JH. Activity-based therapies for repair of the corticospinal system injured during development. *Front Neurol*. 2014;5:229.
20. Martin JH, Chakrabarty S, Friel KM. Harnessing activity-dependent plasticity to repair the damaged corticospinal tract in an animal model of cerebral palsy. *Dev Med Child Neurol*. September 2011;53(suppl 4):9–13.
21. Eyre JA. Development and plasticity of the corticospinal system in man. *Neural Plast*. 2003;10(1–2):93–106.
22. Eyre JA, Smith M, Dabydeen L, et al. Is hemiplegic cerebral palsy equivalent to amblyopia of the corticospinal system? *Ann Neurol*. November 2007;62(5):493–503.
23. Lewis DA. Cortical circuit dysfunction and cognitive deficits in schizophrenia–implications for preemptive interventions. *Eur J Neurosci*. June 2012;35(12):1871–1878.
24. Gonzalez-Burgos G, Fish KN, Lewis DA. GABA neuron alterations, cortical circuit dysfunction and cognitive deficits in schizophrenia. *Neural Plast*. 2011;2011:723184.
25. Lloyd J, Wilson ML, Tenovuo O, Saarijarvi S. Outcomes from mild and moderate traumatic brain injuries among children and adolescents: a systematic review of studies from 2008-2013. *Brain Inj*. March 19, 2015:1–11.
26. Garcia D, Hungerford GM, Bagner DM. Topical review: a review of negative behavioral and cognitive outcomes following traumatic brain injury in early childhood. *J Pediatr Psychol*. October 22, 2014;40(4):391–397.
27. McKinlay A, Grace R, Horwood J, Fergusson D, MacFarlane M. Adolescent psychiatric symptoms following preschool childhood mild traumatic brain injury: evidence from a birth cohort. *J Head Trauma Rehabil*. May–June 2009;24(3):221–227.
28. Gutmann A. Safeguarding children–pediatric research on medical countermeasures. *N. Engl J Med*. March 28, 2013;368(13):1171–1173.
29. Administration OSH. *Occupational Noise Exposure*; 2015. https://http://www.osha.gov/SLTC/noisehearingconservation/. Accessed 04.07.15.
30. Tringali S, Perrot X, Collet L, Moulin A. Repetitive transcranial magnetic stimulation: hearing safety considerations. *Brain Stimul*. July 2012;5(3):354–363.
31. Collado-Corona MA, Mora-Magana I, Cordero GL, et al. Transcranial magnetic stimulation and acoustic trauma or hearing loss in children. *Neurological Res*. June 2001;23(4):343–346.
32. Quintana H. Transcranial magnetic stimulation in persons younger than the age of 18. *J ECT*. 2005;21(2):88–95.

33. Ekici B. Transcranial direct current stimulation-induced seizure: analysis of a case. *Clin EEG Neurosci.* April 2015;46(2):169.
34. Kratz O, Studer P, Barth W, et al. Seizure in a nonpredisposed individual induced by single-pulse transcranial magnetic stimulation. *J ECT.* March 2011;27(1):48–50.
35. Hu SH, Wang SS, Zhang MM, et al. Repetitive transcranial magnetic stimulation-induced seizure of a patient with adolescent-onset depression: a case report and literature review. *J Int Med Res.* 2011;39(5):2039–2044.
36. Holmstrom L, Vollmer B, Tedroff K, et al. Hand function in relation to brain lesions and corticomotor-projection pattern in children with unilateral cerebral palsy. *Dev Med Child Neurol.* February 2010;52(2):145–152.
37. Staudt M, Gerloff C, Grodd W, Holthausen H, Niemann G, Krageloh-Mann I. Reorganization in congenital hemiparesis acquired at different gestational ages. *Ann Neurol.* December 2004;56(6):854–863.
38. Staudt M, Grodd W, Gerloff C, Erb M, Stitz J, Krageloh-Mann I. Two types of ipsilateral reorganization in congenital hemiparesis: a TMS and fMRI study. *Brain.* October 2002;125(Pt 10):2222–2237.
39. Berweck S, Walther M, Brodbeck V, et al. Abnormal motor cortex excitability in congenital stroke. *Pediatr Res.* 2008;63:84–88.
40. Kesar TM, Sawaki L, Burdette JH, et al. Motor cortical functional geometry in cerebral palsy and its relationship to disability. *Clin Neurophysiol (Official J Int Fed Clin Neurophysiol).* July 2012;123(7):1383–1390.
41. Braun C, Staudt M, Schmitt C, Preissl H, Birbaumer N, Gerloff C. Crossed cortico-spinal motor control after capsular stroke. *Eur J Neurosci.* May 2007;25(9):2935–2945.
42. Gillick BT, Krach LE, Feyma T, et al. Safety of primed repetitive transcranial magnetic stimulation and modified constraint-induced movement therapy in a randomized controlled trial in pediatric hemiparesis. *Archives Phys Med Rehabil.* April 2015;96(4 suppl):S104–S113.
42a. Kirton A, Andersen J, Herrero M, Carsolio L, Nettel-Aguirre A, Keess J, Damji O, Carlson H, Mineyko A, Lane C, Hodge J, Hill MD. Brain stimulation and constraint for perinatal stroke hemiparesis: The PLASTIC CHAMPS trial. *Neurology.* 2016, in press.
43. Gillick BT, Krach LE, Feyma T, et al. Primed low-frequency repetitive transcranial magnetic stimulation and constraint-induced movement therapy in pediatric hemiparesis: a randomized controlled trial. *Dev Med Child Neurol.* January 2014;56(1):44–52.
44. Kuhnke N, Juenger H, Walther M, Berweck S, Mall V, Staudt M. Do patients with congenital hemiparesis and ipsilateral corticospinal projections respond differently to constraint-induced movement therapy? *Dev Med Child Neurol.* December 2008;50(12):898–903.
45. Juenger H, Kuhnke N, Braun C, et al. Two types of exercise-induced neuroplasticity in congenital hemiparesis: a transcranial magnetic stimulation, functional MRI, and magnetoencephalography study. *Dev Med Child Neurol.* October 2013;55(10):941–951.
46. Fregni F. Towards novel treatments for paediatric stroke: is transcranial magnetic stimulation beneficial? *Lancet Neurol.* June 2008;7(6):472–473.
47. Brasil-Neto JP, de Araujo DP, Teixeira WA, Araujo VP, Boechat-Barros R. Experimental therapy of epilepsy with transcranial magnetic stimulation: lack of additional benefit with prolonged treatment. *Arq Neuro-Psiquiatria.* March 2004;62(1):21–25.
48. Rotenberg A. Epilepsy. *Handbook of Clinical Neurology.* 2013;116:491–497.
49. Kirton A, Deveber G, Gunraj C, Chen R. Neurocardiogenic syncope complicating pediatric transcranial magnetic stimulation. *Pediatr Neurol.* September 2008;39(3):196–197.
50. Yang XR, Kirton A, Wilkes TC, et al. Glutamate alterations associated with transcranial magnetic stimulation in youth depression: a case series. *J ECT.* September 2014;30(3):242–247.
51. Bridgers SL. The safety of transcranial magnetic stimulation reconsidered: evidence regarding cognitive and other cerebral effects. *Electroencephalogr Clin Neurophysiology Suppl.* 1991;43:170–179.

52. Bloch Y, Harel EV, Aviram S, Govezensky J, Ratzoni G, Levkovitz Y. Positive effects of repetitive transcranial magnetic stimulation on attention in ADHD Subjects: a randomized controlled pilot study. *World J Biol Psychiatry (Official J World Fed Soc Biol Psychiatry)*. August 2010;11(5):755–758.

53. D'Agati D, Bloch Y, Levkovitz Y, Reti I. rTMS for adolescents: safety and efficacy considerations. *Psychiatry Res*. May 30, 2010;177(3):280–285.

54. Walter G, Tormos JM, Israel JA, Pascual-Leone A. Transcranial magnetic stimulation in young persons: a review of known cases. *J Child Adolesc Psychopharmacol*. Spring 2001;11(1):69–75.

55. Mayer G, Faivel N, Aviram S, Walter G, Bloch Y. Repetitive transcranial magnetic stimulation in depressed adolescents: experience, knowledge, and attitudes of recipients and their parents. *J ECT*. June 2012;28(2):104–107.

56. Wall CA, Croarkin PE, Sim LA, et al. Adjunctive use of repetitive transcranial magnetic stimulation in depressed adolescents: a prospective, open pilot study. *J Clin Psychiatry*. September 2011;72(9):1263–1269.

57. Beauchamp TL, Childress JF. *Principles of Biomedical Ethics*. Oxford University Press; 2001.

58. Maslen H, Earp BD, Cohen Kadosh R, Savulescu J. Brain stimulation for treatment and enhancement in children: an ethical analysis. *Front Hum Neurosci*. 2014;8:953.

59. Croarkin PE, Wall CA, Lee J. Applications of transcranial magnetic stimulation (TMS) in child and adolescent psychiatry. *Int Rev Psychiatry*. October 2011;23(5):445–453.

60. Kirton A, Chen R, Friefeld S, Gunraj C, Pontigon AM, Deveber G. Contralesional repetitive transcranial magnetic stimulation for chronic hemiparesis in subcortical paediatric stroke: a randomised trial. *Lancet Neurol*. June 2008;7(6):507–513.

60a. Juenger H, Kuhnke N, Braun C, Ummenhofer F, Wilke M, Walther M, Koerte I, Delvendahl I, Jung NH, Berweck S, Staudt M, Mall V. Two types of exercise-induced neuroplasticity in congenital hemiparesis: a transcranial magnetic stimulation, functional MRI, and magnetoencephalography study. *Dev Med Child Neurol*. 2013;55(10): 941–951.

61. Luber B, Steffener J, Tucker A, et al. Extended remediation of sleep deprived-induced working memory deficits using fMRI-guided transcranial magnetic stimulation. *Sleep*. June 2013;36(6):857–871.

62. Garvey MA, Kaczynski KJ, Becker DA, Bartko JJ. Subjective reactions of children to single-pulse transcranial magnetic stimulation. *J Child Neurol*. December 2001;16(12): 891–894.

NIBS IN PEDIATRIC NEUROLOGICAL CONDITIONS

TMS Applications in ADHD and Developmental Disorders

D.L. Gilbert

Cincinnati Children's Hospital Medical Center, Cincinnati, OH,
United States

OUTLINE

INTRODUCTION

The term "developmental disorders" encompasses a large, diverse number of diagnoses that result in cognitive, emotional, or physical impairments. In some diagnoses, such as fragile X syndrome, knowledge of the genetic etiology has propelled forward research into mechanisms of developmental disorder symptoms and into clinical trials of more rational treatments, through a variety of experimental approaches. In many other diagnoses, such as attention-deficit hyperactivity disorder (ADHD) or Tourette syndrome (TS), there is no unifying genetic basis. Rather, these developmental disorders result from the combined influence of multiple genetic susceptibilities combined with ill-defined environmental factors. Moreover, these diagnoses are categorical terms that encompass a non-specific set of symptoms, evaluated subjectively, without any biomarker. These diagnoses will be the focus in this chapter, primarily ADHD and TS. The research in this area is too broad to cover completely, but several key themes will be emphasized and these could readily be applied to many non-specific, spectrum diagnoses as well as to some genetic ones.

The medical scientific justification for developmental and behavioral diagnoses, while questioned by some due in part to a lack of biomarkers, is clear: these diagnoses occur and co-occur in sufficient numbers of children that epidemiological and clinical studies can be meaningfully conducted. Their clinical heterogeneity, as well as difficulties with diagnostic boundaries and classification, does pose substantial challenges for understanding the neurobiological substrates of these conditions. Despite this, given the high public health impact from both prevalence and disability, research into neural mechanisms is vital.

The purpose of this chapter is to review the use of transcranial magnetic stimulation (TMS) in developmental disorders primarily for the purpose of understanding disease mechanisms. Additional background can be found in Section I (Fundamentals), which reviews the principles and physiology of single- and paired-pulse TMS (Chapter 1: TMS Basics: Single and Paired Pulse Neurophysiology), assessments of normal development (Chapter 2: Assessing Normal Developmental Neurobiology With Brain Stimulation), and protocols for evaluation of neuroplasticity (Chapter 3: Neuroplasticity Protocols: Inducing and Measuring Change). A variety of other topics are covered elsewhere which could fall under this umbrella, including neurological complications of prematurity (Chapter 12: Brain Stimulation in Children Born Preterm—Promises and Pitfalls), cerebral palsy (Chapter 9: TMS Mapping of Motor Development After Perinatal Brain Injury and Chapter 11: Therapeutic Brain Stimulation Trials in Children With Cerebral Palsy), and autism (Chapter 13: Brain Stimulation to Understand and Modulate the Autism Spectrum). Treatment using TMS will not be addressed here as it is addressed elsewhere.

The structure of this chapter will be to first provide a review of the overarching principles of ethics, safety, statistics, study design, and validity with regard to the use of TMS in childhood developmental disorders. Following this, the sections will be organized by measurement, with each section briefly describing the technique and findings in ADHD, which is most studied, as a paradigm for other developmental disorders using that technique in children. Tourette syndrome findings will be discussed and compared.

KEY CONSIDERATIONS FOR TMS RESEARCH IN CHILDREN WITH DEVELOPMENTAL DISORDERS

Ethics, Minimal Risk, and Safety

There are several issues pertinent to TMS in children which deserve specific discussion. These are also addressed in more detail in Chapter 7, "Pediatric Issues in Neuromodulation: Safety, Tolerability and Ethical Considerations"; but are reviewed briefly here. The use of TMS (and the dedication of scarce research resources) for understanding mechanisms of developmental disorders in children can be justified on the basis of the life-long suffering and reduced quality of life related to these disorders coupled with limited available treatments.[1,2] However, as a rule these developmental conditions do not substantially increase the risk of death or imminent morbidity, and most do not require any risky diagnostic testing or treatment as part of the current medical standard of care. Thus, to be ethically acceptable, any biomarker research which has limited possibility for direct benefit for ADHD or other developmental conditions must involve minimal risk. This substantially limits the types of investigation that can be undertaken. For example, attempting to quantify differences in neurotransmitters and receptors using positron emission tomography (PET) scans and radioactive tracers generally must be limited to adults only or, when performed in children, cannot be performed and compared in healthy control children.

There are a number of important factors to consider in regards to what TMS research should be permitted in children with developmental disorders. As these involve intertwined considerations of ethical principles and safety, these are considered together. The application of general ethical principles to TMS research in healthy children and children with developmental impairments involves informed consent, risk/benefit assessment, and subject selection.[3] Informed consent entails ensuring that the parents, and, to the degree possible, the child participants understand the nature and purpose of TMS procedures. At the end of the informed consent process, participants should have a clear understanding of the potential risks and benefits. Laboratories should have clearly written consent and assent documents but also have personnel skilled in providing accurate descriptions of the machines and procedures verbally, and answering questions.[3,4] It can be helpful to compare the TMS procedure to more familiar procedures. Examples include brain MRI scans, which use magnets of comparable strength. Although child participants may never have had an MRI scan, they are often familiar with them and have seen pictures. It's reasonable to explain to parents that TMS employs a magnet like the MRI scanner, but instead of using a large magnet in the shape of a tunnel to take pictures of the brain, the strong hand-held magnet is used to activate the brain and measure its signals to the body or to other brain areas. For most types of TMS procedures, there are risks of side effects like headaches or tingling, but these are nearly always mild and resolve spontaneously.[5] The risk of more severe side effects, such as fainting[6–8] or a seizure[9,10] should also be noted, with the qualification that these are extremely rare. We have collected safety and adverse event data routinely in our laboratory for nearly 15 years, with only mild adverse events in that time.[3,5]

Children, and to an even greater extent children with developmental disabilities, may not as clearly understand descriptions of procedures and their risks. One advantage of TMS research in this regard is that for many procedures, data are only interpretable if the child is comfortable. Surface

EMG data from an anxious or upset child will typically be contaminated by muscle artifact. So the procedure has to be easily tolerated by children. It is helpful in writing protocols to share this information with the local institutional review board (IRB). Sample IRB-approved language from consent and assent documents in our laboratory are presented in Table 8.1.

TABLE 8.1 Sample, IRB-Approved Consent and Assent Language for TMS Studies

PARENT CONSENT FOR TOURETTE TMS STUDY

TMS uses a strong hand-held magnet, such as the strong magnet used for MRI scans. This magnet is placed over the scalp of the person being studied and is triggered to give an impulse. This impulse causes the brain cells underneath to activate and send a message through the brain, just as brain cells activate when a person decides to move part of his or her body. The result depends on where the magnet is placed. If the magnet is placed over the "thumb control area" of the brain, then the magnetic pulse will activate the brain and cause the thumb to twitch. By studying the way a brain responds to a magnetic impulse, we can learn more about how the brain is "wired for movement." This "movement wiring" is affected by Tourette syndrome. By using TMS we can learn how the brain's "movement wiring" develops in children with Tourette syndrome.

TMS does not allow us to diagnose any specific brain diseases. TMS is an investigational device. The research team has received permission from the Cincinnati Children's Hospital to use this device for research purposes only. TMS has been used for research in over 1000 children worldwide, and in many more adults. Dr. Gilbert and his team have already used this TMS in other studies involving over 200 children and adults, without any significant problems.

There are limited risks associated with the use of TMS. No hazards are anticipated based on previous studies involving both children and adults exposed to TMS. Mild and temporary side effects such as scalp discomfort, hand weakness, headache, neck pain, arm pain, hand pain, hand twitching, arm tingling hand pain, a feeling of decreased hand coordination, hearing changes, and tiredness occasionally occur. These side effects usually end by the next day. Rarely, TMS can cause seizure. The research staff will monitor you closely. There may be unknown or unforeseen risks associated with study participation.

CHILD ASSENT FOR TOURETTE TMS STUDY

What Will Happen in the Research?

After you say it is ok, the study staff will do the following things with you:

- Measure the signals of your brain with a magnet called a "TMS" magnet.
  - During this test you will sit in a comfortable chair and the research team will place a strong, hand-sized magnet near your head and tape some wires on your hand.
  - TMS will be used to measure how your brain signals work to move your body.

What Are the Bad Things That Can Happen From This Research?

During the TMS

- Tired
- Bored
- Uncomfortable due to sitting for a while, due to tape on your hands, or due to the feel or noise from the TMS magnet.

TMS Biomarker Research: Statistical Considerations

Biomarkers

TMS applications to date have primarily involved understanding pathophysiology and identifying biomarkers. The term *biomarker* refers to an objectively measurable, biologically based indicator. A number of overlapping, formal definitions have been provided of this term, and essentially TMS physiological measures fit with any of these. A representative definition, from the 1998 National Institutes of Health Biomarkers Definitions Working Group, was "a characteristic that is objectively measured and evaluated as an indicator of normal biological processes, pathogenic processes, or pharmacologic responses to a therapeutic intervention." TMS measures are candidates for clinically meaningful biomarkers, but to function in this regard it is important that they meet a number of criteria.

"Abnormal" TMS Results

Clinicians, researchers, and even parents may expect TMS to identify "abnormal responses" with high accuracy in a binary (normal/abnormal) sense. This does not occur. Rather, taking TMS over M1 as a most typical approach, when a TMS coil is discharged it generates a response from which a series of quantitative, continuous measures can be derived. For a quantitative measure to be "abnormal," a threshold level of abnormality has to be defined. This can be defined empirically in a disorder, based on normative, age-matched data, using receiver operating characteristic analysis.[11,12] However, no highly accurate threshold or "cut-off" for any TMS measure has been identified for any developmental disorder, nor is this likely given the complexity and high variability across individuals and development.

Overlap With Controls—Spectrum Diagnoses

Many developmental and behavioral diagnoses lie on a spectrum with typical development rather than being categorically distinct. The use of TMS to generate quantitative biomarkers may encounter several obstacles. One is that if the range of values of a particular measure within healthy, typically developing children whose nervous systems are maturing at different rates is wide, then there is likely to be significant overlap between affected and unaffected groups. Then for any proposed threshold there will be a large trade-off between sensitivity and specificity (highly sensitive threshold values will have low specificity and vice versa). Another obstacle is that, in children, age or maturation may affect these measures, and therefore "normal" would have to be based on a large body of normative pediatric data. Such data do not exist. Finally, to date even for some of the most common TMS measures like short-interval cortical inhibition

(SICI), there has been little standardization of technique. For example, the selection of pulse intensities and interstimulus intervals affects the magnitude of SICI.[13]

Heterogeneous, Non-specific, Overlapping and Comorbid Diagnoses

Many developmental and behavioral diagnoses co-occur in the same individuals. Even more challenging, characteristic symptoms are sufficiently non-specific that there is the problem of high overlap between diagnoses and high heterogeneity within diagnoses. Taking ADHD as an example, ADHD diagnosis or subthreshold symptoms are common in autism spectrum disorders and fragile X syndrome, absence and other childhood epilepsies, neurofibromatosis type I, Turner syndrome, and Tourette syndrome. ADHD symptoms also commonly occur as acquired sequelae of premature birth, hypoxic ischemic brain injury, traumatic brain injury, stroke, and infectious or inflammatory conditions in the central nervous system. In so-called idiopathic ADHD cases, no single gene exerts a dominant effect in the population. The extreme heterogeneity has several implications for the use of TMS for biomarker research in developmental disorders involving movement, motivation, mood, or cognition. These considerations are described in Table 8.2 using ADHD symptoms as an example, but other individual or clusters of symptoms could be inserted.

Study Designs for TMS Biomarker Research in Developmental Disorders

With the above considerations in mind, designing TMS studies for biomarkers of developmental disorders may use complementary case–control and dimensional approaches. Of the two, the more common approach is the case–control study. With appropriate selection of participants and age-matched controls, this can yield estimates of group differences that are scientifically meaningful. The second approach is a correlational study to incorporate scalar data regarding either severity of core symptoms or severity of associated impairments and determine whether these correlate, in the expected direction, with the quantitative TMS measure.

Case–Control Studies

In this type of design, the disorder is a class variable, indicated as *present* or *absent*. The study hypothesis is that the mean of a particular neurophysiological measure is different when the disorder is present versus absent. This can be tested with a parametric Student's *t*-test or a more conservative non-parametric test, if necessary. If two or more disease groups are compared with normals, an ANOVA would be used. Since there is not

TABLE 8.2   Considerations for Use of TMS Biomarkers in Developmental Diagnoses

Question Regarding Neural Substrate of Behavioral or Cognitive Symptoms	Comment/Implication for Use of TMS to Develop Biomarkers
Is there any "common pathway" within specific structures or neurotransmitter systems which span these many diagnoses and produce core ADHD symptoms?	If yes, then a particular measure might reflect the substrate for ADHD in multiple types of patients with ADHD symptoms. If no, then case–control studies may not identify a biomarker, but principal components or cluster analyses might identify biological subgroups.
Are there multiple distinct (or overlapping) pathways within specific structures or neurotransmitter systems which span these many diagnoses and produce core ADHD symptoms?	If yes, then a particular biomarker might reflect pathophysiology or predict outcomes in some but not all forms of ADHD or in some but not all conditions in which ADHD symptoms are present.
Are any of the subcortical, cerebellar, or other anatomic structures linked to ADHD reflected reliably in physiology of motor cortex?	Motor cortex is the most easily evaluated structure with current TMS protocols. Striato-pallido-thalamo-cortical, transcallosal/transhemispheric, or cerebellar-thalamo-cortical pathways might produce quantifiable motor function differences, which could be linked to both behavioral symptoms and TMS measures in motor cortex.
Can physiological TMS measures captured during tasks reveal mechanisms for performance differences in those tasks?	Children with ADHD may have inefficient response inhibition or impulsive reward responses. If the substrate of these differences is correlated with or reflected in motor cortex function, TMS measures performed during those tasks (functional TMS or "fTMS") might have value in explaining core differences in ADHD versus controls.
Would information from one or more TMS biomarkers be clinically useful?	Any biomarker that appears statistically linked to a diagnosis or diagnostic subgroup should ideally be evaluated for clinical value. This might require a reasonably large population with careful clinical phenotyping and characterization of treatment responses or long term outcomes.

Selected questions for guiding TMS biomarker work in developmental disorders. For simplicity and clarity, ADHD is used as an example of a heterogeneous developmental disorder. Tourette Syndrome, autism spectrum disorder, or other conditions may be substituted.

a large body of normative data obtained using standardized techniques in multiple laboratories, case–control studies should ideally match age, gender, and possibly other demographic factors as closely as possible. It should be borne in mind, however, that most studies involve small convenience samples, so it is important to replicate findings and follow up positive pilot studies with much larger ones.

Co-occurring diagnoses and the current or prior history of treatments also merit some discussion. Many if not most children with a developmental disorder, and particularly those whose families are likely to have sought medical attention, meet criteria for more than one categorical behavioral diagnosis. Interpretation of results may be much easier for studies of pure developmental disorders versus controls. However, the results of such a study may have lower generalizability to clinic populations. Taking Tourette syndrome (TS) as an example, most children presenting clinically also have ADHD and/or obsessive compulsive disorder (OCD). In order to elucidate important mechanisms for tic generation, one should study TS children with tics only. However, this approach might have limited relevance to understanding more clinically relevant pathophysiological mechanisms in the majority of children with TS with impairing, co-occurring ADHD or OCD.

Similarly, medications used to treat symptoms of developmental disorders affect a number of TMS measures. Excluding children with current or prior medication treatment may reduce statistical confounding due to medication effects. On the other hand, this may result in a mildly affected sample, excluding the most severely affected children and failing to represent those suffering most profoundly. In summary, decisions to exclude children based on co-morbid diagnoses or pharmacological treatment from a study may enhance validity, but a filtered group of cases may poorly represent the clinical condition being studied. Similar arguments could be made for studies of children with autism spectrum disorder, for example.

Finally, there is the issue of blinding. Ideally, the TMS operators would perform measures without knowing the diagnostic status of the subjects. Unfortunately, in children with developmental and behavioral disorders, it tends to be quite obvious which children are affected, introducing the possibility of bias in TMS operator performance or data analysis. Potential means to mitigate this possibility include non-informative file names/ study identifiers and offline analysis without knowledge of the diagnosis. In our laboratory, we utilize an independent clinical rater/interviewer. This person obtains information on symptom severity from parents or teachers completely independently, so the laboratory measuring motor physiology does not have access to clinical severity ratings, and vice versa. This at least reduces opportunities for bias in the correlational studies (section: Correlational Studies). Another important step, whenever possible, is to use mathematical analytic techniques that do not depend on visual discrimination or interpretation.[14]

Correlational Studies

A correlational study involves obtaining measures within a sample of affected patients, then determining the relationship between the measure of interest and the severity of the symptom of interest. This may add substantially to the findings of a case–control study. In cases where a technique is new or where ethical considerations do not favor including a healthy control group of children this could also be very helpful. In this type of study, patients with the disorder under investigation may be enrolled without any healthy control subjects. The study hypothesis is that the severity of a disorder's clinical symptoms correlates with the range of a quantitative physiological measure. Depending on the nature and distribution of the scalar and physiological data, this hypothesis can be tested with a parametric or non-parametric correlation statistic. If age, sex, current medications, or other characteristics may be important, linear regression can be performed including these factors in order to identify a more precise estimate of the relationship between symptom severity and physiology.

For this approach to work well, the disorder of interest must have a clearly ratable phenotype, but the distinction between mild cases and controls may not be as important, since a control group is not necessary in this design. On the other hand, if subthreshold symptoms are common in the general population, then in a case–control study a symptom/physiology correlational analysis can be performed including all participants—cases and controls.

Bias in Adult Studies

There are a number of other potential sources of bias that should be taken into consideration. This is particularly true for cross-sectional, case–control studies in adults, which constitute the majority of published studies. Such studies may involve small convenience samples which are not necessarily representative of developmental behavioral disorders at the time symptoms are emerging. Moreover, it is worth noting that adults presenting to clinic may already represent an important but unusual subset of patients. Persistence of clinically significant symptoms of developmental disorders into adulthood may represent in some ways a more severe phenotype. Moreover, given the diverse and complex polypharmacy in use in many clinical settings, current medications and possibly many prior years of medications in childhood create potential for confounding.

Finally, data from a small number of careful adult and longitudinal studies suggest that many quantitative differences between adults, eg, those with TS versus controls, may be compensatory.[15–17] Yet, in a univariate case–control study, these might be falsely assumed to be primary.

TMS IN DEVELOPMENTAL DISORDERS: PRACTICAL CONSIDERATIONS AND REPRESENTATIVE RESULTS

In this section, we will describe results from published studies using TMS to understand the pathophysiology of developmental disorders in childhood. The sections are organized by measurement, rather than disorder. Each section describes briefly the technique, the pitfalls of the technique with regard to the above-described challenges, the findings in childhood developmental disorders using that technique, and the possible implications for those findings.

Getting Children to "Sit Still"

One challenge in utilizing TMS in children is small, active bodies. The "active" part is particularly challenging in developmental disorders. This will be addressed in the technique sections below. Children move around, and for small children there is ample room to move around in most commercial chairs. Labs need to decide how to address this realistically for successful pediatric research, and there is not much published about this. In our laboratory, we find it helpful to do "resting" TMS with children watching quiet videos on a computer monitor in front of them. Educational nature or space videos with no major emotional content (and no pictures of large moving hands to "mirror") work well, in that children generally keep their straight heads toward the screen, are not prone to wiggle or laugh, and do not fall asleep. We have not found it realistic for many children to stare for long periods at a simple neutral target on the screen or wall.

TMS Techniques for Studying Motor Cortex Physiology

Depolarization Thresholds: Resting and Active Motor Thresholds (RMTs and AMTs)

Technique

Resting (RMTs) and active motor thresholds (AMTs) are the amount of energy required to evoke a response in M1 at rest or with activation of the target muscle, expressed as a percentage of the output of the TMS device being used. Methods for evaluation are roughly consistent.[18] Thresholds are likely primarily dependent on ion channel conduction.[19]

A number of experimental decisions may affect the findings. While threshold determination techniques are broadly standardized, there are nonetheless a few differences reported across laboratories' published methods. For most TMS studies, left/dominant motor cortex (M1) is stimulated with a figure-of-eight, double 70-mm flat coil placed lateral to the vertex, tangential to the skull, with the handle pointing posteriorly at 45 degrees from the midsagittal line. In a limited number of studies, a round 90-mm coil placed over the vertex has been used. The muscles evaluated with surface EMG are typically either first dorsal interosseous (FDI), abductor pollicis brevis (APB), or adductor digiti minimi (ADM). Due to the large area of motor cortex controlling the hand, it is usually very simple and quick to identify the "hot spot" for best evoked responses.

To identify the best location and the motor threshold, many laboratories begin by stimulating at a low (subthreshold) intensity, increasing by intervals of 10% until an MEP is identified. After adjusting the location and position of the coil to produce a well-formed MEP, the intensity is decreased sequentially until an intensity is reached where 50% of trials yield no MEP, but the other 50% yield usually small-amplitude MEPs. Most labs report five of 10 trials. Bearing in mind that longer TMS lab times create challenges for restless and inattentive children, we generally utilize three out of six as adequate. In place of this manual/visual observational technique, some commercial software will perform this iterative process automatically. "Rest" can be judged visually by the absence of any surface EMG motion artifact prior to the TMS pulse. An audio amplifier may be used to identify muscle artifact noise as well.

The AMT is measured in muscle activated at some percentage of maximum effort. We prefer just a slight activation so that the MEP onset, and its disappearance as the intensity decreases, can be more clearly visualized. AMT is also approached from above, decreasing the stimulus intensity until the point at 50% of trials yield no MEP visible above slight background EMG activity and the remainder evoke small-amplitude MEPs, visible over background. For AMT, the participant rests between trials to avoid fatigue. The methods of eliciting and maintaining muscle activation vary widely across studies and are not standardized. The intertrial interval for administering TMS single pulses is usually >5s to minimize any modulatory effects.

As young children or those prescribed medications like anticonvulsants may have quite high thresholds, it can be helpful sometimes to identify the AMT first. The generation of the MEP during activation can provide confidence that the coil is in the proper location. If the AMT is greater than 70% of the maximum stimulator output, the RMT may be near to or above the maximum stimulator output.

Pitfalls

RMT and AMT are sensitive to coil position and orientation. These can be difficult to maintain accurately using the figure-of-eight coils in young children without using neuronavigation. The round coil which rests on the vertex is much easier in a hyperkinetic child. There can be interoperator differences as well. Nonetheless, for M1 studies it is common and probably reasonable to mark the scalp, eg, with wax pencil, and position the coil on that basis. Laboratories should conduct test–retest[20,21] and cross-laboratory consistency[22] of this and other measures. A major pitfall involves motor system activity, related to emotion, cognition, or actual movement, affecting cortical excitability. Even activities, such as teeth clenching[23] or thinking about food[24] or rewards[25,26] can affect motor cortex excitability. We sometimes find RMTs appear low initially, because children are nervous and contracting their proximal muscles slightly, even if this is not detected as muscle artifact on the TMS tracing. If this is suspected, it is important to continually encourage children to relax, or to distract them, and to recheck the RMT after obtaining the AMT.

AMT can be challenging because some young children will relax their muscles as soon as they hear and feel the TMS pulse. It's important to train children to "keep pushing until I say stop" and to make sure they are activating the target muscle appropriately until at least 200 ms after the TMS pulse.

Finally, all children but particularly those with developmental disorders, tend to be hyperkinetic. For a hyperactive child who is being told to "sit still," one wonders if there is also an active state involved in suppressing a tendency to move. So, the results in hyperkinetic children may incorporate both trait and an active *state* of complying with the request to sit still. In contrast, for a child for whom sitting still comes more readily, the TMS measures may reflect "trait" only, at passive resting state.

Results and Interpretations

For motor thresholds, differences between children with developmental and behavioral disorders and healthy, typically developing controls have not, to our knowledge, been reported. RMT and AMT decrease with age.[27] No studies have identified statistical differences in RMT or AMT in individuals with TS or comorbid conditions such as ADHD.[28–30]

RMT data from our laboratory and a collaborating laboratory at the Kennedy Krieger Institute in Baltimore, MD, USA, using the Magstim TMS device demonstrate the major challenge in pediatric research which is that younger children have higher motor thresholds and may in fact have RMTs higher than a given TMS device's maximal output (see Fig. 8.1). Similar to other laboratories, in our case–control studies, thresholds were not significantly higher in ADHD or TS children compared to controls ($P = \text{NS}$).

Differences in absolute values between pediatric publications may reflect the type of commercial machine being used in individual laboratories, the coil and its configuration (eg, round versus figure-of-eight), and the target muscle. The main use for RMT and AMT in developmental disorders is to use these values to determine the intensity for paired-pulse or repetitive TMS studies. For example, in paired-pulse TMS, the first subthreshold conditioning pulse may be indexed to an individual's RMT or AMT.[13]

Recruitment Within Motor Cortex: Input–Output Curves (I–O Curve)

Motor cortex TMS input–output curves demonstrate changes in MEP amplitudes as a function of increasing stimulation intensity. A sigmoidal curve with a steep middle phase is expected, indicating that neurons are

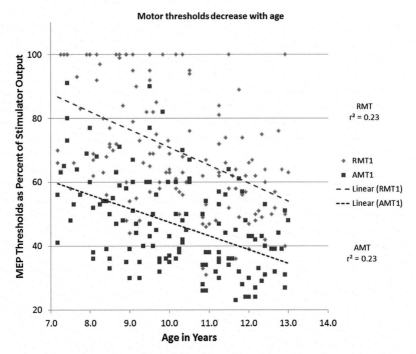

FIGURE 8.1   Resting motor thresholds (RMTs) and active motor thresholds (AMTs) in 195 children ages 7–12 years. Age accounts for about 25% of the variance in thresholds (r^2 for RMT = 0.23; r^2 for AMT = 0.23; $P < 0.001$ for both). *Data from Cincinnati Children's TMS lab.*

readily excited and recruited into a response as the intensity of the TMS pulse increases. In contrast, a flatter curve indicates that cortical networks are less readily recruited to participate in an excitatory response, perhaps because an ongoing inhibitory process is being engaged.

Input–Output Technique

This is a single pulse TMS technique. First, the resting motor threshold is established. Then, for each participant, a series of pulse intensities are used at set percentages above threshold, for example at RMT and then 10%, 20%, 30%, 40%, and 50% above RMT. This gives a measure of the corticospinal pathway excitability for the target muscle.[31] Given the high intertrial variability of MEP amplitudes, sufficient numbers of trials at each intensity must be obtained for the average to be meaningful. In our lab we typically prefer a minimum of 10 trials per intensity. An intertrial interval of 5–6 s may be used.

Pitfalls

This is a straightforward technique and has similar limitations to other MEP amplitude techniques in terms of intertrial variability. As MEP amplitudes are increased in the "premovement" state,[32,33] it may be that persons with hyperkinetic disorders are more often in a "premovement" state and thus this has potential for study in developmental disorders. To date this has not been widely used and published in children.

Motor Cortex Inhibition: The Cortical Silent Period (CSP)

Technique

Five to ten sequential, suprathreshold TMS pulses are administered at 5–10 s intervals to dominant (usually) M1 during voluntary contraction of the contralateral target muscle. Surface EMG is recorded, showing the motor contraction signal, the TMS pulse (difficult to see behind motor activity), continued motor signal, an evoked MEP, a period of silence, then a return of motor activity to baseline (see Fig. 8.2A). We instruct children to "push hard" or "squeeze hard" depending on the target muscle and what device we use. Measuring precisely the amount of effort is optional and we do not verify this in our laboratory as the effect on CSP is minimal. The primary determinant of within-individual variation in CSP duration is the intensity of the pulse,[34] not the effort of the child. Most laboratories index the intensity to the participant's AMT and express the intensity on that basis. For example, CSP may be obtained at 150% of AMT for each participant. Some authors have advocated reporting CSP at two intensities, eg, 130% and 150% of AMT.

For CSP, the tracings are superimposed, rectified, and averaged, reducing the amplitude of the background noise but generating a consistent silent period (see Fig. 8.2B). Most trials have a fairly consistent onset and offset of the silent period.

This period of EMG suppression is known as the cortical silent period (CSP) (see Fig. 8.2A and B).[35,36] The parameter of interest is a time interval, the duration of the EMG suppression, which most labs report identifying visually, although statistical process control can also be used.[14] The specific mechanism that generates the CSP is proposed to be activation of inhibitory neurons projecting onto M1 pyramidal cells. Some of these inhibitory neurons express, or are influenced by other cells which express GABA-B receptors. GABA-B agonism lengthens the CSP.[37,38]

Pitfalls

The CSP is obtained during voluntary activation of a muscle. It is a simple and brief procedure. Because AMT is determined on the background of a noisy surface EMG signal, and CSP intensity is indexed on that basis, the determination of the AMT is less precise than RMT. As a result, the CSP duration may be affected by imprecision in the AMT measure. In addition, as above, some children tend to release at the

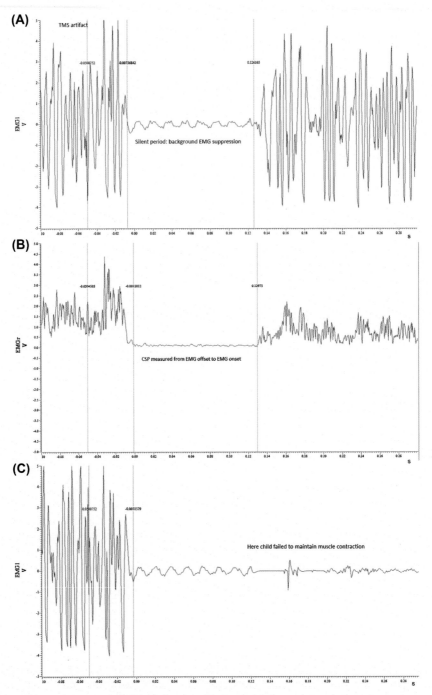

FIGURE 8.2 Surface EMG tracings. (A) Single trial surface EMG tracing showing large-amplitude motor signal obscuring the TMS pulse artifact at −0.05 s. TMS artifact, onset and offset of silent periods are marked using cursors. Regular 60-Hz artifact contaminates the silent period. Obtained using Magstim single pulses at 68% of stimulator output (150% of AMT for this child). (B) EMG tracing of five trials rectified and averaged, clarifying onset and offset. (C) EMG tracing of trial when child "let go" rather than maintaining continuous activity throughout the trial as instructed. This individual trial would be eliminated from the analysis as invalid. *CSP*, Cortical silent period.

instant of the TMS pulse (see Fig. 8.2C). This eliminates the EMG signal re-onset so it is important to encourage children to continuing squeezing until instructed to stop.

Results and Interpretations

CSP as a form of inhibition has not been considered to be a biomarker of idiopathic developmental disorders in children. Case–control studies have not consistently demonstrated any difference in CSP duration between children with ADHD and controls. Two small case–control studies have reported shorter CSP durations in persons with TS. In children with TS, the CSP was found to be shorter irrespective of tic location,[29] whereas in adults with TS, the CSP was shorter predominantly in individuals with "target muscle" (ie, hand) tics.[39] Shorter CSP has not been shown to correlate with greater tic symptom severity.[20] Because baclofen, a GABA-B agonist, may lengthen CSP[37] and is used for tic treatment,[40] this finding may have some clinical relevance. Possibly short CSP could be a biomarker of a subgroup of TS more likely to respond to medications that lengthen the silent period, like baclofen or tiagabine.[38]

Interhemispheric Signaling: Interhemispheric Inhibition (IHI) and Ipsilateral Silent Period (ISP)

Interhemispheric signaling is important for many unimanual and bimanual motor activities. Such signaling may suppress unwanted mirror movements and may also allow resources of both hemispheres to be incorporated into motor planning and execution.[41,42] A vital anatomic substrate for this interaction is the corpus callosum (CC). CC consists of both excitatory and inhibitory neurons, responsible for sequestration, integration, and lateralization of brain activity.[43] In developmental disorders, the extent to which interhemispheric signaling is important in suppression or generation of unwanted behaviors or movements or in developing compensatory or adaptive responses to regulate behavior or movements is under active investigation.[44–46] Imaging studies have probed the structure and connectivity of the CC and suggested this may be relevant to motor control or plasticity in a variety of ways.[47–49] TMS studies have elucidated relationships between callosal transmission and mirror movements or structural callosal abnormalities.[41,50]

Techniques

Several TMS techniques can be used to probe transhemispheric physiology. Interhemispheric inhibition (IHI) is a paired pulse measure where the conditioning pulse is administered through a figure-of-eight coil over one hemisphere and then the test pulse is administered 5–7 ms later with a separate coil over the opposite hemisphere. The placement of the conditioning pulse coil has been described over motor as well as premotor cortex. The resulting inhibition of MEP amplitudes in the paired- vs. single-pulse trials reflects transhemispheric inhibitory signaling.[51] For this procedure, both hands are at rest. Such paired-pulse IHI studies have been limited in children but preliminary data from 22 healthy school-aged children demonstrate inhibition of 40–60% at both short (8 and 10ms) and long (40, 50ms) interstimulus intervals (Kirton, personal communication).

A more widely used technique is the ipsilateral silent period (ISP). Like the CSP, the ISP is defined as a period of suppression of ongoing EMG activity of target muscle after stimulation of motor cortex (see Fig. 8.3). However, in contrast to CSP, the TMS pulse over motor cortex and the EMG measure are on the same side, eg, right brain/right hand. The target muscles for both hands are contracted.[52]

ISP reflects the integrity of transcallosal fibers exerting an inhibitory effect on non-stimulated cortex from the stimulated motor cortex.[53] Experiments usually involve 5–10 trials. Both the duration of the ISP and the onset latency of the ISP have been studied in healthy children as well as in children with ADHD.[28,46,54,55]

Pitfalls

For double-coil IHI measures, the smaller size of the child's head relative to the coil size can make obtaining this measure challenging. For ISP, the main issue is the intensity of stimulation versus the higher motor thresholds in children. ISP is typically elicited at the maximal stimulator output (100%). In adults, this number may be more than double their AMT, and a robust response of longer duration may result. For children with high thresholds, 100% of stimulator output may be much closer to the AMT, possibly affecting duration therefore, as intensity does for CSP. As long as there is some response, this may be less of an issue for ISP onset latency than it is for duration. However, the other issue is tolerance. Children may be less tolerant of TMS at such high intensities, limiting the feasibility and utility of ISP testing in young children.[55] ISPs in children tend to have short duration and can be difficult to identify (see Fig. 8.3).

Results and Interpretations

To date, paired-pulse IHI techniques have not been applied systematically in children, although there are some interesting findings in adults with TS, combining TMS IHI with callosal studies of fractional anisotropy.[44] There is a more substantial literature in typically developing children[46,55] and children with ADHD.[28,45,54,55] The general finding is that as children mature, their latency to ISP onset decreases. This likely reflects in part white matter maturation. ISP durations also increase. These findings lag in children with ADHD. These effects appear not to be solely related to myelination, however, and may include trans-synaptic activity. Administration

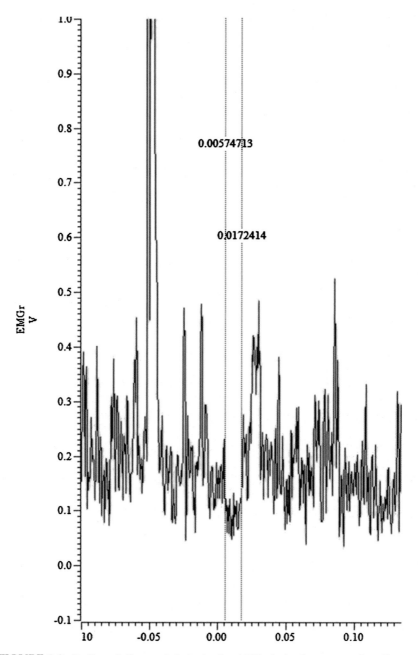

FIGURE 8.3 Ipsilateral silent period obtained at 100% of stimulator output in a 10-year-old child.

of methylphenidate in a cohort of 23 children, while improving clinical symptoms, also shortened ISP latency and lengthened its duration.[54] A study by our research group found that longer ISP latencies may correlate with clinically rated, more severe ADHD, with worse motor development as assessed with the Physical and Neurological Examination for soft signs, and with short-interval cortical inhibition.[55]

IHI Future Directions

This line of inquiry lends itself well to studies combining motor function assessments, imaging modalities (fractional anisotropy, resting connectivity), and TMS.[56,57] It would be most informative if obtained across larger age groups of children, and particularly if longitudinal rather than cross-sectional measures could be obtained. Changes in coil design or intensity may help with obtaining more consistent ISPs in younger children.

Paired Pulse TMS Studies

General Technique in Paired-Pulse Studies

A variety of methods have been developed to test trans-synaptic, neuronal properties using paired, condition/test pulse paradigms. Most of these experiments are performed with a TMS coil stimulating resting, dominant motor cortex, with output in an intrinsic hand muscle. The physiological output of interest is the *amplitude of the MEP* measured automatically, peak to peak or as an area. Because motion or motion preparation can increase cortical excitability, trials with muscle artifact are usually removed, by visual inspection or using an automated threshold criterion.

Generally, paired-pulse studies all involve comparing multiple trials with different interstimulus intervals, to trials having a single, suprathreshold pulse, also called the "test pulse." The test pulse evokes an MEP of size to allow for evaluation of inhibition without a floor effect and facilitation without a ceiling effect. Because the TMS-evoked MEPs reflect instantaneous excitability of several linked cortical and subcortical systems, they vary from trial to trial. A common approach is to identify a stimulus intensity which evokes MEPs in hand muscle of approximately 1 mV peak-to-peak amplitude as the baseline cortical excitability readout.

The basic experimental design is to randomize a number of trials with a conditioning input followed by the test pulse, with intermixed test-pulse only trials. In paired-pulse TMS, the conditioning input is a TMS pulse, however the test pulse can be paired with other inputs, for example electrical sensory stimulation.

The paired-pulse trials evoke responses which are either larger (excitatory) or smaller (inhibitory). There is no standard intensity for a TMS test pulse. Generally, the test pulse intensity is set to produce ~1 mV amplitudes, so that the neurophysiological system will have capacity to detect phenomena which are either inhibitory (tending to reduce the amplitude) or excitatory (tending to increase the amplitude). Amplitudes are usually represented as the average peak-to-peak amplitude of multiple trials, although occasionally researchers report the area under the MEP curve. In our laboratory, mean peak-to-peak amplitudes and mean areas are highly correlated ($r > 0.9$; unpublished data), so we would not generally see any advantage to using areas. Paired-pulse-evoked amplitudes are compared to single-pulse amplitudes and are represented as ratio or percent change. Depending on the protocol, the intensity of the first, conditioning pulse is set and is paired with the test pulse at one or more interstimulus intervals. The use of multiple intervals can yield a nice curve spanning inhibitory effects at multiple interstimulus intervals less than 7 ms, and excitatory effects at multiple longer intervals. However, more intervals means either more pulses, which can tax the patience of the child, or reducing the number of pulses given at each interval, which reduces precision.

It is critical in comparing across studies[58] to review the methods used before concluding that TS or other conditions are associated with a specific finding. Even for the most commonly used paired-pulse paradigms, such as short-interval cortical inhibition (SICI),[59] a variety of conditioning pulse intensities and interstimulus intervals are published. Some careful studies have produced useful data for choosing experimental parameters in adults.[13] Different paired-pulse paradigms have been found to sensitively reflect effects of dopaminergic, GABA-ergic, and cholinergic, and other medications.[60]

General Pitfalls

Measurements with amplitude as the outcome of interest have an important disadvantage. Centrally evoked motor amplitudes have very large intraindividual, intertrial variability. Therefore, whatever paradigm is being used, a sufficient number of trials is needed to understand any disease or drug signal over the background of the intraindividual, intertrial noise. Careful attention to patient phenotype, muscle relaxation, and laboratory conditions is also required.

Motor Cortex Inhibition Biomarker? Short-Interval Cortical Inhibition (SICI)

TMS has been used to evaluate motor cortex physiological biomarkers of many childhood and adult developmental and psychiatric disorders.[61] SICI is probably the most widely evaluated.[59] It is believed to reflect GABAergic tone in cortex, but is also dopamine-sensitive.[62–65]

While single-pulse TMS quantifies a "net" cortical spinal excitability, the SICI paired-pulse TMS protocols are widely used to evaluate inhibitory neuronal populations in order to evaluate disorders involving synaptic transmission as well as to understand and quantify pharmacological effects.[59,60,66] The SICI protocol involves pairing subthreshold (conditioning) with suprathreshold (test) pulses delivered at an interstimulus interval of 1–5 ms. The conditioning pulse is believed to activate the $GABA_A$-mediated cortical inhibitory interneurons which in turn reduce the response to the test pulse.[62] Spacing the same pulses at 10–20 ms intervals tends to result in excitation or facilitation, termed intracortical facilitation or ICF. Commonly, protocols will randomly intermix single- and various paired-pulse TMS trials (see Fig. 8.4).

SICI in ADHD

Research in ADHD is particularly informative and serves as a paradigm for both potential benefits and pitfalls of using paired-pulse measures as

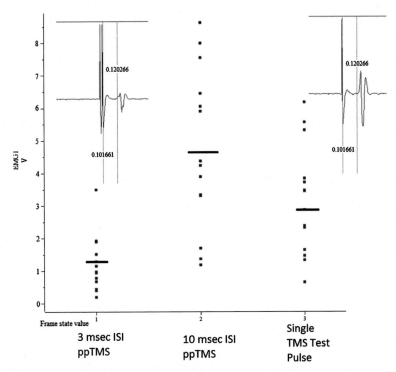

FIGURE 8.4   Short-interval cortical inhibition and intracortical facilitation. Data obtained from an 11-year-old healthy girl. Three types of trials were randomly intermixed with conditioning pulses set at 60% of RMT and test pulses set at 120% of RMT: 3 ms interstimulus interval (ISI) paired pulse (pp) TMS, 10 ms pp TMS, and single (test) pulse TMS. Y-axis is MEP amplitude in mV × 5. Black bars are approximate average MEP amplitudes for each trial type. Examples of surface EMG tracings examples (1 trial) are shown for 3 ms and single pulse, with cursors indicating TMS pulse artifact and onset of MEP approximately 20 ms later.

biomarkers in children. SICI has been described with regard to both the diagnosis and treatment-induced changes in ADHD. Early, small studies[29,67,68] showed reduced SICI in ADHD, and these findings have generally been replicated in larger subsequent studies.[22,55,69]

Rationales for studying ADHD in motor cortex, rather than premotor cortex or elsewhere, include commonly observed impairments in fine motor control in ADHD,[70,71] neuroimaging findings in the frontal cortex and motor control systems,[72,73] and, more recently, magnetic resonance spectroscopy (MRS) research showing reduced motor cortex GABA levels in ADHD.[74] The basic concept is that the MEP amplitude reflects the instantaneous balance of cortical inhibition and excitation and moreover that motor cortex, based on embryological development, shares patterns of neural transmission with other frontal and prefrontal locations. Therefore a motor cortex MEP "readout" can reflect disease processes elsewhere in the brain.

While SICI appears to be altered in an ever-increasing number of neurological, developmental, and psychiatric conditions,[59,61] in ADHD the finding may have the most significance as reduced SICI seems to correlate with hyperactive symptom severity assessed independently with commonly used clinical rating scales,[20,22,69] and can be probed with regard to movement preparation in the context of response inhibition.[69]

SICI and Medications

Pharmacological studies of SICI are believed to be highly informative, but an important caveat is that most studies demonstrating relationships between SICI or other TMS measures and medications are single-dose studies in healthy adults. ADHD is one area where SICI has been evaluated in the context of pharmacological treatment.

Given the studies discussed above demonstrating that children with ADHD have reduced SICI compared to their typically developing peers, it would be reasonable to expect that SICI would increase with ADHD treatment. This expectation was realized in two small studies (see Table 8.3), involving fewer than 20 children each, of the stimulant medication methylphenidate (MPH). Single-dose and short-term (2-week) treatment of children with ADHD with MPH increased or "normalized" SICI.[67,68]

Studies of ADHD in children with Tourette syndrome (TS) have also shown reduced SICI when ADHD is comorbid with TS.[20,29] However, two surprising differences emerged in a study when children with TS plus ADHD were recruited for an open-label study of ADHD treatment with atomoxetine (ATX), the selective norepinephrine reuptake inhibitor. First, the pretreatment SICI in these children was not reduced. Second, in contrast to the SICI *increase* reported for MPH, 4 weeks of ATX treatment *reduced* SICI (see Table 8.3). Corroborating this paradoxical finding, the amount of SICI reduction correlated with the degree of improvement in ADHD.[75]

TABLE 8.3   Results of Case–Control Studies (Not Involving Treatment) and Treatment Studies With the Stimulant Methylphenidate (MPH) and With the Norepinephrine Reuptake Inhibitor Atomoxetine (ATX)

| | | TS + ADHD | | | ADHD Only | | |
| | | SICI Ratio | | | SICI Ratio | | |
		n	Mean	SD	n	Mean	SD
Case–control studies	ADHD (TS,[30] ADHD[22])	6	**0.74**	0.27	38	**0.68**	0.24
	No ADHD	13	0.32	0.21	52	0.49	0.24
	ADHD[29]	16	**0.88**	0.17	16	**0.87**	0.19
	No ADHD	16	0.71	0.26	16	0.79	0.21
MPH treatment studies of ADHD	Pretreatment[67]	–	–	–	18	**0.97**	0.26
	Post single dose 10 mg MPH	–	–	–	18	0.87	0.23
	Pretreatment[68]	–	–	–	18	**0.81**	0.30
	Post 10–14 days 0.8 mg/kg/day MPH	–	–	–	18	0.64	0.44
ATX treatment studies	Pretreatment (TS,[75] ADHD[76])	12	0.33	0.18	61	0.49	0.26
	Post 28 days 1.0 mg/g/day ATX:						
	Non-Responder	7	0.34	0.11	28	0.45	0.38
	Responder	5	0.47*	0.30	33	0.63*	0.36

All studies reported statistically significant results ($P < 0.05$) (see citations in table). Although TMS techniques differed between labs, in case-control studies, SICI ratios are closer to 1.0 in the children with ADHD, indicating less intracortical inhibition (indicated in bold). This was also true in both MPH studies enrolling children with ADHD (indicated in bold). MPH treatment appeared to reduce or "normalize" SICI, as expected. However, in both of the ATX studies, pretreatment, baseline SICI ratios were low—similar to published healthy controls, and after treatment responders' SICI ratios "paradoxically" increased, which *correlated with clinical symptom improvement.

These discrepant findings—reduced SICI in ADHD in a case–control study but not pretreatment in an ATX study—were replicated by two larger studies we have completed in idiopathic ADHD, without TS, in children.[22,76] In our case–control study, children with ADHD had reduced SICI, and this reduction correlated with both ADHD severity and impaired motor function.[22] However, in a distinct cohort of children with ADHD recruited for a study of ATX treatment, the average pretreatment SICI was not reduced. Moreover, as in the TS plus ADHD study, clinical responders to 4 weeks of ATX showed a decrease in SICI, and the magnitude of the SICI decrease correlated with the magnitude of ADHD symptom improvement.[76]

The reasons for this are not clear, but we speculate that the study designs resulted in recruiting children with a different subtype of ADHD. Although clinical rating scale scores and demographics were not different, a majority (62%) of participants in the case–control study were currently prescribed stimulants (which were temporarily discontinued for the TMS).[22] The second, atomoxetine-treatment (AT) study enrolled children whose parents, as part of routine clinical care, were seeking to initiate ADHD treatment with atomoxetine for their children. Current effective stimulant medication use was an exclusion criterion. Most children in the ATX study (67%) had previously been prescribed stimulants but were no longer taking them.[76] Although rigorous data about treatment outcomes prior to the study were not available, it seems reasonable that in the ATX studies, children whose ADHD symptoms were relatively stimulant-resistant, or who at least had stimulant side effects, were over-represented.

Although a large fraction of children may respond reasonably well to either stimulants or ATX, some children respond better to one, and others do not respond to either.[77] This suggests distinct underlying physiological characteristics may underlie beneficial responses to ATX.[78]

Future Directions for SICI and Other Paired-Pulse TMS Measures

Findings to date suggest that TMS SICI may have potential for identifying subtypes within clinically heterogeneous behavioral diagnoses like ADHD. These subtypes may have treatment implications for use of psychostimulants versus norepinephrine reuptake inhibitors. While evaluating and comparing physiological effects in healthy adults, eg, using meta-analysis,[58] is of interest, ultimately using SICI or other paired-pulse measures for pharmacological treatment predictions in affected individuals has greater clinical utility. Using TMS to identify subtypes within various developmental conditions is probably achievable with various TMS measures. However, such biomarker work requires a large sample of clinically well-characterized, affected children. Given the time investment required, multicenter studies may achieve these goals more quickly.

Some promise has also been suggested by combining single- and paired-pulse TMS physiology with genetics[79–82] but the pace of this research, which has primarily occurred in adults, has been slow.

Motor Cortex Inhibition Biomarker? Functional Short-Interval Cortical Inhibition (SICI)

A newly emerging technique has been to combine TMS with various activities which generate states. To the extent that these states differ between persons with different traits, probing motor cortex might result in important functional insights.

For example, in small samples of healthy adults, investigators have probed motor cortex physiology, timing TMS pulses precisely during trials with behavioral and motivational tasks. These include multiple states or tasks which could be altered in a variety of developmental or behavioral disorders. Examples include tasks requiring response inhibition or selective stopping,[83–85] and perception or selection of reward.[24–26] Although possibly creating additional challenges for optimization in hyperkinetic children, there is immense promise for these to be used for biomarker research in developmental studies. One of the few such examples published to date required children with ADHD and healthy children to complete a modified Go–NoGo task. TMS pulses were administered at precise times as children were either in stages of movement preparation or inhibited preprimed responses. The results suggest children with more hyperactive ADHD phenotypes have less SICI than peers both at rest and during movement preparation and, further, do not "engage" motor cortex SICI as robustly while inhibiting responses. Further research validating these findings is needed.

References

1. Hinshaw SP, Arnold LE, Group MTAC. Attention-deficit hyperactivity disorder, multimodal treatment, and longitudinal outcome: evidence, paradox, and challenge. *Wiley Interdiscip Rev Cogn Sci*. 2015;6(1):39–52.
2. Conelea CA, Woods DW, Zinner SH, et al. The impact of Tourette syndrome in adults: results from the Tourette Syndrome impact survey. *Community Ment Health J*. 2013;49(1):110–120.
3. Gilbert DL, Garvey MA, Bansal AS, Lipps T, Zhang J, Wassermann EM. Should transcranial magnetic stimulation research in children be considered minimal risk? *Clin Neurophysiol*. 2004;115(8):1730–1739.
4. Gilbert DL. Design and analysis of motor-evoked potential data in pediatric neurobehavioural disorder investigations. In: Wassermann EM, Epstein CM, Ziemann U, Walsh V, Paus T, Lisanby SH, eds. *The Oxford Handbook of Transcranial Magnetic Stimulation*. Oxford, UK: Oxford University Press; 2008:389–400.
5. Hong YH, Wu SW, Pedapati EV, et al. Safety and tolerability of theta burst stimulation versus single and paired pulse transcranial magnetic stimulation: a comparative study of 165 pediatric subjects. *Front Hum Neurosci*. 2015;9:29.

6. Sczesny-Kaiser M, Hoffken O, Tegenthoff M, Schwenkreis P. Convulsive syncope after single-pulse TMS. *Brain Stimul*. 2013;6(5):830.

7. Hadar AA, Makris S, Yarrow K. Single-pulse TMS related syncopal spell in a healthy subject. *Brain Stimul*. 2012;5(4):652–653.

8. Kirton A, Deveber G, Gunraj C, Chen R. Neurocardiogenic syncope complicating pediatric transcranial magnetic stimulation. *Pediatr Neurol*. 2008;39(3):196–197.

9. Kratz O, Studer P, Barth W, et al. Seizure in a nonpredisposed individual induced by single-pulse transcranial magnetic stimulation. *J ECT*. 2011;27(1):48–50.

10. Bae EH, Schrader LM, Machii K, et al. Safety and tolerability of repetitive transcranial magnetic stimulation in patients with epilepsy: a review of the literature. *Epilepsy Behav*. 2007;10(4):521–528.

11. Sackett DL, Haynes RB, Guyatt G, Tugwell P. *The Interpretation of Diagnostic Data. Clinical Epidemiology: A Basic Science for Clinical Medicine*. 2nd ed. Boston: Little, Brown and Company; 1991:69–152.

12. Connell FA, Koepsell TD. Measures of gain in certainty from a diagnostic test. *Am J Epidemiol*. 1985;121(5):744–753.

13. Orth M, Snijders AH, Rothwell JC, et al. The variability of intracortical inhibition and facilitation. *Clin Neurophysiol*. 2003;114(12):2362–2369.

14. Garvey MA, Ziemann U, Becker DA, Barker CA, Bartko JJ. New graphical method to measure silent periods evoked by transcranial magnetic stimulation. *Clin Neurophysiol*. 2001;112(8):1451–1460.

15. Peterson BS, Choi HA, Hao X, et al. Morphologic features of the amygdala and hippocampus in children and adults with Tourette syndrome. *Arch Gen Psychiatry*. 2007;64(11):1281–1291.

16. Baym CL, Corbett BA, Wright SB, Bunge SA. Neural correlates of tic severity and cognitive control in children with Tourette syndrome. *Brain J Neurol*. 2008;131(Pt 1):165–179.

17. Bloch MH, Leckman JF, Zhu H, Peterson BS. Caudate volumes in childhood predict symptom severity in adults with Tourette syndrome. *Neurology*. 2005;65(8):1253–1258.

18. Mills KR, Nithi KA. Corticomotor threshold to magnetic stimulation: normal values and repeatability. *Muscle Nerve*. 1997;20(5):570–576.

19. Ziemann U, Lonnecker S, Steinhoff BJ, Paulus W. Effects of antiepileptic drugs on motor cortex excitability in humans: a transcranial magnetic stimulation study. *Ann Neurol*. 1996;40(3):367–378.

20. Gilbert DL, Sallee FR, Zhang J, Lipps TD, Wassermann EM. TMS-evoked cortical inhibition: a consistent marker of ADHD scores in Tourette Syndrome. *Biol Psychiatry*. 2005;57:1597–1600.

21. Fleming MK, Sorinola IO, Newham DJ, Roberts-Lewis SF, Bergmann JH. The effect of coil type and navigation on the reliability of transcranial magnetic stimulation. *IEEE Trans Neural Syst Rehabil Eng*. 2012;20(5):617–625.

22. Gilbert DL, Isaacs KM, Augusta M, MacNeil LK, Mostofsky SH. Motor cortex inhibition: a marker of ADHD behavior and motor development in children. *Neurology*. 2011;76(7):615–621.

23. Boroojerdi B, Battaglia F, Muellbacher W, Cohen LG. Voluntary teeth clenching facilitates human motor system excitability. *Clin Neurophysiol*. 2000;111(6):988–993.

24. Gupta N, Aron AR. Urges for food and money spill over into motor system excitability before action is taken. *Eur J Neurosci*. 2011;33(1):183–188.

25. Thabit MN, Nakatsuka M, Koganemaru S, Fawi G, Fukuyama H, Mima T. Momentary reward induce changes in excitability of primary motor cortex. *Clin Neurophysiol*. 2011;122(9):1764–1770.

26. Kapogiannis D, Campion P, Grafman J, Wassermann EM. Reward-related activity in the human motor cortex. *Eur J Neurosci*. 2008;27(7):1836–1842.

27. Muller K, Homberg V, Lenard HG. Magnetic stimulation of motor cortex and nerve roots in children. Maturation of cortico-motoneuronal projections. *Electroencephalogr Clin Neurophysiol*. 1991;81(1):63–70.

28. Garvey MA, Barker CA, Bartko JJ, et al. The ipsilateral silent period in boys with attention-deficit/hyperactivity disorder. *Clin Neurophysiol.* 2005;116(8):1889–1896.
29. Moll GH, Heinrich H, Trott GE, Wirth S, Bock N, Rothenberger A. Children with comorbid attention-deficit-hyperactivity disorder and tic disorder: evidence for additive inhibitory deficits within the motor system. *Ann Neurol.* 2001;49(3):393–396.
30. Gilbert DL, Bansal AS, Sethuraman G, et al. Association of cortical disinhibition with tic, ADHD, and OCD severity in Tourette syndrome. *Mov Disord.* 2004;19(4):416–425.
31. Devanne H, Lavoie BA, Capaday C. Input-output properties and gain changes in the human corticospinal pathway. *Exp Brain Res.* 1997;114(2):329–338.
32. Chen R, Yaseen Z, Cohen LG, Hallett M. Time course of corticospinal excitability in reaction time and self-paced movements. *Ann Neurol.* 1998;44(3):317–325.
33. Stinear CM, Coxon JP, Byblow WD. Primary motor cortex and movement prevention: where Stop meets Go. *Neurosci Biobehav Rev.* 2009;33(5):662–673.
34. Orth M, Rothwell JC. The cortical silent period: intrinsic variability and relation to the waveform of the transcranial magnetic stimulation pulse. *Clin Neurophysiol.* 2004;115(5):1076–1082.
35. Schnitzler A, Benecke R. The silent period after transcranial magnetic stimulation is of exclusive cortical origin: evidence from isolated cortical ischemic lesions in man. *Neurosci Lett.* 1994;180(1):41–45.
36. Fuhr P, Agostino R, Hallett M. Spinal motor neuron excitability during the silent period after cortical stimulation. *Electroencephalogr Clin Neurophysiol.* 1991;81(4):257–262.
37. Siebner HR, Dressnandt J, Auer C, Conrad B. Continuous intrathecal baclofen infusions induced a marked increase of the transcranially evoked silent period in a patient with generalized dystonia. *Muscle Nerve.* 1998;21(9):1209–1212.
38. Werhahn KJ, Kunesch E, Noachtar S, Benecke R, Classen J. Differential effects on motor-cortical inhibition induced by blockade of GABA uptake in humans. *J Physiol.* 1999;517 (Pt 2):591–597.
39. Ziemann U, Paulus W, Rothenberger A. Decreased motor inhibition in Tourette's disorder: evidence from transcranial magnetic stimulation. *Am J Psychiatry.* 1997;154(9):1277–1284.
40. Singer HS, Wendlandt J, Krieger M, Giuliano J. Baclofen treatment in Tourette syndrome: a double-blind, placebo-controlled, crossover trial. *Neurology.* 2001;56(5):599–604.
41. Lepage JF, Beaule V, Srour M, et al. Neurophysiological investigation of congenital mirror movements in a patient with agenesis of the corpus callosum. *Brain Stimul.* 2011;5(2):137–140.
42. Fling BW, Seidler RD. Task-dependent effects of interhemispheric inhibition on motor control. *Behav Brain Res.* 2012;226(1):211–217.
43. Wahl M, Lauterbach-Soon B, Hattingen E, et al. Human motor corpus callosum: topography, somatotopy, and link between microstructure and function. *J Neurosci.* 2007;27(45):12132–12138.
44. Baumer T, Thomalla G, Kroeger J, et al. Interhemispheric motor networks are abnormal in patients with Gilles de la Tourette syndrome. *Mov Disord.* 2010;25(16):2828–2837.
45. Buchmann J, Wolters A, Haessler F, Bohne S, Nordbeck R, Kunesch E. Disturbed trans-callosally mediated motor inhibition in children with attention deficit hyperactivity disorder (ADHD). *Clin Neurophysiol.* 2003;114(11):2036–2042.
46. Garvey MA, Ziemann U, Bartko JJ, Denckla MB, Barker CA, Wassermann EM. Cortical correlates of neuromotor development in healthy children. *Clin Neurophysiol.* 2003;114(9):1662–1670.
47. McNally MA, Crocetti D, Mahone EM, Denckla MB, Suskauer SJ, Mostofsky SH. Corpus callosum segment circumference is associated with response control in children with attention-deficit hyperactivity disorder (ADHD). *J Child Neurol.* 2010;25(4):453–462.
48. Draganski B, Martino D, Cavanna AE, et al. Multispectral brain morphometry in Tourette syndrome persisting into adulthood. *Brain J Neurol.* 2010;133(Pt 12):3661–3675.

49. Fling BW, Benson BL, Seidler RD. Transcallosal sensorimotor fiber tract structure-function relationships. *Hum Brain Mapp*. 2013;34(2):384–395.
50. Reitz M, Muller K. Differences between 'congenital mirror movements' and 'associated movements' in normal children: a neurophysiological case study. *Neurosci Lett*. 1998; 256(2):69–72.
51. Ferbert A, Priori A, Rothwell JC, Day BL, Colebatch JG, Marsden CD. Interhemispheric inhibition of the human motor cortex. *J Physiol*. 1992;453:525–546.
52. Wassermann EM, Fuhr P, Cohen LG, Hallett M. Effects of transcranial magnetic stimulation on ipsilateral muscles. *Neurology*. 1991;41(11):1795–1799.
53. Jung P, Beyerle A, Humpich M, Neumann-Haefelin T, Lanfermann H, Ziemann U. Ipsilateral silent period: a marker of callosal conduction abnormality in early relapsing-remitting multiple sclerosis? *J Neurol Sci*. 2006;250(1–2):133–139.
54. Buchmann J, Gierow W, Weber S, et al. Modulation of transcallosally mediated motor inhibition in children with attention deficit hyperactivity disorder (ADHD) by medication with methylphenidate (MPH). *Neurosci Lett*. 2006;405(1–2):14–18.
55. Wu SW, Gilbert DL, Shahana N, Huddleston DA, Mostofsky SH. Transcranial magnetic stimulation measures in attention-deficit/hyperactivity disorder. *Pediatr Neurol*. 2012; 47(3):177–185.
56. Dirlikov B, Nebel MB, Barber AD, et al. A multimodal examination of interhemispheric connectivity and mirror overflow in children with ADHD. In: *The 20th Annual Meeting of the Organization for Human Brain Mapping; June 8–12, 2014*. Hamburg, Germany; 2014.
57. Crocetti D, Dirlikov B, Peterson D, Gilbert DL, Mostofsky SH. Interhemispheric motor inhibition is associated with callosal structural integrity in children with ADHD. In: *The 20th Annual Meeting of the Organization for Human brain Mapping; June 8–12, 2014*. Hamburg, Germany; 2014.
58. Gilbert DL, Ridel KR, Sallee FR, Zhang J, Lipps TD, Wassermann EM. Comparison of the inhibitory and excitatory effects of ADHD medications methylphenidate and atomoxetine on motor cortex. *Neuropsychopharmacology*. 2006;31(2):442–449.
59. Rothwell JC, Day BL, Thompson PD, Kujirai T. Short latency intracortical inhibition: one of the most popular tools in human motor neurophysiology. *J Physiol*. 2009;587 (Pt 1):11–12.
60. Ziemann U, Reis J, Schwenkreis P, et al. TMS and drugs revisited 2014. *Clin Neurophysiol*. 2014;126(10):1847–1868.
61. Bunse T, Wobrock T, Strube W, et al. Motor cortical excitability assessed by transcranial magnetic stimulation in psychiatric disorders: a systematic review. *Brain Stimul*. 2014;7(2):158–169.
62. Kujirai T, Caramia MD, Rothwell JC, et al. Corticocortical inhibition in human motor cortex. *J Physiol*. 1993;471:501–519.
63. Di Lazzaro V, Oliviero A, Meglio M, et al. Direct demonstration of the effect of lorazepam on the excitability of the human motor cortex. *Clin Neurophysiol*. 2000;111(5):794–799.
64. Di Lazzaro V, Oliviero A, Saturno E, et al. Effects of lorazepam on short latency afferent inhibition and short latency intracortical inhibition in humans. *J Physiology*. 2005;564 (Pt 2):661–668.
65. McDonnell MN, Orekhov Y, Ziemann U. The role of GABA(B) receptors in intracortical inhibition in the human motor cortex. *Exp Brain Res*. 2006;173(1):86–93.
66. Ziemann U. TMS and drugs. *Clin Neurophysiol*. 2004;115(8):1717–1729.
67. Moll GH, Heinrich H, Trott G, Wirth S, Rothenberger A. Deficient intracortical inhibition in drug-naive children with attention-deficit hyperactivity disorder is enhanced by methylphenidate. *Neurosci Lett*. 2000;284(1–2):121–125.
68. Buchmann J, Gierow W, Weber S, et al. Restoration of disturbed intracortical motor inhibition and facilitation in attention deficit hyperactivity disorder children by methylphenidate. *Biol Psychiatry*. 2007;62(9):963–969.

69. Hoegl T, Heinrich H, Barth W, Losel F, Moll GH, Kratz O. Time course analysis of motor excitability in a response inhibition task according to the level of hyperactivity and impulsivity in children with ADHD. *PLoS One*. 2012;7(9):e46066.
70. Macneil LK, Xavier P, Garvey MA, et al. Quantifying excessive mirror overflow in children with attention-deficit/hyperactivity disorder. *Neurology*. 2011;76(7):622–628.
71. Cole WR, Mostofsky SH, Larson JC, Denckla MB, Mahone EM. Age-related changes in motor subtle signs among girls and boys with ADHD. *Neurology*. 2008;71(19):1514–1520.
72. Mostofsky SH, Cooper KL, Kates WR, Denckla MB, Kaufmann WE. Smaller prefrontal and premotor volumes in boys with attention-deficit/hyperactivity disorder. *Biol Psychiatry*. 2002;52(8):785–794.
73. Castellanos FX, Proal E. Large-scale brain systems in ADHD: beyond the prefrontal-striatal model. *Trends Cognit Sci*. 2012;16(1):17–26.
74. Edden RA, Crocetti D, Zhu H, Gilbert DL, Mostofsky SH. Reduced GABA Concentration in attention-deficit/hyperactivity DisorderReduced GABA Concentration in ADHD. *Arch Gen Psychiatry*. 2012;69(7):750–753.
75. Gilbert DL, Zhang J, Lipps TD, et al. Atomoxetine treatment of ADHD in Tourette Syndrome: reduction in motor cortex inhibition correlates with clinical improvement. *Clin Neurophysiol*. 2007;118:1835–1841.
76. Chen TH, Wu SW, Welge JA, et al. Reduced short interval cortical inhibition correlates with atomoxetine response in children with attention-deficit hyperactivity disorder (ADHD). *J Child Neurol*. 2014;29(12):1672–1679.
77. Newcorn JH, Kratochvil CJ, Allen AJ, et al. Atomoxetine and osmotically released methylphenidate for the treatment of attention deficit hyperactivity disorder: acute comparison and differential response. *Am J Psychiatry*. 2008;165(6):721–730.
78. Schulz KP, Fan J, Bedard AC, et al. Common and unique therapeutic mechanisms of stimulant and nonstimulant treatments for attention-deficit/hyperactivity disorder. *Arch Gen Psychiatry*. 2012;69(9):952–961.
79. Gilbert DL, Wang ZW, Sallee FR, et al. Dopamine transporter genotype influences the physiological response to medication in ADHD. *Brain*. 2006;129:2038–2046.
80. Menzler K, Hermsen A, Balkenhol K, et al. A common SCN1A splice-site polymorphism modifies the effect of carbamazepine on cortical excitability–a pharmacogenetic transcranial magnetic stimulation study. *Epilepsia*. 2014;55(2):362–369.
81. Eichhammer P, Langguth B, Wiegand R, Kharraz A, Frick U, Hajak G. Allelic variation in the serotonin transporter promoter affects neuromodulatory effects of a selective serotonin transporter reuptake inhibitor (SSRI). *Psychopharmacology*. 2003;166(3):294–297.
82. Langguth B, Sand P, Marek R, et al. Allelic variation in the serotonin transporter promoter modulates cortical excitability. *Biol Psychiatry*. 2009;66(3):283–286.
83. Badry R, Mima T, Aso T, et al. Suppression of human cortico-motoneuronal excitability during the Stop-signal task. *Clin Neurophysiol*. 2009;120(9):1717–1723.
84. Coxon JP, Stinear CM, Byblow WD. Stop and go: the neural basis of selective movement prevention. *J Cogn Neurosci*. 2009;21(6):1193–1203.
85. Majid DS, Cai W, George JS, Verbruggen F, Aron AR. Transcranial magnetic stimulation reveals dissociable mechanisms for global versus selective corticomotor suppression underlying the stopping of action. *Cereb Cortex*. 2012;22(2):363–371.

TMS Mapping of Motor Development After Perinatal Brain Injury

M. Staudt

Schön Klinik, Vogtareuth, Germany; University Children's Hospital, Tübingen, Germany

INTRODUCTION

In 1993, Lucinda Carr et al.[1] published a study in *Brain* which changed our understanding of sensorimotor reorganization in children with early unilateral brain lesions dramatically. Using focal transcranial magnetic stimulation (TMS), the authors found that in many, but not all, children with early unilateral lesions resulting in hemiparesis, TMS of the contralesional hemisphere elicited short-latency motor-evoked potentials (MEP) not only in the (contralateral) non-paretic hand, but also in the (ipsilateral) paretic hand. This finding provided clear evidence that such children possess fast-conducting, monosynaptic corticospinal projections connecting the contralesional hemisphere to alpha-motoneurons in the ipsilateral cervical spinal cord, controlling the paretic hand. This phenomenon is apparently specific for lesions to the developing brain, since it has never been described after adult hemiparetic stroke.

> It is important to differentiate the TMS phenomenon in this group of patients from ipsilateral MEPs that can be elicited in other conditions as in patients with unilateral lesions acquired in adulthood (Benecke et al., 1991; Turton et al., 1996; Eyre et al., 2001) or later childhood (Nezu et al., 1999), in healthy children (Müller et al., 1997a; Eyre et al., 2001), or from precontracted forearm muscles in healthy adults (Wassermann et al., 1994; Ziemann et al., 1999). These ipsilateral MEPs usually have longer latencies, which has been interpreted as evidence for oligosynaptic, corticoreticulospinal pathways (Benecke et al., 1991; Ziemann et al., 1999) or, alternatively, as indicating slower conduction velocities of such ipsilaterally projecting fibers (Eyre et al., 2001). *Staudt et al.*[6]

In this chapter, I will review the current knowledge on this fascinating pathway, including information on the ontogenesis of the corticospinal tract, on functional aspects of these ipsilateral corticospinal pathways, on typical activation patterns in functional neuroimaging studies, and finally on the clinical relevance of this finding for individual patients.

DEVELOPMENTAL ASPECTS

In the human embryo, the corticospinal tract develops in a corticofugal direction, with axons growing out of pyramidal cells in the primary motor cortex (M1) and finding their way through the cerebral white matter, the posterior limb of the internal capsule and the anterior part of the brainstem to the spinal cord. By the 24th week of gestation, projections for the upper extremity have already reached their segmental level in the cervical spinal cord, and enter a process of synaptogenesis with the alpha-motoneurons.[2] During this time, each hemisphere still possesses bilateral projections, ie, projections to both sides of the spinal cord – so that ipsilateral and contralateral projections compete for synaptic space on the alpha-motoneurons. For the following steps of development, neuronal activity becomes an important determinant[3,4]: more active projections will be strengthened, while less active projections will become weaker and weaker. Under normal conditions, the contralateral projections are more active than the ipsilateral projections (and, to my knowledge, it is not known yet why it is the contralateral), so that the ipsilateral projections are weakened and eventually disappear (Fig. 9.1, top). This process takes place during the third trimester of pregnancy

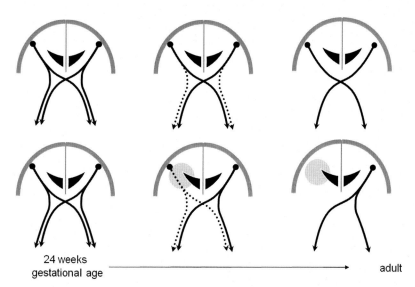

24 weeks
gestational age ⟶ adult

FIGURE 9.1 (Top) Normal development: Each hemisphere initially possesses bilateral corticospinal projections, ie, to left and right extremities. During ongoing development, the ipsilateral projections are gradually withdrawn (*dotted lines*), while the contralateral projections persist. (Bottom) When a brain lesion (*gray circle*) affects the corticospinal projections of one hemisphere during this period of development (→ lines become dotted), the ipsilateral projections from the contralesional hemisphere can persist into adulthood.

and the first months of life; already at the age of 2 years, fast-conducting ipsilateral projections to the upper extremity can no longer be detected.[5]

These mechanisms can nicely explain the development of ipsilateral corticospinal control over the paretic hand in children with early unilateral lesions (Fig. 9.1, bottom): when these lesions weaken or disrupt the contralateral projections from the lesioned hemisphere, the ipsilateral projections from the contralesional hemisphere become stronger than the contralateral projections from the lesioned hemisphere, so that these ipsilateral projections are not retracted, but persist into adulthood – and can thus allow the contralesional hemisphere to "take over" some portion of motor control of the paretic hand.

This possibility of "re-organization" with ipsilateral corticospinal pathways leads to three different constellations along a spectrum of lateralization for focal TMS investigating hand muscles in children with early unilateral brain lesions (Fig. 9.2 – from Staudt et al.[6]):

1. MEPs in the paretic hand can be elicited only by TMS of the (contralateral) affected hemisphere.
2. MEPs in the paretic hand can be elicited both by TMS of the affected hemisphere and by TMS of the contralesional hemisphere.
3. MEPs in the paretic hand can be elicited only by TMS of the (ipsilateral) contralesional hemisphere.

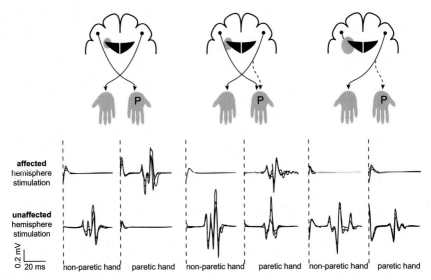

FIGURE 9.2   (Top) Schematic drawing of the three possibilities of corticospinal projections to the paretic hand (P) in patients with unilateral spastic cerebral palsy (*gray circle* = lesion). Corticospinal projections to the paretic hand can originate only in the lesioned hemisphere (left), in both hemispheres (middle) or only in the contralesional hemisphere (right). (Bottom) MEP tracings from representative patients (recorded in the first dorsal interosseus muscles of both hands after focal motor cortex single-pulse TMS). *From Staudt M, Grodd W, Gerloff C, Erb M, Stitz J, Krägeloh-Mann I. Two types of ipsilateral reorganization in congenital hemiparesis: a TMS and fMRI study. Brain. 2002;125:2222–2237, with permission.*

And in line with the disappearance of ipsilateral projections in healthy children during the first months to years of life, the "latest" lesion published so far which had led to the persistence of ipsilateral fast-conducting corticospinal projections occurred at the age of 2 years; when a lesion disrupts the corticospinal tract in one hemisphere later during childhood, no short-latency MEPs can be elicited from this hand any longer from either hemisphere.[7]

The phenomenon of bilateral projections to the paretic hand (Fig. 9.2, middle) has been reported for many patients.[1,6,8] It is still not known, however, whether this phenomenon can be explained by bilateral corticospinal projections to single alpha motoneurons, or simply by the fact that the MEP is composited of responses from many thousands of motor units, some of which might receive input exclusively from contralateral projections, others exclusively from ipsilateral projections.

Furthermore, at the cortical level, it is not yet clear whether, in a situation with ipsilateral projections (Fig. 9.2 right), these ipsilateral projections are axonal branches (or "collaterals") of the axons projecting contralaterally (ie, to the non-paretic hand), or whether there are distinct subpopulations of pyramidal neurons in M1 (Fig. 9.3), some with ipsilateral projections,

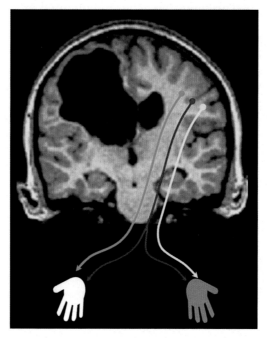

FIGURE 9.3 Theoretically, there are two possibilities for the anatomical correlate of ipsilateral corticospinal projections: pyramidal neurons could project bilaterally (*red*), ie, with branched corticospinal axons; alternatively, pyramidal neurons could project either only ipsilaterally (*yellow*) or only contralaterally (*green*).

some with contralateral projections – or whether all three types of pyramidal neurons exist in M1 of the contralesional hemisphere (ie, only ipsilateral, only contralateral, and bilateral projections).

Most studies looking topographically at the cortical origin of these projections found overlapping representations for the two hands,[6,8,9] but there seem to be exceptions to this rule, with topographically distinct representations for the two hands within the contralesional hemisphere.[9]

Large-scale reorganization within the lesioned hemisphere, on the other hand, is apparently extremely rare: only two single cases have been reported with a shift of the primary motor representation of the paretic hand, one with a shift to the parietal lobe,[10] the other with a shift to the occipital lobe.[11]

FUNCTIONAL ASPECTS

When a hand receives corticospinal input only from the ipsilateral hemisphere, this hand will in most cases develop a clear functional motor deficit,[12–14] but show some preserved active motor function ranging from an active opening/closing of the hand to individual finger movements. There is a good correlation with the age of the fetus/child at the time when the lesion occurred: in our own study,[12] we found that "ipsilateral" children with hemiparesis resulting from malformations (1st and 2nd trimester) could perform individual finger movements, "ipsilateral" children with hemiparesis resulting from unilateral periventricular lesions (early 3rd trimester) could move individual fingers or at least open/close the paretic hand actively, while some "ipsilateral" children with cortico-subcortical infarcts (late 3rd trimester) could not move the fingers of the paretic hand actively at all. A "normal" function of a hand under ipsilateral control, on the other hand, is quite unusual – despite many studies reporting on patients with ipsilateral projections, this phenomenon has, to date, only been observed in a single patient.[15]

On a group level, children controlling their paretic hands via only contralateral projections perform better than children controlling their paretic hands via only ipsilateral projections – and the group of children with bilateral projections is in-between.[6,16] This is not surprising, since a complete disruption or retraction of all contralateral projections from the lesioned hemisphere will occur in children with large lesions,[6] and a takeover of motor control by the contralesional hemisphere may only be the "second best" solution. Looking more closely, however, this statement holds true only for children with brain malformations (acquired during the 1st or 2nd trimesters of pregnancy) and for children with periventricular lesions (acquired during the early 3rd trimester of pregnancy), but not for children with corticosubcortical infarcts (typically acquired during the late 3rd trimester of pregnancy or around term birth). In these children, active hand function (opening/closing the fingers of the paretic hand) can be absent although preserved contralateral projections from the infarcted hemisphere can be detected by TMS.[12]

The fact that all these exceptions (poor hand function despite ipsilateral projections/poor hand function despite contralateral projections) can be found only in the group of children with the "latest" lesion[12] could indicate that, in these children, the time window for successful reorganization with ipsilateral projections has already been closed. Accordingly, to my knowledge, all reported children with lesions acquired later in childhood and a complete disruption (as defined by absence of stroke-side MEP) of the corticospinal tract have no or minimal active hand function.

TMS evidence for ipsilateral corticospinal projections is usually associated with another peculiarity in hand function: most of these children develop strong *mirror movements* (MMs), ie, involuntary movements of the other hand during voluntary unimanual movements, which are often stronger than those observed in some healthy children, and also do not disappear after the age of 10 years, but persist into adulthood.[12,17,18] These MMs not only occur during voluntary movements of the paretic hand (a non-specific finding, which can also be observed after hemiparetic stroke acquired in adulthood[19]), but also during voluntary movements of the non-paretic hand: apparently, the "structural abnormality" of the contralesional hemisphere in these children (ie, the presence of ipsilateral corticospinal projections) is also reflected in a "functional abnormality" of the contralesional hemisphere: it can no longer activate "its own" contralateral hand without co-activating the paretic hand. Therefore, during neurological examination, especially of older children with hemiparesis, MMs in the paretic hand should be assessed ("Open and close your good hand!") and, when present, be interpreted as an indicator of the presence of ipsilateral projections to the paretic hand.

FUNCTIONAL IMAGING

Functional MRI studies in hemiparetic children with ipsilateral corticospinal projections have added informative, and sometimes surprising aspects complementary to TMS neurophysiology. Not surprisingly, when such children were studied during active movements with their paretic hands, cortical activation can be observed in the "hand knob" area of the contralesional hemisphere – the "reorganized" primary motor representation of the paretic hand; the area where TMS had elicited bilateral projections. The surprise was to see activation also in the central (Rolandic) region of the lesioned hemisphere – an area from where TMS had been unable to elicit MEPs.[20–23] And since the same activation could also be elicited by passive movement of the paretic hand, and since MEG studies demonstrated the first cortical response to a tactile stimulus in the paretic hand in the same area,[22] it became clear that, in many of these children, the lesioned hemisphere still harbored the primary somatosensory (S1) representation of the paretic hand. This suggests a resulting "hemispheric dissociation" between an (ipsilaterally reorganized) M1 representation and a (contralaterally preserved) S1 representation of the paretic hand (Fig. 9.4 – from Staudt et al.[22]), and it is attractive to speculate that this "hemispheric M1 S1 dissociation" itself might be responsible, at least in part, for the functional impairment despite a "normal" (though ipsilateral) corticospinal projection from a "healthy hemisphere" and often good somatosensory functions in these subjects.

MR TRACTOGRAPHY

To date, no convincing evidence has been provided yet that ipsilateral corticospinal projections can be visualized (separately from contralateral corticospinal tracts) using MR tractography techniques. A case

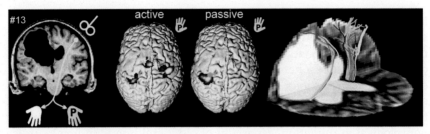

FIGURE 9.4   Example of a patient with a unilateral periventricular brain lesion (left) resulting in a "hemispheric dissociation" between the primary motor (M1) representation of the paretic hand in the contralesional hemisphere and the primary somatosensory (S1) representation in the lesioned hemisphere. (Left) TMS of the lesioned hemisphere elicited no response, while TMS of the contralesional hemisphere elicited bilateral responses. *Yellow arrows* symbolize corticospinal projections. fMRI (center) during active movement of the paretic hand (P) showed activation in both Rolandic cortices, while fMRI during passive movement of the paretic hand showed activation only in the lesioned hemisphere. Magnetoencephalography of repetitive tactile stimulation of the paretic thumb demonstrated the first cortical response (corresponding to S1) in the Rolandic cortex of the lesioned hemisphere (*blue circle*). (Right) MR diffusion tractography visualized axonal "bypasses" of somatosensory projections around the lesion. *From Staudt M, Braun C, Gerloff C, et al. Developing somatosensory projections bypass periventricular brain lesions. Neurology. 2006;67:522–525, with permission.*

report[24] describing a "double crossover" of corticospinal projections at the brainstem level in a hemiparetic patient with schizencephaly still needs to be reproduced. On the other hand, the ascending thalamocortical projections delivering somatosensory information to S1 in the lesioned hemisphere, thereby often "bypassing" the lesion, can easily be visualized using MR tractography[22] (Fig. 9.4 right). And finally, diffusion tractography techniques have successfully been applied to document asymmetry of corticospinal projections at the brainstem level, with marked asymmetry facilitating early prognosis for the development of hemiparesis.[25–28]

CLINICAL RELEVANCE I: FUNCTIONAL THERAPY

Whether children controlling their paretic hand via ipsilateral projections require a special type of functional therapy is still a matter of debate.[29,30] In our group, we have investigated two well-defined samples of hemiparetic children, one with ipsilateral and one with contralateral corticospinal projections, and could demonstrate that these groups respond differently to constraint-induced movement therapy (CIMT), both on a behavioral level[31] and regarding cortical neuromodulation induced by this therapy.[32] These results indeed suggest that the same therapeutic approach may not necessarily be optimal for each type of childhood hemiparesis. Along the same lines, we recently demonstrated that mirror movements have a specific negative impact on the performance of asymmetrical bimanual activities,[17] as already suggested previously.[33] Therefore, this phenomenon, when present, should specifically be addressed in functional therapeutic approaches. Whether the TMS information of the presence/absence of ipsilateral projections is helpful for "designing" an individual functional therapy for an individual child beyond the simple clinical observation of MM is not known.

CLINICAL RELEVANCE II: EPILEPSY SURGERY

The clinical importance of this information comes also from a related field of pediatric neurology. When children with early unilateral brain lesions develop epilepsy, this is a significant threat for the neuroplastic potential of the one good hemisphere of such children. Therefore, epilepsy in these children must be diagnosed early and treated vigorously. And since the epilepsies in such children are often difficult to control with antiepileptic medication, the option of epilepsy surgery must be discussed early – especially since many of these children are excellent candidates for hemispherotomy, ie, the complete surgical disconnection of the epileptic hemisphere.

When this procedure is performed in patients without hemiparesis or with hemiparesis due to lesions acquired in late childhood or adulthood, the contralateral hand will lose all active function.[34,35] In many children with hemiparesis due to early lesions, however, this procedure will not lead to a deterioration of hand function. That the paretic hand can still perform active movements immediately after the hemispherotomy is convincing evidence that this hand receives (and had received preoperatively) functional cortical input from the (ipsilateral) contralesional hemisphere. Consequently, TMS has been applied to look for ipsilateral projections in such children preoperatively, hoping to predict that paretic hand function will be preserved postoperatively. And indeed, all published cases to date have confirmed that when grasp function of the paretic hand is present preoperatively, and when TMS detects corticospinal projections to the paretic hand only from the contralesional hemisphere, that these children were still able to grasp after the hemispherotomy.[10,23,36,37] The function of the paretic hand is often unchanged, though in some cases it may even improve compared to preoperatively.[38,39] Deterioration has also been observed but still with a preserved grasp function (personal experience, unpublished). This deterioration might, at least in some cases, be explained by the above-mentioned situation of an "M1–S1 dissociation," in which hemispherotomy leads to a disconnection of the thalamocortical projections to S1 in the affected hemisphere.

In this rare and dramatic situation (considering hemispherotomy in a hemiparetic child with preserved grasp function), it is, in my view, absolutely necessary to perform TMS preoperatively whenever possible, and to collect and publish the postoperative outcome with respect to hand motor function – in order to gain more confidence in predicting a preservation of hand function in more children, which should eventually result in an increasing number of children to whom epilepsy surgery is offered.

PERSPECTIVE: BRAIN STIMULATION

Brain stimulation techniques like repetitive transcranial magnetic stimulation (rTMS) or transcranial direct current stimulation (tDCS) have been applied in several studies for adult hemiparetic stroke, with the purpose of enhancing the recovery process either by facilitating the affected hemisphere (controlling the paretic hand) or by inhibiting negative influences from the contralesional hemisphere.[39a] First results are also being published for hemiparetic children with early brain lesions.[39b] For these techniques, children who control both hands with only one (the contralesional)[39c] hemisphere are especially challenging, and I hypothesize that these children again react differently to brain-stimulating therapies from those controlling their paretic hands via crossed projections from the affected hemisphere[39d]. In any case, all previous research calls for a stratification of hemiparetic children in intervention studies with respect to their underlying type of corticospinal reorganization.[40]

References

1. Carr LJ, Harrison LM, Evans AL, Stephens JA. Patterns of central motor reorganization in hemiplegic cerebral palsy. *Brain.* October 1993;116(Pt 5):1223–1247.
2. Eyre JA, Miller S, Clowry GJ, Conway EA, Watts C. Functional corticospinal projections are established prenatally in the human foetus permitting involvement in the development of spinal motor centres. *Brain.* January 2000;123(Pt 1):51–64.
3. Martin JH, Friel KM, Salimi I, Chakrabarty S. Activity- and use-dependent plasticity of the developing corticospinal system. *Neurosci Biobehav Rev.* 2007;31(8):1125–1135.
4. Salimi I, Friel KM, Martin JH. Pyramidal tract stimulation restores normal corticospinal tract connections and visuomotor skill after early postnatal motor cortex activity blockade. *J Neurosci.* July 16, 2008;28(29):7426–7434.
5. Eyre JA, Taylor JP, Villagra F, Smith M, Miller S. Evidence of activity-dependent withdrawal of corticospinal projections during human development. *Neurology.* November 13, 2001;57(9):1543–1554.
6. Staudt M, Grodd W, Gerloff C, Erb M, Stitz J, Krägeloh-Mann I. Two types of ipsilateral reorganization in congenital hemiparesis: a TMS and fMRI study. *Brain.* 2002;125:2222–2237.
7. Maegaki Y, Maeoka Y, Ishii S, et al. Mechanisms of central motor reorganization in pediatric hemiplegic patients. *Neuropediatrics.* June 1997;28(3):168–174.
8. Kesar TM, Sawaki L, Burdette JH, et al. Motor cortical functional geometry in cerebral palsy and its relationship to disability. *Clin Neurophysiol.* July 2012;123(7):1383–1390.

9. Vandermeeren Y, Davare M, Duque J, Olivier E. Reorganization of cortical hand representation in congenital hemiplegia. *Eur J Neurosci.* February 2009;29(4):845–854.

10. Kamida T, Baba H, Ono K, Yonekura M, Fujiki M, Kobayashi H. Usefulness of magnetic motor evoked potentials in the surgical treatment of hemiplegic patients with intractable epilepsy. *Seizure.* September 2003;12(6):373–378.

11. Basu A, Graziadio S, Smith M, Clowry GJ, Cioni G, Eyre JA. Developmental plasticity connects visual cortex to motoneurons after stroke. *Ann Neurol.* January 2010;67(1): 132–136.

12. Staudt M, Gerloff C, Grodd W, Holthausen H, Niemann G, Krägeloh-Mann I. Reorganization in congenital hemiparesis acquired at different gestational ages. *Ann Neurol.* 2004;56(6):854–863.

13. Koudijs SM, Leijten FS, Ramsey NF, van Nieuwenhuizen O, Braun KP. Lateralization of motor innervation in children with intractable focal epilepsy–a TMS and fMRI study. *Epilepsy Res.* June 2010;90(1–2):140–150.

14. Eyre JA, Smith M, Dabydeen L, et al. Is hemiplegic cerebral palsy equivalent to amblyopia of the corticospinal system? *Ann Neurol.* November 2007;62(5):493–503.

15. Fiori S, Staudt M, Pannek K, et al. Is one motor cortex enough for two hands? *Dev Med Child Neurol.* 2015;57(10):977–980

16. Holmström L, Vollmer B, Tedroff K, et al. Hand function in relation to brain lesions and corticomotor-projection pattern in children with unilateral cerebral palsy. *Dev Med Child Neurol.* February 2010;52(2):145–152.

17. Adler C, Berweck S, Lidzba K, Becher T, Staudt M. Mirror movements in unilateral spastic cerebral palsy: specific negative impact on bimanual activities of daily living. *Eur J Paediatr Neurol.* 2015;19(5):504–509.

18. Klingels K, Jaspers E, Staudt M, et al. Do mirror movements relate to hand function and timing of the brain lesion in children with unilateral cerebral palsy? *Dev Med Child Neurol.* 2015. http://dx.doi.org/10.1111/dmcn.12977

19. Nelles G, Spiekramann G, Jueptner M, et al. Evolution of functional reorganization in hemiplegic stroke: a serial positron emission tomographic activation study. *Ann Neurol.* 1999;46:901–909.

20. Thickbroom GW, Byrnes ML, Archer SA, et al. Differences in sensory and motor cortical organization following brain injury early in life. *Ann Neurol.* 2001;49:320–327.

21. Staudt M, Krägeloh-Mann I, Holthausen H, et al. Searching for motor functions in dysgenic cortex: a clinical TMS and fMRI study. *J Neurosurg.* 2004;32:159–161.

22. Staudt M, Braun C, Gerloff C, et al. Developing somatosensory projections bypass periventricular brain lesions. *Neurology.* 2006;67:522–525.

23. Zsoter A, Pieper T, Kudernatsch M, Staudt M. Predicting hand function after hemispherotomy: TMS versus fMRI in hemispheric polymicrogyria. *Epilepsia.* June 2012;53(6):e98–e101.

24. Chang WH, Kim YB, Ohn SH, Park CH, Kim ST, Kim YH. Double decussated ipsilateral corticospinal tract in schizencephaly. *Neuroreport.* October 28, 2009;20(16):1434–1438.

25. Khong PL, Zhou LJ, Ooi GC, Chung BH, Cheung RT, Wong VC. The evaluation of Wallerian degeneration in chronic paediatric middle cerebral artery infarction using diffusion tensor MR imaging. *Cerebrovasc Dis.* 2004;18(3):240–247.

26. De Vries LS, Van der Grond J, Van Haastert IC, Groenendaal F. Prediction of outcome in new-born infants with arterial ischaemic stroke using diffusion-weighted magnetic resonance imaging. *Neuropediatrics.* February 2005;36(1):12–20.

27. van der Aa NE, Verhage CH, Groenendaal F, et al. Neonatal neuroimaging predicts recruitment of contralesional corticospinal tracts following perinatal brain injury. *Dev Med Child Neurol.* August 2013;55(8):707–712.

28. Bleyenheuft Y, Grandin CB, Cosnard G, Olivier E, Thonnard JL. Corticospinal dysgenesis and upper-limb deficits in congenital hemiplegia: a diffusion tensor imaging study. *Pediatrics.* December 2007;120(6):e1502–e1511.

29. Islam M, Nordstrand L, Holmström L, Kits A, Forssberg H, Eliasson AC. Is outcome of constraint-induced movement therapy in unilateral cerebral palsy dependent on corticomotor projection pattern and brain lesion characteristics? *Dev Med Child Neurol.* March 2014;56(3):252–258.

30. Staudt M, Berweck S. Is constraint-induced movement therapy harmful in unilateral spastic cerebral palsy with ipsilateral cortico-spinal projections? *Dev Med Child Neurol.* March 2014;56(3):202–203.

31. Kuhnke N, Juenger H, Walther M, Berweck S, Mall V, Staudt M. Do patients with congenital hemiparesis and ipsilateral corticospinal projections respond differently to constraint-induced movement therapy? *Dev Med Child Neurol.* 2008;50:898–903.

32. Juenger H, Kuhnke N, Braun C, et al. Two types of exercise-induced neuroplasticity in congenital hemiparesis: a transcranial magnetic stimulation, functional MRI, and magnetoencephalography study. *Dev Med Child Neurol.* October 2013;55(10):941–951.

33. Kuhtz-Buschbeck JP, Sundholm LK, Eliasson AC, Forssberg H. Quantitative assessment of mirror movements in children and adolescents with hemiplegic cerebral palsy. *Dev Med Child Neurol.* November 2000;42(11):728–736.

34. Gardner WJ, Karnosh LJ, McClure JR, Gardner AK. Residual function following hemispherectomy for tumour and for infantile hemiplegia. *Brain.* 1955;78:487–502.

35. Holthausen H, Strobl K. Modes of reorganization of the sensorimotor system in children with infantile hemiplegia and after hemispherectomy. *Adv Neurol.* 1999;81: 201–220.

36. Shimizu T, Nariai T, Maehara T, et al. Enhanced motor cortical excitability in the unaffected hemisphere after hemispherectomy. *Neuroreport.* September 28, 2000;11(14): 3077–3084.

37. Sun W, Fu W, Wang D, Wang Y. Ipsilateral responses of motor evoked potential correlated with the motor functional outcomes after cortical resection. *Int J Psychophysiol.* September 2009;73(3):377–382.

38. Pascoal T, Paglioli E, Palmini A, Menezes R, Staudt M. Immediate improvement of motor function after epilepsy surgery in congenital hemiparesis. *Epilepsia.* August 2013;54(8):e109–e111.

39. van der Kolk NM, Boshuisen K, van Empelen R, et al. Etiology-specific differences in motor function after hemispherectomy. *Epilepsy Res.* February 2013;103(2–3):221–230.

39a. Hsu WY, Cheng CH, Liao KK, Lee IH, Lin YY. Effects of repetitive transcranial magnetic stimulation on motor functions in patients with stroke: a meta-analysis. *Stroke.* 2012;43(7):1849–1857.

39b. Gillick BT, Krach LE, Feyma T, et al. Primed low-frequency repetitive transcranial magnetic stimulation and constraint-induced movement therapy in pediatric hemiparesis: a randomized controlled trial. *Dev Med Child Neurol.* 2014;56(1):44–52.

39c. Kirton A, Andersen J, Herrero M, et al. Brain stimulation and constraint for perinatal stroke hemiparesis: the PLASTIC CHAMPS trial. *Neurology* [in press].

39d. Staudt M, Gordon AM. Combining rTMS and CIMT: a "One-Size-Fits-All" Therapy for Congenital Hemiparesis? *Neurology* [in press].

40. Staudt M. The role of transcranial magnetic stimulation in the characterization of congenital hemiparesis. *Dev Med Child Neurol.* 2010;52(2):113–114.

10

The Right Stimulation of the Right Circuits: Merging Understanding of Brain Stimulation Mechanisms and Systems Neuroscience for Effective Neuromodulation in Children

J.B. Carmel, K.M. Friel

Weill Cornell Medical College, New York, NY, United States; Burke-Cornell Medical Research Institute, White Plains, NY, United States

INTRODUCTION

A tremendous appeal of brain stimulation, as opposed to drug therapy, is the ability to target therapy to parts of the brain in need of repair. However, a challenge is that the effects of brain stimulation on brain circuits are not well understood. The overall hypothesis of this chapter is that for stimulation protocols to be effective, knowledge of functional circuits must be combined with the understanding of stimulation mechanisms. In some cases, the deranged circuits can be targeted directly, as in deep brain stimulation (DBS) for movement disorders, such as Parkinson's disease. Other stimulation approaches seek to enhance brain plasticity generally and then apply task-specific training in order to gain the circuit specificity. In either case, pediatric brain stimulation presents the further challenge of a moving target: as brain anatomy and physiology change with development, the response to brain stimulation is likely to change as well.

The critical experiments to address our central hypothesis, that matching the right stimulation to the right circuits will produce the most effective brain stimulation, alter either the target of stimulation or the stimulation parameters, while keeping the other condition the same. Since experimental evidence is largely missing, the chapter begins with an example of effective brain stimulation – deep brain stimulation (DBS) for Parkinson's disease – and describes how the success lies in directing known stimulation mechanisms to alter specific neural circuits. Next, we discuss pediatric hemiplegia, in which the circuits that control the impaired side of the body may differ depending on the nature of the brain injury, allowing comparison of similar brain stimulation for different neural circuits. We will report on the few experiments that compare different stimulation for the same circuits: motor cortex stimulation with focal versus diffuse direct-current stimulation. Finally, we suggest ways of closing the gap in understanding how electrical stimulation and neural circuitry interact, using both animal models and mechanistic studies in children.

FOCAL STIMULATION TO KNOWN CIRCUITS: LESSONS FROM DEEP BRAIN STIMULATION FOR PARKINSON'S DISEASE

Few medical interventions are as dramatic as placement of DBS electrodes for movement disorders. With the patient awake, the head is shaved, the scalp numbed, and a craniotomy performed. Then pins are placed into the skull to hold a stereotaxic frame, which will drive the stimulating electrode. Using the patient's magnetic resonance image (MRI) to target the brain structure that will be stimulated, an electrode the size of a knitting needle is advanced many centimeters into the brain to reach a pea-sized nucleus. When the electrode reaches its target, stimulation is initiated. When effective, the person's tremor is immediately calmed and movement of the affected limb becomes fluid and well-controlled, often for the first time in many years.

This procedure has produced remarkable long-term results. For people with Parkinson's disease, deep brain stimulation improves motor function and reduces the need for pharmacological therapy.[1] Patients with DBS have greater independence in their daily activities and better quality of life.[2] In addition to providing symptomatic relief, there is evidence that DBS might actually slow neurodegeneration and alter neuroplasticity.[3]

Why has DBS for Parkinson's disease been so effective? We argue that this application is an example of targeting the right circuit with the right stimulation. Discovering the basal ganglia circuits affected by the disease took decades. Parkinson initially identified the disease as degeneration of the substantia nigra. Later, the neurotransmitter dopamine was discovered as critical to the pathophysiology of Parkinson's, paving the way for dopamine repletion as therapy.[4] The circuits of the basal ganglia were subsequently identified using physiology studies in monkeys. These studies created a working model of basal ganglia pathways involved in both direct and indirect motor control.

An important advance in understanding the circuits of Parkinson's disease came after an unfortunate outbreak of parkinsonism in people taking a synthetic heroin. The development of parkinsonism was found to be related to the contaminating neurotoxin, 1-methyl-4-phenyl-1,2,3,6-tetrahydropyridine (MPTP). Mahlon DeLong, an expert in basal ganglia physiology, used MPTP in monkeys to create a large animal model of the disease. In monkeys with MPTP-induced parkinsonism, targeting the subthalamic nucleus with a blocking drug calmed the tremor and broke the bradykinesia of the condition.[5] This experiment, as well as the large number of previous physiology experiments, identified the two main targets of DBS for Parkinson's disease, the subthalamic nucleus and the internal portion of the globus pallidus. These experiments identified the circuits for which modulation with DBS might be effective.

Identification of the right stimulation frequency involved a serendipitous experiment. Alim Louis Benabid, a French neurosurgeon who had been performing ablations of the basal ganglia and thalamus for movement disorders, was testing an awake patient to determine the proper site. Usually, he used 50-Hz stimulation to activate a region in order to know if it had been properly targeted. In one patient with tremor he tried 100-Hz stimulation in ventral intermediate nucleus (Vim) of thalamus, and the patient's tremor calmed. When the stimulation was stopped, the tremor returned, indicating that the 100-Hz stimulation caused reversible dysfunction in the target. Benabid then began using high-frequency (HF, >100 Hz) DBS rather than lesions to treat patients.[6]

The most effective protocol for PD treatment involved targeting the subthalamic nucleus with HF DBS.[7] Lesions of the subthalamic nucleus cause the movement disorder known as ballism, involuntary, high amplitude, flinging movements of the limbs typically occurring proximally. Stimulation of the subthalamic nucleus allowed targeted therapy, without causing ballistic movement. By implanting DBS leads on both sides of the brain, PD symptoms on both sides of the body can be ameliorated.

While the HF stimulation protocol was found empirically, the effects on the brain are now better understood. Recent evidence indicates that subthalamic stimulation may prevent the intense synchronization of motor cortex rhythms found in patients with PD. Using motor cortex surface recordings, subthalamic HF DBS was found to disrupt locking of phase and amplitude in the beta frequency in motor cortex. There was a strong correlation between disruption of the pathological physiology and improvement in motor symptoms for both rest and reaching.[8]

An advantage of invasive brain stimulation is that electrodes positioned in or on the brain can be used both to record and to stimulate. This can be used to better understand the circuits that are being modulated and also to modulate based on the state of the target, a closed loop system. Stimulation based on the current state of the circuit, or adaptive DBS (aDBS), has the potential to deliver therapy only at the time that it is needed. In a test of this approach, investigators at Oxford University stimulated the subthalamic nucleus either with conventional deep brain stimulation (cDBS) or aDBS. For aDBS, local field potentials recorded from the stimulating electrodes in the subthalamic nucleus were processed and the information was used to trigger HF DBS. The investigators found a substantial improvement in function in subjects with aDBS compared with cDBS, even though stimulation was given less than half of the time in the aDBS group.[9]

What are the lessons to be learned for pediatric brain stimulation? There is the direct translation of DBS for diseases with deranged basal ganglia circuitry, such as dystonia.[10,11] However, more importantly, the field of DBS for PD demonstrates an important process for developing effective brain stimulation. First, the circuits underlying PD were delineated with the help of a highly homologous primate model. Second, the type of stimulation was optimized to target the specific circuit derangements. Finally, studies of targeted neuromodulation were used to optimize the therapy in an iterative fashion.

BRAIN STIMULATION IN PEDIATRIC HEMIPARESIS: WHICH CIRCUITS?

Hemiplegia is the most common pediatric condition for which brain stimulation has been employed. There are three likely reasons for this. First, like many investigations in pediatrics, methods and approaches are borrowed from advances in adult studies. Since stroke is highly prevalent in adults, brain stimulation has been used extensively to study and modulate circuits and improve function after stroke. The study of brain stimulation for pediatric hemiparesis, from stroke or other etiologies, stems naturally from adult studies, similar to methods such as constraint therapy and robotic therapy. Second, hemiparesis is relatively common in children, affecting approximately one per 1000.[12] Finally, the circuits involved in pediatric hemiparesis and recovery have been relatively well-studied compared to other common pediatric conditions.

Like adults with stroke, symptoms of pediatric hemiparesis are largely due to injury to the corticospinal tract (CST). The CST connects motor cortex directly to the spinal cord and is the principal pathway for voluntary movement.[13] Since the CST decussates, when it is injured there is impairment of the contralateral half of the body, largely the arm and hand. In several studies, there is a tight correlation between the degree of injury to the CST and the degree of hand impairment.[14,15]

The circuits that develop and persist after CST damage in children often differ substantially from those that remain in adults. Early in development each motor cortex sends terminations to both sides of the spinal cord. With experience, the terminations are pruned to mostly innervate the spinal cord on the side opposite to the motor cortex from which the CST originated (crossed). As detailed by animal studies, when injury or loss of activity from one half of the CST occurs early in development, the bilateral terminations from the uninjured hemisphere persist.[16,17] In addition to the age of injury, the degree of CST injury can also determine whether the injured (small lesion) or the uninjured hemisphere (large lesion) controls the impaired hand (Fig. 10.1).[16] Similar patterns of CST control of the impaired hand are observed in children with hemiparesis; some have preserved crossed connections and others have bilateral control from the uninjured hemisphere.[18,19]

A key to understanding which circuits to target for brain stimulation is identifying which circuits mediate motor recovery. This has been studied in animal models and in children with hemiplegia. In a cat model of pediatric hemiplegia, visuomotor skill training early in life can reroute aberrant CST connections to their appropriate spinal targets, can reshape the motor representation of the impaired limbs in the brain, and can improve

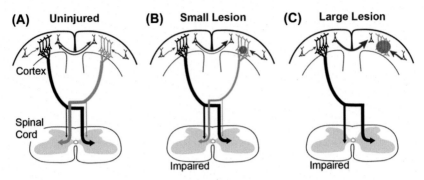

FIGURE 10.1   **Motor circuits in pediatric hemiparesis.** (A) The uninjured motor system in maturity. The corticospinal tract CST connects motor cortex directly to the spinal cord, largely through crossed connections (*black and dark gray*). Ipsilateral CST connections present at birth (*thin lines*) are withdrawn in early development. Transcallosal connections between homotopic areas of cortex are largely inhibitory (*red*). Similarly, local interneurons within or adjacent to motor cortex can also be inhibitory. (B) Changes to motor circuits after minor early injury to motor pathways. Small injury causes (1) weak CST control from the injured hemisphere to the spinal cord on the impaired side (*thinned gray line*), (2) increased transcallosal inhibition (*thicker red arrow*), and (3) increased perilesional inhibition (*thicker red arrow*). (C) Changes to motor circuits after severe injury to motor cortex. The balance of descending motor control switches further toward the uninjured hemisphere (*thicker black line* to spinal cord on impaired side). The transcallosal and perilesional inhibition of the injured hemisphere may also be altered.

motor skill.[17] In children with hemiplegia, intensive bimanual training can improve skillful use of the hand[20] and can expand the motor representation of the impaired hand (unpublished data). Interestingly, motor maps show plasticity independent of the hemisphere of origin: both crossed and uncrossed motor pathways have the capacity for plastic strengthening with training. Understanding the circuits that are modulated during motor rehabilitation provides targets that can be stimulated.

In order to effectively apply brain stimulation to children with hemiparesis, two other circuits must be considered. The first is the connection between the two motor cortices via the corpus callosum. The net effect of these connections in adults is largely inhibitory (Fig. 10.1, red), allowing the two hands to function independent of one another, although developmental studies in normal children have been limited.[21] After hemispheric injury, not only is descending motor control affected, but there may also be a loss of inhibitory transcallosal connections. With less inhibitory tone from the injured hemisphere, the uninjured hemisphere can exert excessive inhibition. This imbalance in inhibitory tone between the two hemispheres causes the uninjured hemisphere to "bully" the injured one.[22] Transcallosal inhibition is measured physiologically as interhemispheric inhibition (IHI) using paired pulses of transcranial magnetic stimulation (TMS) delivered with a stimulating coil held over each hemisphere and an interstimulus interval on the order of milliseconds.[23] Recent data demonstrate that paired-pulse IHI principles in school-aged children are comparable to adults (Adam Kirton; personal communication).

In adult hemiparesis, this "IHI imbalance" model has driven most therapeutic brain stimulation approaches with many targeting the uninjured hemisphere to reduce excessive transcallosal inhibition. Inhibition is mediated by GABAergic interneurons.[24] Since both TMS and transcranial direct current stimulation (tDCS) may affect GABA tone,[25] this offers a possible target for modulation. The most effective neuromodulation paradigm published so far doubled down on this approach, applying cathodal tDCS, which can decrease excitability, to the uninjured hemisphere and anodal tDCS, which can increase excitatability, to the injured hemisphere.[26] The large improvement in hand function of adults with chronic subcortical stroke, as measured by the Fugl–Meyer and Wolf motor function tests, strongly correlated with a decrease in IHI, providing correlative evidence for rebalancing of transcallosal inhibition as a mechanism.

The final circuit that is known to play a role in limiting function after unilateral brain injury is inhibition from perilesional cortex (Fig. 10.1, red). After focal injury to cortex in mice, the adjacent cortex develops increased GABA tone, as determined physiologically.[27] Decreasing this tone using a selective GABA antagonist lowered the perilesional inhibition and restored function in mice. Whether this strategy operates in humans after focal injury has not been tested. But one putative mechanism of excitatory stimulation of the injured hemisphere after focal injury is a decrease in perilesional inhibitory tone.

Given our current understanding of the important circuits in pediatric hemiparesis, how should neuromodulation be applied? Thus far, two published trials have tested the functional effects of brain stimulation on hand function in children with hemiparesis, and each used inhibitory (low-frequency) TMS over the uninjured hemisphere. Kirton and colleagues first applied this protocol to a small sample ($n = 10$) of children with subcortical stroke and described possible increases in hand function.[28] This protocol may have also changed IHI, increasing IHI from the injured to the uninjured hemisphere.[29] Gillick and colleagues combined constraint therapy and inhibitory TMS to the uninjured hemisphere and suggested that the combination (but not constraint alone) improved hand function.[30] The children in the Gillick study had congenital hemiparesis, and were, therefore, more likely to have control over the affected hand from the uninjured hemisphere, although those with strong ipsilateral organization patterns (no stroke side evoked potentials) were excluded from this trial.

To date, no study has determined the interaction between the pattern of CST connections and response to neuromodulation, nor has a study changed the site of stimulation depending on which hemisphere controls the affected arm. The reasons for this are simple: it raises the complexity of the study and subsequent interpretation. While modifying the stimulation site or type depending on the pattern of connections is a long-term goal of neuromodulation, this will require larger studies that are sufficiently powered to account for multiple types of connectivity or stimulation.

TARGETING MOTOR CORTEX: WHICH STIMULATION MODALITY?

The effects of brain stimulation on neural circuits are complex and poorly understood. This section explores how the biophysics of the stimulation interacts with the targeted circuits. TMS and tDCS have wholly different neurophysiological effects. Whereas TMS produces a phasic depolarization of neurons, tDCS shifts the membrane potential very slightly and holds that field continuously, causing neurons to fire more or less often, depending on the polarity of stimulation.[31] TMS causes synchronous depolarization of large populations of neurons, while tDCS likely alters the gain of neural circuits by placing them in a constant electrical field. From the perspective of activity-dependent plasticity, TMS causes neurons to fire, similar to endogenous activity, and this depolarization is critical for many forms of neural plasticity. The effect of tDCS may be to make motor training or sensory experience more effective in modulating neural circuits than they would be without stimulation. This could be accomplished through regional plasticity mechanisms such as local secretion of BDNF, a key regulator of neural plasticity.[32]

For motor cortex, the choice of TMS versus tDCS depends on how one wants to modulate the system. To explore the possibilities, we can turn our attention back to hemiplegia. After unilateral brain injury, the injured hemisphere exerts weak descending control. To strengthen those circuits, other regions of cortex that send descending connections could be recruited. High-frequency TMS applied to reorganized motor areas would likely have an effect similar to the epidural stimulation that was used in both animal and human studies to improve motor function. The mechanisms mediating motor recovery with excitatory stimulation could include strengthening synapses, promoting axon sprouting at subcortical targets, or increasing dendritic plasticity. In a cat model of pediatric hemiplegia, stimulation of the pyramidal tract from the injured side of the brain redirected corticospinal connections to their appropriate target in the spinal cord, increased synaptic strength, and improved visuomotor skill.[33] To improve motor function with anodal tDCS, it is likely that training would be needed. tDCS can enhance motor learning,[34] which is a fundamental process underlying motor recovery.

Two inhibitory circuits may be targeted with brain stimulation. The most popular target is the uninjured hemisphere. As detailed earlier, low-frequency TMS appears effective for motor recovery in chronic adult stroke hemiparesis with possible early evidence in pediatric hemiparesis. In contrast, cathodal tDCS applied to the uninjured hemisphere may require concomitant training in order to be effective. Another possible target for stimulation is the perilesional cortex. Focal brain injury has been shown to increase GABAergic signaling in the surrounding cortex.[27] Since non-invasive brain stimulation can modulate GABA, the excessive inhibition surrounding the lesion could be a promising target. As opposed to pharmacological manipulation of GABA, which affects the whole brain, including both excitatory and inhibitory circuits, brain stimulation may be able to modulate inhibitory circuits with some selectivity.

Practical considerations, rather than the mechanisms of neuromodulation, often lead clinicians or investigators to choose TMS or tDCS. In this respect, tDCS has advantages including moderate cost and increased portability and ease of use. But given the paucity of data on how stimulation affects the brain, and potential difference in pediatric tDCS effects,[35] to focus on these practical elements may be premature. Instead, it makes sense to use all of the tools at our disposal—including TMS as a probe of brain function as well as a modulator of that function—to understand the effects of brain stimulation. It is likely easier to engineer brain stimulation to achieve a targeted modulation than it is to understand what that modulation should be.

FOCAL STIMULATION VERSUS GENERAL PLASTICITY AND TASK-SPECIFIC TRAINING

Finally, in considering which stimulation is right for the target of interest, we can consider focused versus more diffuse stimulation paradigms. This section first addresses this question through discussion of two different tDCS stimulation paradigms: the typical tDCS electrode placement, which directs current widely through the brain, versus so-called high-definition tDCS (HD-tDCS) (see Chapter 5: Transcranial Direct-Current Stimulation (tDCS): Principles and Emerging Applications in Children), which produces more focal current density. Second, we consider achieving circuit specificity not by focal stimulation, but by pairing a stimulation paradigm that affects global plasticity with task-specific training. We discuss stimulation of the vagus nerve as a proplasticity treatment that can be combined with either sensory experience or motor training in order to achieve circuit specificity.

There have been few studies that address whether stimulation should be applied focally or more diffusely. One study compared the typical montage for tDCS against HD-tDCS on the ability to alter TMS-evoked muscle responses in healthy adults.[36] Both montages caused increases in responses under the anode and decreased responses under the cathode. However, the peak effect for HD-tDCS effects was delayed (at 30 min), and the effects lasted longer (over 2 h). These data suggest different effects based on the stimulation montage and more persistent changes with focal stimulation. Similar studies have not yet been completed in children.

Vagal nerve stimulation (VNS) has been used as a clinical tool to treat medication-refractory seizures since 1989 years (see Chapter 21: Invasive Neuromodulation in Pediatric Epilepsy: VNS and Emerging Technologies) and is also US FDA-approved for depression; its effects on brain plasticity have been described more recently. To stimulate the vagus nerve, a cuff electrode is placed around the left vagus nerve (has a higher proportion of afferent fibers and less efferent cardiac innervation than the right) and a pulse generator placed subcutaneously in the upper chest to deliver stimulation. VNS stimulates the brain by providing a biphasic current that cycles between on and off periods, activating axons projecting to the autonomic nuclei in the brain stem's nucleus of the solitary tract. How the brain stem centers might relate to cortical plasticity is not completely understood, but likely involves ascending signals, either noradrenergic from the locus ceruleus or cholinergic from the nucleus basalis.[37,38] These ascending projections are diffuse throughout cortex.

To augment plasticity, VNS must be combined with sensory experience or motor training. Studies in the auditory system and the motor system have demonstrated how VNS can drive adaptive cortical plasticity in health and disease. In the auditory system pairing VNS with sounds of a certain frequency can shift tonotopic maps in auditory cortex and improve responses to the trained tones.[39] A common hearing problem is tinnitus, or ringing in the ears, that is most often caused by hearing loss. In rats with noise-induced tinnitus, tones can be used to restore normal hearing and the tonotopic map in auditory cortex. These remarkable results have spurred clinical trials using VNS paired with tones outside of the tinnitus frequency.[40] The goal is to make auditory cortex more sensitive to the trained tones so that the hypersensitivity to the tinnitus tone is reduced. Preliminary results show that the treatment is safe and well-tolerated.

Similarly, VNS can change maps in motor cortex when paired with motor training. Rats were trained on a task involving the distal forelimb (spinning a wheel with their paw and digits) or the proximal forelimb (pressing a lever). Rats receiving VNS had a larger representation of the trained part of the forelimb in motor cortex; rats without VNS did not have changes in motor cortex maps.[41] Following these plasticity studies in uninjured rats, VNS has been used to promote recovery of motor function after brain injury, including both ischemic and hemorrhagic stroke. VNS enabled large-scale recovery of function associated with cortical plasticity in adult rat stroke models.[42] Like the tinnitus studies, the motor recovery work has been translated to human trials of motor recovery after stroke (ClinicalTrials.gov NCT01669161).

How might VNS modulate cortical plasticity? This work follows plasticity studies with invasive stimulation of the nucleus basalis,[43] which sends cholinergic projections diffusely throughout cortex and enables plasticity. To determine if the effects of VNS are mediated by the cholinergic system, auditory plasticity was measured with and without the antimuscarinic drug scopolamine.[44] The plasticity effects of VNS were blocked in the presence of scopolamine, suggesting that its effects are mediated by activation of muscarinic receptors. Ongoing work is exploring the system neuroscience mechanisms through which VNS acts.

CHALLENGES AND OPPORTUNITIES

The central hypothesis of this chapter, that effective brain stimulation requires the right stimulation of the right circuits, can only be tested when the effects of brain stimulation are better understood. Our focus should be on advancing that understanding, particularly in the developing brain. One challenge that has been partially addressed is how current moves through a child's head compared to that of an adult. Studies have estimated current flow using computational models based on children's anatomy as measured with MRI.[35,45,46] These modeling studies show that a smaller head, thinner CSF space, and less subcutaneous tissue result in much larger effective brain currents in children than in adults. In addition, lesions of the brain cause significant shifts in current flow, with important implications, particularly for targets near the brain lesion.

While the density of current can be estimated, how that current may affect the developing brain is also not well known. The developing brain differs from the mature brain in several ways that are likely to affect responses to brain stimulation. The pediatric brain undergoes changes in myelination through adolescence, and responses to phasic brain stimulation are likely to be muted in neural systems that are incompletely myelinated. For example, both the corticospinal tract[47] and corpus callosum[48] do not become fully myelinated until adolescence. This may explain why muscle responses to TMS stimulation of motor cortex are difficult to elicit in young children. In addition, development is a time of great change in various neurotransmitter systems and the responsiveness of neurons to those chemicals. Insofar as brain stimulation modifies neurotransmitter systems differently, this could significantly alter the response. Finally, there is evidence that lasting modulation of neural circuits occurs due to changes at the synapse. During development, there is an active remodeling of synaptic connections. Plasticity is likely altered during this time, with significant implications for brain stimulation.

How then can the field close the gaps in understanding about how stimulation may affect the developing brain? Brain stimulation has largely relied on human studies to understand mechanisms, using measures of brain physiology, imaging, and pharmacology. Mechanistic trials in children can also make significant advances with these tools, although the use of pharmacology is likely to be much more restricted. Non-invasive neurophysiology tools, including TMS and EEG, can assay changes in brain maps and rhythms. Improvements in neuroimaging of brain function and structure, including measurement of metabolites with magnetic resonance spectroscopy, could also yield important insights. It will also be important to have very sensitive tools to measure behavior in order to measure possible therapeutic application and also to monitor for possible deleterious effects.

Finally, animal models will be important to better define the effects of brain stimulation on developing neural circuits. The vital work that has been done with whole-animal studies and brain slices in adult animals should also be performed in young animals. Studies of animals and brain slices allow the invasive monitoring of neural circuits and can specifically measure physiology, metabolism, intracellular signaling, and genetic changes. These studies are critical for building understanding of stimulation from cells, to synapses, to circuits, and ultimately to behavior.

Brain stimulation uses the language of the nervous system, electricity, to alter neural circuits. Stimulation achieves modulation, at least in part, by altering neural activity. Neural activity is crucial for patterning the nervous system during development and for allowing experience to alter neural circuits in maturity. We have explored how stimulation alters neural circuits under a variety of conditions. Stimulation can be highly focal, targeting the pea-sized subthalamic nucleus in PD, or diffuse, with conventional tDCS or VNS. Knowing which of these approaches is best for modulation in children with brain injury or disease will involve an iterative process of advancing our basic understanding of brain stimulation with knowledge of practical application to children.

Acknowledgment

The authors thank Anil Sindhurakar for preparing the figure.

References

1. Follett KA, et al. Pallidal versus subthalamic deep-brain stimulation for Parkinson's disease. *N Engl J Med.* 2010;362:2077–2091.
2. Odekerken VJ, et al. Subthalamic nucleus versus globus pallidus bilateral deep brain stimulation for advanced Parkinson's disease (NSTAPS study): a randomised controlled trial. *Lancet Neurol.* 2013;12:37–44.
3. Schuepbach WM, et al. Neurostimulation for Parkinson's disease with early motor complications. *N Engl J Med.* 2013;368:610–622.
4. Hornykiewicz O. Dopamine miracle: from brain homogenate to dopamine replacement. *Mov Disord.* 2002;17:501–508.

5. Baron MS, Wichmann T, Ma D, DeLong MR. Effects of transient focal inactivation of the basal ganglia in Parkinsonian primates. *J Neurosci*. 2002;22:592–599.
6. Benabid AL, et al. Long-term suppression of tremor by chronic stimulation of the ventral intermediate thalamic nucleus. *Lancet*. 1991;337:403–406.
7. Moro E, et al. Long-term results of a multicenter study on subthalamic and pallidal stimulation in Parkinson's disease. *Mov Disord*. 2010;25:578–586.
8. de Hemptinne C, et al. Therapeutic deep brain stimulation reduces cortical phase-amplitude coupling in Parkinson's disease. *Nat Neurosci*. 2015;18(5):779–786.
9. Little S, et al. Adaptive deep brain stimulation in advanced Parkinson disease. *Ann Neurol*. 2013;74:449–457.
10. Marks WA, Honeycutt J, Acosta F, Reed M. Deep brain stimulation for pediatric movement disorders. *Semin Pediatr Neurol*. 2009;16:90–98.
11. Kumar R, Dagher A, Hutchison WD, Lang AE, Lozano AM. Globus pallidus deep brain stimulation for generalized dystonia: clinical and PET investigation. *Neurology*. 1999;53:871–874.
12. Himmelmann K, Hagberg G, Beckung E, Hagberg B, Uvebrant P. The changing panorama of cerebral palsy in Sweden. IX. Prevalence and origin in the birth-year period 1995–1998. *Acta Paediatr*. 2005;94:287–294.
13. Porter R, Lemon RN. *Corticospinal Function and Voluntary Movement*. Oxford, England: Oxford University Press; 1993.
14. Friel KM, Kuo HC, Carmel JB, Rowny SB, Gordon AM. Improvements in hand function after intensive bimanual training are not associated with corticospinal tract dysgenesis in children with unilateral cerebral palsy. *Exp Brain Res*. 2014;232:2001–2009.
15. Bleyenheuft Y, Grandin CB, Cosnard G, Olivier E, Thonnard JL. Corticospinal dysgenesis and upper-limb deficits in congenital hemiplegia: a diffusion tensor imaging study. *Pediatrics*. 2007;120:e1502–e1511.
16. Martin J, Friel K, Salimi I, Chakrabarty S. Corticospinal development. In: Squire L, ed. *Encyclopedia of Neuroscience*. vol. 3. Oxford: Academic Press; 2009:302–314.
17. Friel KM, Williams PT, Serradj N, Chakrabarty S, Martin JH. Activity-based therapies for repair of the corticospinal system injured during development. *Front Neurol*. 2014;5:229.
18. Staudt M. Reorganization after pre- and perinatal brain lesions. *J Anat*. 2010;217:469–474.
19. Eyre JA, et al. Is hemiplegic cerebral palsy equivalent to amblyopia of the corticospinal system? *Ann Neurol*. 2007;62:493–503.
20. Gordon AM, et al. Bimanual training and constraint-induced movement therapy in children with hemiplegic cerebral palsy: a randomized trial. *Neurorehabil Neural Repair*. 25:692–702.
21. Garvey MA, et al. Cortical correlates of neuromotor development in healthy children. *Clin Neurophysiol*. 2003;114:1662–1670.
22. Perez MA, Cohen LG. Interhemispheric inhibition between primary motor cortices: what have we learned? *J Physiol*. 2009;587:725–726.
23. Di Lazzaro V, et al. Direct demonstration of interhemispheric inhibition of the human motor cortex produced by transcranial magnetic stimulation. *Exp Brain Res*. 1999;124:520–524.
24. Lee H, Gunraj C, Chen R. The effects of inhibitory and facilitatory intracortical circuits on interhemispheric inhibition in the human motor cortex. *J Physiol*. 2007;580:1021–1032.
25. Stagg CJ, et al. Polarity-sensitive modulation of cortical neurotransmitters by transcranial stimulation. *J Neurosci*. 2009;29:5202–5206.
26. Lindenberg R, Renga V, Zhu LL, Nair D, Schlaug G. Bihemispheric brain stimulation facilitates motor recovery in chronic stroke patients. *Neurology*. 2010;75:2176–2184.
27. Clarkson AN, Huang BS, Macisaac SE, Mody I, Carmichael ST. Reducing excessive GABA-mediated tonic inhibition promotes functional recovery after stroke. *Nature*. 2010;468:305–309.

28. Kirton A, et al. Contralesional repetitive transcranial magnetic stimulation for chronic hemiparesis in subcortical paediatric stroke: a randomised trial. *Lancet Neurol.* 2008;7(6):507–513.

29. Kirton A, Deveber G, Gunraj C, Chen R. Cortical excitability and interhemispheric inhibition after subcortical pediatric stroke: plastic organization and effects of rTMS. *Clin Neurophysiol.* 2010;121:1922–1929.

30. Gillick BT, et al. Primed low-frequency repetitive transcranial magnetic stimulation and constraint-induced movement therapy in pediatric hemiparesis: a randomized controlled trial. *Dev Med Child Neurol.* 2014;56:44–52.

31. Purpura DP, McMurtry JG. Intracellular activities and evoked potential changes during polarization of motor cortex. *J Neurophysiol.* 1965;28:166–185.

32. Fritsch B, et al. Direct current stimulation promotes BDNF-dependent synaptic plasticity: potential implications for motor learning. *Neuron.* 2010;66:198–204.

33. Salimi I, Friel KM, Martin JH. Pyramidal tract stimulation restores normal corticospinal tract connections and visuomotor skill after early postnatal motor cortex activity blockade. *J Neurosci.* 2008;28:7426–7434.

34. Reis J, et al. Noninvasive cortical stimulation enhances motor skill acquisition over multiple days through an effect on consolidation. *Proc Natl Acad Sci USA.* 2009;106:1590–1595.

35. Kessler SK, et al. Dosage considerations for transcranial direct current stimulation in children: a computational modeling study. *PLoS One.* 2013;8:e76112.

36. Kuo HI, et al. Comparing cortical plasticity induced by conventional and high-definition 4 × 1 ring tDCS: a neurophysiological study. *Brain Stimul.* 2013;6:644–648.

37. Fornai F, Ruffoli R, Giorgi FS, Paparelli A. The role of locus coeruleus in the antiepileptic activity induced by vagus nerve stimulation. *Eur J Neurosci.* 2011;33:2169–2178.

38. Reed A, et al. Cortical map plasticity improves learning but is not necessary for improved performance. *Neuron.* 2011;70:121–131.

39. Engineer ND, et al. Reversing pathological neural activity using targeted plasticity. *Nature.* 2011;470:101–104.

40. Engineer ND, Moller AR, Kilgard MP. Directing neural plasticity to understand and treat tinnitus. *Hear Res.* 2013;295:58–66.

41. Porter BA, et al. Repeatedly pairing vagus nerve stimulation with a movement reorganizes primary motor cortex. *Cereb Cortex.* 2012;22:2365–2374.

42. Khodaparast N, et al. Vagus nerve stimulation delivered during motor rehabilitation improves recovery in a rat model of stroke. *Neurorehabil Neural Repair.* 2014;28(7):698–706.

43. Kilgard MP, Merzenich MM. Cortical map reorganization enabled by nucleus basalis activity. *Science.* 1998;279:1714–1718.

44. Nichols JA, et al. Vagus nerve stimulation modulates cortical synchrony and excitability through the activation of muscarinic receptors. *Neuroscience.* 2011;189:207–214.

45. Minhas P, Bikson M, Woods AJ, Rosen AR, Kessler SK. Transcranial direct current stimulation in pediatric brain: a computational modeling study. *Conf Proc IEEE Eng Med Biol Soc.* 2012;2012:859–862.

46. Gillick BT, Kirton A, Carmel JB, Minhas P, Bikson M. Pediatric stroke and transcranial direct current stimulation: methods for rational individualized dose optimization. *Front Hum Neurosci.* 2014;8:739.

47. ten Donkelaar HJ, et al. Development and malformations of the human pyramidal tract. *J Neurol.* 2004;251:1429–1442.

48. Olivares R, Montiel J, Aboitiz F. Species differences and similarities in the fine structure of the mammalian corpus callosum. *Brain Behav Evol.* 2001;57:98–105.

CHAPTER

11

Therapeutic Brain Stimulation Trials in Children With Cerebral Palsy

B.T. Gillick

University of Minnesota, Minneapolis, MN, United States

K.M. Friel

Burke-Cornell Medical Research Institute, White Plains, NY, United States;
Weill Cornell Medical College, New York, NY, United States

J. Menk, K. Rudser

University of Minnesota, Minneapolis, MN, United States

OUTLINE

INTRODUCTION

Therapeutic brain stimulation trials involving children with cerebral palsy (CP) suggest potential for neuroplastic change in what historically has been defined as a permanent or static diagnosis of motor deficits.[1] Since the application of brain stimulation in CP is in a nascent phase, investigation of safety, feasibility, and efficacy of such applications is paramount. Results should be considered preliminary when advocating for clinical adoption until further validation is obtained. In an optimal research environment, we would facilitate comparisons between studies by creating synchrony in how we develop trials and report outcomes. This would lead to a more complete understanding of findings and a clearer assessment of the benefits of such interventions in children with CP.

"Therapeutic" in the context of brain stimulation refers to an intervention of one or more sessions using brain stimulation through the power of electricity, magnetism, or implantation (whether done invasively through surgery or implantation in the brain or non-invasively through the skull). As discussed in other chapters in this book, the forms of pediatric brain stimulation currently in use involve interventions that are either surgical (eg, deep brain stimulation, see Chapter 19: Deep Brain Stimulation in Children: Clinical Considerations and Chapter 20: Deep Brain Stimulation Children—Surgical Considerations) or non-surgical (eg, Chapter 4: Therapeutic RTMS in Children and Chapter 5: Transcranial direct current stimulation (tDCS): Principles and emerging applications in children). Few brain stimulation trials specific to children with CP have been reported to date (Table 11.1). However, the incidence of children born with

TABLE 11.1 Brain Stimulation Trials in Children With CP to Date

Brain Stimulation Studies in Cerebral Palsy

Studies	Study Design
	Stated Diagnosis
Repetitive Transcranial Magnetic Stimulation	
Valle et al.[6]	Cerebral palsy with spastic quadraplegia
Kirton et al.[8]	Arterial ischemic stroke with congenital hemiparesis
Gillick et al.[9]	Ischemic stroke or periventricular leukomalacia with congenital hemiparesis
Transcranial Direct Current Stimulation	
Young et al.[15]	Dystonia/cerebral palsy
Aree-uea et al.[21]	Spastic cerebral palsy
Duarte et al.[18]	Spastic cerebral palsy
Grecco et al.[14,17]	Spastic diparetic cerebral palsy
Young et al.[16]	Dystonia/cerebral palsy
Grecco et al.[14,17]	Spastic cerebral palsy
Bhanpuri et al.[27]	Secondary dystonia/cerebral palsy
Bhanpuri et al.[27]	Secondary dystonia/cerebral palsy
Grecco et al.[20]	Spastic diparetic cerebral Palsy
Lazzari et al.[19]	Cerebral palsy
Gillick et al.[13]	Hemispheric stroke or periventricular leukomalacia with congenital hemiparesis

Continued

TABLE 11.1 Brain Stimulation Trials in Children With CP to Date—cont'd

Study Criteria

Inclusion: Ages 5–18, upper limb plasticity score of ≧1 on Ashworth, exclusion: rTMS contraindications, Ashworth score 5 and contractures, uncontrolled epilepsy

Inclusion: Age at stroke between 1 month and 17 years, single arterial ischemic stroke of posterior limb internal capsule, formal neurological follow-up for 2 years since stroke, pediatric stroke outcome measure score of >/–1 Current age <7 years. Excluded if neonatal or unknown age of stroke, seizures for more than 1 month poststroke, movement disorder, use of drugs which could alter cortical excitability, disease with ongoing risk of stroke

Inclusion: Ages 8–17, ≧10 degrees of active finger movement, presence of a motor-evoked potential from ipsilesional motor cortex. Exclusion: Contraindications to TMS, seizures in last 2 years, Botox within last 6 months, current therapies

Study Criteria

Inclusion: Primary or secondary dystonia affecting one or both hands. Exclusion: Upper extremity spasticity

Inclusion: CP and spasticity of right upper limb, Gross Motor Function Classification System (GMFCS) levels II–IV, age 8–18, Modified Ashworth Scale score 1–3. Exclusion: Severe spasticity and contractures, mental disorders, concomitant therapies, pregnancy, orthopedic surgery on upper limb, initiation or change in dosage of oral antispastic drug in last 5 days, Botox 3 months prior.

Inclusion: GMFCS I–III, 12 months independent gait, ages 5–10 years, able to comprehend examination. Exclusion: Surgery or neurolytic block in previous 12 months, orthopedic deformities, epilepsy, metal implants in skull or hearing aids

Inclusion: GMFCS I–III, 12 months independent gait, ages 5–10 years, able to comprehend examination. Exclusion: Surgery or neurolytic block in previous 12 months, orthopedic deformities, epilepsy, metal implants in skull or hearing aids

Inclusion: Primary or secondary dystonia affecting one or both hands. Exclusion: Upper extremity spasticity

Inclusion: GFMCS I–III, 12 months of independent gait, age 6–10 years, able to comprehend examination. Exclusion: Surgery or neurolytic block in previous 12 months, orthopedic deformities, epilepsy, metal implants in skull or hearing aids

Inclusion: Primary or secondary dystonia affecting one or both hands. Exclusion: None stated

Inclusion: GMFCS I-III, 12 months independent gait, ages 5-10 years, able to comprehend examination. Exclusion: Surgery or neurolytic block in previous 12 months, orthopedic deformities, epilepsy, metal implants in skull or hearing aids

Inclusion: GMFCS I-III, 12 months independent gait, ages 4-12 years, able to comprehend examination. Exclusion: Surgery or neurolytic block in previous 12 months, orthopedic deformities, epilepsy, metal implants in skull or hearing aids

Inclusion: Ages 7-18, 10 degrees of active finger movement, cognitively able to follow commands, no seizures within 2 previous years. Exclusion: TMS contraindications, skin disease or abnormalities, Botox/phenol block in previous 6 months

| Sample Size (n) | Mean [SD]Age (Years, Months) | Sham Controlled | Protocol | | Location | Total Sessions |
			Blinding	Intensity		
17	9.1 [3.2]	Yes	Double blind	90% RMT	Motor cortex	5
10	Median 13.25 IQR 10.08–16.78	Yes	Partially	100% RMT	Contralesional Motor cortex	8
19	10.10 [2.1]	Yes	Double blind	90% RMT	Contralesional Motor cortex	5

| Sample Size (n) | Mean [SD]Age (Years, Months) | Sham Controlled | Protocol | | Montage | Electrode Site |
			Blinding	Intensity		
11	13.1 [4.0]	No	None	1.0 mA	M1-SO	M1 cathodal
46	Active: 13.0 [3.2], sham: 14.0 [3.0]	Yes	Double blind	1.0 mA	M1-SO	M1 anodal
24	Experimental: 7.8 [2.0], control: 8.1 [1.5]	Yes	Double blind	1.0 mA	M1-SO	M1 anodal
24	Experimental: 7.8 [3.0], control: 8.0 [2.2]	Yes	Double blind	1.0 mA	M1-SO	M1 anodal

Continued

TABLE 11.1   Brain Stimulation Trials in Children With CP to Date—cont'd

Sample Size (n)	Mean [SD]Age (Years, Months)	Sham Controlled	Protocol			
			Blinding	Intensity	Location	Total Sessions
14	12.6 [3.8]	Yes	Double blind	1.0 mA	M1-SO	M1 cathodal
20	Experimental: 7.2 [1.8], control: 7.8 [1.5]	Yes	Double blind	1.0 mA	M1-SO	M1 anodal
9	15.3 [4.2]	Yes, cross-over	Double blind	2.0 mA	M1-SO	Experiment#1: M1 cathodal
9	15.3 [4.2]	Yes, cross-over	Double blind	2.0 mA	M1-SO	Experiment#2: M1 anodal
20	Experimental 8.2 (1.6), control: 8.8 (1.1)	Yes	Double blind	1.0 mA	M1-SO	M1 anodal
20		Yes	Double blind	1.0 mA	M1-SO	M1 anodal
13	Intervention 16.9 [2.6], control 10.7 [3.2]	Yes	Double blind	0.7 mA	C3-C4	Ipsilesional M1 anodal, contralesional M1 Cathodal

Frequency	Session Duration (Min)	Pulses	Primary Outcome
5 Hz, 1 Hz or sham	15, 25, 15 respectively	1500	Passive range of Motion of upper extremity
1 Hz	20	1200	Melbourne assessment of upper extremity function, grip strength
6-Hz priming followed by 1 Hz	20 total (10 priming: 10 1 Hz)	1200 total	Assisting hand assessment

Total Sessions	Time (min)		
1	18 total (9 on: 20 off: 9 on)		Electromyogram tracking task
5	20		Ashworth scale of spasticity
10	20		Pediatric balance scale (PBS)
10	20		Gait analysis
1	18 (9 on: 20 off: 9 on)		Electromyogram tracking task
1	20		Spatiotemporal gait variables
5	9		Electromyogram tracking tasks
5	9		
10	20		Gait analysis
1	20		Stabilometric analysis
1	10		Symptom survey

Secondary Outcome

Ashworth Scale muscle tone of upper extremity and subjective report

Purdue Pedg Board test, Halstead–Reitan finger tapping, in-hand manipulation

Canadian Occupational performance measure, 12-object stereognosis, finger extension force

Passive range of motion

Stabilometric analysis, Pediatric Evaluation of Disability Inventory (PEDI)

6-min Walk test (6WT), Gross Motor Function Measure (GMFM), treadmill test, motor-evoked potential

Stabiliometric analysis

Barry Albright Dystonia Scale

Continued

II. NIBS IN PEDIATRIC NEUROLOGICAL CONDITIONS

TABLE 11.1   Brain Stimulation Trials in Children With CP to Date—cont'd

Secondary Outcome

Gross Motor Function Measure (GMFM), Pediatric Evaluation Disability Inventory (PEDI), motor-evoked potentials (MEPs)

Mobility training

Participant survey, vital signs, Modified Pediatric Stroke Outcome Measure

Results

(Clarify primary finding, additional findings, effect size)

Significant reduction of spasticity after 5-Hz rTMS, but not 1 Hz or sham, as indexed by degree of passive movement. No significant difference between groups in Ashworth scores. No significant differences in subjective reports

Significant differences in MAUEF scores, grip strength. Data not fully assessed for secondary outcomes due to range in severity

Significant differences in AHA, no significant differences noted between groups in secondary outcomes

Outcomes

Decreased non-task muscle overflow in three of 10 participants, non-significant

Significant decrease in upper limb spasticity and shoulder PROM

Significantly improved PBS and stabilometric values, and PEDI mobility and self-care scores

Significant improvements in Gait velocity, cadence, gait profile scores, gait variability scores, 6'WT, and MEPs

Significant reduction in overflow

Gait improvements, reduction in oscillatory sway

Cathodal group tracking error/overflow: three subjects improved, two worsened. Anodal group tracking error/overflow: none improved, five worsened. No changes in dystonia scale.

Significant improvements in gait, GMFM scores and increase in MEP which was not sustained in 7-week follow-up

Significant improvements in sway velocity

No significant improvements nor decline in function between or within groups

Effect Size

Velocity (0.2), cadence (19.3), GPS (−4.8), GVS (−3.5 and −6.4), 6'WT (87.8), MEP (0.3)

Estimate 0.36% of maximum voluntary contraction; 95% confidence interval: 0.62%, 0.10%; $F(1, 525) = 7.34$, $P = 0.007$

Cathodal 1% maximum voluntary contraction.

Reported Adverse Events

"All patients were able to tolerate the TMS treatments, with no incidence of seizures"

"No serious adverse events reported. Two patients had single episodes of neurocardiogenic syncope with their initial exposure to TMS, two patients reported mild headache, one nausea and neck stiffness"

"No serious adverse event were reported; the most common minor adverse events were self-limiting headache, which resolved within 24h after rTMS, and cast irritation."

Reported Adverse Events

Statements "One child complained of discomfort and withdrew from study" "One child was comfortable after the experimenters reduced the stimulator current"

One drop out, one child with rash and mild skin burn which resolved in 3 days

Redness ($n = 3$), tingling ($n = 18$)

Statement "No adverse effects were found"

No serious adverse events reported. Report of "discomfort" in three participants which abided after adjusted intensity.

Redness and tingling ($n = 3$) in experimental group

No serious adverse events reported. Tingling for "most patients". Two cases Reduced stimulation to 1.5 due to skin discomfort, mild headache, $n = 1$

Statement "No child experienced any serious adverse event throughout the study. Four children reported mild tingling with anodal tDCS"

None mentioned.

Statement "No serious adverse events, including seizure, occurred." Most common side effects itchiness, burning, sleepiness, difficulty concentrating.

CP is approximately 2 per 3000 live births, each incurring estimated lifetime healthcare costs of over 1 million dollars.[2] With such a prevalent and costly diagnosis, the potential benefit of both improving motor function and decreasing burden of care underscores brain stimulation as a promising treatment for children with CP.

A comprehensive discussion of the design, monitoring, and analysis of therapeutic trials and dosing parameters lies outside of the purview of this chapter. An overview of such clinical trials in general may be found elsewhere,[3] along with guidance documents from regulatory governing bodies and international harmonization efforts such as the U.S. Food and Drug Administration (FDA) and The International Conference on Harmonisation of Technical Requirements for Registration of Pharmaceuticals for Human Use (ICH).[4,5] Not all trials applying brain stimulation in CP are randomized, controlled, behavioral clinical trials and other forms of research can also be informative. Non-randomized trials, such as observational or "naturalistic" studies, may ask fundamentally different questions and still, through responsible rigor, yield valuable and reliable information about the safety, feasibility, and efficacy of brain stimulation applications. Decisions relating to this nascent phase of brain stimulation, however, should be based on the best available scientific evidence. Therefore, the purpose of this chapter, focusing specifically on brain stimulation intervention trials in CP, is to discuss current trials by investigating study design as a framework for future research in CP and related pediatric conditions.

SCIENTIFIC EVIDENCE IN THERAPEUTIC BRAIN STIMULATION IN CHILDREN WITH CP: SUMMARY OF CURRENT INTERVENTIONAL TRIALS

In this chapter, trials of the application of therapeutic brain stimulation in CP as serial interventions using rTMS or tDCS are primarily discussed (Table 11.1, Fig. 11.1). Although investigated in children with CP, incorporating TMS as a cortical excitability measure, as well investigating deep brain stimulation and transcranial electric stimulation have been discussed in other chapters in greater detail.

rTMS

The application of repetitive trains of electromagnetic current, rTMS, has been investigated in children with CP both with and without accompanying motor rehabilitation interventions. In a randomized, double-blinded, sham-controlled study, Valle et al.[6] investigated the influence of five sessions of rTMS on spasticity in 17 children (mean [SD] age 9

(A)   **(B)**

FIGURE 11.1   Non-invasive brain stimulation applications in children. (A) Child receiving transcranial direct-current stimulation in an M1-SO montage while performing a behavioral activity. (B) Child during a transcranial magnetic stimulation testing session with stereotactic neuronavigation.

[3] years) with CP and spastic quadriplegia. The authors found that passive movement assessment of spasticity was significantly reduced after 5-Hz excitatory rTMS over the motor cortex. The side of stimulation was not reported. The authors stated that no adverse events occurred.

In another small randomized, controlled trial, eight sessions of 1-Hz low-frequency rTMS were applied to the contralesional motor cortex in 10 children (median age 13 years, range 8–20 years) with hemiparesis due to childhood arterial ischemic stroke. Significant improvements in paretic hand function lasting beyond the treatment period were suggested with favorable tolerability.[7] In addition, exploration of cortical neurophysiology using TMS suggested imbalances in interhemispheric inhibition may be present and altered with rTMS treatment. Stimulus–response curves could only be obtained in seven of the 10 children. As the contralesional hemisphere was stimulated with low-frequency (inhibitory) rTMS, the unaffected hand was also evaluated and no changes were noted. No serious adverse events were reported although two participants had episodes of neurocardiogenic syncope during their first exposure to single-pulse TMS.[8] Of these two participants, one withdrew from the study and recovered completely within 24 h. The other recovered almost immediately after a meal and recommenced involvement in the intervention. Both were found to have previous history of vasovagal syncope.

In a double-blind, randomized controlled trial of 19 children between the ages of 8 and 17 years with congenital hemiparesis due to perinatal stroke or periventricular leukomalacia, Gillick et al.[9] employed five sessions of 6-Hz primed 1-Hz low-frequency contralesional rTMS combined with constraint-induced movement therapy. Statistically significant differences were found between the two groups on the primary outcome measure of the Assisting Hand Assessment. No serious adverse events occurred, and all participants completed the study. Minor adverse events included headache and constraint-induced movement therapy (CIMT) cast irritation, which resolved within 24 h.[10]

As evidenced by the paucity of studies surrounding rTMS in children with cerebral palsy, further studies in larger populations and with defined dosing parameters are indicated to define the effects of these interventions. As has been found in reports of the effects of rTMS in adults with stroke,[11] the limited rTMS sample sizes reported earlier (range 10–19), with limited power and variable reported effect sizes support the incorporation of well-designed studies set to elucidate the potential role of rTMS in neurorecovery.

tDCS

tDCS as a mode of delivery of a constant direct current has recently been applied in children with CP to investigate its potential to reduce or increase motor cortex excitability for therapeutic effect.[12] Depending upon the type of CP and clinical manifestations, protocols and outcomes have varied.

Studies have reported analysis of outcomes of single sessions of tDCS in children with CP. In a pilot safety study of 13 children, ages 7–18 years, with congenital hemiparesis due to hemispheric stroke, a single session of 10 min, 0.7-mA tDCS was applied in a bihemispheric (anodal ipsilesional, cathodal contralesional) montage over both motor cortices.[13] Two children withdrew from the study, one due to complaints of discomfort in the area of the electrodes and one due to a lack of ipsilesional motor threshold presence as this was required for continuation in the study. No significant functional (Grip-Strength Dynamometry and Box and Blocks Test) or cognitive (TOKEN test of intelligence) declines in performance were noted. No major adverse events were reported. Minor adverse events included itchiness ($n=1$, intervention group) and burning ($n=1$, control group) sensations under the tDCS electrodes. The authors concluded that performing future tDCS studies is feasible and safe within the investigated parameters in this CP population. A study assessing a single tDCS session on balance and gait and balance in 20 children, ages 6–10 was performed.[14] Participants received M1-SO anodal tDCS of 1.0 mA over M1 of the dominant hemisphere for a 20-min session. Significant reductions in oscillatory sway were found in the experimental group with significant intraexperimental group improvements in gait velocity, cadence, and oscillations in standing.

Single-session analysis has led to multiday/session investigations within the same lab. In an open-label trial of 11 children, ages 7–18 years, with "dystonia," specific types of CP included varied from "dyskinetic" to "right hemiplegic."[15] This study applied tDCS to the side of the scalp contralateral to the hand most affected by dystonia in a primary motor cortex/supraorbital (M1-SO) montage for two 9-min sessions at 1.0 mA, with a 20-min break between sessions. Although the results evaluating tracking error and overflow muscle activity did not reach statistical significance, three children showed a reduction in dystonic symptoms and one more showed a reduction in tracking error within a testing session 5 min following the stimulation. No serious adverse events were reported, although two children reported electrode site discomfort with one withdrawing from the study and the other reporting comfort after reduction of the tDCS stimulator to 0.65 mA. In a subsequent double-blinded sham-controlled crossover study of 14 children, ages 7–19 years, with dystonia and related diagnoses, an M1-SO tDCS montage was also employed with the same timing and intensity protocol.[16] A significant reduction in dystonic symptoms was reported in the "real stimulation" group, with improved hand task performance contralateral to the cathode. No serious adverse events were reported although three participants required the stimulator intensity to be lowered due to "discomfort" at the electrode site. Building upon these results, a multiday study in 10 children, ages 10–21 years, with primary or secondary dystonia, received 10 total tDCS sessions over 2 weeks. In an M1-SO montage, participants received either cathodal or anodal stimulation, on the hemisphere contralateral to the affected hand, for five consecutive visits, and then received sham stimulation for five visits on each day of the other week. An electromyogram tracking task measured tracking error and overflow to the muscles not involved in the tracking. In the cathodal group of seven total participants, an improvement of tracking error was found, yet there was a worsening of overflow. In the anodal group of six participants, a worsening of tracking error was found but an improvement in overflow. The effects in both groups were found to be small, of less than 1% maximum voluntary contraction. Individual analysis revealed changes deemed too small to be clinically relevant.

tDCS has also been reported to be coupled with rehabilitation therapies in real time. A double-blind, randomized controlled trial of 24 children (no range reported, mean age 7.8 years experimental group, 8.0 years control) with heterogeneous spastic diplegic CP affecting the lower limbs employed tDCS in conjunction with treadmill training.[17] Participants received M1-SO anodal tDCS of 1.0 mA over M1 of the dominant hemisphere during 10 treadmill training sessions of 20 min each over 2 weeks. The intervention group was reported to make significantly greater gains in gait velocity, cadence, gait profile scores, gait variability scores, 6-min walk test distances, and motor-evoked potential amplitudes in the quadriceps musculature. Changes were reported to be maintained after 1 month. No adverse effects were reported. From this same lab, another 2014 study of 24 children, ages 5–12 years, with CP were investigated with similar coupling of 10 sessions of 20 min of 1.0 mA anodal M1 tDCS and treadmill training.[18] The authors reported significantly improved stabilometric values, Pediatric Balance Scale (PBS) scores, Pediatric Evaluation of Disability Inventory (PEDI) mobility, and self-care scores. This same lab has reported both a single session and multiple sessions of tDCS and combined with virtual reality training.[19,20] In a study involving 20 children with diagnosis of cerebral palsy (age range or mean not reported), participants were randomly assigned to an active or placebo tDCS group while participating in virtual reality training.[19] The tDCS parameters were 1.0 mA in an M1-SO montage for 20 min with the anode placed over the primary motor cortex. The designated hemisphere was not described. Static balance measured by a force place revealed significant increases in sway velocity in both groups. In the most recent tDCS publication from this lab, 20 children, mean ages 8.2 (experimental group) and 8.8 (control group), received 10 sessions of tDCS combined with a training program with virtual reality.[20] The tDCS protocol involved an M1-SO montage set to 1.0 mA for 20 min over the primary motor cortex which the authors deemed "responsible" for controlling the limb which the child felt displayed greater deficits during gait. Outcomes revealed significant improvements in gait, Gross Motor Function Measure scores, and an increase in motor-evoked potential amplitude which were not sustained at 7-week follow-up.

Another lab coupling tDCS with a behavioral intervention recently completed a double-blind, randomized control trial of 46 children, with cerebral palsy, mean ages 13.0 (experimental group) and 14.0 (control group).[21] Participants received five sessions of tDCS at 1.0 mA with the anode over the left M1 and the cathode in the supraorbital region for 20 min during simultaneous physical therapy. The experimental group revealed a significant decrease in upper-limb spasticity and shoulder passive range of motion. Adverse event details were not assessed by formal clinical evaluation but rather by subjective family reporting. No serious adverse events occurred and minor adverse events included acute mood changes, irritability, tingling, itching, headache, burning sensation, sleepiness, and trouble concentrating, all of which were described as mild and self-limiting. The authors concluded that in this population performing a study with tDCS is feasible with potentially mild adverse events.

From the tDCS trials mentioned earlier, variability in both the dosing and the outcomes measured are explored. As with all forms of NIBS, investigations surrounding the unknown effects, side effects, optimal dosing parameters, and translational studies between adult and children are indicated.[22]

TRIAL DESIGN

As neuromodulation technologies continue in their development in children with CP, a multifaceted design of such trials is imperative. As the studies noted above suggest both feasibility and minimal adverse events, they invite exploration of brain stimulation as an intervention to influence function. Specific to the pediatric population with CP, various considerations need to be addressed in designing optimal efficacy trials. Beginning with consideration of the child, the success of the study hinges on incorporation of each subsequent component (Fig. 11.2). These components are discussed in the remaining sections in this chapter.

Randomized clinical trials (RCTs) are complex endeavors with numerous moving parts to plan, implement, monitor, and adjust. The goal is to conduct trials of potentially beneficial treatments in a manner which maintains scientific credibility, ethical integrity, and efficiency in both time and cost. A challenge in the design of such trials is to strike the appropriate balance among the competing goals from scientists, clinicians, participants

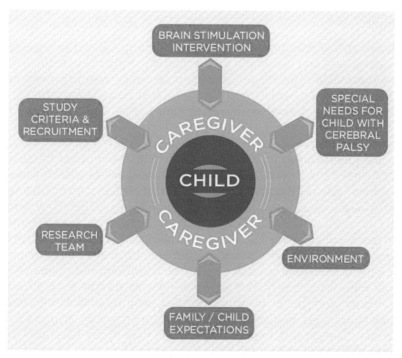

FIGURE 11.2 The caregiver/child with CP interaction in brain stimulation trials. With the child at the center of the trial, caregiver support is central to trial interactions with the stimulation intervention, special needs, environment, expectations, research team, study criteria, recruitment, and subsequent scientific and clinical/individualized outcomes.

and their families, and regulatory groups, all under ethical, logistical, and economic pressures that need to be simultaneously accommodated. Components central to establishing appropriate therapeutic brain stimulation trials in children with CP include the following.

Environment

Although the translational goal of this research is to establish clinical usage, current trials of brain stimulation in children with CP will occur within academic centers or hybrid rehabilitation/academic settings. Access to advanced neuromodulation equipment, trained users, neuro-imaging, plus appropriate research event-related medical care is essential to investigating the effectiveness along with safety and feasibility of brain stimulation interventions. Given limitations of time and funding, the pursuit of multicenter research to pool participant numbers and resources may be of tremendous benefit. Although there may be added variability from inclusion of different sites, it is outweighed by higher sample size and greater generalizability.

Timing

Although the critical period for intervention in CP and other neurologic diagnoses may occur at a young age, exploring the potential effect of brain stimulation on the younger brain demands recognition of potential lifelong developmental impact.[23] The immediate and long-term effects of brain stimulation intervention must therefore be explored, beginning with older children and working into earlier developmental timeframes as safety and tolerability evidence accumulates. In addition, as these children may be involved in other medical care, it is important to include assessment of concurrent rehabilitation interventions as well as surgical procedures, spasticity management, and pharmacologic seizure management. These aspects of care of a child with CP potentially affect behavioral outcomes, as well as cortical excitability. As an example of a research team recognizing these potentially confounding interventions, Valle et al.[6] reported in their rTMS study that the number of patients taking centrally acting medications did not differ across the 1-Hz, 5-Hz and sham groups. Future trials could incorporate such components such as a stratification factor in randomization to protect against potential imbalances across treatment arms. As another example, botulinum toxin, as used in CP for spasticity management, has been reported to have negligible effects after 6 months.[24] Consideration of this wash-out period may indicate the appropriate time at which to enroll a child with CP in a brain stimulation trial. For example, Gillick et al.[10] excluded children who had received botulinum toxin or phenol blocks within 6 months before enrolling in the study.

Study Criteria and Recruitment

When considering which children may be appropriate for inclusion in a study, safety is the paramount consideration. Criteria must acknowledge not only the type of brain stimulation to be investigated, but also the past medical history of the child and potential for severe or minor adverse events. Previous trials, detailing reports of adverse events, benefit future trials by pointing options for optimal design and considerations of appropriateness.

Various inclusion and exclusion criteria play different roles in the design of such clinical trials. The justification for a given criterion may have multiple sources—scientific, ethical, statistical, or economic. In general, each criterion will be motivated by at least one of the following: safety, compliance, larger magnitude of treatment effect, avoiding protected or vulnerable population(s), heteroscedasticity, and likelihood to obtain outcomes (eg, events for event-based analyses, avoiding missing data). There is a balance to strike between being extremely restrictive and thus homogeneous albeit narrowly generalizable, and being less restrictive and more generalizable to broader populations albeit with perhaps a more heterogeneous set of participants being evaluated. Though being restrictive may require fewer people in the study for the same power, it can take longer to complete the study due to smaller populations of patients meeting the more strict criteria.[25] Specific considerations in the context of trials in children with CP are highlighted in the following.

Inclusion Criteria

Diagnosis

CP is considered a secondary diagnosis to a primary lesion such as perinatal stroke.[26] Determination of the primary diagnosis is central not only to establishing targeted homogeneity within a sample representing children with CP but also to targeting the parameters of a therapeutic trial intervention. As mentioned earlier, Young et al. investigated the impact of tDCS on a sample of children with dystonia. Studies from Kirton et al. and Gillick et al. investigated children with unilateral spastic hemiparesis, yet a more heterogeneous sample was included in the Gillick study as compared to a more narrowed diagnosis of ischemic stroke in the Kirton study.[7,9] Conclusions as to feasibility, impact, and responders may be influenced by such variability.

Baseline Level of Function

In order to categorize children by level of function, a targeted level of function should be established. Use of a classification system assists with this determination. Are severely affected children allowed in the study or does the study seek to influence higher levels of function with potentially less disability and less potential for seizure and/or adverse event? For example, in rTMS, Kirton et al. reported Pediatric Stroke Outcome Measure motor scores, and Gillick et al. reported classification using the Manual Ability Classification Scale Levels.[7,9] In tDCS trials, Duarte et al. reported use of the Gross Motor Functional Classification System in children with spastic CP, while Bhanpuri et al. rated children with dystonic CP using the Barry–Albright Dystonia rating.[18,27] Understanding the differing impacts of brain stimulation interventions, while at the same time incorporating the severity of involvement, can guide future applications in responsive populations.

Communication Ability

In research with children, specifically children with disabilities, an understanding of both age-appropriate developmental abilities and the impact of the disability on cognition and communication is central to reporting change, side effects, or impact. When evaluating children with CP and the myriad of presentations and levels of disability, an understanding of each child's ability to comprehend instructions and procedures as well as their ability to express questions, feedback, and symptoms, needs to be considered before undertaking a brain stimulation intervention. Evaluation of cognitive and language function as well as the ability to follow commands provides the greatest assurance of the child's ability to assent to a study, to understand what will be asked of him or her, and to report not only his or her understanding of the trial but also any adverse effects, which may be subtle.[22] Trained pediatric investigators, who can recognize both developmental abilities and disability-related deficits, are indispensable to facilitate accurate and complete reporting in this population. Incorporating the communication level and style of these children can therefore allow appropriate understanding of child and caregiver expectations while correspondingly minimizing attrition.

Cortical Excitability

TMS, used as a means to elicit motor-evoked potentials, can, through corticospinal tract and motor neuron activation, help assess the functional integrity of central motor pathways, but motor threshold analysis is variable.[28] The more immature brain may not be sufficiently responsive to motor threshold assessments,[29] and may limit detection of corticospinal tract integrity and excitability change. The presence or absence of a motor-evoked potential therefore may in part be age-related, or it may indicate cortical reorganization. The presence of a motor-evoked potential through a corresponding resting or active motor threshold should ideally be established in both hemispheres if cortical excitability and pathway organization is considered essential to the therapeutic trial. Stereotactic neuronavigation may increase the precision of this assessment. Kirton et al.[30] reported the inability to measure an ipsilesional resting motor threshold (RMT) in four of the 10 participants, yet were able to find active motor thresholds (AMT) in two of the four participants without RMTs. However, based on their protocol, they did not exclude these children. Conversely, Gillick et al. excluded 11 of an enrolled 36 children who did not present with ipsilesional motor thresholds, whether active or resting in attempts to control for heterogeneity of the population and assess the potential influence of a combined rTMS/CIMT intervention on ipsilesional cortical excitability.[9]

Age

Determination of the optimal age for stimulation trials in children with CP is multifaceted. Positing a critical period—during which development dictates future development, with the potential for optimal neuroplastic change—interventions at a younger age may have greater impact.[31,32] However, the brain of a child, and more specifically the brain of a child with CP, is not the same as the brain of an adult, and differences in architecture alone can alter the effect of the stimulation on a child's brain.[12,33] In addition, such aspects as the lesion (unilateral, bilateral) and the extent of development and myelination, along with the potential for interhemispheric inhibition, must be taken into consideration. An additional concern is that researchers may set out to understand the impact on the brain in younger children only to find that those children cannot tolerate the sensation or sit still long enough for the study. These issues affect approaches for recruitment, enrollment, and retention of participants. Finally, in this population with CP, consideration not only of physical age but also of developmental age is indicated. For example, a 10 year-old child with CP may not communicate or function at the same level as a 10-year-old typically developing child. Adaptations to the environment, outcome measures, and expectations of the child would then need to be changed to appropriately match the developmental age of the child.

Exclusion Criteria

Diagnoses

Many diagnoses coexisting with CP arise from stroke or periventricular leukomalacia.[34,35] In addition, with regard to the effects on the developing brain and related tracts, acquired traumatic brain injury and non-congenital (>1 year) stroke have been investigated less than congenital hemiparesis.[36] The influence of brain stimulation in children has not yet been documented in enzymatic disorders, mitochondrial diseases, disorders of cellular migration and proliferation, nor in neoplasm or hypoxic ischemic encephalopathy.

Skin Integrity

Brain stimulation can result in minor adverse events in skin, such as tDCS electrodes causing redness or burning sensations.[37] As eczema and related skin conditions can occur in children with CP, skin assessments should directly address the potential for an adverse event such as irritation or injury. In the aforementioned tDCS trials mentioning minor adverse events, electrode site discomforts were reported.[13,16,17,27] Careful screening and a risk mitigation plan can address these issues.

Contraindications

The unknown effects of NIBS on the developing fetus and the possible effects of indwelling metal or incompatible medical devices, call for careful exclusion in view of potential risks. In addition, it is unknown whether oral hardware, such as braces, non-removable retainers, and expanders, may influence the current field or directionality of the signal. If the brain stimulation protocol incorporates stereotactic neuronavigation using magnetic resonance images (MRIs), a distortion of the image may occur. Additionally, where MRI is part of the protocol, previous surgeries related to CP may have inserted hardware (rods, screws, plates), which may not be compatible.

Sample Size, Effect Size, and Power Considerations

The sample size for a trial should reflect the objectives of the trial and the phase of scientific investigation. When conducting a study to evaluate the effectiveness of a potential treatment, the sample should be of sufficient size to distinguish between the null of no treatment difference and an alternative that is clinically meaningful. The sample size can be determined based on a specified magnitude of treatment effect between groups, the variability of the statistic used for estimating the treatment effect, and level of power given a specified level of type I error for the study (most often set at 0.05). A common level of power used for design is 80% (higher power requires larger sample size). While this may be sufficient to be awarded funding and receive regulatory approval (eg, by IRB), upon completion results may not distinguish between meaningful benefit and the null. Having higher power is often difficult due to resource constraints. More complete discussion as well as specific formulas can be found elsewhere.[3] At this juncture, trials have included small numbers of children with CP. Establishing the feasibility and safety of recruiting a specific population within a diagnosis can be challenging as well as the ability to determine an effect size or statistically significant impact. For example, primarily due to our exclusion criteria, our own recent rTMS/CIMT trial only enrolled 19 children, while 206 families originally expressed interest in the study.[9]

Data Sharing/Multicenter Trials

As therapeutic trials in brain stimulation for children with CP often have small numbers of participants, with limited statistical power and fewer still long-term assessments of the potential impact of the intervention, expanded data sharing among investigators may facilitate our scientific understanding and advance the field by continued refinement of trial designs and by replication or reproduction of study results. Because the primary diagnoses in CP are highly heterogeneous, such data sharing would allow further understanding of the impact of our interventions in what could be more homogeneous groups. More importantly, running multicenter trials, which facilitate larger sample sizes and can decrease regional bias or ethnic inequities in representation, would allow for expedited trial completion and subsequent translation of results to clinical applications. The continued execution of small, underpowered studies that could then be combined for analysis after they are all complete (eg, through a meta-analysis framework) is not ideal. To make substantive, robust conclusions from brain stimulation trials, an appropriately powered study, which will most likely require a multicentered effort with the same outcome measures or endpoints, is indicated.

Control Groups, Blinding, and Placebo/Sham Interventions

Integrating control groups greatly advances the validity and reliability of a study. Unfortunately, the presence of a control group may deter both caregivers and children from participating, on the presumption that the child may miss a "beneficial" aspect of the brain stimulation under investigation. Although many potential participants assume that being allocated to the treatment group is superior to being allocated to the control group, participants must be educated about the reason for a clinical trial: neither treatment arm has proven superiority. Being allocated to the sham group does not necessarily decrease a participant's chance of benefitting from being in the trial, particularly if other forms of therapy are included. However, some study designs, such as a delayed crossover for the control group or open-label extensions where all patients are offered the experimental intervention, may be employed, thereby guaranteeing participants an opportunity to receive the brain stimulation intervention. Although brain stimulation modalities have "placebo" or "sham" settings, such settings have their limitations in terms of sensation at the scalp and sound of the stimulator. Both the child and the investigator may be able to detect such variation, which becomes a greater problem when investigating multiple children who are able to interact and discuss their sensations, such as in a camp setting. Assessing blindedness after completion of a trial by asking questions of both the child and the caregivers contributes not only to an understanding of the integrity of concealment but also to the ability to create a more sound trial in the future.

Institutional Review Board (IRB)

Children are considered a vulnerable population when applying for regulatory approval from an applicable IRB. Heightened scrutiny may also arise for this developing population in terms of long-term outcomes and impact of brain stimulation in children with a neurologic disorder. Clearly outlining the protocol and study design allows the IRB to be educated on both the background of these interventions in children and to the potential risk/benefit of these interventions.

Medical Monitor and Data Safety and Monitoring Board (DSMB)

Incorporation, as appropriate, of a medical monitor or a DSMB can guide the investigative team through various concerns during the conduct of a trial while optimizing safety. A medical monitor should be a person who is not an investigator on the study but has experience in review of symptoms (eg pediatric neurologist), and could serve not only to review child status before the study, at a defined interim point, and after the study is completed, but also to review on a case-by-case basis when an adverse event occurs. A DSMB, consisting of professionals independent from the study who can speak to the safety of the study in terms of disease, treatment, and managing side effects (ie, physician, biostatistician, and academicians familiar with brain stimulation and pediatric research) will meet at designated intervals (eg, quarterly) to review trial progress and data integrity along with the emerging safety profile to provide interim feedback as to potential study modification and continuation or stopping. Depending on the intervention, size, and scope of the study design, an independent DSMB may be required by an IRB or FDA.

Adverse Events Management/Risk Mitigation/Stopping Rules

Brain stimulation interventions, such as rTMS and tDCS, have been reviewed for safety in both adults and children.[38–41] However adverse events, from serious (seizure) to minor (tingling) have occurred. This has been discussed in an earlier chapter on safety. Within a report or publication, a simple statement of safety (eg, "No serious adverse events occurred") does not describe measures used to assess such safety—questionnaire, participant report of symptoms, or neurophysiologic measure such as electroencephalogram or electromyography. Specific to children with CP, questioning specifically adapted to this population will help in recognition of rare and subtle symptoms and adverse events, which are potentially most important. Collecting data on adverse events, even if a participant drops out of treatment provides important feedback on how well a child tolerates the intervention. *Before the study begins,* the protocol should incorporate a plan on how and what to collect for adverse events reporting. Identifying age-appropriate questions (eg, open- or closed-ended questions, the manner in which the questions are ordered and the timing of the follow-up on adverse events) is essential to the pediatric study population. Safety outcomes may be specific to the intervention, such as monitoring unaffected hand function while delivering inhibitory stimulation to the contralesional hemisphere. In addition, a clear statement and understanding of stopping rules, which direct how the study may or may not proceed, can guide both the team and the caregiver/child. For example, if a child has a seizure during an rTMS session, what action occurs after initial management? Does the study then terminate or does a medical monitor/DSMB review the case and assess the possibility to proceed? Such questions should be defined and answered a priori in order to plan the brain stimulation protocol and manage adverse events.

Endpoints

The choice of primary endpoint will be dictated by the objectives of the trial and the phase of scientific investigation. For preliminary investigations, the primary objectives of the trial will be safety and feasibility, though clinical endpoints will be evaluated secondarily in order to inform the design of a subsequent, larger trial. Because the majority of current trials have been performed in adults, a recommended primary endpoint persisting in trials with children is safety. As detailed in other chapters, in addition to a thorough account of the approach used to assess adverse events (ranging from serious to minor), risk mitigation techniques should be incorporated into each trial. Even if the study is investigating improvements in motor function, safety remains a primary concern. Explicitly stating the investigation parameters, dosing, single versus serial assessments and endpoints will facilitate replicating the study. Efficacy outcome measures should be validated with evidence of sensitivity to change and may include subjective measures of goal achievement (eg Canadian Occupational Performance Measure) in addition to objective tools.

Analysis and Assessment of Intervention Fidelity

In reporting on a trial, whether or not the trial will include an interim analysis a priori is important to incorporate into the study design, eg, through group sequential trial designs.[42,43] Unplanned reporting on a trial midway through can lead to erroneous conclusions based on incomplete, unreliable data. Scenarios such as (1) the results are showing a difference between groups and suggest no need to continue when indeed the differences observed between groups may drastically attenuate or disappear entirely or (2) there is no difference between groups, when indeed the study may develop to find one. As is the case with all trials, not only those involving children with CP, such premature interpretation could erroneously bias clinical applications of brain stimulation. Similarly, missing data in clinical trials jeopardize the integrity and validity of results. Due to potential bias of unknown magnitude and direction, merely increasing the sample size will result in a more precise, yet still biased estimate of unknown magnitude and direction. Because all methods for handling missing data have some level of untestable assumptions, the issue is quite serious and best avoided altogether.[44] Avoiding missing data can be addressed in part through study design, though development of strong rapport with the child, parent, and primary caregiver can have the biggest impact.

Replication, Trial Registration, and Consolidated Standards of Reporting Trials (CONSORT)

In order to advance the science of these trials and build upon the findings of previous studies, transparency in protocols is necessary. Trial registration and submitting results to clinicaltrials.gov is required by law (cf. Food and Drug Administration Amendments Act Section 801) and provides detailed descriptions of each trial, which allows for replicating the study to either support or disprove findings. The European Medicines Agency has an analogous site for registration: ema.europa.eu. Caregivers and children alike have increasingly been using clinicaltrials.gov to find studies, and researchers refer to these sites to learn what trials are currently underway. Development of a CONSORT diagram allows for a detailed account of a brain stimulation trial, with reported decisions, such as whether or not to include a motorevoked potential, possibly helpful to others in designing their own future trials.

Research Team

The multidisciplinary team typically involved in the medical management of the child with CP is central to the success of a research trial using brain stimulation. As already mentioned, pediatric specialists have specialized skills essential to the success of the trial. Study coordinators who typically interface with the caregiver and child from initial recruitment onward can be essential in providing a consistent point of contact and educational resource during a trial. Incorporating a biostatistician, with expertise and experience in clinical trials can assist not only in analysis but also key aspects of design and monitoring. With caregiver consent and a thorough age-appropriate understanding and assent to participation, the child cannot only take an educated viewpoint during the trial, but also participate actively in assessing comfort with the intervention and giving feedback on tolerability. If reported, such feedback improves future study designs for child satisfaction with the study. Importantly, both the caregiver's and child's education and feedback are essential in establishing research rapport with the child. At the initial point of contact is the caregiver. The manner in which he or she learned of the trial, the recruitment method and the retention of the child participant will largely be guided by the caregiver's understanding of the workings of brain stimulation and the data on safety and efficacy. As a diagnosis of CP entails acknowledgment of both a non-progressive disorder and a lifetime disability with special needs, discussions addressing the possible impact (or lack thereof) of brain stimulation are essential. The team should make clear the possibility of the child being placed in a control or sham brain stimulation group, and that, as these are *trials*, neither group may experience a meaningful benefit to motor or cognitive function. As longitudinal and long-term studies continue to be developed, the impact of brain stimulation trials will be better understood. As a result, children and caregivers may recognize that their participation in brain stimulation trials, with or without any immediate impact on their own long-term outcomes living with CP, will have a great impact on the development of future interventions.

Unlike an adult with an acute injury or stroke impacting their function after years of typical development and growth, a child with CP likely has been involved in interventions since early infancy. Brain stimulation for this population appears to be the next intervention on the horizon, holding out the possibility not only to change current levels of function, but also to improve function across the lifespan. Involvement in a study of such stimulation can provide the caregiver and the child with new hope. At the same time, as investigators in a new frontier, although we are posting that we may now have a new means of *improving* function, we must be cautious in our accounts of the potential impact of our interventions. Training all team members—study coordinators, technicians, and others such as graduate students and postdoctoral fellows—to consider this special component of working with children with CP and to present our investigations as opportunities to "explore change" instead of to "improve function" may lead to improved study participation and success. Investigators who incorporate feedback from caregivers and families as they progress through brain stimulation trials may find greater effectiveness in recruitment and retention of participants and increased support from their families.

CONCLUSIONS

While we as researchers in brain stimulation continue to develop our trials, whether our child participants have a diagnosis of CP or autism, epilepsy or a host of other conditions currently under investigation, we act as stewards not only to our science but also to the population researched. We are representatives who should integrate into our work both caution and acknowledgment of responsibility for the safety and behavioral outcomes we investigate. Well-constructed studies are paramount to furthering research in this area, garnering continued support from the community, and building a knowledge base to advance therapies in children with CP. Poorly considered or designed studies dismantle support from the community, make future directions in this area less clear, and waste valuable resources: money, investigators' time, participants' time, and everyone's hopes. Uniquely in the pediatric population, we address development as well as diagnosis in our interventions. Although CP may be defined at present as a "static" diagnosis, the promise of plasticity provides a window of opportunity, especially for children, which we may discover a means to open. By building responsibly upon our collective findings in this field we may safely and effectively discover the opportunity to make the static, dynamic.

References

1. Kirton A. Modeling developmental plasticity after perinatal stroke: defining central therapeutic targets in cerebral palsy. *Pediatr Neurol.* 2013;48(2):81–94. http://dx.doi.org/10.1016/j.pediatrneurol.2012.08.001.
2. CDC. http://www.cdc.gov/ncbddd/cp/index.html.
3. Friedman L, Furber C, DeMets D. 4th ed. *Fundamentals of Clinical Trials.* vol. xviii. New York: Springer; 2010. ISBN: 9781441915863:445. ISBN: 1441915869.
4. Food and Drug Administration (FDA). http://www.fda.gov/RegulatoryInformation/Guidances/ucm122046.htm.
5. *The International Conference on Harmonisation of Technical Requirements for Registration of Pharmaceuticals for Human Use.* http://www.ich.org/products/guidelines.html.
6. Valle A, Dionisio K, Pitskel N, et al. Low and high frequency repetitive transcranial magnetic stimulation for the treatment of spasticity. *Dev Med Child Neurol.* 2007;49(7):534–538. http://dx.doi.org/10.1111/j.1469-8749.2007.00534.x.
7. Kirton A, Chen R, Friefeld S, Gunraj C, Pontigon AM, Deveber G. Contralesional repetitive transcranial magnetic stimulation for chronic hemiparesis in subcortical paediatric stroke: a randomised trial. *Lancet Neurol.* 2008;7(6):507–513.
8. Kirton A, Deveber G, Gunraj C, Chen R. Neurocardiogenic syncope complicating pediatric transcranial magnetic stimulation. *Pediatr Neurol.* 2008;39(3):196–197.
9. Gillick B, Krach L, Feyma T, et al. Primed low-frequency repetitive transcranial magnetic stimulation and constraint-induced movement therapy in pediatric hemiparesis: a randomized controlled trial. *Dev Med Child Neurol.* 2014;56(1):44–52. http://dx.doi.org/10.1111/dmcn.12243.
10. Gillick B, Krach L, Feyma T, et al. Safety of primed repetitive transcranial magnetic stimulation and modified constraint-induced movement therapy in a randomized controlled trial in pediatric hemiparesis. *Arch Phys Med Rehabil.* 2014. http://dx.doi.org/10.1016/j.apmr.2014.09.012.

11. Hsu W-Y, Cheng C-H, Liao K-K, Lee I-H, Lin Y-Y. Effects of repetitive transcranial magnetic stimulation on motor functions in patients with stroke: a meta-analysis. *Stroke.* 2012;43(7):1849–1857.

12. Kessler S, Minhas P, Woods A, Rosen A, Gorman C, Bikson M. Dosage considerations for transcranial direct current stimulation in children: a computational modeling study. *PLoS One.* 2013;8(9):e76112.

13. Gillick B, Feyma T, Menk J, et al. Safety and feasibility of transcranial direct current stimulation in pediatric hemiparesis: randomized controlled preliminary study. *Phys Ther.* 2015;95(3):337–349.

14. Grecco LAC, Duarte NAC, Zanon N, Galli M, Fregni F, Oliveira C. Effect of a single session of transcranial direct-current stimulation on balance and spatiotemporal gait variables in children with cerebral palsy: a randomized sham-controlled study. *Braz J Phys Ther.* 2014;18(5):419–427.

15. Young S, Bertucco M, Sheehan Stross R, Sanger T. Cathodal transcranial direct current stimulation in children with dystonia: a pilot open-label trial. *J Child Neurol.* 2013;28(10):1238–1244.

16. Young S, Bertucco M, Sanger T. Cathodal transcranial direct current stimulation in children with dystonia: a sham-controlled study. *J Child Neurol.* 2014;29(2):232–239.

17. Grecco LAC, de Almeida Carvalho Duarte N, Mendonça M, et al. Transcranial direct current stimulation during treadmill training in children with cerebral palsy: a randomized controlled double-blind clinical trial. *Res Dev Disabil.* 2014;35(11):2840–2848.

18. de Almeida Carvalho Duarte N, Grecco LAC, Galli M, Fregni F, Oliveira C. Effect of transcranial direct-current stimulation combined with treadmill training on balance and functional performance in children with cerebral palsy: a double-blind randomized controlled trial. *PLoS One.* 2014;9(8):e105777.

19. Lazzari RD, Politti F, Santos CA, et al. Effect of a single session of transcranial direct-current stimulation combined with virtual reality training on the balance of children with cerebral palsy: a randomized, controlled, double-blind trial. *J Phys Ther Sci.* 2015;27(3):763–768.

20. Collange Grecco L, de Almeida Carvalho Duarte N, Mendonça M, Galli M, Fregni F, Oliveira C. Effects of anodal transcranial direct current stimulation combined with virtual reality for improving gait in children with spastic diparetic cerebral palsy: a pilot, randomized, controlled, double-blind, clinical trial. *Clin Rehabil.* 2015;29(12):1212–1223.

21. Aree uea B, Auvichayapat N, Janyacharoen T, et al. Reduction of spasticity in cerebral palsy by anodal transcranial direct current stimulation. *J Med Assoc Thai.* 2014;97(9):954–962.

22. Davis N. Transcranial stimulation of the developing brain: a plea for extreme caution. *Front Hum Neurosci.* 2014;8:600.

23. Eyre JA. Corticospinal tract development and its plasticity after perinatal injury. *Neurosci Biobehav Rev.* 2007;31(8):1136–1149.

24. Mooney J, Koman LA, Smith B. Pharmacologic management of spasticity in cerebral palsy. *J Pediatr Orthop.* 2003;23(5):679–686.

25. Rudser K, Bendert E, Koopmeiners J. Sample size and screening size trade-off in the presence of subgroups with different expected treatment effects. *J Biopharm Stat.* 2014;24(2):344–358.

26. Kirton A, deVeber G. Cerebral palsy secondary to perinatal ischemic stroke. *Clin Perinatology.* 2006;33(2):367–386.

27. Bhanpuri N, Bertucco M, Young S, Lee A, Sanger T. Multiday transcranial direct current stimulation causes clinically insignificant changes in childhood dystonia: a pilot study. *J Child Neurol.* 2015;30(12):1604–1615.

28. Wasserman E. *The Oxford Handbook of Transcranial Magnetic Stimulation.* vol. viii. New York: Oxford; 2008. Contributor: Eric M Wassermann (Eric Michael), 1956–:747.

29. Eyre J, Smith M, Dabydeen L, et al. Is hemiplegic cerebral palsy equivalent to amblyopia of the corticospinal system? *Ann Neurol.* 2007;62(5):493–503.

30. Kirton A, Deveber G, Gunraj C, Chen R. Cortical excitability and interhemispheric inhibition after subcortical pediatric stroke: plastic organization and effects of rTMS. *Clin Neurophysiol.* 2010;121(11):1922–1929.
31. Kennard MA. Age and other factors in motor recovery from precentral lesions in monkeys. *Am J Physiol.* 1936;115(1):138.
32. Anderson V, Spencer Smith M, Wood A. Do children really recover better? neurobehavioural plasticity after early brain insult. *Brain.* 2011;134(8):2197–2221.
33. Gillick B, Kirton A, Carmel J, Minhas P, Bikson M. Pediatric stroke and transcranial direct current stimulation: methods for rational individualized dose optimization. *Front Hum Neurosci.* 2014;8:739.
34. Rosenbaum P. A report: the definition and classification of cerebral palsy April 2006. *Dev Med Child Neurol.* 2007;109(suppl 109):8.
35. Rosenbaum P, Eliasson AC, Hidecker MJC, Palisano R. Classification in childhood disability: focusing on function in the 21st century. *J Child Neurol.* 2014;29(8):1036–1045.
36. Chung M, Lo W. Noninvasive brain stimulation: the potential for use in the rehabilitation of pediatric acquired brain injury. *Arch Phys Med Rehabil.* 2015;96(4 suppl):S129–S137.
37. Loo CK, Martin DM, Alonzo A, Gandevia S, Mitchell PB, Sachdev P. Avoiding skin burns with transcranial direct current stimulation: preliminary considerations. *Int J Neuropsychopharmacol.* 2011;14(3):425–426.
38. Wassermann EM. Risk and safety of repetitive transcranial magnetic stimulation: report and suggested guidelines from the international workshop on the safety of repetitive transcranial magnetic stimulation, June 5–7, 1996. *Electroencephalogr Clin Neurophysiol.* 1998;108(1):1–16.
39. Rossi S, Hallett M, Rossini PM, Pascual-Leone A. Safety of TMS consensus group. Safety, ethical considerations, and application guidelines for the use of transcranial magnetic stimulation in clinical practice and research. *Clin Neurophysiol.* 2009;120(12):2008–2039.
40. Quintana H. Transcranial magnetic stimulation in persons younger than the age of 18. *J ECT.* 2005;21(2):88–95.
41. Krishnan C, Santos L, Peterson M, Ehinger M. Safety of noninvasive brain stimulation in children and adolescents. *Brain Stimul.* 2015;8(1):76–87.
42. Fleming T, Sharples K, McCall J, Moore A, Rodgers A, Stewart R. Maintaining confidentiality of interim data to enhance trial integrity and credibility. *Clin Trials.* 2008;5(2):157–167.
43. Jennison C, Turnbull B. *Group Sequential Methods with Applications to Clinical Trials.* vol. xix. Boca Raton: Chapman and Hall; 2000:390.
44. Little R, D'Agostino R, Cohen M, et al. The prevention and treatment of missing data in clinical trials. *N Engl J Med.* 2012;367(14):1355–1360.

12

Brain Stimulation in Children Born Preterm—Promises and Pitfalls

J.B. Pitcher

The University of Adelaide, Adelaide, SA, Australia

INTRODUCTION

In the developed world, 6–12% of all infants born annually are preterm (ie, born before 37 completed weeks of gestation). According to the 2014 March of Dimes Premature Birth Report Card (http://www.marchofdimes.org/mission/prematurity-reportcard.aspx), in 2013, the overall rate for the United States was 11.4% of all births, ranging between 8.8% (California) and 16.6% (Mississippi). Despite advances in neonatal care, many of these children will experience suboptimal neurodevelopmental outcomes that, while often subtle, nonetheless negatively impact upon their physical, educational, and psychosocial opportunities. Relatively few of these neurodevelopmental difficulties are explained by perinatal brain lesions, which are experienced by fewer than 1% of preterm children (<10% of children born <32 completed weeks gestational age; GA).[1] Increasing evidence suggests that the culprits are microstructural brain abnormalities (white matter, short- and long-range connectivity), not readily detected with standard magnetic resonance imaging (MRI).[2-4] To date, this has contributed to limiting our understanding of the underlying mechanisms, our ability to accurately predict long-term outcomes, and our ability to design and administer efficacious interventions that might improve short- and long-term neurodevelopmental outcomes in individuals born preterm.

A number of other factors have limited this understanding. There are a plethora of studies in the field, but many are confounded by not differentiating the effects on brain development of GA from those of low birthweight, and have included infants or children with known perinatal brain lesions (eg, periventricular and intraventricular hemorrhages, leukomalacia, etc.) and/or neurosensory impairments (eg, retinopathy/visual loss, hearing loss, etc.). Significantly, the vast majority of studies are on those children born most preterm and rarely include children born 33–36 completed weeks GA, who comprise approximately 70% of all preterm births in developed countries. Until very recently, these children were classified as "near term," were considered "normal," and rarely if ever received neonatal follow up that differed from their term-born peers. However, in the past decade, an increasing number of studies have shown that these children have a higher incidence of neonatal morbidity and mortality, are rehospitalized more often in the first 3 years,[5] and experience greater respiratory and metabolic dysfunction than their term-born peers (reviewed in Ref. 6). This has led to the emergence of the classifications moderate (32–33 weeks GA) and late preterm (34–36 weeks GA),[7] although often they are simply referred to as late-preterm births. While the late preterm are less likely than their more preterm peers to suffer severe brain dysfunction, emerging evidence shows a high prevalence of low-severity motor, cognitive, and behavioral dysfunction in these children. They are more likely to require access to special education resources[8-10] and, as adults, have a lower net income and are less likely to complete a university education.[11] Despite this, the data on the characteristics of neurodevelopmental dysfunction in moderate and late preterm children (both short and long term) remain sparse and our understanding of the altered underlying physiology responsible is almost non-existent.

The problem of high-prevalence, low/moderate-severity neurodevelopmental dysfunction in moderate/late preterm children also raises another issue. The vast majority of neurodevelopmental outcome studies, particularly in infants and children, have used instruments that are specifically designed to detect relatively frank disability and, arguably, lack the specificity or sensitivity to detect subtle dysfunction. The advent of non-invasive brain stimulation (NIBS) techniques, such as transcranial magnetic stimulation (TMS), transcranial direct-current stimulation (tDCS) either alone, or coupled with electroencephalography (TMS-EEG) or magnetic resonance imaging (TMS-MRI), has offered the potential to detect and characterize more subtle neurodevelopmental abnormalities in preterm (and term-born children). In addition, many of these NIBS techniques offer potential as interventional therapies, particularly in the domain of inducing neuroplasticity. But the field is in its infancy and there are few data on NIBS and neurodevelopment, either in preterm or term-born children. This chapter reviews a number of common TMS techniques that have been used in preterm children (including single-, paired-, transcallosal and repetitive TMS paired-pulse techniques), common pitfalls that currently limit their use in these children, and the potential they promise both as screening and therapeutic tools. As the use of TMS (and indeed other non-invasive brain stimulation techniques, such as tDCS) in preterm children and adolescents is presently a nascent field, the chapter reviews the little that has been done, but highlights more what the gaps in our knowledge are. I have confined the discussion to preterm and term-born children without any history of perinatal or pediatric brain lesions, as these conditions are covered in the preceding chapters.

THE MOTOR THRESHOLD

TMS is generally applied at the scalp overlying the cortical region of interest. To date, the primary motor cortex has been by far the most popular target of investigation, largely because it is one of few cortical areas that allows relatively direct measurement of its output. When applied to the scalp overlying the hand area of the primary motor cortex at sufficient intensity, TMS activates corticospinal output neurons and results in a response in the hand muscles on the opposite, or contralateral, side of the body, known as the motor-evoked potential (MEP). This MEP is easily measured using surface electrodes on the skin overlying the muscle. The size and latency of the MEP reflect the excitability of neurons and impulse conduction characteristics of the motor cortex, corticospinal tract, and spinal motor centers in real time. The resting motor threshold is the lowest TMS intensity required to excite corticospinal fibers indirectly via excitatory input from corticocortical projections and to evoke a small MEP response in the relaxed target muscle.[12] Operationally, it is most commonly defined as the lowest stimulator output intensity at which an MEP of 50 µV minimum peak-to-peak amplitude is elicited in at least 50% of 10 consecutive trials.[13]

Not surprisingly, the motor threshold reflects the excitability of various structures modulated by the TMS pulse, and a series of pharmacological studies has specifically implicated corticocortical axons and their excitatory synapses with corticospinal output neurons, the voltage-gated sodium channels responsible for modulating axonal excitability, and ionotropic non-NMDA glutamate receptors that control fast excitatory synaptic neurotransmission in neocortex (reviewed in Ref. 12). Magnetic resonance imaging (MRI) studies in healthy adults have shown that a lower motor threshold correlates with better maturation, myelination, and structural integrity of the white matter in a range of brain structures including the primary, premotor and prefrontal cortices, the internal capsule, corpus callosum, corona radiata, and cerebral peduncles.[14] Individuals with low motor thresholds have better white matter structural integrity, evident as higher fractional anisotropy in the neural connections involved in corticospinal output.

The threshold to TMS is high in term-born neonates, then decreases over the next decade or so.[15] When exactly the resting motor threshold reaches adult levels is still not clear; early work indicated that the most rapid lowering of threshold is in the first 8–10 years of life,[15] but we recently showed that the resting motor threshold continues to decline well into adulthood, in a pattern similar to that reported for age-related trajectories of white matter development previously reported by imaging studies.[16–18]

To date, there are no published studies of motor threshold in preterm neonates. However, increased motor thresholds have been reported in children and adolescents who were born preterm.[19,20] In adolescents born 24–41 weeks, GA was linearly associated with resting motor threshold, such that the most preterm children had the highest thresholds. There was no "critical" GA after which no further gains in cortical excitability were made, suggesting that every week of gestation contributes to cortical development.

Many preterm infants also experience intrauterine growth restriction or are small for gestational age, but few studies differentiate these effects. We have differentiated the effects of GA from those of fetal growth, by calculating a birth weight centile (BW%) whereby an individual's birth weight is expressed as a percentage of their predicted weight for a given GA, based on their sex, parity, maternal anthropometry, and ethnicity.[20–22] Low BW% is linearly associated with increased threshold independently of GA, although this appears to be confined to the right hemisphere.[20] Similarly, in young adults born at term (≥37 completed weeks GA), low BW% was associated with increased right versus left hemisphere resting motor threshold,[22] and this difference was amplified in those individuals who were discharged to more socially disadvantaged homes after birth.

The observation that low BW% preferentially affects the right hemisphere resting motor threshold is consistent with imaging studies showing a reduction in right hemisphere cortical surface area due to reduced gyrification in adolescents born small for GA at term, compared with that in their appropriately grown peers.[23]

In preterm children (and younger term-born children) in whom the resting motor threshold is too high to be measured, the alternative is to record the active motor threshold, since voluntary activation of the target muscle will lower its threshold to stimulation. The most common definition of active motor threshold is the lowest stimulus intensity at which an MEP of at least 200-μV peak-to-peak amplitude is evoked in at least five out of 10 consecutive trials, while the subject holds an isometric contraction of the target muscle at 10% of their maximal voluntary contraction (MVC). In our original study of preterm motor and cognitive development, the PREMOCODE-1 study, resting and active motor thresholds for the first dorsal interosseous muscle (FDI) correlated highly in both motor hemispheres (right hand; $r = 0.87$; $P \leq 0.0001$; $N = 132$) (left hand; $r = 0.81$; $P \leq 0.0001$; $N = 126$) in 12-year-old children, and the relationships between active motor threshold and GA and BW% were similar to those for resting motor threshold.[20] This suggests the active motor threshold might be a useful tool in younger preterm children in whom resting motor thresholds are unobtainable. However, there are two potential pitfalls that require recognition. First, it is physiologically incorrect to interpret the two thresholds as manifestations of the same phenomenon. The difference between the resting and active motor thresholds is not simply a reflection of a greater proportion of the motor neuron pool being brought closer to its firing threshold with voluntary activation. Compared to the resting state, even a minimal voluntary activation significantly reduces the amount of GABA$_A$-ergic short-interval intracortical inhibition (SICI) acting on the corticomotor output neurons[24,25] and probably increases the degree of superimposed short-interval intracortical facilitation.[26] The second pitfall is a methodological one; assuming an accurate MVC can be obtained, the child must then hold a steady 10% MVC while the active threshold is determined. In children aged 5 years and over, this is relatively straightforward, particularly if the visual feedback of their actual and required contraction force is presented to the child as a game and they are aware that the TMS stimulus will interfere with their contraction momentarily. In children under 5 years, significantly more time and patience is often needed to obtain compliance; in infants it is not feasible.

The motor thresholds appear to correlate reasonably well with a range of functional motor and cognitive abilities in preterm and term-born children. For example, general intellectual ability (GIA) as measured by the Woodcock–Johnson III Tests of Cognitive Abilities, and overall motor skills development (Movement Assessment Battery for Children, second edition) both correlate positively with GA and negatively with resting and active motor thresholds,[20,27] although, when different cognitive (eg, verbal ability, spatial relations) and motor abilities (manual dexterity, balance) are examined, resting motor threshold is a much better predictor than active motor threshold. However, regression modeling of the factors influencing cognitive and motor abilities in preterm children at the end of the first decade of life show a much more complex interplay between GA, BW%, cortical development, parental factors, and social disadvantage. In short, preterm birth and low BW% confer an increased risk of suboptimal cortical development, evident using TMS as reduced cortical excitability. This reduced cortical excitability is associated with poorer motor (particularly manual dexterity) and cognitive abilities. However, the degree to which this increased risk is manifest depends critically upon a range of other factors, particularly the level of social disadvantage the child goes home to at birth; the greater the degree of disadvantage experienced in early life, the poorer the outcome. Conversely, a significant number of preterm children (without brain lesions) who experience little or no early disadvantage are largely indistinguishable from their term-born peers, in terms of their cortical excitability and their functional abilities. So while motor thresholds measured with TMS can certainly contribute to assessing cortical excitability and functional development in preterm children, they are also likely to reflect the child's postnatal environment as much, if not more, than their preterm birth.

SHORT-INTERVAL INTRACORTICAL INHIBITION AND FACILITATION

Normal cortical functioning relies on a balance between facilitatory and inhibitory influences acting upon neurons, largely mediated via networks of interneurons. Paired-pulse TMS studies, where a conditioning stimulus is followed by a test stimulus, have been utilized extensively to assess these interneuronal circuits.[28] When a subthreshold stimulus is followed 1–5ms later by a suprathreshold test stimulus, the size of the evoked response is smaller (ie, inhibited) than if the test stimulus was delivered alone. This has been termed short-interval intracortical inhibition (SICI). At interstimulus intervals (ISIs) around 1ms, the inhibition is thought to be largely due to corticocortical axonal refractoriness,[25] but between 1.5 and 5ms the effect is likely to be due to synaptic inhibition.[25,29] Unless assessed at ISIs at which the influence of intracortical facilitation is minimized, SICI reflects the combined net inhibition produced by activation by the conditioning stimulus of low-threshold inhibitory circuits as well as those higher-threshold excitatory influences activated by the stimulus.[30] Pharmacological studies show SICI most likely reflects postsynaptic inhibition mediated via $GABA_A$ receptors.[31] Pharmacological studies have also shown that ICF is glutamatergically mediated, since NMDA receptor agonists reduce ICF.[31,32] The role of these short-interval circuits in motor function is not entirely clear. However, current evidence suggests roles for SICI in preventing inappropriate muscle activation, including volitional inhibition of prepared motor responses[33,34] and focusing excitatory drive to optimize movement production,[35] fractional activation particularly of intrinsic hand muscles,[36] and in motor learning and neuroplasticity.[37–39] Changes in these short-latency interneuronal circuits are observed in a range of neurological and neuropsychiatric disorders in children, adolescents, and adults.[40–45]

Relatively little is known about the normal developmental trajectory of cortical inhibitory and excitatory processes in healthy children and adolescents. During brain development, there is a developmental shift in GABA from being excitatory to being inhibitory, associated with a progressive decline in neuronal intracellular chloride, a reduction in the glutamic acid decarboxylase GAD27 to GAD67, thereby increasing GABA synthesis,[44] a decrease in the ratio of the chloride cation cotransporter NKCC1 to the KCC2 chloride cation cotransporter, and a corresponding change in GABA polarity from depolarizing to hyperpolarizing.[46] Moreover, the current evidence is that GABAergic synapses form before glutamatergic synapses, suggesting that early in development all excitatory drive is GABAergic.[47] This developmental switch in GABA polarity is thought to play a key role in the activity-dependent development of functional neural circuits.[48] The exact timing of this shift in the human brain is unknown, but postmortem studies of human prefrontal cortex and hippocampus suggest that the process is relatively complex, beginning in the second trimester, and continuing well into the first decade of life.[44,46] Consistent with this is the emerging evidence that the density of $GABA_A$ receptors declines progressively from early postnatal life into adulthood, and the relative expression of the different subunits varies throughout development.[49,50] Similarly, a handful of studies have suggested SICI (measured with TMS) may have a protracted maturation into at least late adolescence.[51,52] However, recent findings by our group suggest that adult levels of SICI and ICF are apparent by the end of the first decade of life in term-born children.[53]

To date, only three studies have reported SICI and/or intracortical facilitation (ICF) in preterm children, and all have reported no apparent effect of GA.[19,20,54] However, one of these studies failed to report the ISIs at which SICI and ICF were evoked.[54] Using a conditioning stimulus of 80% of active motor threshold and a test stimulus of 120% active motor threshold, Flamend and colleagues[19] reported reduced SICI in 8.5-year-old children born less than 32 weeks GA on the "dominant" side when compared with their term-born peers, and an absence of SICI on the non-dominant side in all children regardless of GA. However, this study included only seven preterm and seven term-born children. In children aged approximately 13 years of age who were born 24–41 weeks GA, we assessed SICI at 3 ms ISI and ICF at 10 ms ISI in the left ($N = 95$) and right hemispheres ($N = 72$; both hemispheres, $N = 70$).[20] We found no differences due to GA (or BW%) on either SICI or ICF. However, even with a relatively large sample, the data were highly variable and the statistical power of the sample was less than 30%. So, as with the findings of Flamend and colleagues,[19] we would urge a cautious interpretation of these data until an appropriately powered study is completed. As with all TMS studies in preterm children, the main barrier to assessing SICI and/or ICF prior to approximately 10–11 years of age remains their relatively high thresholds to stimulation. Moreover, if studies only report the results from children in whom responses were present, without detailing those children in whom responses could not be obtained due to high thresholds, the likelihood of confounded and spurious findings is high.

LONG-INTERVAL INTRACORTICAL INHIBITION AND THE CORTICAL SILENT PERIOD

If a suprathreshold conditioning stimulus is applied to the motor cortex 50–200 ms prior to a suprathreshold test stimulus, the amplitude of the resultant MEP is inhibited when compared to the MEP evoked by the test stimulus alone. This type of inhibition has been termed long-interval intracortical inhibition.[55,56] At shorter ISIs (~50 ms) inhibition of the MEP is significantly influenced by spinal excitability.[57,58] However, when Long-interval Intracortical Inhibition (LICI) circuits are probed at longer ISIs (100–200 ms) using low stimulus intensities the resultant inhibition is predominantly due to G-protein-coupled $GABA_B$ receptors acting postsynaptically to increase long-lasting inhibitory postsynaptic potentials in cortical motor neurons.[59,60] There are few reports regarding LICI in children or adolescents[42,43] and none in the preterm population. In both the aforementioned studies, the youngest age tested was 9 years.

Similarly, the cortical silent period reflects $GABA_B$-mediated motor cortical inhibition (except for the early spinal component), and is assessed by applying a single suprathreshold TMS pulse over the cortical representation of the target muscle, while the child holds a steady background contraction. The cortical silent period is evident as a period of "silence" in the electromyogram following the evoked MEP.[61] When measured in the muscle contralateral to the hemisphere being stimulated, the cortical silent period most likely predominantly reflects $GABA_B$-ergic intracortical inhibition,[31,62] whereas when measured in the muscle ipsilateral to the side of stimulation, the cortical silent period reflects predominantly interhemispheric (transcallosal) inhibitory processes.[63]

There are no studies of the cortical silent periods in preterm children, and no developmental studies of either LICI or the cortical silent periods. However, reduced ipsilateral (but not contralateral) silent periods have been reported in children with attention-deficit hyperactivity disorder.[64,65] In these studies, ipsilateral silent periods in the normally developing control children were shown to shorten in onset latency (from 55 to 30 ms) and increase in duration (from less than 10 ms to approximately 30 ms) between the ages of 8 and 14 years, but contralateral silent periods were unchanged. This suggests that $GABA_B$-mediated inhibitory processes (ie, mediating contralateral silent periods) are essentially mature by the end of the first decade and unaffected in ADHD, but transcallosal inhibitory processes (ie, mediating ipsilateral silent periods) are still developing at age 14 years in normally developing children, and are adversely affected in ADHD. Interestingly, the risk for developing ADHD appears to be high in very preterm children, decreases exponentially with increasing GA, but is not related to cerebral palsy (in very preterm children) or growth restriction.[66,67] This suggests that very preterm (but perhaps not late preterm–early term) children are likely to have delayed or aberrant maturation of the transcallosal inhibitory circuits, but normal $GABA_B$-mediated intracortical inhibitory circuits. However, this remains to be determined experimentally. Theoretically, measuring cortical silent periods should be more feasible in preterm children, particular those aged under 10 years, than LICI or transcallosal inhibition (see next section), since contraction of the muscles lowers the threshold for stimulation.

TRANSCALLOSAL INHIBITION

The corpus callosum is the largest bundle of nerve fibers in the nervous system, with over 200 million axons, and the largest white matter structure. It plays critical roles in the transfer and integration of motor, sensory (vision, hearing, touch), and higher cognitive information between the left and right hemispheres.[68] It begins developing around GA week 10–11, but undergoes rapid growth at 20–40 weeks GA, spanning virtually the entire window for survivable preterm and term birth. It reaches adult length of approximately 10 cm at 24 months postnatal age, but myelination is protracted, beginning around GA week 16, and continuing into the third decade.[69,70] Not surprisingly, the corpus callosum is particularly vulnerable in the preterm neonate, partly because of its rapid growth and partly because of its periventricular anatomical location; a common site for perinatal hemorrhage after preterm birth.

A mainstay of predicting neurodevelopmental outcome after preterm birth is the severity of perinatal and neonatal white matter injury.[71] A large number of studies have correlated structural changes of the corpus callosum (particularly thinning and reduced white matter volume) of individuals born preterm with reduced cognitive performance in childhood and adolescence. However, these findings have not always been consistent, suggesting size per se is a relatively poor predictor of function. It is increasingly being recognized that diffuse microstructural white matter abnormalities that adversely affect connectivity are likely to account for the greatest proportion of adverse neurodevelopmental outcomes in this population.[71] A recent magnetic resonance spectroscopy study has also shown that the biochemical maturation of white matter metabolic processes is adversely disrupted by preterm birth, and that this may underlie these changes in white matter volume and microstructure.[3] Taken together, these findings highlight some of the limitations of descriptive correlative studies linking structural brain changes to functional behavioral outcomes.

In addition to interhemispheric transfer, the corpus callosum also has an important role in selectively allowing neural activity in one hemisphere while selectively inhibiting neural activity in the other.[72] There are also excitatory connections, but mutual inhibition appears to predominate between the motor hemispheres.[73] In addition to the ipsilateral silent period, the strength of these interhemispheric inhibitory processes between homologous areas of the motor cortices can be functionally determined using two magnetic stimulators, with coils held over the motor representations for the left and right target muscles in each hemisphere. First described by Ferbert et al.,[63] the effect of a suprathreshold stimulus over one hemisphere (ie, the conditioning stimulus) on the MEP evoked by stimulation of the opposite hemisphere (ie, the test stimulus) is examined. When the conditioning stimulus precedes the test stimulus at an ISI between 8 and 20 ms, the MEP evoked by the test stimulus is inhibited, when compared to the MEP evoked when the test stimulus is applied without the preceding conditioning stimulus. Later studies confirmed that the effect was cortical in origin[74] and mediated by transcallosal pathways,[72,75] although some transmission via subcortical pathways also probably contributes.[76] Transcallosal inhibition at the shortest ISIs is likely to reflect transmission in the fastest conducting axons. When measured using the ipsilateral silent period technique, transcallosal inhibition is absent in normally developing children under 6 years of age.[77] A recent diffusion tensor imaging study showed that transcallosal motor fibers are completely absent in children under 5 years of age, develop slowly between 6 and 10 years, before becoming more prevalent between 11 and 20 years (Fig. 12.1).[78]

We obtained preliminary data from 38 preterm and term-born adolescents (mean = 12 years, 6 months) that suggest that motor cortically evoked transcallosal inhibition is reduced by preterm birth, but only inhibition of the left hemisphere by the right, and not vice versa (Fig. 12.2).[79]

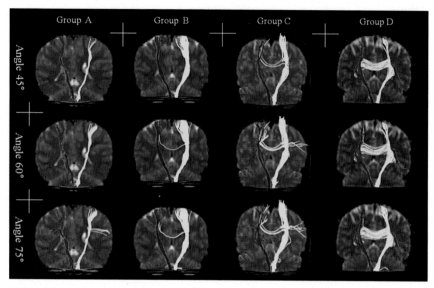

FIGURE 12.1 Results of diffusion tensor tractography for transcallosal motor fiber originating from the corticospinal tract with three angles: group A—the age range was 0–5 years, group B—the age range was 6–10 years, group C—the age range was 11–15 years, and group D—the age range was 16–20 years. *From Kwon HG, Son SM, Jang SH. Development of the transcallosal motor fiber from the corticospinal tract in the human brain: diffusion tensor imaging study.* Front Hum Neurosci. *18 March 2014;8.* http://dx.doi.org/10.3389/fnhum.2014.00153.

Interestingly, this reduced right-to-left hemisphere inhibition was strongly associated with poorer general intellectual ability ($r = -0.51$, $P = 0.001$), verbal comprehension ($r = -0.52$, $P = 0.001$), concept formation ($r = -0.49$, $P = 0.001$), and working memory ($r = -0.38$, $P = 0.02$), but not motor skills development. We have previously shown that hemispheric motor asymmetry is reduced in preterm adolescents; the more preterm their birth, the less lateralized toward the right hand (left hemisphere) an individual is.[20] This is evident as less hemispheric asymmetry in motor function, rather than a greater likelihood of left-handedness.

Whether the underlying mechanism involves altered neural development in the right or left hemisphere, both, or the corpus callosum itself is not yet clear. However, hemispheric motor lateralization and language lateralization are correlated,[80,81] and the failure to develop a dominant hemisphere has been shown to significantly increase the risk of language, reading, and speech difficulties.[82] Our preliminary results suggest that functional neural transmission across interhemispheric pathways is strongly associated with some cognitive abilities in early adolescence, particularly verbal ability. We are currently investigating whether reduced right-to-left hemisphere transcallosal inhibition (assessed with two-coil TMS) in preterm children is associated with reduced development of

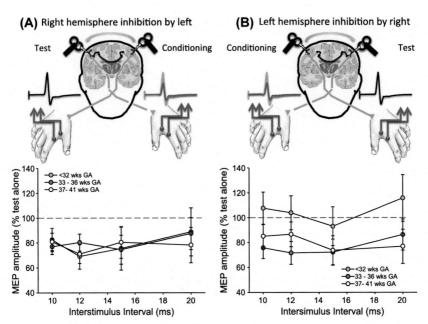

FIGURE 12.2   Inhibition of the motor potential evoked from the right motor cortex by conditioning stimulation of the left motor cortex (A), and of the left hemisphere by conditioning of the right hemisphere (B) in adolescents. Early preterm birth may reduce the strength of transcallosal inhibition that the right hemisphere exerts on the left. Left-to-right transcallosal inhibition appears preserved. *From Pitcher JB, Schneider LA, Harris RJ, Ridding MC, Burns NR. Reduced interhemispheric inhibition and neurodevelopment in adolescents born preterm.* J Paediatr Child Health. *2015;51(Suppl. 1):42; A126.*

hemispheric lateralization toward the left hemisphere, and the physiological mechanisms underlying how this might contribute to cognitive development, particularly verbal abilities. To our knowledge, there are currently no other published data of transcallosal inhibition in preterm individuals of any age group.

As with all other "at rest" TMS techniques, the application of transcallosal inhibition protocols in preterm children is limited by their relatively high resting motor thresholds. Our experience has been that these techniques have limited success in children born before 32 weeks completed GA and who are younger than approximately 12 years of age. In term-born children, we have recorded transcallosal inhibition in children as young as 10 years. But the use of a coil on each hemisphere simultaneously is limited by small head size. We have used customized 50-mm coils (and smaller), but the smaller the coil the more sensitive it is to movements on the scalp. This can be minimized using one of the commercially available stereotactic coil positioning and guidance systems. The obvious choice is to measure ipsilateral silent periods, as muscle activation and a single coil largely resolve these issues with the interhemispheric two-coil technique. But while the ipsilateral silent period is most likely mediated at least in part by transcallosal pathways, the lack of correlation between the degree of inhibition induced by the ipsilateral silent period and that induced by interhemispheric inhibition at short ISIs (ie, when interhemispheric inhibition is greatest), suggests that while related, these two techniques may reflect activity in different populations of cortical neurons or transcallosal pathways.[83]

NEUROPLASTICITY INDUCED WITH NON-INVASIVE BRAIN STIMULATION

A significant proportion of all preterm children will experience difficulties with motor, cognitive, and behavioral development that, while often subtle, adversely impact on their physical development, educational achievements, social adjustment, and income in adulthood.[11,84] This strongly suggests that preterm birth interferes with the brain's capacity to reorganize the strength of its synaptic connections in response to activity or experience; its neuroplasticity. TMS can be used to induce and measure neuroplasticity, particularly in the motor cortex. There are a variety of repetitive TMS protocols, as well as protocols that pair peripheral stimuli with single-pulse TMS, such as paired associative stimulation,[85] that can be used to selectively induce long-term potentiation-like (LTP; ie, synaptic strengthening) or long-term depression-like (LTD; ie, synaptic weakening) changes in the motor cortex. LTP and LTD are generally recognized as the major synaptic mechanisms underlying neuroplasticity, and that neuroplastic processes underlie learning and memory. Commonly, single-pulse TMS is used to compare changes in the amplitude of the motor-evoked potential before and after the plasticity induction, and the magnitude of the MEP change used as a marker of the neuroplasticity induced. However, changes in SICI and other processes have also been used as markers, and there is considerable debate in the literature regarding which is the more informative marker to use.

We used continuous theta burst stimulation (cTBS), which induces an LTD-like plasticity, to assess neuroplastic capacity in young adolescents born at a range of GAs.[86] In even only mildly preterm children, LTD-like neuroplasticity was blunted or absent when compared with their term-born peers, and was linearly associated with GA ($R^2 = 0.30$, $P = 0.004$); the more preterm, the smaller the LTD-like changes that could be evoked. However, a stronger predictor of the neuroplastic response was the salivary cortisol level immediately preceding neuroplasticity induction; children with the lowest cortisol levels had the poorest LTD-like neuroplasticity ($R^2 = 0.43$, $P = 0.001$). While preterm children tended to have the lowest cortisol levels, this was not statistically significant. However, these findings are consistent with the idea that preterm children often have altered hypothalamic–pituitary–adrenal (HPA) axis dysfunction, due to exposure to excess endogenous (maternal) and/or exogenous glucocorticoids in utero.[87] We are currently expanding these data in a blinded study to determine, amongst other things, whether LTP-like neuroplasticity is similarly affected in preterm young adolescents. In animal hippocampal slice studies, prenatal exposure to the commonly used glucocorticoid, betamethasone, selectively blocked the ability of cortisol to inhibit LTP induction in the hippocampi of female, but not male, guinea pigs, but otherwise, prenatal exposure to betamethasone had no effect on LTP induction.[88] So, on this basis, we might expect that LTP-like neuroplasticity is unaffected by preterm birth. However, to our knowledge, there is not a complementary animal study examining LTD. The majority of the preterm children in our original study were exposed to antenatal betamethasone and it was not powered to differentiate the effects of GA from antenatal glucocorticoid exposure. So whether or not LTD is blunted by GA or antenatal glucocorticoid exposure, or a combination of both, remains to be determined in our new larger cohort.

OTHER CONSIDERATIONS

This chapter has been confined to the use of TMS in preterm children without perinatal or other brain lesions. In the original PREMOCODE-1 study,[20,27] we also excluded children with sensory disabilities, congenital conditions, and children with a history of seizures, among other criteria; however, this led to an interesting bias in our sample. All the children born less than 27 completed weeks GA who fulfilled the criteria for study entry had BW% greater than 70%, with a mean of $83.7 \pm 8.2\%$, compared with $26.8 \pm 27.0\%$ in those born at 28–32 weeks. Apart from BW% and GA, these two GA groups did not differ significantly when their outcomes (including responses to TMS) were compared. There was insufficient power to detect any relationship with the indication for their preterm delivery; normal fetal growth might be expected prior to preterm delivery due to cervical insufficiency, but poor fetal growth was more likely prior to preterm birth due to maternal infection or placental insufficiency. However, this highlights several issues when studying preterm birth outcomes, with or without TMS. First, GA and BW% not only interact to influence brain development, but each can have quite profound, independent effects, that should be differentiated.[22] Second, multiple factors in the postnatal environment interact with prenatal (and intergenerational) factors to determine neurodevelopmental outcome, including seemingly discrete physiological measures, such as resting motor threshold.

WHERE TO NOW?

As is clear from the small number of studies performed to date on preterm children using non-invasive brain stimulation techniques such as TMS, there are significant physiological limitations regarding the age of children than can be assessed and the types of techniques that can be used. However, emerging techniques such as TMS-EEG and TMS-MRI potentially offer the opportunity to use brain stimulation techniques that not only bypass the need to obtain and record an MEP, but also investigate the effect of preterm birth on brain areas and functional networks other than the motor areas. While technically challenging, these techniques offer tangible options for elucidating the physiological mechanisms underlying the motor, cognitive, and behavioral difficulties faced by preterm children. Whether or not plasticity induction paradigms can be developed as targeted therapies, or to prime the preterm brain to be more receptive to conventional remedial motor, cognitive, and behavioral therapies, remains to be determined.

References

1. Volpe JJ. Brain injury in premature infants: a complex amalgam of destructive and developmental disturbances. *Lancet Neurol.* 2009;8(1):110–124.
2. Counsell SJ, Edwards AD, Chew ATM, et al. Specific relations between neurodevelopmental abilities and white matter microstructure in children born preterm. *Brain.* 2008;131(12):3201–3208.
3. Blüml S, Wisnowski JL, Nelson Jr MD, Paquette L, Panigrahy A. Metabolic maturation of White matter is altered in preterm infants. *PLoS One.* 2014;9(1):e85829.
4. Feldman HM, Lee ES, Loe IM, Yeom KW, Grill-Spector K, Luna B. White matter microstructure on diffusion tensor imaging is associated with conventional magnetic resonance imaging findings and cognitive function in adolescents born preterm. *Dev Med Child Neurol.* 2012;54(9):809–814.
5. Boyle EM, Poulsen G, Field DJ, et al. Effects of gestational age at birth on health outcomes at 3 and 5 years of age: population based cohort study. *BMJ.* 2012;344:e896.
6. Boyle JD, Boyle EM. Born just a few weeks early: does it matter? *Arch Dis Child Fetal Neonatal Ed.* 2011;98(1):F85–F88.
7. Raju TN, Higgins RD, Stark AR, Leveno KJ. Optimizing care and outcome for late-preterm (near-term) infants: a summary of the workshop sponsored by the National Institute of Child Health and Human Development. *Pediatrics.* 2006;118(3):1207–1214.
8. Chyi LJ, Lee HC, Hintz SR, Gould JB, Sutcliffe TL. School outcomes of late preterm infants: special needs and challenges for infants born at 32 to 36 weeks gestation. *J Pediatr.* 2008;153(1):25–31.
9. van Baar AL, Vermaas J, Knots E, de Kleine MJK, Soons P. Functioning at school age of moderately preterm children born at 32 to 36 weeks' gestational age. *Pediatrics.* 2009;124(1):251–257.
10. Anderson P, Doyle LW, The Victorian Infant Collaborative Study G. Neurobehavioral outcomes of school-age children born extremely low birth weight or very preterm in the 1990s. *JAMA.* 2003;289(24):3264–3272.
11. Lindstrom K, Winbladh B, Haglund B, Hjern A. Preterm infants as young adults: a Swedish National cohort study. *Pediatrics.* 2007;120(1):70–77.

12. Ziemann U. TMS and drugs. *Clin Neurophysiol.* 2004;115(8):1717–1729.
13. Rossini PM, Barker AT, Berardelli A, et al. Non-invasive electrical and magnetic stimulation of the brain, spinal cord and roots – basic principles and procedures for routine clinical application – report of an IFCN committee. *Electroencephalogr Clin Neurophysiol.* 1994;91(2):79–92.
14. Kloppel S, Baumer T, Kroeger J, et al. The cortical motor threshold reflects microstructural properties of cerebral white matter. *NeuroImage.* 2008;40(4):1782–1791.
15. Eyre JA, Miller S, Clowry GJ, Conway EA, Watts C. Functional corticospinal projections are established prenatally in the human foetus permitting involvement in the development of spinal motor centres. *Brain.* 2000;123(Part 1):51–64.
16. Paus T, Collins DL, Evans AC, Leonard G, Pike B, Zijdenbos A. Maturation of white matter in the human brain: a review of magnetic resonance studies. *Brain Res Bull.* 2001;54(3):255–266.
17. Imperati D, Colcombe S, Kelly C, et al. Differential development of human brain White matter tracts. *PLoS One.* 2011;6(8):e23437.
18. Pitcher JB, Doeltgen SH, Goldsworthy MR, et al. A comparison of two methods for evoking 50% of the maximal motor evoked potential. *Clin Neurophysiol.* 2015;126(12):2337–2341.
19. Flamand VH, Nadeau L, Schneider C. Brain motor excitability and visuomotor coordination in 8-year-old children born very preterm. *Clin Neurophysiol.* 2012;123(6):1191–1199.
20. Pitcher JB, Schneider LA, Burns NR, et al. Reduced corticomotor excitability and motor skills development in children born preterm. *J Physiol.* 2012;590(22):5827–5844.
21. Gardosi J, Figueras F, Clausson B, Francis A. The customised growth potential: an international research tool to study the epidemiology of fetal growth. *Paediatr Perinat Epidemiol.* 2011;25(1):2–10.
22. Pitcher JB, Robertson AL, Cockington RA, Moore VM. Prenatal growth and early postnatal influences on adult motor cortical excitability. *Pediatrics.* 2009;124(1):e128–e136.
23. Martinussen M, Fischl B, Larsson HB, et al. Cerebral cortex thickness in 15-year-old adolescents with low birth weight measured by an automated MRI-based method. *Brain.* 2005;128(11):2588–2596.
24. Ridding MC, Taylor JL, Rothwell JC. The effect of voluntary contraction on cortico-cortical inhibition in human motor cortex. *J Physiol.* 1995;487(Pt 2):541–548.
25. Fisher R, Nakamura Y, Bestmann S, Rothwell J, Bostock H. Two phases of intracortical inhibition revealed by transcranial magnetic threshold tracking. *Exp Brain Res.* 2002;V143(2):240–248.
26. Ortu E, Deriu F, Suppa A, Tolu E, Rothwell JC. Effects of volitional contraction on intracortical inhibition and facilitation in the human motor cortex. *J Physiol.* 2008;586(21):5147–5159.
27. Schneider LA, Burns NR, Giles LC, et al. Cognitive abilities in preterm and term-born adolescents. *J Pediatr.* 2014;165(1):170–177.
28. Rothwell JC, Day BL, Thompson PD, Kujirai T. Short latency intracortical inhibition: one of the most popular tools in human motor neurophysiology. *J Physiol.* 2009;587(1):11–12.
29. Strafella AP, Paus T. Cerebral blood-flow changes induced by paired-pulse transcranial magnetic stimulation of the primary motor cortex. *J Neurophysiol.* 2001;85(6):2624–2629.
30. Peurala SH, Muller-Dahlhaus JFM, Arai N, Ziemann U. Interference of short-interval intracortical inhibition (SICI) and short-interval intracortical facilitation (SICF). *Clin Neurophysiol.* 2008;119(10):2291–2297.
31. Ziemann U, Reis J, Schwenkreis P, et al. TMS and drugs revisited 2014. *Clin Neurophysiol.* 2015;126(10):1847–1868.
32. Schwenkreis P, Witscher K, Pleger B, Malin J-P, Tegenthoff M. The NMDA antagonist memantine affects training induced motor cortex plasticity – a study using transcranial magnetic stimulation. *BMC Neurosci.* 2005;6:35.

33. Stinear CM, Byblow WD. Role of intracortical inhibition in selective hand muscle activation. *J Neurophysiol.* 2003;89(4):2014–2020.
34. Coxon JP, Stinear CM, Byblow WD. Intracortical inhibition during volitional inhibition of prepared action. *J Neurophysiol.* 2006;95(6):3371–3383.
35. Ridding MC, Rothwell JC. Afferent input and cortical organisation: a study with magnetic stimulation. *Exp Brain Res.* 1999;126(4):536–544.
36. Rosenkranz K, Rothwell JC. The effect of sensory input and attention on the sensorimotor organization of the hand area of the human motor cortex. *J Physiol (Lond).* 2004;561(1):307–320.
37. Rosenkranz K, Seibel J, Kacar A, Rothwell J. Sensorimotor deprivation induces interdependent changes in excitability and plasticity of the human hand motor cortex. *J Neurosci.* 2014;34(21):7375–7382.
38. Rosenkranz K, Kacar A, Rothwell JC. Differential modulation of motor cortical plasticity and excitability in early and late phases of human motor learning. *J Neurosci.* 2007;27(44):12058–12066.
39. Kidgell DJ, Goodwill AM, Frazer AK, Daly RM. Induction of cortical plasticity and improved motor performance following unilateral and bilateral transcranial direct current stimulation of the primary motor cortex. *BMC Neurosci.* 2013;14.
40. Rogasch NC, Daskalakis ZJ, Fitzgerald PB. Cortical inhibition, excitation, and connectivity in schizophrenia: a review of insights from transcranial magnetic stimulation. *Schizophr Bull.* 2014;40(3):685–696.
41. Chen R, Cros D, Curra A, et al. The clinical diagnostic utility of transcranial magnetic stimulation: report of an IFCN committee. *Clin Neurophysiol.* 2008;119(3):504–532.
42. Reis J, Cohen LG, Pearl PL, et al. GABAB-ergic motor cortex dysfunction in SSADH deficiency. *Neurology.* 2012;79(1):47–54.
43. Croarkin PE, Nakonezny PA, Lewis CP, et al. Developmental aspects of cortical excitability and inhibition in depressed and healthy youth: an exploratory study. *Front Hum Neurosci.* 2014;8.
44. Hyde TM, Lipska BK, Ali T, et al. Expression of GABA signaling molecules KCC2, NKCC1, and GAD1 in cortical development and schizophrenia. *J Neurosci.* 2011;31 (30):11088–11095.
45. Hoegl T, Heinrich H, Barth W, Lösel F, Moll GH, Kratz O. Time course analysis of motor excitability in a response inhibition task according to the level of hyperactivity and impulsivity in children with ADHD. *PLoS One.* 2012;7(9):e46066.
46. Morita Y, Callicott JH, Testa LR, et al. Characteristics of the cation cotransporter NKCC1 in human brain: alternate transcripts, expression in development, and potential relationships to brain function and schizophrenia. *J Neurosci.* 2014;34(14):4929–4940.
47. Ben-Ari Y. Basic developmental rules and their implications for epilepsy in the immature brain. *Epileptic Disord.* 2006;8(2):91–102.
48. Ben-Ari Y, Woodin MA, Sernagor E, et al. Refuting the challenges of the developmental shift of polarity of GABA actions: GABA more exciting than ever! *Front Cell Neurosci.* 2012;6.
49. Fillman SG, Duncan CE, Webster MJ, Elashoff M, Weickert CS. Developmental co-regulation of the β and γ GABAA receptor subunits with distinct α subunits in the human dorsolateral prefrontal cortex. *Int J Dev Neurosci.* 2010;28(6):513–519.
50. Duncan CE, Webster MJ, Rothmond DA, Bahn S, Elashoff M, Shannon Weickert C. Prefrontal GABAA receptor α-subunit expression in normal postnatal human development and schizophrenia. *J Psychiatric Res.* 2010;44(10):673–681.
51. Mall V, Berweck S, Fietzek UM, et al. Low level of intracortical inhibition in children shown by transcranial magnetic stimulation. *Neuropediatrics.* 2004;35(02):120–125.
52. Walther M, Berweck S, Schessl J, et al. Maturation of inhibitory and excitatory motor cortex pathways in children. *Brain Dev.* 2009;31(7):562–567.

53. Schneider LA, Goldsworthy MR, Cole JP, Ridding MC, Pitcher JB. Maturation of short-interval inhibitory and excitatory circuits in the human primary motor cortex. *Brain Stimul*. 2015. [Under review].
54. Schneider C, Charpak N, Ruiz-Peláez JG, Tessier R. Cerebral motor function in very premature-at-birth adolescents: a brain stimulation exploration of kangaroo mother care effects. *Acta Paediatr*. 2012;101(10):1045–1053.
55. Hammond G, Garvey C-A. Asymmetries of long-latency intracortical inhibition in motor cortex and handedness. *Exp Brain Res*. 2006;172(4):449–453.
56. Hanajima R, Furubayashi T, Iwata NK, et al. Further evidence to support different mechanisms underlying intracortical inhibition of the motor cortex. *Exp Brain Res*. 2003;151(4):427–434.
57. Di Lazzaro V, Oliviero A, Mazzone P, et al. Direct demonstration of long latency cortico-cortical inhibition in normal subjects and in a patient with vascular parkinsonism. *Clin Neurophysiol*. 2002;113(11):1673–1679.
58. Nakamura H, Kitagawa H, Kawaguchi Y, Tsuji H. Intracortical facilitation and inhibition after transcranial magnetic stimulation in conscious humans. *J Physiol*. 1997;498 (Pt 3):817–823.
59. McDonnell M, Orekhov Y, Ziemann U. The role of GABA$_B$ receptors in intracortical inhibition in the human motor cortex. *Exp Brain Res*. 2006;173(1):86–93.
60. Werhahn KJ, Kunesch E, Noachtar S, Benecke R, Classen J. Differential effects on motorcortical inhibition induced by blockade of GABA uptake in humans. *J Physiol (London)*. 1999;517(2):591–597.
61. Fuhr P, Agostino R, Hallett M. Spinal motor neuron excitability during the silent period after cortical stimulation. *Electroencephalogr Clin Neurophysiol/Evoked Potentials Sect*. 1991;81(4):257–262.
62. Inghilleri M, Berardelli A, Cruccu G, Manfredi M. Silent period evoked by transcranial stimulation of the human cortex and cervicomedullary Junction. *J Physiol*. 1993;466:521–534.
63. Ferbert A, Priori A, Rothwell JC, Day BL, Colebatch JG, Marsden CD. Interhemispheric inhibition of the human motor cortex. *J Physiol*. 1992;453:525–546.
64. Buchmann J, Wolters A, Haessler F, Bohne S, Nordbeck R, Kunesch E. Disturbed trans-callosally mediated motor inhibition in children with attention deficit hyperactivity disorder (ADHD). *Clin Neurophysiol*. 2003;114(11):2036–2042.
65. Garvey MA, Barker CA, Bartko JJ, et al. The ipsilateral silent period in boys with attention-deficit/hyperactivity disorder. *Clin Neurophysiol*. 2005;116(8):1889–1896.
66. Lindström K, Lindblad F, Hjern A. Preterm birth and attention-deficit/hyperactivity disorder in schoolchildren. *Pediatrics*. 2011;127(5):858–865.
67. Van Hus JW, Potharst ES, Jeukens-Visser M, Kok JH, Van Wassenaer-Leemhuis AG. Motor impairment in very preterm-born children: links with other developmental deficits at 5 years of age. *Dev Med Child Neurol*. 2014;56(6):587–594.
68. Gazzaniga MS. Forty-five years of split-brain research and still going strong. *Nat Rev Neurosci*. 2005;6(8):653–659.
69. Sacco S, Moutard M-L, Fagard J. Agenesis of the corpus callosum and the establishment of handedness. *Dev Psychobiol*. 2006;48(6):472–481.
70. Giedd JN, Blumenthal J, Jeffries NO, et al. Brain development during childhood and adolescence: a longitudinal MRI study. *Nat Neurosci*. 1999;2:861–863.
71. Back SA, Miller SP. Brain injury in premature neonates: a primary cerebral dysmaturation disorder? *Ann Neurol*. 2014;75(4):469–486.
72. Meyer B-U, Röricht S, Woiciechowsky C. Topography of fibers in the human corpus callosum mediating interhemispheric inhibition between the motor cortices. *Ann Neurol*. 1998;43(3):360–369.
73. De Gennaro L, Cristiani R, Bertini M, et al. Handedness is mainly associated with an asymmetry of corticospinal excitability and not of transcallosal inhibition. *Clin Neurophysiol*. 2004;115(6):1305–1312.

74. Di Lazzaro V, Oliviero A, Profice P, et al. Direct demonstration of interhemispheric inhibition of the human motor cortex produced by transcranial magnetic stimulation. *Exp Brain Res*. 1999;124(4):520–524.

75. Meyer BU, Roricht S, Voneinsiedel HG, Kruggel F, Weindl A. Inhibitory and excitatory interhemispheric transfers between motor cortical areas in normal humans and patients with abnormalities of the corpus callosum. *Brain*. 1995;118(Part 2):429–440.

76. Gerloff C, Cohen LG, Floeter MK, Chen R, Corwell B, Hallett M. Inhibitory influence of the ipsilateral motor cortex on responses to stimulation of the human cortex and pyramidal tract. *J Physiol (London)*. 1998;510(1):249–259.

77. Heinen F, Glocker F-X, Fietzek U, Meyer B-U, Lücking C-H, Korinthenberg R. Absence of transcallosal inhibition following focal mangnetic stimulation in preschool children. *Ann Neurol*. 1998;43(5):608–612.

78. Kwon HG, Son SM, Jang SH. Development of the transcallosal motor fiber from the corticospinal tract in the human brain: diffusion tensor imaging study. *Front Hum Neurosci*. 2014;8.

79. Pitcher JB, Schneider LA, Harris RJ, Ridding MC, Burns NR. Reduced interhemispheric inhibition and neurodevelopment in adolescents born preterm. Paper presented at: perinatal Society of Australia and New Zealand, 19th Annual Conference 2015. Melbourne, Australia.

80. Pujol J, Deus J, Losilla JM, Capdevila A. Cerebral lateralization of language in normal left-handed people studied by functional MRI. *Neurology*. 1999;52(5):1038–1043.

81. Knecht S, Drager B, Deppe M, et al. Handedness and hemispheric language dominance in healthy humans. *Brain*. 2000;123(12):2512–2518.

82. Orton ST. Word-blindness. by Samuel Torrey Orton, MD. In: *School children and other papers on strephosymbolia (specific language disability—dyslexia) 1925–1946*. Vol 2. Pomfret, Conneticutt: Orton Society; 1966.

83. Chen R, Yung D, Li J-Y. Organization of ipsilateral excitatory and inhibitory pathways in the human motor cortex. *J Neurophysiol*. 2003;89(3):1256–1264.

84. Bennet L, Van Den Heuij L, Dean JM, Drury P, Wassink G, Jan Gunn A. Neural plasticity and the Kennard principle: does it work for the preterm brain? *Clin Exp Pharmacol Physiol*. 2013;40(11):774–784.

85. Stefan K, Kunesch E, Cohen LG, Benecke R, Classen J. Induction of plasticity in the human motor cortex by paired associative stimulation. *Brain*. 2000;123(Part 3):572–584.

86. Pitcher JB, Riley AM, Doeltgen SH, et al. Physiological evidence consistent with reduced neuroplasticity in human adolescents born preterm. *J Neurosci*. 2012;32(46):16410–16416.

87. Moisiadis VG, Matthews SG. Glucocorticoids and fetal programming part 1: outcomes. *Nat Rev Endocrinol*. 2014;10(7):391–402.

88. Setiawan E, Jackson MF, MacDonald JF, Matthews SG. Effects of repeated prenatal glucocorticoid exposure on long-term potentiation in the juvenile guinea-pig hippocampus. *J Physiol*. 2007;581(3):1033–1042.

Brain Stimulation to Understand and Modulate the Autism Spectrum

L.M. Oberman

Brown University, Providence, RI, United States

A. Pascual-Leone, A. Rotenberg

Harvard Medical School, Boston, MA, United States

INTRODUCTION

Autism spectrum disorder (ASD) is a behaviorally defined complex neurodevelopmental syndrome characterized by impairments in social communication, restricted and repetitive behaviors, interests and activities, and abnormalities in sensory reactivity.[1] While ASD diagnostics are principally based upon behavioral assessments, neuropathologic, genetic and neuroimaging data indicate that the behavioral ASD phenotype is the product of aberrant brain development.[2] More specifically, numerous recently identified ASD-related genes implicate abnormal synaptic function as the core of ASD pathophysiology.[3–7] These data also indicate that ASD pathophysiology likely is not limited to dysfunction of a single brain region, but rather involves a breakdown in functioning of distributed neural circuits.

Some published reports suggest an imbalance in cortical excitation/ inhibition favoring excess excitation and/or deficient in inhibition. This is underscored by a vulnerability to seizures and epilepsy in the ASD population. ASD research also points to a role for aberrant synaptic plasticity in ASD pathogenesis.[8–12] Consistent with this hypothesis, many of the genes associated with ASD regulate aspects of synaptic development and basic mechanisms of long-term potentiation (LTP) and long-term depression (LTD) of excitatory synaptic strength.[5] Indeed, monogenic ASD animal models routinely exhibit deficits in LTP or LTD as part of their phenotype, and ASD can be conceptualized as primarily a "synaptopathology."[13]

Though most evidence for an increase in the excitation/inhibition ratio and altered synaptic plasticity in humans with ASD is indirect, direct measures of circuit level abnormalities in excitability and plasticity in humans can be obtained by transcranial magnetic stimulation (TMS).[14–17] Given its capacity to modulate brain excitability and network activity beyond the duration of a stimulation train, repetitive TMS (rTMS) has also been explored as both a measure of cortical synaptic plasticity and as a therapeutic intervention for ASD.

TMS MEASURES OF ASD PATHOPHYSIOLOGY

TMS paradigms have been developed to probe cortical reactivity, intracortical excitation/inhibition balance, and plasticity, and have been used to explore the neurophysiology of ASD. Most studies to date have focused on relatively older individuals with ASD who do not have other intellectual disability. These studies, summarized below by TMS modality, show a number of atypical responses in the ASD population (findings summarized in Table 13.1).

TABLE 13.1 Published Reports of Investigational Use of TMS in ASD

Publication	Diagnosis	Intellectual Disability	n (ASD)	Age Range (Years)	Gender (Male: Female)	Medication	Site	Protocol	Intensity
Theoret et al.[18] Current Biology	ASD	None	10	23–58	7:3	–	L M1	Single pulse and paired pulse (1, 2, 3, 6, 9, 12, 15 ms ISI)	Single pulse: 90%, 100%, 05%, 110%, 115%, 120%, 130%, 140%, 150%, 160% RMT Paired pulse: 80% and 120% RMT
Minio-Paluello et al.[19] Biological Psychiatry	AS	None	16	21–35	16:0	–	L M1	Single pulse	120% RMT
Oberman et al.[28] Frontiers in Synaptic Neuroscience	AS	None	5	26–54	3:2	–	L M1	cTBS	80% AMT
Enticott et al.[20] Developmental Medicine and Child Neurology	ASD (11 HFA, 14 AS)	None	25	9–26	20:5	Clomipramine, risperidone, quetiapine, venlafaxine, flouxetine, valproate	L and R M1	Single pulse and paired pulse (2 and 15 ms ISI)	Single pulse: 115% RMT Paired pulse: 90% and 115% RMT
Jung et al.[27] Developmental Medicine and Child Neurology	ASD (7 HFA, 6 AS, 2 PDD-NOS)	None	15	15–29	14:1	None	R M1 (PAS)	PAS, single pulse and paired pulse (2 and 3 ms ISI)	Lowest intensity producing average 1-mV motor-evoked potential

Continued

TABLE 13.1 Published Reports of Investigational Use of TMS in ASD—cont'd

Publication	Duration	Trains	Pulses Delivered	Sessions	Blinding	Assessment Times	Reported Effects	Side Effects
Theoret et al.[18] Current Biology	n/a	n/a	264	1	None	Online	No group difference in RMT or response to ppTMS. Impaired corticospinal facilitation in response to finger movements viewed from the egocentric point of view in the ASD group	Not indicated
Minio-Paluello et al.[19] Biological Psychiatry	n/a	n/a	72	1	None	Online	No modulation of corticospinal excitability in response to the observation of painful stimuli affecting another individual in the Asperger's group	Not indicated
Oberman et al.[28] Frontiers in Synaptic Neuroscience	190s (iTBS),47s (cTBS)	iTBS: 20×2s, 10s ISI; cTBS1 train of 47s	600	2 (one cTBS, one iTBS)	None	Before, 5, 10, 20, 30, 40, 50, 60, 75, 90, 105, 120min after cTBS	Longer-lasting facilitation (enhanced MEP) following iTBS in ASD Longer-lasting-inhibition (suppressed MEP) following cTBS in ASD No effect the following day using opposite protocol for ASD (suggesting enhanced metaplasticity)	None
Enticott et al.[20] Developmental Medicine and Child Neurology	n/a	n/a	120	1	None	Online	Reduced intracortical inhibition in the HFA group as compared to the AS or control group	Not indicated
Jung et al.[27] Developmental Medicine and Child Neurology	13.3min	n/a	200	1	None	Before, after, 30min after, 60min after PAS	No LTP-like MEP facilitation in ASD (group difference significant at 60min). No group difference in response to ppTMS	–

Publication	Diagnosis	Intellectual Disability	n (ASD)	Age Range (Years)	Gender (Male: Female)	Medication	Site	Protocol	Intensity
Oberman et al.[23] European Journal of Neuroscience	AS	None	35 (cTBS), 9 also had iTBS	18–64	30:5	–	L M1	cTBS, iTBS	80% AMT
Enticott et al.[26] Biological Psychiatry	ASD	None	34	14–49	26:8	SSRI, atypical antipsychotic, benzodiazepine, antidepressant	L M1	Single pulse	120% RMT
Enticott et al.[21] Frontiers in Human Neuroscience	ASD	None	32	14–40	24:8	SSRI, atypical antipsychotic, benzodiazepine, antidepressant	L M1	Single pulse	120% RMT
Enticott et al.[22] Neuropharmacology	ASD	None	36	14–49	28:8	Fluoxetine, citalopram, sertraline, lorazepam, olanzapine, venlafaxine, risperidone, mirtazapine, quetiapine	L and R M1	Single pulse and paired pulse (2, 15, and 100 ms ISI)	Single pulse = 115% and 130% RMT; 115% and 130% AMT Paired pulse = 90% RMT and 120% RMT
Oberman et al.[31] Frontiers in Human Neuroscience	ASD	None	19	9–18	19:0	Citalopram, atomoxetine, buspirone, sertraline, methylphenidate, risperidone, guanfacine	L M1	cTBS	80% AMT
Oberman et al.[29] Medical Hypotheses	ASD	None	35	18–64	30:5	–	L M1	cTBS	80% AMT

Continued

TABLE 13.1   Published Reports of Investigational Use of TMS in ASD—cont'd

Publication	Duration	Trains	Pulses Delivered	Sessions	Blinding	Assessment Times	Reported Effects	Side Effects
Oberman et al.[23] European Journal of Neuroscience	190s (iTBS)40s (cTBS)	iTBS: 20×2s, 10s ISI; cTBS1 train of 40s	600	1 or 2	Data analysis (patient group)	Before, 5, 10, 20, 30, 40, 50, 60, 75, 90, 105, 120min after cTBS	Longer-lasting facilitation (enhanced MEP) following iTBS in ASD Longer-lasting inhibition (suppressed MEP) following TBS in ASD	None
Enticott et al.[26] Biological Psychiatry	n/a	n/a	50	1	None	Online	No group difference in degree of corticospinal excitability in response to observation of single static hand stimuli. Impaired corticospinal facilitation in response to single hand transitive hand actions in the ASD group	Not indicated
Enticott et al.[21] Frontiers in Human Neuroscience	n/a	n/a	50	1	None	Online	No group difference in degree of corticospinal excitability in response to single static hand stimuli or two person interactive hands	Not indicated
Enticott et al.[22] Neuropharmacology	n/a	n/a	170	1	None	Online	No group difference in RMT. Heterogeneous response to paired-pulse TMS in the ASD group	Not indicated
Oberman et al.[21] Frontiers in Human Neuroscience	40s	1 train of 40s	600	1	Data analysis (patient group)	Before, 5, 10, 20, 30, 40, 50, 60, 75, 90, 105, 120min after cTBS	Positive linear relationship between age and duration of modulation of TBS after effects in children and adolescents with ASD. A subgroup of the ASD participants showed paradoxical facilitation	Mild headache, mild fatigue
Oberman et al.[29] Medical Hypotheses	40s	1 train of 40s	600	1	Data analysis (patient group)	Before, 5, 10, 20, 30, 40, 50, 60, 75, 90, 105, 120min after cTBS	Longer-lasting inhibition (suppressed MEP) and greater degree of inhibition (area under the curve) following TBS in ASD. Age did not significantly contribute to the model	None

All studies used TMS-evoked MEP amplitude as outcome measures. All studies used figure-of-eight coils.
ASD, autism spectrum disorder; HFA, high-functioning autism; PDD-NOS, pervasive developmental disorder, not otherwise specified; AS, Asperger's syndrome; M1, primary motor cortex; ISI, interstimulus interval; ppTMS, paired-pulse TMS; PAS, paired associative stimulation; iTBS, intermittent theta burst stimulation; cTBS, continuous theta burst stimulation; RMT, resting motor threshold; AMT, active motor threshold; MEP, motor-evoked potential; LTP, long-term potentiation; SSRI, selective serotonin reuptake inhibitor.

Single-Pulse TMS

Single-pulse TMS has been applied to the primary motor cortex (M1) to probe corticospinal excitability as measured by baseline motor threshold, the lowest intensity of stimulation required to induce a motor-evoked potential (MEP) or size of MEP in response to a suprathreshold pulse of TMS. Single-pulse TMS has also been used to investigate the modulation of corticospinal excitability in response to visually presented stimuli.

Six studies have found no difference in either motor threshold or MEP size of response to a suprathreshold TMS pulse between individuals with ASD and neurotypical individuals.[18–23] These results suggest that baseline primary motor cortex excitability is not affected in ASD. However, these results may be biased by: (1) the inclusion of autistic individuals on psychoactive medications; (2) the restriction to subjects without intellectual disability; and (3) the exclusion of autistic individuals with a history of seizures or abnormal electroencephalography (EEG) findings, and thereby a hyperexcitable cerebral cortex (see Ref. 24).

In protocols that combine a behavioral task with single-pulse TMS, several studies have assessed the modulation of M1 excitability following the observation of another person's actions. In neurotypical individuals, the observation of another person's actions results in a simultaneous activation of the observer's sensorimotor system. This phenomenon is termed interpersonal motor resonance (IMR), and may be an index of mirror neuron system activity.[25] These studies report mixed results in subjects with ASD that appear to be dependent on the properties of the stimuli. Theoret et al.[18] published the first of these studies and reported a reduction in modulation of corticospinal excitability in the ASD group, but only when the action (simple finger movement) was presented from an egocentric point of view. Another study, conducted by Minio-Paluello et al.[19] used stimuli depicting either a painful pinprick or a cotton swab touch of a static hand. Individuals with ASD, as compared with neurotypical controls, had a significantly reduced modulation in corticospinal excitability in response to observation of the painful stimulus.[19] Two additional studies[21,26] have been published exploring IMR in ASD. In one, the investigators found impaired modulation of corticospinal excitability in response to a transitive (goal- or object-directed) hand movement, but intact modulation in response to a static hand stimulus.[26] In a second study, the same group reported intact modulation of corticospinal excitability in response to both a static hand stimulus and a stimulus involving two hands interacting. Such findings suggest that differences in visual processing or attention may account for the aberrant IMR responses to certain stimuli, but typical responses to others.[21]

Paired-Pulse TMS

Studies using paired-pulse TMS (ppTMS) have investigated intracortical inhibition and facilitation in ASD. These studies, despite their somewhat mixed results, are relevant to the hypothesis that an increased cortical excitation/inhibition ratio contributes to ASD pathophysiology. Two studies report no significant difference in response to the short intracortical inhibition (SICI) paradigm between ASD and neurotypical individuals.[18,27] Three studies have employed the intracortical facilitation (ICF) paradigm and found no significant difference between ASD and neurotypical controls.[18,20,22] Three studies have also reported mixed results with some ASD individuals showing impaired intracortical inhibition and others showing typical responses.[20,22,28] One caveat is that all of these findings were from studies that excluded individuals with comorbid epilepsy, which is very prevalent in ASD, thus the generalizability to the overall population is unclear. Enticott et al.[20] report that SICI is absent in subjects with high-functioning autism (with language delay), while not distinct from neurotypical levels in subjects with Asperger's syndrome, an ASD subgroup that is not associated with language delay. Similarly, the same group reported in a later study an absence of SICI in a subgroup of ASD participants with early language delay, but not in those with typical language acquisition development.[22] Oberman et al. also reported paradoxical facilitation in response to the long intracortical inhibition (LICI) and SICI paradigms in a small sample of participants with ASD.[28] Thus, abnormal intracortical inhibition may be present in a subgroup of individuals with ASD, but this alteration of cortical physiology is not consistently demonstrable in all individuals with ASD.

rTMS

Two rTMS paradigms – paired-associative stimulation (PAS) and theta burst stimulation (TBS) – have been employed to study corticospinal plasticity mechanisms in ASD. Jung et al.[27] found abnormally reduced LTP-like facilitation of MEPs following the PAS (25 ms) paradigm, suggesting an impairment in Hebbian plasticity mechanisms. Seemingly counter to these findings, Oberman et al. published a series of studies where high-functioning adults with ASD showed an *increased* duration of response to the TBS paradigm that appears to be present from age 18 to 65 and reliable across studies and cohorts.[23,28,29] The latter findings led the authors to suggest that the symptoms of ASD may result from an underlying abnormality in cortical synaptic plasticity mechanisms.[8,30] An additional study, where TBS was applied to children with ASD, demonstrated an increase in the duration of response across childhood, and revealed a subgroup of children who showed paradoxical facilitation to the typically suppressive cTBS paradigm.[31] The authors suggested that their findings may reflect abnormalities in GABAergic inhibitory control in those individuals who showed paradoxical facilitation. Thus, both PAS and TBS applied to ASD suggest abnormalities in cortical plasticity mechanisms; however, initial studies have yielded conflicting findings, with PAS suggesting impaired and TBS enhanced plasticity. These differences may reflect the small sample sizes of the studies, etiologic heterogeneity within ASD, or paradigmatic differences (Hebbian vs. non-Hebbian plasticity).[32]

In summary, the findings from the above-mentioned literature using TMS as an investigational device partially support the molecular, animal model, and imaging literature suggesting an increase in the excitation/inhibition ratio and aberrant plasticity mechanisms. However, what these studies reveal most clearly is the variability of the findings. Other than no abnormality in baseline corticospinal excitability, all other indexes of response to TMS vary both within and across studies. One should note that the sample sizes in the studies are relatively small (ranging from 5 to 36) and all of the studies enrolled high-functioning individuals. Thus, a number of unanswered questions related to the use of TMS as an investigative device in ASD remain, including whether the aberrant physiological findings are causal or a consequence of ASD pathology, developmental effects, effects related to intellectual disability, comorbidities, or verbal abilities, and what underlying mechanisms are driving the observed heterogeneity in the population. Despite the limitations of the studies to date, the results suggest that TMS measures of altered and variable brain physiology among individuals with ASD have the potential to serve as biomarkers to guide the search for ASD subtypes. Importantly, all these TMS measures can be applied to animal models[33–37] and thus might serve as valuable translatable phenotypes to bridge model systems and human applications.

TMS AS A THERAPEUTIC INTERVENTION IN ASD

A small number of studies[38–48] and case reports[49–51] have investigated whether neuromodulation via rTMS can induce neurophysiological and possible clinical improvements in ASD (findings summarized in Table 13.2). A range of symptom domains have been targeted including: irritability, repetitive behaviors, executive functioning, motor behavior, speech production, mentalizing, eye–hand coordination. These studies have targeted a number of brain regions and circuits including: dorsal lateral prefrontal cortex (DLPFC), medial prefrontal cortex, supplementary motor, premotor, and primary motor cortices. Finally, both low-frequency and high-frequency stimulation have been applied.

TABLE 13.2 Published Reports of Therapeutic Use of TMS in ASD

Publication	Study Type	Diagnosis	Intellectual Disability	n (ASD)	Age Range (Years)	Gender (Male: Female)	Medication	Site	Coil	Protocol	Intensity	Duration	Trains	Pulses Delivered
Sokhadze et al.[44] Journal of autism and developmental disorders	RCT with waitlist control	Autism	None	13 (8 rTMS, 5 waitlist)	12–27	13:0	–	L dlPFC (5 cm anterior to M1)	Fo8	0.5 Hz	90% RMT	10 min	15×10 s, 20–30 s ISI	150
Baruth et al.[42] Journal of Neurotherapy	RCT with waitlist control	ASD	2	25 (16 rTMS, 9 waitlist)	9–26	21:4	–	L & R dlPFC (5 cm anterior to M1)	Fo8	1 Hz	90% RMT	10 min	15×10 s, 20–30 s ISI	150
Sokhadze et al.[43] Applied Psychophysiology and Biofeedback	Clinical trial with no control	Autism	None	13	8–27	12:1	–	L dlPFC (5 cm anterior to M1)	Fo8	0.5 Hz	90% RMT	10 min	15×10 s, 20–30 s ISI	150

Publication	Sessions	Blinding	Measures	Assessment Times	Reported Effects	Side Effects
Sokhadze et al.[44] Journal of Autism and Developmental Disorders	6	None	EEG Gamma power Accuracy, RT (Kanizsa) Abberant Behavior Checklist (ABC) Social Responsiveness Scale (SRS) (Caregiver report) Repetitive Behavior Scale-Revised (RBS-R)	Before and 2 weeks after treatment course	Decreased frontal EEG P3a amplitude To non-targets following TMS decreased centroparietal latency EEG P3b to non-target and non-Kanizsa following TMS Decrease in gamma power for non- target and non-Kanizsa following TMS Reduced repetitive behavior (RBS-R) following TMS	–
Baruth et al.[42] Journal of Neurotherapy	12 (6L, 6R)	None	EEG Gamma power Accuracy, RT (Kanizsa) Abberant Behavior Checklist (ABC) Social Responsiveness Scale (SRS) (Caregiver report) Repetitive Behavior Scale-Revised (RBS-R)	Before and 2 weeks after treatment course	Increased EEG gamma power to targets, decrease to non-targets Reduced repetitive behavior (RBS-R) Reduced irritability (ABC)	Itching sensation at nose (5) Mild headache (1)
Sokhadze et al.[43] Applied Psychophysiology and Biofeedback	6	None	EEG event-related potentials (visual oddball) Accuracy, RT (Kanizsa) Abberant Behavior Checklist (ABC) Social Responsiveness Scale (SRS) (Caregiver report) Repetitive Behavior Scale – Revised (RBS-R)	Before and 2 weeks after treatment course	Reduced error rate Increased frontal EEG P50 amplitude to targets Increased frontal EEG P50 latency to targets Decreased frontal EEG N200 latency to novel distractors Decreased parieto-occipital EEG P50 to novel distractor Increased centroparietal EEG P50 to targets and decreased to standard distractor Centro-parietal EEG P3b amplitude increase to targets and decrease to standard distractors Centro-parietal EEG P200 increased latency to targets Reduced repetitive behaviors (RBS-R)	–

Continued

TABLE 13.2 Published Reports of Therapeutic Use of TMS in ASD—cont'd

Publication	Study Type	Diagnosis	Intellectual Disability	n (ASD)	Age Range (Years)	Gender (Male: Female)	Medication	Site	Coil	Protocol	Intensity	Duration	Trains	Pulses Delivered
Enticott et al.[50] Journal of ECT	Single case study	AS	None	1	20	0:1	None	Bilateral dmPFC (7 cm anterior to M1)	H-coil	5 Hz	100% RMT	15 min	30 × 10s, 20s ISI	1500
Fecteau et al.[46] European Journal of Neuroscience	Crossover trial with sham control	AS	None	10	18–62	7:3	None	L & R pars opercularis, L & R par triangularis (MRI neuronavigation), sham (central lobe midline)	Fo8	1 Hz	70% Of stimulator output	30 min	1	1800
Casanova et al.[38] Translational Neuroscience	RCT with waitlist control	ASD	None	45 (25 rTMS, 20 waitlist)	9–19	39:6	–	L & R dlPFC (5 cm anterior to M1)	Fo8	1 Hz	90% RMT	10 min	15 × 10s, 20–30s ISI	150
Enticott et al.[45] Brain Stimulation	Crossover trial with sham control	ASD (6 HFA, 5 AS)	None	11	14–26	10:1	–	SMA (15% of nasion to inion anterior to Cz), L M1, sham (M1)	Fo8	1 Hz	100% RMT	15 min	1	900
Niederhofer[51] Clinical Neuropsychiatry	Single case study with placebo	Autism	Not reported	1	42	0:1	None	M1	–	1 Hz	–	1h	1	1200
Sokhadze et al.[52] Applied Psychophysiology and Biofeedback	RCT with waitlist control	ASD (36 HFA, 4 AS)	None	40 (20 rTMS, 20 waitlist)	9–21	32:8	–	L & R dlPFC (5 cm anterior to M1)	Fo8	1 Hz	90% RMT	10 min	15 × 10s, 20–30s ISI	150

Publication	Sessions	Blinding	Measures	Assessment Times	Reported Effects	Side Effects
Enticott et al.[50] Journal of ECT	10	Double	Interpersonal Reactivity Index (IRI) Autism Spectrum Quotient (AQ) Ritvo Autism-Aspergers Diagnostic Scale (RAADS)	Before, after, and 1 month after treatment course	Reduction on all measures Anecdotal reports of improvement from patient and relatives	None
Fecteau et al.[46] European Journal of Neuroscience	1 per site	Double	Response latency on Boston Naming Test	Before and after TMS	Increased response latency after L pars opercularis Decreased response latency after L pars triangularis	Many reported, including; Sleepy Trouble concentrating Improved mood Headache Dizziness
Casanova et al.[38] Translational Neuroscience	12 (6 L, 6 R)	None	EEG event-related potentials (ERP) Accuracy, RT (Kanizsa) Aberrant Behavior Checklist (ABC) Social Responsiveness Scale (SRS) (Caregiver report) Repetitive Behavior Scale – Revised (RBS-R)	Before and 2 weeks after treatment course	Reduced error rate Increased frontal EEG N200 to targets Reduced frontal EEG N200 latency increased frontal RHEEG P300 to target Increased parietal EEG N200 to targets Reduced repetitive behavior (RBS) Reduced irritability (ABC)	–
Enticott et al.[48] Brain Stimulation	1 per site	Single	EEG movement-related cortical potentials motor response time	Before and after TMS	SMA: Increased early EEG component PMC: Increased EEG negative slope	–
Niederhofer[51] Clinical Neuropsychiatry	5	Single	Aberrant Behavior Checklist (ABC)	Before and after treatment course	ABC Irritability: Active 40–33, Sham 39–35 ABC Sterotypy: Active 18–12, sham 16–15	-
Sokhadze et al.[52] Applied Psychophysiology and Biofeedback	12 (6 L, 6 R)	None	EEG event-related potentials (ERPs) RT Accuracy (Kanizsa)	Before and after treatment course	Reduced omission error rates increased EEG ERN amplitude reduced EEG ERN latency	–

Continued

TABLE 13.2 Published Reports of Therapeutic Use of TMS in ASD—cont'd

Publication	Study Type	Diagnosis	Intellectual Disability	n (ASD)	Age Range (Years)	Gender (Male: Female)	Medication	Site	Coil	Protocol	Intensity	Duration	Trains	Pulses Delivered
Panerai et al.[47] Autism	(i) Crossover trial with sham (ii) RCT with sham (iii) Crossover trial with sham (iv) RCT (TMS, EHI training, TMS+EHI training)	Autism	Severe to profound	(i) 9 (ii) 17 (iii) 4 (iv) 13	11–16	43:0	–	(i) left and right PrMC (2.5 cm anterior to M1) (ii), (iii), (iv) left PrMC	Fo8	(i) 8 Hz, 1 Hz (ii) 8 Hz, 1 Hz (iii), (iv) 8 Hz	90% RMT	8 Hz 30 mins, 1 Hz 15 min;	8 Hz 30×3.6 s, 56.4 s ISI	900
Enticott et al.[45] Brain Stimulation	RCT with sham control	ASD (4 HFA, 24 AS)	None	28 (15 active, 13 sham)	18–59	23:5	Yes (39%)	dmPFC (7 cm anterior to M1)	H-coil	5 Hz	100% RMT	15 min	30×10 s, 20 s ISI	1500
Cristancho et al.[49] Journal of ECT	Single case study	Autism, depression	Not reported	1	15	1:0	Olanzapine, fluoxetine, guanfacine, clonazepam	(i) R DLPFC (6 cm anterior to M1), (ii) L DLPFC (6 cm anterior to M1)	Fo8	1 Hz	90% RMT	Variable (between 5 and 25 min)	(i) 15×10 s, 10–30 s ISI (week 1), 30×10 s, 10–30 s ISI (week 2), (ii) 30–60×10 s, 10–15 s ISI	(i) 150–300, (ii) 300–600

Publication	Sessions	Blinding	Measures	Assessment Times	Reported Effects	Side Effects
Panerai et al.[47] Autism	(i) 3; (ii) 10; (iii) 5 active, 5 sham, 5 active, 5 sham; (iv) 10	Double	(i), (ii), (iii), (iv) Psychoeducational profile – revised (PEP-R) eye–hand coordination	Before and after treatment course	(i), (ii), (iii) improved eye–hand coordination score following lPrMC HF TMS (iv) Improved eye–hand coordination score following combined TMS + EHI training compared to each technique alone	—
Enticott et al.[45] Brain Stimulation	10	Double	Interpersonal Reactivity Index (IRI) Autism Spectrum Quotient (AQ) Ritvo Autism-Aspergers Diagnostic Scale (RAADS) Reading the Mind in the Eyes Test (RMET) Mentalising Animations Task	Before, after, and 1 month after treatment course	Reduced social relatedness (RAADS) Reduced personal distress (IRI)	1 "Light-headedness" 2 facial discomfort during rTMS
Cristancho et al.[49] Journal of ECT	(i) 10, (ii) 26	None	Mental status examination	Before and after treatment course	Anecdotal reports of improve mood, eye contact, interpersonal communication, verbal expression, focus, activity	Mild headaches, jaw twiching, transient dizziness

Continued

TABLE 13.2 Published Reports of Therapeutic Use of TMS in ASD—cont'd

Publicaiton	Study Type	Diagnosis	Intellectual Disability	n (ASD)	Age Range (Years)	Gender (Male: Female)	Medication	Site	Coil	Protocol	Intensity	Duration	Trains	Pulses Delivered
Casanova et al.[39] Frontiers in Human Neuroscience	Proof of feasibility study, no control	ASD	None	18	9–21	14:4	–	L & R dlPFC (5 cm anterior to M1)	Fo8	5 Hz	90% RMT	10 min	8 × 10 s, 20 s ISI	160
Sokhadze et al.[40] Frontiers in Systems Neuroscience	RCL with waitlist control	ASD	None	54	9–21	44:10	–	L & R dlPFC (5 cm anterior to M1)	Fo8	1 Hz	90% RMT	10 min	9 × 20 s, 20–30 s ISI	180
Sokhadze et al.[41] Appl Psychophysiol Biofeedback	Clinical trial with waitlist control	ASD	None	42	10–21	34:8	-	L & R dlPFC (5 cm anterior to M1)	Fo8	1 Hz	90% RMT	10 min	9 × 20 s, 20–30 s ISI	180

Publication	Sessions	Blinding	Measures	Assessment Times	Reported Effects	Side Effects
Casanova et al.[39] Frontiers in Human Neuroscience	18	None	Aberrant Behavior Checklist, restricted Behavior Pattern, Time-domain measures of HRV (R–R interval, SDNN, RMSSD, pNN50), Frequency – domain measures of HRV (LV and HF of HRV, LF/HF ratio index), SCL	Before and 2 weeks after treatment	Increase in R–R interval, SDNN, and HF power. Significant decrease in the LF/HF ratio and SCL. Significant improvements in RBS-R and ABC rating scores	None
Sokhadze et al.[40] Frontiers in Systems Neuroscience	18	None	Aberrant Behavior Checklist (ABC), Repetitive Behavior Scale (RBS-R), EEG event-related potentials (ERPs), RT and post-error RT	Before and after	Decreased irritability and hyperactivity on the Aberrant Behavior Checklist (ABC), and decreased stereotypic behaviors on the Repetitive Behavior Scale (RBS-R). Decreased amplitude and prolonged latency in the frontal and fronto-central N100, N200 and P300 (P3a) ERPs to non-targets. Increased centroparietal P100 and P300 (P3a) ERPs to non-targets.	None
Sokhadze et al.[41] Appl Psychophysiol Biofeedback	18	None	Aberrant Behavior Checklist, Repetitive Rehavior Scale-Revised, EEG Gamma power, Theta/Beta ratio, RT and Accuracy, Post-error RT, EEG event-related potentials (ERPs)	Before and after	Integrated TMS-NFB treatment enhanced the process of target recognition. Significant improvements in RBS-R and ABC rating scores. Improvement in both early and later stage ERP indices.	None

ASD, autism spectrum disorder; HFA, high-functioning autism; AS, Asperger's syndrome; M1, primary motor cortex; dlPFC, dorsolateral prefrontal cortex; dmPFC, dorsomedial prefrontal cortex; Fo8, figure-of-eight coil; ISI, interstimulus interval; rTMS, repetitive transcranial magnetic stimulation; RMT, resting motor threshold.

Low-Frequency Prefrontal Cortex Stimulation

The earliest and majority of these studies have been reported by Casanova et al., who have employed low-frequency, subthreshold rTMS to the DLPFC (left or sequential bilateral) in ASD. This paradigm is designed to address cortical inhibition deficits in ASD resulting from suspected minicolumnar abnormalities. Statistically significant improvements have been reported in irritability and repetitive behaviors.[38–42] Though the effect sizes were large in these studies, the trial designs were open-label, thus could not account for placebo effects. Following this paradigm, improvements have also been reported in EEG components related to target detection and error monitoring.[38,40,42–44,52] These executive functions are of importance for the adaptive and flexible interactions to changing environmental conditions.[53] These results have been further corroborated and improved in a pilot trial using EEG neurofeedback in combination with rTMS.[41] Recently, Casanova et al. evaluated electrocardiogram parameters and electrodermal activity in response to low-frequency DLPFC stimulation and showed an increase in cardiac interval variability and a decrease in tonic skin conductance level, indicating enhanced autonomic balance, arguably through frontal inhibition of limbic activity.[39] These changes were not seen in the waitlist control group, but again, the study was open label, so results may be confounded by placebo effects or adaptation to the environment and protocol.

Although not examining clinical improvement, Fecteau et al.[46] discovered that 1-Hz rTMS to individuals with ASD enhanced object naming when applied to left pars triangularis, but reduced object naming when applied to left pars opercularis. Enticott et al.[48] reported changes in a different EEG-evoked potential, specifically, movement-related cortical potentials (MRCPs) that are involved in preparation and execution of movements. Following a single session of 1-Hz rTMS to supplementary motor area and M1 (relative to M1 sham) ASD participants showed increases in these components. There were, however, no observable changes in motor behavior.

High-Frequency Prefrontal Cortex Stimulation

Other investigators have tested high-frequency stimulation in ASD to enhance excitability within presumably underactive cortical regions and associated networks. Enticott et al.[45] applied high-frequency (5 Hz) rTMS to bilateral medial prefrontal cortex among adults with ASD. This study was designed to have an excitatory effect on networks devoted to mentalizing, which have shown reduced activation in ASD in neuroimaging studies. Participants received either active or sham stimulation each weekday for 2 weeks. The authors reported a significant improvement in the Social Relatedness Subscale of the Ritvo Autism Asperger Diagnostic Scale (RAADS) with a medium effect size, but no effect on other behavioral scales including the Autism Spectrum Quotient (AQ), Interpersonal Reactivity Index (IRI), or experimental measures of mentalizing (reading the mind in the eyes test and animations mentalizing test). High-frequency (8 Hz) rTMS to premotor cortex was also employed by Panerai et al.[47] in a series of small, sham-controlled studies among children with ASD with intellectual disability. There were significant improvements in eye–hand coordination following left premotor cortex HF-rTMS, and these effects were accentuated when paired with behavioral eye–hand integration training. This is the only study in the published literature that included participants with intellectual disability.

CONCLUSION

Despite the relatively large number of studies ($N = 14$) assessing rTMS as a therapeutic intervention in ASD, few studies followed strict clinical trial protocols (eg, randomization, identification of clear and objective primary endpoints, double-blinding with appropriate sham conditions, sufficient power, etc.). Recently Lefaucheur et al.[54] proposed guidelines for evaluation of the efficacy of TMS for treatment of neurological and psychiatric conditions based on the number of successful double-blind randomized placebo-controlled clinical trials and the sample size of those trials. Though ASD was not specifically mentioned in these guidelines, the evidence published thus far would likely lead to therapeutic rTMS for ASD being classified as Level C: "possibly effective." Level C is defined as indications where two or more controlled trials have been conducted on small samples without adherence to strict clinical trial protocols. Based on these guidelines, Lefaucheur et al.[54] classified neuropathic pain and major depression as Level A "definitely effective." Level B "probably effective" was given to chronic motor stroke, low-frequency stimulation for major depression, depression in the context of Parkinson's disease, and the negative symptoms of schizophrenia, and Level C "possibly effective" was given to tinnitus, auditory hallucinations, motor symptoms of Parkinson's disease, acute stroke, complex regional pain syndrome, hemispatial neglect, epilepsy, PTSD, cigarette consumption, and Broca's non-fluent aphasia.

There remain major gaps in knowledge with regard to rTMS for ASD treatment, particularly with respect to stimulus parameters, application sites, and targeted symptoms. Though the existing literature has some limitations, many of the concerns and challenges are not unique to this population or this intervention, and have been addressed by other researchers and clinicians in the development of novel therapeutic interventions. Through collaboration across disciplines and across labs, researchers and clinicians can begin to address these challenges (both common to all interventions and conditions and unique to TMS in ASD) and develop valid and reliable uses of TMS to both study the pathophysiology and develop novel treatments for ASD. This type of collaborative effort is underway through the establishment of the "TMS in ASD Consensus Group." This group of researchers, clinicians, regulatory affairs officers, and community partners meets annually prior to the International Meeting for Autism Research (IMFAR) and convenes monthly through teleconference to facilitate ongoing discussion, establish consensus on investigative and therapeutic protocols and encourage collaboration across centers.

The studies published thus far have significant limitations, however, we contend that TMS holds significant promise for autism research and treatment and should be properly evaluated using scientific rigor. For therapeutic applications, future trials should be adequately data-supported, randomized, double-blind and placebo-controlled. Studies should also strive to enroll a representative sample large enough to be powered to account for dropouts and still detect a significant effect in a predetermined primary outcome measure. Thus far no study has met all of these criteria. Until at least two such trials have been conducted, it would be premature to offer rTMS for the treatment of ASD in clinical settings.

Acknowledgments

The authors would like to acknowledge the TMS in ASD Consensus Group whose discussion contributed to the ideas expressed in this chapter. This work was primarily funded by the NIH (R01MH100186). Dr. A. Pascual-Leone was also supported in part by the Sidney R. Baer Jr. Foundation, the NIH (R21 NS082870, R01HD069776, R01NS073601, R21 MH099196, R21 NS085491, R21 HD07616), and Harvard Catalyst, The Harvard Clinical and Translational Science Center (NCRR and the NCATS NIH, UL1 RR025758). Dr. A. Rotenberg's efforts related to this work are supported by the Boston Children's Hospital Translation Research Program, the NIH (R01NS088583), and the Autism Speaks Preclincal Autism Consortium.

The content is solely the responsibility of the authors and does not necessarily represent the official views of Harvard Catalyst, Harvard University and its affiliated academic healthcare centers, or any of the listed granting agencies.

Conflict of Interest Disclosures

APL serves on the scientific advisory boards for Nexstim, Neuronix, Starlab Neuroscience, Neuroelectrics, and Neosync; and is listed as an inventor on several issued and pending patents on the real-time integration of transcranial magnetic stimulation (TMS) with electroencephalography (EEG) and magnetic resonance imaging (MRI). AR has served as advisor for Nexstim Inc., Sage Therapeutics Inc., NeuroRex Inc., and has joint grants with Brainsway Inc. and Vivonics Inc. He is a founder and consultant to Neuro'motion Labs and is an inventor on a patent pending for technologies to enhance the development of emotional regulation.

References

1. APA. *Diagnostic and Statistical Manual of Mental Disorders*. 5th ed. Arlington: American Psychiatric Publishing; 2013.
2. Ameis SH, Catani M. Altered white matter connectivity as a neural substrate for social impairment in autism spectrum disorder. *Cortex*. January 2015;62:158–181.

3. Mefford HC, Batshaw ML, Hoffman EP. Genomics, intellectual disability, and autism. *N Engl J Med.* February 23, 2012;366(8):733–743.
4. Huguet G, Ey E, Bourgeron T. The genetic landscapes of autism spectrum disorders. *Annu Rev Genomics Hum Genet.* 2013;14:191–213.
5. Murdoch JD, State MW. Recent developments in the genetics of autism spectrum disorders. *Curr Opin Genet Dev.* June 2013;23(3):310–315.
6. Lai MC, Lombardo MV, Baron-Cohen S. Autism. *Lancet.* March 8, 2014;383(9920):896–910.
7. Ronemus M, Iossifov I, Levy D, Wigler M. The role of de novo mutations in the genetics of autism spectrum disorders. *Nat Rev Genet.* February 2014;15(2):133–141.
8. Oberman L, Rotenberg A, Pascual-Leone A. Altered brain plasticity as the proximal cause of autism spectrum disorders. In: Tracy J, Hampstead B, Sathian K, eds. *Plasticity of Cognition in Neurologic Disorders.* New York: Oxford University Press; 2014.
9. Markram H, Rinaldi T, Markram K. The intense world syndrome – an alternative hypothesis for autism. *Front Neurosci.* November 2007;1(1):77–96.
10. Dolen G, Bear MF. Fragile x syndrome and autism: from disease model to therapeutic targets. *J Neurodev Disord.* June 2009;1(2):133–140.
11. Rubenstein JL, Merzenich MM. Model of autism: increased ratio of excitation/inhibition in key neural systems. *Genes Brain Behav.* October 2003;2(5):255–267.
12. Casanova MF, Buxhoeveden D, Gomez J. Disruption in the inhibitory architecture of the cell minicolumn: implications for autism. *Neuroscientist.* December 2003;9(6):496–507.
13. Zoghbi HY, Bear MF. Synaptic dysfunction in neurodevelopmental disorders associated with autism and intellectual disabilities. *Cold Spring Harb Perspect Biol.* March 2012;4(3).
14. Huerta PT, Volpe BT. Transcranial magnetic stimulation, synaptic plasticity and network oscillations. *J Neuroeng Rehabil.* 2009;6:7.
15. Thickbroom GW. Transcranial magnetic stimulation and synaptic plasticity: experimental framework and human models. *Exp Brain Res.* July 2007;180(4):583–593.
16. Ziemann U. TMS induced plasticity in human cortex. *Rev Neurosci.* 2004;15(4):253–266.
17. Huang YZ, Edwards MJ, Rounis E, Bhatia KP, Rothwell JC. Theta burst stimulation of the human motor cortex. *Neuron.* January 20, 2005;45(2):201–206.
18. Theoret H, Halligan E, Kobayashi M, Fregni F, Tager-Flusberg H, Pascual-Leone A. Impaired motor facilitation during action observation in individuals with autism spectrum disorder. *Curr Biol.* February 8, 2005;15(3):R84–R85.
19. Minio-Paluello I, Baron-Cohen S, Avenanti A, Walsh V, Aglioti SM. Absence of embodied empathy during pain observation in Asperger syndrome. *Biol Psychiatry.* January 1, 2009;65(1):55–62.
20. Enticott PG, Rinehart NJ, Tonge BJ, Bradshaw JL, Fitzgerald PB. A preliminary transcranial magnetic stimulation study of cortical inhibition and excitability in high-functioning autism and Asperger disorder. *Dev Med Child Neurol.* August 2010;52(8): e179–e183.
21. Enticott PG, Kennedy HA, Rinehart NJ, et al. Interpersonal motor resonance in autism spectrum disorder: evidence against a global "mirror system" deficit. *Front Hum Neurosci.* 2013;7:218.
22. Enticott PG, Kennedy HA, Rinehart NJ, Tonge BJ, Bradshaw JL, Fitzgerald PB. GABAergic activity in autism spectrum disorders: an investigation of cortical inhibition via transcranial magnetic stimulation. *Neuropharmacology.* May 2013;68:202–209.
23. Oberman L, Eldaief M, Fecteau S, Ifert-Miller F, Tormos JM, Pascual-Leone A. Abnormal modulation of corticospinal excitability in adults with Asperger's syndrome. *Eur J Neurosci.* September 2012;36(6):2782–2788.
24. Badawy RA, Strigaro G, Cantello R. TMS, cortical excitability and epilepsy: the clinical impact. *Epilepsy Res.* February 2014;108(2):153–161.
25. Uithol S, van Rooij I, Bekkering H, Haselager P. Understanding motor resonance. *Soc Neurosci.* 2011;6(4):388–397.

26. Enticott PG, Kennedy HA, Rinehart NJ, et al. Mirror neuron activity associated with social impairments but not age in autism spectrum disorder. *Biol Psychiatry*. March 1, 2012;71(5):427–433.

27. Jung NH, Janzarik WG, Delvendahl I, et al. Impaired induction of long-term potentiation-like plasticity in patients with high-functioning autism and Asperger syndrome. *Dev Med Child Neurol*. January 2013;55(1):83–89.

28. Oberman L, Ifert-Miller F, Najib U, et al. Transcranial magnetic stimulation provides means to assess cortical plasticity and excitability in humans with fragile x syndrome and autism spectrum disorder. *Front Synaptic Neurosci*. 2010;2:26.

29. Oberman LM, Pascual-Leone A. Hyperplasticity in autism spectrum disorder confers protection from Alzheimer's disease. *Med Hypotheses*. September 2014;83(3):337–342.

30. Oberman LM, Pascual-Leone A. Cortical plasticity: a proposed mechanism by which genomic factors lead to the behavioral and neurological phenotype of autism spectrum and psychotic spectrum disorders. *Behav Brain Sci*. 2008;31:241–320.

31. Oberman LM, Pascual-Leone A, Rotenberg A. Modulation of corticospinal excitability by transcranial magnetic stimulation in children and adolescents with autism spectrum disorder. *Front Hum Neurosci*. 2014;8:627.

32. Enticott PG, Oberman LM. Synaptic plasticity and non-invasive brain stimulation in autism spectrum disorders. *Dev Med Child Neurol*. January 2013;55(1):13–14.

33. Rotenberg A, Muller PA, Vahabzadeh-Hagh AM, et al. Lateralization of forelimb motor evoked potentials by transcranial magnetic stimulation in rats. *Clin Neurophysiol*. January 2010;121(1):104–108.

34. Vahabzadeh-Hagh AM, Muller PA, Pascual-Leone A, Jensen FE, Rotenberg A. Measures of cortical inhibition by paired-pulse transcranial magnetic stimulation in anesthetized rats. *J Neurophysiol*. February 2011;105(2):615–624.

35. Hsieh TH, Dhamne SC, Chen JJ, Pascual-Leone A, Jensen FE, Rotenberg A. A new measure of cortical inhibition by mechanomyography and paired-pulse transcranial magnetic stimulation in unanesthetized rats. *J Neurophysiol*. February 2012;107(3): 966–972.

36. Muller PA, Dhamne SC, Vahabzadeh-Hagh AM, Pascual-Leone A, Jensen FE, Rotenberg A. Suppression of motor cortical excitability in anesthetized rats by low frequency repetitive transcranial magnetic stimulation. *PloS One*. 2014;9(3):e91065.

37. Hsieh TH, Huang YZ, Chen JJJ, et al. Novel use of theta burst cortical electrical stimulation for modulating motor plasticity in rats. *J Med Biol Eng*. 2015;35(1):62–68.

38. Casanova MF, Baruth JM, El-Baz A, Tasman A, Sears L, Sokhadze E. Repetitive transcranial magnetic stimulation (rTMS) modulates event-related potential (ERP) indices of attention in autism. *Transl Neurosci*. June 1, 2012;3(2):170–180.

39. Casanova MF, Hensley MK, Sokhadze EM, et al. Effects of weekly low-frequency rTMS on autonomic measures in children with autism spectrum disorder. *Front Hum Neurosci*. 2014;8:851.

40. Sokhadze EM, El-Baz AS, Sears LL, Opris I, Casanova MF. rTMS neuromodulation improves electrocortical functional measures of information processing and behavioral responses in autism. *Front Syst Neurosci*. 2014;8:134.

41. Sokhadze EM, El-Baz AS, Tasman A, et al. Neuromodulation integrating rTMS and neurofeedback for the treatment of autism spectrum disorder: an exploratory study. *Appl Psychophysiol Biofeedback*. December 2014;39(3–4):237–257.

42. Baruth JM, Casanova MF, El-Baz A, et al. Low-frequency repetitive transcranial magnetic stimulation (rTMS) modulates evoked-gamma frequency oscillations in autism spectrum disorder (ASD). *J Neurother*. July 1, 2010;14(3):179–194.

43. Sokhadze E, Baruth J, Tasman A, et al. Low-frequency repetitive transcranial magnetic stimulation (rTMS) affects event-related potential measures of novelty processing in autism. *Appl Psychophysiol Biofeedback*. June 2010;35(2):147–161.

44. Sokhadze EM, El-Baz A, Baruth J, Mathai G, Sears L, Casanova MF. Effects of low frequency repetitive transcranial magnetic stimulation (rTMS) on gamma frequency oscillations and event-related potentials during processing of illusory figures in autism. *J Autism Dev Disord*. April 2009;39(4):619–634.

45. Enticott PG, Fitzgibbon BM, Kennedy HA, et al. A double-blind, randomized trial of deep repetitive transcranial magnetic stimulation (rTMS) for autism spectrum disorder. *Brain Stimul*. March–April 2014;7(2):206–211.

46. Fecteau S, Agosta S, Oberman L, Pascual-Leone A. Brain stimulation over Broca's area differentially modulates naming skills in neurotypical adults and individuals with Asperger's syndrome. *Eur J Neurosci*. July 2011;34(1):158–164.

47. Panerai S, Tasca D, Lanuzza B, et al. Effects of repetitive transcranial magnetic stimulation in performing eye-hand integration tasks: four preliminary studies with children showing low-functioning autism. *Autism Int J Res Pract*. October 10, 2013;18(6):638–650.

48. Enticott PG, Rinehart NJ, Tonge BJ, Bradshaw JL, Fitzgerald PB. Repetitive transcranial magnetic stimulation (rTMS) improves movement-related cortical potentials in autism spectrum disorders. *Brain Stimul*. January 2012;5(1):30–37.

49. Cristancho P, Akkineni K, Constantino JN, Carter AR, O'Reardon JP. Transcranial magnetic stimulation in a 15-year-old patient with autism and comorbid depression. *J ECT*. December 2014;30(4):e46–e47.

50. Enticott PG, Kennedy HA, Zangen A, Fitzgerald PB. Deep repetitive transcranial magnetic stimulation associated with improved social functioning in a young woman with an autism spectrum disorder. *J ECT*. March 2011;27(1):41–43.

51. Niederhofer H. Effectiveness of the repetitive transcranial magnetic stimulation (rTMS) of 1 Hz for autism. *Clin Neuropsychiatry*. 2012;9(2):107.

52. Sokhadze EM, Baruth JM, Sears L, Sokhadze GE, El-Baz AS, Casanova MF. Prefrontal neuromodulation using rTMS improves error monitoring and correction function in autism. *Appl Psychophysiol Biofeedback*. June 2012;37(2):91–102.

53. Garavan H, Ross TJ, Murphy K, Roche RA, Stein EA. Dissociable executive functions in the dynamic control of behavior: inhibition, error detection, and correction. *NeuroImage*. December 2002;17(4):1820–1829.

54. Lefaucheur JP, Andre-Obadia N, Antal A, et al. Evidence-based guidelines on the therapeutic use of repetitive transcranial magnetic stimulation (rTMS). *Clinical Neurophysiol*. November 2014;125(11):2150–2206.

CHAPTER

14

Non-Invasive Brain Stimulation in Pediatric Epilepsy: Diagnostic and Therapeutic Uses

S.K. Kessler

Children's Hospital of Philadelphia, Philadelphia, PA, United States

Pediatric Brain Stimulation
http://dx.doi.org/10.1016/B978-0-12-802001-2.00014-X

INTRODUCTION

Epilepsy is one of the most common neurologic disorders in children, affecting nearly 1 in 100 children.[1] The pathophysiologic hallmarks of epilepsy include abnormal neuronal excitability and impaired neuronal network structure or dynamics, resulting in recurrent unprovoked seizures as well as the potential for impaired cognitive development.[2] The non-invasive brain stimulation methods, transcranial magnetic stimulation (TMS) and transcranial direct-current stimulation (TDCS), have a number of applications in pediatric epilepsy, many still investigational and some for clinical practice. The goal of this chapter is not a comprehensive review of studies of non-invasive brain stimulation (NIBS) related to epilepsy, but a discussion of key clinical applications of NIBS in pediatric epilepsy, currently and in the near future.

DIAGNOSTIC USES OF TMS

Evaluating Cortical Excitability

TMS provides a non-invasive method for probing cortical excitability. Traditionally, measures of cortical excitability using TMS have been restricted to motor cortex, where a single stimulus of adequate intensity over motor cortex induces a muscle response called a motor-evoked potential (MEP). MEP measurement can be used to evaluate the integrity of corticospinal tracts as well as the inhibitory and excitatory properties of motor cortex. With the advent of technology to record EEG while administering TMS, the inhibitory and excitatory properties of other regions of cortex may be measured, along with other cortical network properties.

Several measures of cortical excitability can be obtained with MEPs. The first is *motor threshold* (MT), the stimulus intensity needed to reliably produce MEPs above a specified amplitude (most frequently, 50 μV).[3] Motor threshold reflects membrane excitability of pyramidal neurons.[4] Another measure of cortical physiology which uses single-pulse stimulation is the *cortical silent period* (CSP), which is the interruption of ongoing voluntary muscle activity by a TMS pulse, measured as the duration of "silence" on the electromyographic tracing. The CSP has been proposed as a marker of $GABA_B$ receptor-mediated intracortical inhibition because the typical CSP duration mirrors the duration of the IPSP elicited by $GABA_B$ receptor activation in animal slice preparations of pyramidal cells.

The other commonly used measures of cortical excitability employ responses to paired stimuli. In paired-pulse TMS, a conditioning stimulus, by triggering a series of excitatory and inhibitory influences on pyramidal neurons, modulates the MEP amplitude elicited by a subsequent test stimulus.[5] Paired stimuli at short intervals (2–5 ms) and long intervals (100–300 ms) produce inhibition (lower MEP amplitude), and intervals of 6–20 ms produce facilitation (higher MEP amplitude) in healthy individuals. The ratio of the average conditioned MEP amplitude over the average unconditioned MEP amplitude is the recovery ratio. The mechanism for the observed inhibitory effects is likely related to the activation of inhibitory interneurons resulting in biphasic (fast and slow) hyperpolarization of target pyramidal cells.[6] The fast inhibitory postsynaptic component is due to activation of ionotropic $GABA_A$ receptors[7] and is the process that likely underlies TMS induced *short intracortical inhibition* (SICI).[8,9] The slow IPSP is due to opening of K+ channels linked to metabotropic $GABA_B$ receptors, correlating with TMS-induced *long intracortical inhibition* (LICI).[10] The mechanism of *intracortical facilitation* (ICF), which occurs at interstimulus intervals of 6–20 s, is less clear, but likely involves both γ-aminobutyric acid (GABA)-mediated inhibition and excitatory neurotransmission mediated by NMDA receptors.[11]

In the early days of TMS research, the underlying physiologic mechanisms underlying each of these measures was hypothesized based on correlations between the timing and characteristics of the TMS measure, and timing and characteristics of known inhibitory or excitatory processes involving pyramidal neurons and interneurons. Supporting evidence for these hypothesized mechanisms came from experiments which evaluated the immediate effect of drugs with well-characterized mechanisms of action on TMS measures in healthy subjects. Subsequently, these TMS measures are being used as biomarkers to characterize drugs or treatments whose mechanisms are less clear, to elucidate defects in cortical excitability or inhibition in neuropsychiatric disorders, or to evaluate subgroups of individuals with disorders (such as those who are drug-responsive versus drug-resistant). Though many of these studies have not included children, an overview of cortical excitability studies in epilepsy is presented here to illustrate potential applications of TMS in children with epilepsy.

Using TMS to Characterize Epilepsy

Idiopathic Generalized Epilepsies

Idiopathic generalized epilepsies (IGE), now often called genetic generalized epilepsies,[12] are a group of epilepsy syndromes characterized by seizures that have non-focal mechanisms of onset (such as absence, myoclonic, or primary generalized tonic clonic seizures) and typical EEG findings (generalized spike wave discharges, provoked by hyperventilation or photic stimulation), associated with diffuse cortical and subcortical hyperexcitability, particularly in thalamocortical circuits. They occur in otherwise healthy individuals with no identifiable cause other than a genetic predisposition. Most IGE syndromes have childhood onset, though some begin in adulthood. The most common IGE syndromes are childhood absence epilepsy (CAE), juvenile absence epilepsy (JAE), juvenile myoclonic epilepsy (JME), and generalized epilepsy with generalized tonic clonic seizures alone (GE-GTC). Although the underlying genetic causes of IGE syndromes are not yet well characterized, defects in GABA receptor subunits have been implicated.[13]

Early studies using TMS to characterize the physiology of IGE syndromes often yielded mixed results, perhaps due to confounding factors such as the effects of antiepileptic drugs (AEDs), or lack of controlling for testing in pre- or postictal states.[14] A clearer electrophysiological profile of IGE has emerged from studies of newly diagnosed (drug-naïve) patients, though large discrepancies in some parameters are still apparent. The most consistent finding is that SICI is decreased (that is, there is less inhibition) in patients with IGE syndromes compared to healthy controls.[15–17] Studies which included paired-pulse stimulation at long interstimulus intervals, where inhibition is typically seen (LICI), have shown facilitation instead.[15] Both SICI and LICI are abnormal in the dominant and non-dominant hemispheres. ICF appears unaffected in IGE, except in one study of eight drug-naïve IGE patients, six of whom were adolescents ages 15–17, with JAE (three patients), JME (two patients), or GE-GTC (one patient), where ICF was increased at baseline compared to controls, and decreased after chronic treatment with valproate.[18] The inclusion of adolescents without age-matched controls was a possible confounding factor leading to apparent ICF abnormalities not seen in other studies.

MT is the parameter which appears to have the greatest inconsistency across studies of IGE, with some studies reporting increases, and others decreases, in IGE patients compared to healthy controls. A meta-analysis of TMS studies evaluating IGE patients older than age 12 years concluded that MT is decreased in patients with JME compared to healthy controls, but not in patients with other IGE syndromes.[19] In non-JME patients, MT was increased in the meta-analysis, but without statistical significance. The analysis was limited by the small number of non-JME IGE patients, but a later study comparing adolescent- and adult-onset IGE syndromes replicated the finding that MT is more abnormal in JME than in other IGE syndromes.[16] A key feature which distinguishes JME from the other IGE syndromes is the prominence of myoclonic seizures, which represent a brief massive activation of bilateral motor networks—so perhaps it is not surprising that unlike other syndromes, JME is particularly characterized by decreased MT, reflecting increased membrane excitability of pyramidal neurons of the corticospinal tract. The large degree of variability in the findings between studies, even when limited to drug-naïve patients, may also reflect the possibility that MT is not a stable trait in patients with epilepsy, fluctuating with different degrees of seizure susceptibility. In fact, a classic feature of JME specifically is the increased occurrence of seizures in the morning hours, which was investigated by Badawy et al. in a study of diurnal variation in

MT and ppTMS parameters in drug-naïve JME and other IGE patients.[20] SICI and LICI, which were decreased in all patients compared to healthy controls, were more abnormal in the morning than in the afternoon, and the effect was greater among JME patients.

Thus, the overall picture of IGE physiology is that MT is variably abnormal, but generally decreased in JME patients; intracortical inhibition is symmetrically decreased and perhaps even reversed at long interstimulus intervals; and ICF is still unclear. Other TMS-based measures of cortical physiology, within motor cortex or outside of motor cortex using TMS paired with EEG or other modalities, may emerge as useful tools in the future. For example, based on observations of polyphasic MEPs in IGE patients, and to a lesser degree in their unaffected siblings, compared to healthy controls, MEP polyphasia has been proposed as a marker of IGE,[21] although replication and further study are needed.

Because TMS studies of IGE physiology have largely excluded children, there are essentially no studies evaluating cortical excitability in drug-naïve patients with CAE. Though several barriers to the inclusion of children were likely present in early studies, the rationale for exclusion of children in many later studies is the problem of the influence of age on MT and intracortical inhibition. Young children have greater MT than adolescents or adults. Thus, age-matched controls are necessary, in both cross-sectional studies, and in longitudinal studies evaluating changes in MT or ppTMS parameters over time. It is also not clear that motor cortex physiology, although reflective of diffuse impairments in cortical excitability in adults with focal or generalized epilepsies, is an equally reliable marker of seizure susceptibility in children with less mature brain networks. Finally, abnormalities in TMS markers of cortical excitability are not specific to epilepsy. Abnormalities have been noted in a large number of other disorders, including ADHD,[22] a common comorbid condition in children with epilepsy.[23] Future studies utilizing TMS measures in children with epilepsy will have to take this into account.

Recent studies, often employing functional neuroimaging paired with EEG, investigating the pathophysiologic underpinnings of the IGEs have focused on the role of bilateral distributed cortical and subcortical networks as the sources of seizures and of the specific behavioral and psychiatric comorbidities that characterize the disorders as a group.[24] The advent of technology that allows recording of EEG signal during TMS stimulation, with minimal obscuration of the EEG tracing by TMS artifact, may offer new methods for examining the interplay between cortical excitability in connectivity in JME, and other developmental epilepsies.

Focal Epilepsies

Focal epilepsies are characterized by abnormal neuronal excitability in one or more focal brain regions, resulting in seizures arising from that region with the potential for spread more broadly. Epilepsy is now understood to be a network disorder, in which one area of neuronal dysfunction can cause dysfunction across a network. Studies of the neurophysiology of focal epilepsies using TMS pivot around the question, is motor cortex excitability influenced by epileptogenic zones outside of the motor cortex, and how?

As in studies of IGE, early studies yielded conflicting results about which TMS motor cortex parameters were abnormal in focal epilepsies, probably because of differences in stimulation methods, and heterogeneous patient populations. An often-cited study by Hamer et al. looking more carefully at this question included 23 adult patients with well-defined focal epilepsy not directly involving motor cortex based on imaging and electrophysiologic data.[25] The key finding was that CSP in the hemisphere ipsilateral to the epileptogenic zone was shortened compared with the contralateral hemisphere, a finding more pronounced in extratemporal epilepsy compared to temporal epilepsy, and in neocortical temporal epilepsy compared to mesial temporal epilepsy. No interhemispheric differences were seen in MT, SICI, or ICF. ICF was enhanced in extratemporal epilepsy, in both hemispheres. Thus, CSP and ICF were proposed as potential tools for lateralization and localization of epilepsy. The additional finding that MT was higher in patients, in both hemispheres, compared to healthy controls, can be explained as a medication effect, as the majority of patients were chronically exposed to at least one sodium-channel-blocking AED. Of note, all of the patients in this sample had treatment-resistant epilepsy with one or more seizures monthly, and almost all were on multiple AEDs, key factors which may influence the findings.

Further studies using TMS to evaluate focal epilepsies in drug-naïve patients revealed greater degrees of interhemispheric differences, and in different TMS parameters. In multiple studies in non-overlapping populations by Badawy et al.,[15,26] SICI at the 2 and 4 ms interstimulus interval and LICI, at the 250 and 200 ms interval, were decreased in the ipsilateral hemisphere compared with the contralateral hemisphere, and compared to healthy control subjects. Interestingly, MT in these drug-naïve patients was increased in the ipsilateral hemisphere compared to the contralateral hemisphere and to healthy controls—raising the possibility that the elevated MT seen in the Hamer study was not purely a medication effect, but perhaps an interictal adaptation to a state of decreased intracortical inhibition.

Further research is clearly needed before TMS can be reliably employed as a focal epilepsy lateralization and localization tool, in conjunction with the current armamentarium of non-invasive methods (including EEG, MRI, PET, MEG) for identifying epileptogenic cortex. Development of the tool may significantly benefit from TMS paired with EEG. TMS-EEG offers the potential for probing abnormalities in cortical excitability in cortical regions outside of motor cortex, where TMS pulses induce EEG-detected evoked potentials.[27] TMS-EEG may also be used to observe how perturbation of a focal brain region reverberates throughout a network, and thus has the potential for identification of abnormal networks in a new and dynamic way.

Using TMS to Characterize Mechanism of Action in Epilepsy Treatments

Based largely on the work of Ulf Ziemann and colleagues,[28,29] the effects of neuropsychiatric drugs on MEP-based single- and paired-pulse TMS parameters have been extensively characterized in convenience samples of healthy adults. Early studies employed drugs with well-defined mechanisms of action to characterize different TMS measures. From these studies came the now widely accepted idea that specific TMS measures reflect axonal membrane excitability (MT), cortical inhibition mediated by GABA-A receptors (SICI), cortical inhibition mediated by GABA-B receptors (LICI, CSP), or excitatory synaptic excitability (motor-evoked potential amplitude, intracortical facilitation). Thus AEDs, such as phenytoin or carbamazepine, which affect axonal membrane excitability through action on voltage-gated sodium channels, generally increase MT. AEDs, which are GABA-A agonists, such as the benzodiazepines, lorazepam and diazepam, increase SICI and decrease ICF. AEDs which are GABA-B agonists, such as pregabalin, increase LICI. CSP is increased by GABA-B agonists, but decreased by GABA-A agonists.

TMS measures of cortical excitability can also be used to evaluate antiepileptic treatments whose mechanisms of action are unknown. One example with particular relevance for children is the ketogenic diet. The ketogenic diet (KD) is a treatment for medication-resistant epilepsy consisting of a high-fat, low-carbohydrate, low-protein diet, which shifts the major source of brain energy from glucose to ketone bodies. The KD likely has multiple mechanisms of anticonvulsant action, including increased production and action of the inhibitory neurotransmitter GABA.[30] In a small study of healthy adult volunteers, paired-pulse TMS was used to study the physiologic effects of short-term ketogenic diet use.[31] All eight subjects, (mean age 36 ± 8.5 years), achieved the expected biochemical changes typical of the KD, including ketonuria, decreased blood glucose, and decreased insulin levels. Resting motor threshold (RMT) and the duration of the CSP did not change. SICI showed a substantial and statistically significant elevation 2 weeks after KD initiation, with subsequent return to baseline after resumption of a normal diet. This pattern of effects on ppTMS parameters suggests that the KD enhances GABAergic transmission, particularly type A GABA receptor-mediated action, but not voltage-gated sodium channel function. These findings are consistent with the observation that the KD suppresses seizures induced by GABA-A antagonists, but not by other mechanisms,[30] and lend further evidence to the very small number of studies in humans suggesting that the KD increases brain GABA levels.[32] However, correlating changes in one TMS parameter to a single cellular mechanism is likely an oversimplification. Larger studies employing TMS to study drug response in patients with epilepsy, discussed in detail below, support that AEDS with differing cellular mechanisms may evoke common changes in TMS measures of cortical excitability.

Although some caution is needed in extrapolating these findings to a population of children with refractory epilepsy, they raise the interesting possibility of using TMS as a non-invasive biomarker of KD treatment response. Future studies might also include TMS parameters not studied here, including LICI. Potential pitfalls for such a study include the effects of age on cortical excitability measures, the effect of concomitant antiepileptic medications, proximity to the previous or subsequent seizure, and the tolerability of the TMS in young children. Studies using ppTMS in children with epilepsy are particularly susceptible to the problem of intrinsically high motor thresholds, further increased by sodium-channel-blocking antiepileptic drugs. In this situation, some children cannot be evaluated when MT is increased beyond maximal machine output or cannot tolerate testing with high-intensity stimuli.

Using TMS as a Biomarker of Drug Response

Despite a substantial increase in the number of antiepileptic drugs (AEDs) over the past two decades, sustained seizure freedom remains elusive for one in three patients with epilepsy. Progress in predicting treatment at the individual level has been slow—common tools such as EEG are not helpful in either predicting treatment response at baseline, or shortly after treatment initiation. Because of the electrophysiologic abnormalities demonstrated by single- and paired-pulse TMS in patients with epilepsy, TMS methods show promise as the eagerly sought but elusive biomarker of treatment response.

The largest and most promising study using TMS to predict epilepsy treatment response included adolescents and adults with IGE and focal epilepsies.[26] Median age at study entry for IGE patients was 18 years (range 14–68), and for focal epilepsy patients was 23 years (range 14–70 years); the number of patients between the ages of 14 and 18 was not specified. IGE patients were treated with either valproate or lamotrigine, and patients with focal epilepsy were treated with either carbamazepine or lamotrigine. Baseline measurements of MT and cortical recovery curves at short and long interstimulus intervals were measured in drug-naïve patients at least 24 h outside of a peri-ictal period. Follow-up cortical excitability measurements were made 4–16 weeks after AED initiation, and seizure outcomes were assessed at 1 year. At baseline, both SICI and LICI were decreased in IGE patients. In focal epilepsy patients, in the hemisphere ipsilateral to the epileptogenic zone, at some short interstimulus intervals, paired pulse stimulation resulted in facilitation instead of inhibition. In patients with either epilepsy type who remained seizure-free 1 year after initiating medication, cortical recovery curves essentially normalized. In contrast, in AED-treated patients with ongoing seizures, normalization at 2, 3, and 250 ms interstimulus intervals tended not to occur. In IGE patients, the positive predictive values for 1 year seizure freedom of a 100% reduction in recovery ratio at the 250 ms interval was 0.97 (95% CI 0.82–0.99), while the negative predictive value was just 0.42 (95% CI 0.23–0.63). For focal epilepsy patients the predictive values were less robust: the positive predictive value was 0.69 (95% CI 0.39–0.91) and the negative predictive value was only 0.45 (95% CI 0.27–0.64). In other words, a positive normalization result predicted seizure freedom, but failed normalization did not predict treatment resistance very well. Thus, while MEP-based paired-pulse TMS measures are able to separate patients into treatment-responsive and treatment-resistant groups, they are not quite robust enough for individual prediction of response.

Nevertheless, a tool for testing treatment response at the group level that can be carried out easily and non-invasively in an office setting could be employed as a surrogate marker of outcome in preliminary efficacy trials of AEDs, reducing the cost and burden of seizure count based on outcome measures currently employed in AED studies. The idea of a physiologic surrogate marker in AED trials is particularly salient in AED studies involving children, where exposure to the potential adverse effects of a new agent can be minimized in study designs where early exit is allowed if no effect on TMS measures of cortical excitability are seen.

Among the many interesting findings in this longitudinal study of drug response is that changes in TMS parameters for treatment responders were similar across medications with different mechanisms of action. Thus, the electrophysiological responses may be more closely tied to an overall change in the balance of excitation and inhibition affecting seizure susceptibility than to the effect of a drug on a specific neurotransmitter, receptor, or ion channel.

Functional Mapping

Identifying anatomic cortical regions responsible for motor and language functions is another key diagnostic use of TMS. For children with medication-resistant epilepsy, resective surgery to remove the epileptogenic zone is a potentially curative treatment option. Although the potential benefits of epilepsy surgery are substantial—seizure freedom, reduced chronic exposure to AEDs, improved or preserved cognitive function over time, improved quality of life—these benefits must be weighed against the potential for harm—permanent functional deficits when eloquent areas of cortex are part of the epileptogenic zone. Because epileptogenic lesions, whether developmental (such as focal cortical dysplasia) or acquired (such as perinatal stroke) can induce functional plasticity—remapping of motor or cognitive functions to areas less affected by pathology—identification of functional areas based on normative anatomy alone is inadequate. Locating functional cortex—typically motor and language areas—in relation to the epileptogenic zone is critical for making decisions to proceed, or not, with invasive EEG monitoring or resection, and can aid in surgical planning.

While imaging methods, such as functional MRI (fMRI) or magnetoen-cephalography (MEG), can be used to identify cortical regions involved in specific functions, both have two key limitations. The first limitation is that both record changes in brain activity (blood oxygen level in the case of fMRI and electrical oscillations in the case of MEG), but both detect brain regions that participate in the activity being examined, without the means to make a direct causal inference. TMS, on the other hand, evokes a response with stimulation over a specific cortical site that can then be causally linked to that site. The second limitation, with special relevance in pediatrics, is that many children may not be able to tolerate MRI imaging without sedation—that is, they may not be able to maintain adequate stillness and be able to complete the tasks required for functional mapping. Mapping using navigated TMS does not require the prolonged positional stability that MRI requires. Functional mapping can also be performed by direct electrical cortical stimulation either intraoperatively or extraoperatively during intracranial EEG monitoring. While direct cortical stimulation is the gold standard in terms of accuracy of localization, the clear limitation is that it is invasive—it cannot be performed prior to the decision to proceed with at least the first step of epilepsy surgery.

Functional mapping using TMS is carried out by combining TMS—single pulse for motor mapping, and brief trains of stimuli for language mapping—with a frameless stereotactic navigation system. The components of these systems include software which creates a three-dimensional (3D) reconstruction of the head and brain from high-resolution MRI images of the patient, and a tracking system which allows registration of the patient's head in actual space to the 3D reconstruction using anatomic landmarks, such as the tip of the nose and the tragus of the ears, and surface matching. The tracking system uses a dual infrared camera which tracks optical sensors worn by the patient (using a headband, glasses, or more recently, adhesives). The TMS coil also carries a sensor detectable by the camera, which allows the software to indicate where the coil is stimulating in relation to the patient's cortex, visualized on the MRI-based 3D reconstruction. For optimal use of the images, dicom files marked with the locations identified for motor and language functions can be exported and uploaded for use in OR navigation systems, to guide the neurosurgical approach.

Motor mapping is straightforward: single TMS pulses over motor cortex elicit motor-evoked potentials in the target muscle. Sites eliciting an MEP for the target muscle are marked on the images to create a somatotopic map. Several studies have validated the accuracy of TMS motor maps compared to direct cortical stimulation to the millimeter level, including in epilepsy patients with lesions involving motor cortex.[33] Typically, motor mapping focuses on hand muscle function, because hand motor deficits have the greatest functional implication for patients, but sometimes there is a need to map the proximal arm or leg. Cortical malformations or early life injury may lead to functional reorganization of motor functions within or beyond the anatomic motor cortex.[34] In addition, some patients may experience strengthening of the ipsilateral motor pathways when contralateral pathways are poorly functioning. Thus, mapping of both hemispheres when the epileptogenic lesion involves motor cortex may provide information critical for decision making about epilepsy surgery. Fig. 14.1 shows mapping results from our center for an 8-year-old patient with intractable epilepsy due to a dysplastic precentral gyrus in the right hemisphere (*red arrow*). The child had only mild weakness in the left hand. Hand motor functions for abductor pollicis brevis and first dorsal interosseous were localized to the postcentral gyrus (*green and yellow markers*). Subsequent resection of the dysplastic cortex (type IIb focal cortical dysplasia) resulted in seizure freedom (observed for 1 year), without additional left hand weakness. Mapping was performed with a Magstim

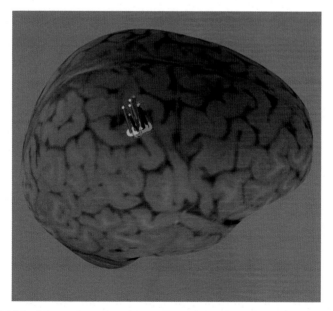

FIGURE 14.1   Motor mapping results in a patient with motor cortex dysplasia. *Blue pins* represent sites of stimulation, which elicited motor-evoked potentials (MEPs) from abductor digiti minimi, and *orange pins* represent sites, which elicited MEPs from abductor pollicis brevis. Also note the abnormal width and shape of the precentral gyrus, anterior to the marked sites.

200 stimulator, 70-mm figure-of-eight coil (Magstim Company Ltd, UK), and Brainsight Neuronavigation System (Rogue Research Inc., Canada).

Speech/language mapping is more challenging than motor mapping for a number of reasons, but there is a rapidly growing body of studies validating methods and exploring how to optimize stimulation parameters and tasks to achieve reliable results. Some of these challenges include:

- Language tasks: the most straightforward task, and the one most often reported, is a picture-naming paradigm, in which a series of pictures of common objects are presented on a screen. Subjects should be instructed to name the object as part of a phrase like "this is a ..." so that speech arrest (complete inability to produce language) can be distinguished from anomia (patient can speak the lead in phrase without the ability to name the object). Other error types that may be seen are semantic paraphasias (for example, calling a *spoon* a *pencil*), or phonemic paraphasias (such as calling a *spoon* a *foon*). As is true for other methods of language mapping like fMRI or MEG, a key issue in using this task in children is to ensure that the child can name the picture being presented at baseline. Thus, having a child go through the picture stack before the day of the test, and/or a practice run prior to testing with stimulation are helpful. Customization of the picture sequence includes removing pictures of objects that the child does not know, either because they are outside the child's experience or because the child has cognitive limitations. The procedure is performed with video, optimally with automated correlation of the individual trial to the site of stimulation, for later off-line analysis of the errors. Sites with errors should be tested again for confirmation.
- Timing of picture presentation: The optimal parameters for the picture-to-TMS-stimulus interval (or picture-to-trigger interval), or the length of time the picture is presented, or the time between pictures are not clear—and in practice may have to be customized to each patient based on knowledge of the patient's cognitive abilities, including processing speed. A number of studies have explored different picture-to-trigger intervals, with onset latency varying from −100 to 400 ms, but with different tasks and other stimulation parameters (see Krieg[35] for a review). In adult brain tumor patients the optimal timing of stimulus to pulse onset was compared to the gold standard of direct cortical stimulation.[35] The researchers reported that for target sites around Broca's area, TMS concurrent with picture onset and delayed TMS after picture onset had comparable sensitivity (100% for both). However, TMS at picture onset had much better specificity (67%) compared to delayed TMS (28%)—that is, there was a greater proportion of false-positive trials with delayed TMS. The effect was

even stronger with posterior language areas. Based on this and other studies,[36] 0 s latency between picture presentation and stimulus onset appears to be optimal. However, these results may not hold true for children, or children with epilepsy who may have delayed processing speeds or other cognitive impairments. For practical purposes, starting with a 0 s latency is probably best, but with an increase in latency if speech arrest or other errors are not occurring.

- TMS stimulus intensity: Stimulus intensity for language mapping is based on motor threshold, though the optimal intensity is still uncertain. Some studies stimulate at 100% of resting MT, and some at 110%.[36,37] Because language mapping requires stimulation over the frontal temporal region, the temporalis muscle is stimulated as well, a source of discomfort for some patients, particularly because TMS pulses are given as short trains of repetitive stimuli, unlike the single pulses used for motor mapping. Some studies report reducing stimulus intensity to 80–90% of RMT to reduce discomfort. Because MT tends to be higher in children, and higher in patients taking sodium-channel-blocking antiepileptic medication, this point is particularly salient. A reasonable approach may be stimulating at 100% MT, but decreasing stimulus intensity when there is intolerable discomfort, or increasing the intensity if no speech errors are induced. As additional reports of language mapping in pediatric populations appear in the literature, the parameters with the highest yield and greatest tolerability should start to become clearer.
- TMS stimulus characteristics: Other parameters that can be varied include the pulse frequency and the number of pulses per train, but there are no studies specifically evaluating which settings are optimal. A typical paradigm in recent publications is 5-Hz stimulation with 5 or 10 pulses per train. When the picture to stimulus interval is 0, a longer pulse train should be given to encompass the time required for the word to be spoken.
- Location of stimulation: the bulk of studies published to date on preoperative language mapping have largely focused on adults with tumors, and have mapped the dominant hemisphere. In children with epilepsy, where atypical language representation is not uncommon,[38] bilateral mapping may be needed to identify atypical laterality or bilaterality.

Whenever possible, multimodal mapping of motor and language functions using TMS, along with fMRI or MEG, is probably best for building a complete and reliable picture of functional cortex for surgical planning, particularly in children where the gold standard of direct cortical stimulation may not be performed. That is, TMS may not be a replacement for other methods so much as an enhancement of noninvasive preoperative testing.

At the time of this writing, functional mapping is the only application of TMS in epilepsy for which the US Food and Drug Administration (FDA) has approved clearance, a regulatory designation which allows the device to be marketed for the specified purpose. Specifically, the Nexstim Navigated Brain Stimulation (NBS) system (Nexstim Plc, Finland), an integrated TMS and frameless stereotaxy system, gained FDA clearance for motor and speech mapping for presurgical planning across all ages and indications. Thus, while the other applications of TMS in epilepsy presented here are still in the investigational realm, motor and speech mapping is being performed for clinical purposes at a growing number of centers. As with any procedure, rates of reimbursement by insurance in the United States for motor mapping procedures likely vary by market, but the CPT codes for motor mapping have been bundled with other neurophysiology procedures and are thus covered by many carriers.

THERAPEUTIC USES OF TMS

The ability of low-frequency repetitive transcranial magnetic stimulation (rTMS) to induce a sustained decrease in cortical excitability has been explored as a potential therapeutic option for epilepsy. The underlying mechanism of the effect is not clear but may involve or be similar to the cellular process termed long-term depression (LTD), in which low-frequency electrical direct cortical stimulation results in weakening of postsynaptic responses. There are two potential applications of rTMS for seizures. One is the interictal, chronic application of rTMS to reduce seizure susceptibility in patients with medication-resistant epilepsy. The second is the use of rTMS to interrupt a seizure in progress, particularly in the context of status epilepticus. Although initial investigations have shown promise, there are still few controlled trials of rTMS for epilepsy, and none have included children under the age of 14 years. The discussion here includes studies of rTMS for epilepsy in adults to illustrate the range of potential applications for future investigation in children.

Repetitive TMS for Treatment of Refractory Focal Epilepsy

A number of case reports, case series, and several small controlled trials have evaluated rTMS treatment for epilepsy, with variable results that cannot easily be compared because of the large degree of variability in patient selection and stimulation parameters. As is true across many applications of non-invasive stimulation, the optimal stimulation parameters for treatment are not clear, and individual studies evaluating every possible permutation are not feasible. However, there are several emerging themes from the existing evidence.

Patient Selection

The best candidates for rTMS treatment are persons with neocortical epilepsy on the cortical convexity.[39] Because the direct effects of TMS have limited spatial penetration (the field decays with the square of distance), epileptogenic zones that are primarily mesial temporal, orbital frontal, or interhemispheric are less likely to receive stimulation of sufficient intensity to incur change. Although stimulation may have an indirect impact on other areas through networks connected to the target region, it is unclear whether the LTD-like process expected with low-frequency stimulation occurs anywhere but the region directly receiving stimulation. While there are reports of rTMS treatment success in patients with multifocal or poorly localized epileptogenic zones, the general consensus is that patients with a single, circumscribed epileptogenic focus are most likely to respond. rTMS may be particularly effective for patients with focal cortical dysplasia (FCD), though there is too little evidence to confidently conclude that one pathology is better suited than another. Whether there is something intrinsically susceptible to rTMS about the abnormal neurons and there abnormal connections in FCD, or whether it is simply easier to target the target region because of visible abnormalities on MRI is not clear. FCD and other malformations of cortical development are among the more common reasons for treatment-resistant focal epilepsy, and patients with these lesions should be evaluated for resective epilepsy surgery. However, when the dysplasia involves motor, language, or visual cortex, surgery may be avoided because of the potential for functional deficits with resection. These patients are perhaps the ideal candidates for rTMS.

One of the key questions about rTMS for treatment of refractory focal epilepsy is whether children might be better candidates than adults, because the capacity for cortical plasticity (and thus susceptibility to neuromodulation) is greater at younger ages. Future trials of rTMS for epilepsy treatment should consider stratification by age to evaluate the influence of this parameter on treatment response.

Stimulation Parameters

There are no studies systematically and directly comparing the efficacy of different stimulation parameters in epilepsy, and such studies are unlikely to be forthcoming. However, based on the existing evidence, some observations can be made about parameters more likely to be effective.

- *Coil shape*: focal coils (figure-of-eight, bat-wing) are probably better than a circular coil, particularly when targeting a unifocal lesion, such as the epileptogenic zone.[40] Weak results in early trials may have been largely due to circular coil use.[41]
- *Neuronavigation*: navigation is necessary, particularly with the use of a focal coil, and when there is an MRI-apparent lesion being targeted

for stimulation. Although an argument could be made that it is not needed when choosing a target based on the 10-10 EEG electrode placement system, MRI-based stereotactic navigation adds precision to the application of stimulation.[42]

- *Stimulation intensity*: high intensity (90–110% of MT) is probably more efficacious than low-intensity stimulation.[40,43] Because refractory epilepsy patients who are on sodium-channel-blocking AEDs may have high motor thresholds, using an air-cooled coil is best, and even then special procedures may be needed to cool the coil in between stimulation blocks.[44] In controlled trials, masking the treatment arm when using a sham coil can be difficult,[45] so at least one study has employed a high-dose/low-dose strategy, comparing stimulation at 90% MT to stimulation at 20% MT, with a clinically substantial and statistically significant treatment effect in the high-dose arm.[43]
- *Stimulation frequency*: unclear. Most studies have employed 0.5 or 1 Hz stimulation. Recent studies have given stimulation in 10-min blocks with a 1–10-min break between trains.
- *Session number and frequency*: unclear. Early studies with less robust effects employed few treatment sessions (1 week or less),[40,41] whereas recent reports with more robust effects have employed two sessions per day at initiation or early in the course followed by daily sessions for 10–14 days.[43,44,46]

Barriers to conducting therapeutic trials of rTMS for pediatric epilepsy merit consideration. A major barrier is that rTMS is resource- and labor-intensive. Unlike AEDs, which can be dispensed and taken at home, or an implantable stimulation device, rTMS requires that the patient come to the hospital or clinic, and requires an operator to deliver the treatment daily for weeks. Thus, the treatment incurs costs not seen in other types of treatment trials. Another barrier is the tolerability of the procedure. Although single- and paired-pulse TMS are generally well tolerated, even by younger children,[47] rTMS causes greater discomfort, particularly with frontal or temporal targets where local stimulation of underlying muscle can cause pain. There are also several aspects of stimulation that may complicate trial design. One is the issue of sham stimulation—though several methods of sham stimulation are available, including coils which mimic the sound of a TMS pulse, the experiences of verum and sham stimulation are still different enough that true masking of the treatment arm is difficult. The second issue is that in crossover designs, the washout period is not clearly known, risking contamination (lingering treatment effects) of the arm which receives sham second. While none of these issues prevents further research, some care is needed to address each of them in research proposals.

Repetitive TMS for Seizure Cessation

Using rTMS to interrupt a seizure in progress has been explored to a limited extent in animal models, and in specific clinical scenarios in patients with seizures. The basis of its use is the ability of rTMS to immediately affect or interrupt neuronal activity. Dr. Rotenberg et al. described seizure suppression with rTMS in the rat intraperitoneal kainite seizure model, using EEG to guide manual triggering of the stimulator at onset of individual seizures.[48] When administered at 0.5 and 0.75 Hz, but not at 0.25 Hz, rTMS shortened the average duration of kainite-induced seizures, with a trend toward improved effect when rTMS was delivered closer to seizure onset. In addition to providing proof of concept data to support studies in humans, this animal model also provides a way to optimize stimulation parameters.

Epilepsia partialis continua (EPC) is a term that describes persistent simple partial seizures with focal motor symptoms arising either from epileptic activity in primary motor cortex, or from primary sensory cortex in the post-central gyrus.[49] EPC presents as intermittent or continuous focal clonic jerks lasting hours, and sometimes weeks or months. It is almost always AED-resistant. EPC may arise from a number of etiologies, including peri-Rolandic focal cortical dysplasia (FCD), Rasmussen encephalitis, focal brain injury such as stroke, or the metabolic disorder associated with mutations in POLG1. The theoretical goals for applying low-frequency rTMS are (1) to interrupt ongoing neuronal activity, stopping the EPC acutely, and (2) to dampen the excitability of the involved cortical region in a lasting way, leading to a state in which focal motor seizures are less likely to occur.

Descriptions of rTMS for EPC include a case series of seven patients treated at the Berenson–Allen Center for Noninvasive Brain Stimulation.[50] The cohort included one child, an 11-year-old with Rasmussen encephalitis, and six adults, ages 18–79 years, whose etiologies included hypoglycemia, stroke, and resection of a cortical vascular malformation. Stimulation parameters varied, though all underwent 1-Hz stimulation either alone or primed first with high-frequency stimulation (6, 20, or 100 Hz). Neuronavigation to target anatomic motor cortex was not used. In five of the seven patients, rTMS disrupted seizures, but in three the effect lasted only 20–30 min after rTMS ended. In two patients, the anticonvulsive effect lasted for months. The 11-year-old child with Rasmussen encephalitis experienced seizure cessation during stimulation, but seizures resumed 30 min after each session. No adverse events occurred—notably, none of the seven patients had exacerbation of seizures, even during high-frequency stimulation. Other published cases of rTMS for EPC[51,52] include children and adolescents treated using varying stimulus intensity, rtMS frequency, train duration, and number of trains and sessions. Based on these limited data, it would not be unreasonable to recommend a trial of rTMS therapy in a child experiencing debilitating focal motor seizures with a focal cortical dysplasia over motor cortex. In this scenario, rTMS, with its low risk of adverse effects, may be preferable to surgical resection, with its high risk of permanent limb weakness.

In some ways, patients with EPC are the ideal population for conducting a clinical trial of rTMS for epilepsy treatment. Unlike the semiology of other seizure types, the focal clonic jerking of EPC is easily observable and measurable with surface EMG recording. Instead of seizure frequency counting, which is subject to error in many seizure types, outcome measures in an EPC trial might include immediate seizure cessation and time to return of EPC. The data from such a trial might also inform our understanding of the stimulation factors (frequency, intensity, stimulation duration, number of sessions, use of neuronavigation) and patient factors (etiology, age, concurrent medications), which affect response in a broader population of epilepsy patients. In the absence of a trial, further observational reports may nonetheless be informative.

Status epilepticus (SE) is the other scenario where rTMS for acute seizure cessation attracts interest. Convulsive status epilepticus is one of the more common life-threatening neurologic emergencies in children and adults. It is now often defined in stages: impending status epilepticus (a generalized tonic clonic seizure persisting for longer than 5 min), established status epilepticus (seizure persists after benzodiazepine administration), refractory status epilepticus (seizures persists after administration of second-line anticonvulsant medications), or super-refractory status epilepticus (seizures continue or recur 24 h or more after the onset of anesthetic therapy). Because the mortality and morbidity rates in refractory and super-refractory SE are high, and medication-induced coma carries significant medical risk, novel interventions to prevent SE from becoming refractory, and to treat super-refractory SE are actively being sought.[53] As discussed above, the rationale for low-frequency rTMS in SE is that, in most circumstances, a continuous train of timed TMS pulses at low frequencies (1 Hz or less) results in inhibition of long tract output of the underlying cortex, an effect that can last beyond the stimulation period.

To our knowledge, only two reports of rTMS for refractory SE have been published, which include three patients in total, all adults. The first report is of a 68-year-old woman with herpes simplex encephalitis in super-refractory SE for 38 days prior, despite multiple antiepileptic medications, high-dose corticosteroids, and four periods of medication-induced burst suppression.[44] RTMS was delivered using a figure-of-eight coil at 100% machine output (which was the motor threshold) at 0.5 Hz for 60 min (1800 stimuli) targeted to the region where seizures localized (right posterior temporal in all but three sessions), with continuous EEG monitoring. After 2 days of once-daily sessions, twice-daily sessions were undertaken.

After the start of twice-daily rTMS, EEG improvement was seen, and after the fourth stimulation (day 3), electrographic seizures stopped. Five further days of stimulation were given, with continued sustained improvement of the EEG and the patient's mental status. A common query in this "open-label" situation is whether the outcome was coincidental—ie, due to time, not treatment. However, fast, spontaneous resolution is not part of the natural history of super-refractory SE. Thus, the temporal relationship of the rTMS and the patient's improvement suggest an rTMS effect. The authors make the important point that the refractoriness of SE increases over time, and thus the efficacy of rTMS may have been greater if used earlier.

The other report[54] describes two adult patients, both with a history of treatment-resistant focal epilepsy prior to presentation with increased seizure frequency punctuated by altered mental status, but not a single unrelenting GTC seizure. Also of note, one had a VNS which was deactivated years prior for lack of efficacy, and the other had an active VNS that was deactivated immediately prior to TMS therapy. Both received a single 30-min session of 1-Hz rTMS with a figure-of-eight coil over the presumed epileptogenic zone. In one patient stimulus intensity was 100% hand MT, and in the other patient, 70% of machine output, because MT was not measurable. In both patients, seizure frequency declined, but the effect was short-lived. The authors further observed that rTMS did not induce secondary generalization of seizures in either patient, and did not interfere with the functioning of ICU equipment or EEG monitoring.

THERAPEUTIC USES OF TRANSCRANIAL DIRECT-CURRENT STIMULATION

Transcranial direct-current stimulation (tDCS) is a newer method of non-invasive brain stimulation for modulating cortical excitability whose mechanisms are different from TMS. tDCS utilizes weak electrical currents applied to the scalp which penetrate underlying tissues. By imposing a voltage gradient on the cortical neurons along the path of current, tDCS induces region changes in cortical excitability that persist beyond the stimulation period.[55] The effects of tDCS depend on the intensity and spatial extent of induced electrical fields, which in turn depend on both the stimulation dose (applied current intensity, arrangement of electrodes) and neuroanatomic factors. Based on effects in healthy subjects on motor cortex, anodal stimulation was thought to be excitatory and cathodal stimulation inhibitory—but subsequent research in other brain regions, in patient populations, and in pediatric patients, shows that this is a likely oversimplification of the physiologic processes affected by tDCS. Nevertheless, based on these ideas, reduced excitability through hyperpolarization of neurons underlying the cathode in tDCS has been proposed as a method to reduce susceptibility to seizures, or to suppress seizures acutely. As a non-invasive brain stimulation method applied for therapeutic purposes, tDCS has several key advantages over repetitive TMS—including cost, ease of use, portability, safety, and tolerability.

A number of preclinical animal studies have begun to explore the potential of tDCS as a seizure therapy, but currently there are few clinical investigations in human patients with epilepsy (see San-Juan et al.[56] for a review). Interestingly, of the few human studies, the majority report on pediatric patients. The earliest clinical study of tDCS in epilepsy, conducted by Fregni et al., was a randomized sham controlled trial of 19 patients (mean age 24 years, ±7.9) who underwent a single 20-min session of sponge-electrode-based tDCS with montages customized to the presumed epileptogenic zone.[57] The key findings were that the density of epileptiform discharges on EEG immediately after the session was substantially lower than at baseline, the sessions were well tolerated, and no seizures were induced during stimulation. No effect of stimulation was seen on seizure counts. Similar findings of a short-term reduction in epileptiform discharges after a single session of cathodal tDCS were obtained in a randomized sham controlled study of 36 children, ranging in age from 6 to 15 years.[58] Both studies highlight the possibility that tDCS may be useful in epileptic encephalopathies, in which the burden of interictal epileptiform activity contributes to cognitive and functional impairment.

A study in five children with the syndrome of continuous spike wave during slow-wave sleep (CSWS) with focal discharges showed no impact on spike index of cathodal stimulation during wakefulness for one session,[59] but a later case report of two children, ages 11 and 7 years, with CSWS and frequent focal epileptiform discharges who received multiple sessions of cathodal tDCS during sleep showed a large reduction in interictal epileptiform discharges during, and for a short while after, stimulation.[60] These preliminary studies raise some important points about factors which may contribute to the effectiveness of the therapy. The first is the number of sessions—based on experience in brain stimulation in general, and on these preliminary studies in particular, it seems reasonable to hypothesize that a single session may have a short-term effect, but is unlikely to have a sustained effect. A sustained effect will likely depend on accumulation of LTD-like effects on cortex, and thus will likely require repeated sessions. An advantage of tDCS over TMS in this respect is the capacity for delivering tDCS at home—family members can be taught how to affix electrodes and turn on stimulation, particularly if the system is preprogrammed with the desired parameters. The second factor which must be considered in tDCS therapy for epilepsy is state dependence—just as experience with tDCS in cognitive neuroscience or motor rehabilitation indicates that tDCS paired with an activity is needed to achieve an effect, it is possible that for tDCS to reduce spike frequency in CSWS, it may need to be applied in sleep, when the spike discharges are occurring, in order to affect the cortical oscillations specific to sleep that are inducing frequent discharges. Preclinical animal model research, when possible, will be key to optimizing parameters to move the field forward in applying tDCS as a therapy for medication-resistant epilepsy.

Acute seizure suppression with tDCS is also gaining interest. There is a single case report of tDCS for acute seizure suppression in a 20-year-old patient with epilepsia partialis continua.[61] Stimulation was given with a battery-driven stimulator with 35-cm^2 saline-soaked sponges, with the cathode over the location of the C4 electrode in a 10–20-electrode system (the site thought to best correlate with motor cortex in most patients) and the anode over the contralateral supraorbital region. Current intensity of 2mA was given for 20min. EPC immediately ceased with the start of stimulation, and reappeared gradually after the completion of stimulation. After five other sessions in the subsequent 2 weeks, the immediate effects were replicated, and the patient reported a subject improvement between sessions. Effect on EEG discharges was not reported. Whether more robust or longer-lasting improvement might occur with further sessions is an interesting and still open question. However, the possibility of immediate clinical seizure suppression with tDCS was illustrated. Seizure cessation by tDCS was also illustrated recently in a rat pentylenetetrazol (PTZ) status epilepticus model, with the key finding that tDCS administered with the benzodiazepine lorazepam was more effective than either intervention delivered alone.[62] This finding raised the interesting possibility of a new multimodal therapy for acute status epilepticus in patients who do not respond to benzodiazepine treatment alone.

CONCLUSION

Non-invasive brain stimulation in the forms of transcranial magnetic stimulation and transcranial direct-current stimulation have emerging promising applications in childhood epilepsy. Increasingly, these applications are based on rationales from animal research and rigorously conducted human studies. An increasing future role in a variety of diagnostic and therapeutic settings is anticipated.

References

1. Hirtz D, Thurman DJ, Gwinn-Hardy K, Mohamed M, Chaudhuri AR, Zalutsky R. How common are the "common" neurologic disorders? *Neurology*. 2007;68(5):326–337.
2. Engel Jr J. ILAE classification of epilepsy syndromes. *Epilepsy Res*. 2006;70(Suppl. 1):S5–S10.
3. Rossini PM, Berardelli A, Deuschl G, et al. Applications of magnetic cortical stimulation. The International Federation of Clinical Neurophysiology. *Electroencephalogr Clin Neurophysiol Suppl*. 1999;52:171–185.

4. Rossini PM, Barker AT, Berardelli A, et al. Non-invasive electrical and magnetic stimulation of the brain, spinal cord and roots: basic principles and procedures for routine clinical application. Report of an IFCN committee. *Electroencephalogr Clin Neurophysiol.* 1994;91(2):79–92.
5. Kujirai T, Caramia MD, Rothwell JC, et al. Corticocortical inhibition in human motor cortex. *J Physiol.* 1993;471:501–519.
6. Benardo LS. Separate activation of fast and slow inhibitory postsynaptic potentials in rat neocortex in vitro. *J Physiol.* 1994;476(2):203–215.
7. Connors BW, Malenka RC, Silva LR. Two inhibitory postsynaptic potentials, and GABAA and GABAB receptor-mediated responses in neocortex of rat and cat. *J Physiol.* 1988;406:443–468.
8. Di Lazzaro V, Pilato F, Dileone M, et al. GABAA receptor subtype specific enhancement of inhibition in human motor cortex. *J Physiol.* 2006;575(Pt 3):721–726.
9. Davies CH, Davies SN, Collingridge GL. Paired-pulse depression of monosynaptic GABA-mediated inhibitory postsynaptic responses in rat hippocampus. *J Physiol.* 1990;424:513–531.
10. McDonnell MN, Orekhov Y, Ziemann U. The role of GABA(B) receptors in intracortical inhibition in the human motor cortex. *Exp Brain Res.* 2006;173(1):86–93.
11. Ziemann U. Intracortical inhibition and facilitation in the conventional paired TMS paradigm. *Electroencephalogr Clin Neurophysiol Suppl.* 1999;51:127–136.
12. Berg AT, Berkovic SF, Brodie MJ, et al. Revised terminology and concepts for organization of seizures and epilepsies: report of the ILAE Commission on Classification and Terminology, 2005–2009. *Epilepsia.* 2010;51(4):676–685.
13. Jones-Davis DM, Macdonald RL. GABA(A) receptor function and pharmacology in epilepsy and status epilepticus. *Curr Opin Pharmacol.* 2003;3(1):12–18.
14. Tassinari CA, Cincotta M, Zaccara G, Michelucci R. Transcranial magnetic stimulation and epilepsy. *Clin Neurophysiol.* 2003;114(5):777–798.
15. Badawy RA, Curatolo JM, Newton M, Berkovic SF, Macdonell RA. Changes in cortical excitability differentiate generalized and focal epilepsy. *Ann Neurol.* 2007;61(4):324–331.
16. Badawy RA, Vogrin SJ, Lai A, Cook MJ. Patterns of cortical hyperexcitability in adolescent/adult-onset generalized epilepsies. *Epilepsia.* 2013;54(5):871–878.
17. Manganotti P, Tamburin S, Bongiovanni LG, Zanette G, Fiaschi A. Motor responses to afferent stimulation in juvenile myoclonic epilepsy. *Epilepsia.* 2004;45(1):77–80.
18. Cantello R, Civardi C, Varrasi C, et al. Excitability of the human epileptic cortex after chronic valproate: a reappraisal. *Brain Res.* 2006;1099(1):160–166.
19. Brigo F, Storti M, Benedetti MD, et al. Resting motor threshold in idiopathic generalized epilepsies: a systematic review with meta-analysis. *Epilepsy Res.* 2012;101(1–2):3–13.
20. Badawy RA, Macdonell RA, Jackson GD, Berkovic SF. Why do seizures in generalized epilepsy often occur in the morning? *Neurology.* 2009;73(3):218–222.
21. Chowdhury FA, Pawley AD, Ceronie B, Nashef L, Elwes RD, Richardson MP. Motor evoked potential polyphasia: a novel endophenotype of idiopathic generalized epilepsy. *Neurology.* 2015;84(13):1301–1307.
22. Gilbert DL, Isaacs KM, Augusta M, Macneil LK, Mostofsky SH. Motor cortex inhibition: a marker of ADHD behavior and motor development in children. *Neurology.* 2011;76(7):615–621.
23. Reilly C, Atkinson P, Das KB, et al. Parent- and teacher-reported symptoms of ADHD in school-aged children with active epilepsy: a population-based study. *J Atten Disord.* 2014; pii:1087054714558117. [Epub ahead of print].
24. Wolf P, Yacubian EM, Avanzini G, et al. Juvenile myoclonic epilepsy: a system disorder of the brain. *Epilepsy Res.* 2015;114:2–12.
25. Hamer HM, Reis J, Mueller HH, et al. Motor cortex excitability in focal epilepsies not including the primary motor area–a TMS study. *Brain.* 2005;128(Pt 4):811–818.
26. Badawy RA, Macdonell RA, Berkovic SF, Newton MR, Jackson GD. Predicting seizure control: cortical excitability and antiepileptic medication. *Ann Neurol.* 2010;67(1):64–73.

27. Farzan F, Barr MS, Levinson AJ, et al. Reliability of long-interval cortical inhibition in healthy human subjects: a TMS-EEG study. *J Neurophysiol.* 2010;104(3):1339–1346.
28. Ziemann U. TMS and drugs. *Clin Neurophysiol.* 2004;115(8):1717–1729.
29. Ziemann U, Reis J, Schwenkreis P, et al. TMS and drugs revisited 2014. *Clin Neurophysiol.* 2014;126(10):1847–1868.
30. Bough KJ, Rho JM. Anticonvulsant mechanisms of the ketogenic diet. *Epilepsia.* 2007;48(1):43–58.
31. Cantello R, Varrasi C, Tarletti R, et al. Ketogenic diet: electrophysiological effects on the normal human cortex. *Epilepsia.* 2007;48(9):1756–1763.
32. Dahlin M, Elfving A, Ungerstedt U, Amark P. The ketogenic diet influences the levels of excitatory and inhibitory amino acids in the CSF in children with refractory epilepsy. *Epilepsy Res.* 2005;64(3):115–125.
33. Vitikainen AM, Salli E, Lioumis P, Makela JP, Metsahonkala L. Applicability of nTMS in locating the motor cortical representation areas in patients with epilepsy. *Acta Neurochir (Wien).* 2013;155(3):507–518.
34. Makela JP, Vitikainen AM, Lioumis P, et al. Functional plasticity of the motor cortical structures demonstrated by navigated TMS in two patients with epilepsy. *Brain Stimul.* 2013;6(3):286–291.
35. Krieg SM, Tarapore PE, Picht T, et al. Optimal timing of pulse onset for language mapping with navigated repetitive transcranial magnetic stimulation. *NeuroImage.* 2014;100: 219–236.
36. Ille S, Sollmann N, Hauck T, et al. Combined noninvasive language mapping by navigated transcranial magnetic stimulation and functional MRI and its comparison with direct cortical stimulation. *J Neurosurg.* 2015;123(1):212–225.
37. Tarapore PE, Findlay AM, Honma SM, et al. Language mapping with navigated repetitive TMS: proof of technique and validation. *NeuroImage.* 2013;82:260–272.
38. Berl MM, Zimmaro LA, Khan OI, et al. Characterization of atypical language activation patterns in focal epilepsy. *Ann Neurol.* 2014;75(1):33–42.
39. Hsu WY, Cheng CH, Lin MW, Shih YH, Liao KK, Lin YY. Antiepileptic effects of low frequency repetitive transcranial magnetic stimulation: a meta-analysis. *Epilepsy Res.* 2011;96(3):231–240.
40. Fregni F, Otachi PT, Do Valle A, et al. A randomized clinical trial of repetitive transcranial magnetic stimulation in patients with refractory epilepsy. *Ann Neurol.* 2006;60(4):447–455.
41. Theodore WH, Hunter K, Chen R, et al. Transcranial magnetic stimulation for the treatment of seizures: a controlled study. *Neurology.* 2002;59(4):560–562.
42. Ruohonen J, Karhu J. Navigated transcranial magnetic stimulation. *Neurophysiol Clin/ Clin Neurophysiol.* 2010;40(1):7–17.
43. Sun W, Mao W, Meng X, et al. Low-frequency repetitive transcranial magnetic stimulation for the treatment of refractory partial epilepsy: a controlled clinical study. *Epilepsia.* 2012;53(10):1782–1789.
44. Thordstein M, Constantinescu R. Possibly lifesaving, noninvasive, EEG-guided neuromodulation in anesthesia-refractory partial status epilepticus. *Epilepsy Behav.* 2012; 25(3):468–472.
45. Bae EH, Theodore WH, Fregni F, Cantello R, Pascual-Leone A, Rotenberg A. An estimate of placebo effect of repetitive transcranial magnetic stimulation in epilepsy. *Epilepsy Behav.* 2011;20(2):355–359.
46. VanHaerents S, Herman ST, Pang T, Pascual-Leone A, Shafi MM. Repetitive transcranial magnetic stimulation; A cost-effective and beneficial treatment option for refractory focal seizures. *Clin Neurophysiol.* 2015;126(9):1840–1842.
47. Garvey MA, Kaczynski KJ, Becker DA, Bartko JJ. Subjective reactions of children to single-pulse transcranial magnetic stimulation. *J Child Neurol.* 2001;16(12):891–894.
48. Rotenberg A, Muller P, Birnbaum D, et al. Seizure suppression by EEG-guided repetitive transcranial magnetic stimulation in the rat. *Clin Neurophysiol.* 2008;119(12):2697–2702.

49. Nakken KO, Server A, Kostov H, Haakonsen M. A patient with a 44-year history of epilepsia partialis continua caused by a perirolandic cortical dysplasia. *Epilepsy Behav.* 2005;6(1):94–97.
50. Rotenberg A, Bae EH, Takeoka M, Tormos JM, Schachter SC, Pascual-Leone A. Repetitive transcranial magnetic stimulation in the treatment of epilepsia partialis continua. *Epilepsy Behav.* 2009;14(1):253–257.
51. Misawa S, Kuwabara S, Shibuya K, Mamada K, Hattori T. Low-frequency transcranial magnetic stimulation for epilepsia partialis continua due to cortical dysplasia. *J Neurol Sci.* 2005;234(1–2):37–39.
52. Morales OG, Henry ME, Nobler MS, Wassermann EM, Lisanby SH. Electroconvulsive therapy and repetitive transcranial magnetic stimulation in children and adolescents: a review and report of two cases of epilepsia partialis continua. *Child Adolesc Psychiatr Clin N Am.* 2005;14(1):193–210. [viii–ix].
53. Shorvon S, Ferlisi M. The treatment of super-refractory status epilepticus: a critical review of available therapies and a clinical treatment protocol. *Brain.* 2011;134(Pt 10):2802–2818.
54. Liu A, Pang T, Herman S, Pascual-Leone A, Rotenberg A. Transcranial magnetic stimulation for refractory focal status epilepticus in the intensive care unit. *Seizure.* 2013;22(10):893–896.
55. Nitsche MA, Paulus W. Sustained excitability elevations induced by transcranial DC motor cortex stimulation in humans. *Neurology.* 2001;57(10):1899–1901.
56. San-Juan D, Morales-Quezada L, Orozco Garduno AJ, et al. Transcranial direct current stimulation in epilepsy. *Brain Stimul.* 2015;8(3):455–464.
57. Fregni F, Thome-Souza S, Nitsche MA, Freedman SD, Valente KD, Pascual-Leone A. A controlled clinical trial of cathodal DC polarization in patients with refractory epilepsy. *Epilepsia.* 2006;47(2):335–342.
58. Auvichayapat N, Rotenberg A, Gersner R, et al. Transcranial direct current stimulation for treatment of refractory childhood focal epilepsy. *Brain Stimul.* 2013;6(4):696–700.
59. Varga ET, Terney D, Atkins MD, et al. Transcranial direct current stimulation in refractory continuous spikes and waves during slow sleep: a controlled study. *Epilepsy Res.* 2011;97(1–2):142–145.
60. Faria P, Fregni F, Sebastiao F, Dias AI, Leal A. Feasibility of focal transcranial DC polarization with simultaneous EEG recording: preliminary assessment in healthy subjects and human epilepsy. *Epilepsy Behav.* 2012;25(3):417–425.
61. Grippe TC, Brasil-Neto JP, Boechat-Barros R, Cunha NS, Oliveira PL. Interruption of epilepsia partialis continua by transcranial direct current stimulation. *Brain Stimul.* 2015;8(6):1227–1228.
62. Dhamne SC, Ekstein D, Zhuo Z, et al. Acute seizure suppression by transcranial direct current stimulation in rats. *Ann Clin Transl Neurol.* 2015;2(8):843–856.

Brain Stimulation in Pediatric Depression: Biological Mechanisms

P. Croarkin
Mayo Clinic College of Medicine, Rochester, MN, United States

S.H. Ameis
Campbell Family Mental Health Research Institute, CAMH; University of Toronto, Toronto, ON, Canada

F.P. MacMaster
University of Calgary, Calgary, AB, Canada

OUTLINE

Pediatric Brain Stimulation
http://dx.doi.org/10.1016/B978-0-12-802001-2.00015-1

INTRODUCTION

Non-invasive brain stimulation (NIBS) modalities, such as transcranial direct-current stimulation (tDCS) and repetitive transcranial magnetic stimulation (rTMS), have demonstrated early promise as therapeutic interventions for youth with neurologic or psychiatric disorders. Techniques which directly modulate dysfunctional neurocircuitry with a low side-effect burden are appealing for this population. Recent systematic reviews bolster this idea. To date 48 studies, including 513 youth (18 years of age), have been completed[1] and collectively suggest that tDCS and rTMS are safe[1] and potentially effective[2] treatments in the developing brain.

Research with tDCS is in the midst of a renaissance as it is tolerable, safe, inexpensive, and has potential positive effects for a wide range of neurologic conditions (see Chapter 5: Transcranial Direct Current Stimulation (tDCS): Principles and Emerging Applications in Children). The biological effects of tDCS likely vary based on the delivery of anodal (excitatory) and cathodal (inhibitory) stimulation.[3] Prior work suggests that tDCS sessions modulate neural resting membrane potential and cortical excitability through changes in glutamatergic and GABAergic synaptic activity.[1,3] The physiologic and behavioral effects of one tDCS session are brief in duration. However, multiple sessions[4] or sessions coupled with other interventions[5] may have a more significant impact on synaptic plasticity and behavior. A deeper understanding of the relevant neurobiology could inform future therapeutic work in youth.

Over the past decade, therapeutic trials of repetitive transcranial magnetic stimulation (rTMS) for depression have focused on low frequency (1 Hz) or high frequency (5–20 Hz) over the dorsolateral prefrontal cortex. More contemporary work examines theta burst stimulation (TBS) pulse sequences as potentially more efficient and durable dosing strategies for the modification of cortical activity and treatment. TBS sequences deliver groups of three high-frequency pulses (50 Hz) with intervals of 200 ms (5 Hz). There are two primary TBS patterns that are thought to have discordant neurophysiological effects.[6] Continuous theta burst stimulation (cTBS) provides TBS pulses without interruption (typically 20–40 s 300–600 pulses) and is thought to decrease cortical excitability. Intermittent theta burst stimulation (iTBS) delivers 2-s trains of TBS (30 pulses) every 10 s and is thought to increase cortical excitability.[7]

The antidepressant effects of rTMS[8,9] and tDCS[10] were serendipitously discovered. Since that time, clinical trials have advanced the understanding and delivery of these techniques for the treatment of psychiatric disorders such as depression. However, a definitive understanding of the biologic effects and mechanisms of these NIBS modalities is lacking.[2] The number of relevant variables encompassing location of stimulation, dosing parameters, frequency of sessions, and individual neurobiology present considerable challenges in this area of inquiry. An increased understanding would have utility in study design, optimizing stimulus delivery, and personalizing treatments. Arguably, this is an ethical mandate for studies of NIBS in vulnerable populations such as youth with psychiatric disorders. This chapter reviews prior research on the mechanisms of therapeutic tDCS and rTMS for depression. As much of this prior work focuses on adults it must be considered in a neurodevelopmental context to inform work with youth.

PRECLINICAL

Therapeutic trials of tDCS and rTMS for depression are promising. Four magnetic stimulators have been cleared by the United States Food and Drug Administration (FDA) for the treatment of adults with major depressive disorder (MDD) that has not responded to antidepressant medications. Therapeutic research with tDCS is promising and rTMS is increasingly used in clinical settings to treat depression. However, it is difficult to identify patients most likely to benefit. Executing adequately powered clinical and mechanistic studies is also challenging due to ethical, pragmatic, and financial constraints. Treatments are time-consuming and potentially costly for patients. Biomarkers to identify underlying neurochemical mechanisms of action and patients most likely to benefit from rTMS could revolutionize clinical practice. Investigations with animal models could optimize rTMS delivery and catalyze discovery in humans. Animal studies have unique methodologic challenges that include the careful consideration of coil or electrode size, confounds of anesthesia, and the impact of handling during stimulus delivery.[11]

Early animal research demonstrated the basic neurophysiological effects of tDCS.[3] Anodal stimulation over the cortex increased neuronal activity, while cathodal stimulation reduced activity.[12–14] Further, stimulation impacted cortical neurons differently depending on the location and orientation of the electrodes. Thresholds also vary among groups of neurons. For example, pyramidal neurons required higher charges for stimulation as compared to other neurons.[3] Recent work examining a mouse model of tDCS treatment for nicotine addiction demonstrated that anodal stimulation, twice daily, for 20 min, over five successive days at 0.2 mA, had antidepressant properties in naïve mice and normalized depressive behaviors in mice with prolonged nicotine exposure.[15] Further work focused on animal models of tDCS treatment for depression will be invaluable for mechanistic understanding and improving therapeutic protocols.

Initial animal studies of rTMS suggested that its behavioral and neuroendocrine effects parallel that of antidepressant medications.[16] For example, repeated sessions of rTMS in rodent models exert quantifiable antidepressant effects, such as increases in apomorphine-induced stereotypy, increased electroconvulsive shock thresholds, upregulation of beta-adrenergic receptors, and decreased immobility time with the forced swim test.[17] Other early work demonstrated putative neuroprotective and neuroplastic effects of rTMS sessions, such as increased hippocampal brain-derived neurotrophic factor.[18] Recently, studies have also examined the differential effects of low- and high-frequency rTMS sessions on brain-derived neurotrophic factor and GluR1. It appears that high-frequency rTMS sessions potentiate these two markers of neuroplasticity in awake rats.[19] Fifteen daily sessions of rTMS in rats also increased hippocampal glutamate and gamma-aminobutyric acid concentrations as assessed by high-performance liquid chromatography.[20] Electrophysiological studies have demonstrated that rTMS modulates cortical excitability in a frequency-dependent fashion in rodents.[21,22] Further work examining protein expression and electrophysiology sought to demonstrate the differential effects of iTBS and cTBS in rat cortical excitability. These two TBS stimulation patterns have variable effects on neuronal inhibitory functioning with iTBS most likely impacting the output of pyramidal cells, while cTBS appears to modulate dendritic input to pyramidal cells via interneurons.[23] Recently, another elegant line of research examined the impact of various low-intensity rTMS dosing patterns on neuronal cell cultures. Findings suggested that varying stimulus frequency and patterns of TMS had differential effects on proapoptotic genes, antiapoptotic genes, and neurite development. Complex dosing patterns, such as TBS, appeared to have favorable effects on cell survival, leading the authors to contemplate the central importance of the rhythm of stimulus delivery in mechanistic effects.[24] Given the inherent ethical and pragmatic challenges of examining mechanisms of rTMS during human neurodevelopment, preclinical research will continue to play a central role.

NEUROPHYSIOLOGY

Electroencephalography (EEG), cortical excitability, and cortical inhibition measures have potential for the study of mechanisms of NIBS in depression. Previous studies have demonstrated that EEG biomarkers, such as quantitative indices, evoked potentials, alterations in frequency bands, hemispheric alpha asymmetry, and theta concordance, may predict antidepressant response.[25] Pre- and post-treatment EEG studies of tDCS and rTMS may prove useful for examining early predictors of response, identifying relevant cortical changes with intervention, and elucidating mechanisms of action. For example, recent efforts with synchronized transcranial magnetic stimulation (TMS) have postulated that low field strength, sinusoidal stimulation matching a patient's unique alpha frequency may be effective for MDD.[26] Other work has examined changes in cortical excitability and inhibition with measures of resting motor threshold, motor-evoked potential (MEP) amplitude, cortical silent period (CSP), and paired-pulse TMS measures such as short-interval intracortical inhibition (SICI), intracortical facilitation (ICF), and long-interval intracortical inhibition (LICI) (see Fig. 15.1 and Chapter 1: TMS Basics: Single and Paired Pulse Neurophysiology).[27]

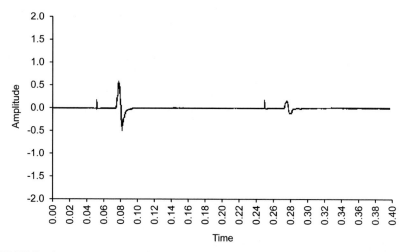

FIGURE 15.1 The long-interval cortical inhibition (LICI) paradigm is an index of $GABA_B$-mediated inhibitory neurotransmission. Two suprathreshold stimulations are applied to the motor cortex with interstimulus intervals of 50–200 ms. Paired-pulse measures of cortical inhibition and facilitation have promise as non-invasive measures for mechanistic studies of brain stimulation treatments and baseline predictors of response. For example, LICI measures of $GABA_B$ mediated inhibitory function may have utility in predicting response to fluoxetine. *Reproduced with permission from: Croarkin PE, Nakonezny PA, Husain MM, et al. Evidence for pretreatment LICI deficits among depressed children and adolescents with nonresponse to fluoxetine. Brain Stimul. 2014;7(2):243–251.*

Paired-pulse measures of ICF may have utility in pediatric depression, as one cross-sectional study demonstrated that depressed children and adolescents have increased ICF compared to healthy controls. These findings suggest that depressed children and adolescents may have excessive NMDA-mediated glutamatergic neurotransmission.[28] However, at present it is unclear if this finding is a trait or state characteristic. A more sophisticated approach interleaves EEG and TMS measures for neurophysiology studies,[29] providing more dynamic measures of the prefrontal cortex. Electrophysiological measures of neuroplasticity, such as paired associative stimulation (PAS), may also have utility in examining the mechanisms of NIBS for depression. The PAS paradigm involves repetitive median nerve stimulation coupled with TMS stimulation of the contralateral motor cortex. Cortical changes related to PAS are thought to index long-term potentiation. Prior work with PAS suggests that depressed adults have reductions in this measure of neuroplasticity.[30]

Powell and colleagues recently examined the neurophysiological effects related to a single session of tDCS for depressed patients with cortical EEG. Eighteen depressed adults participating in a sham-controlled study of tDCS (2 mA for 20 min or sham stimulation applied to the left dorsolateral prefrontal cortex) had postsession EEG activity and event-related potentials during a visual working memory task. After active stimulation, participants had decreased N2 amplitude and reduced, task-related, frontal theta activity. The authors proposed that these findings suggest a lasting, indirect effect from left frontal stimulation.[31] A recent study of healthy adult participants examined the impact of a 20-min session of anodal and cathodal tDCS over the primary motor cortex on $GABA_B$-mediated inhibitory function. LICI and CSP measures were collected immediately prior and 20 min after the tDCS session. Anodal stimulation increased the amplitude of MEPs, reduced CSP duration, and had no effect on LICI. Cathodal stimulation had no effect on MEPs, CSP duration, or LICI.[32] Of note, while work with tDCS and depressed youth is lacking, LICI measures have shown early promise as an index of responsiveness to standard antidepressant treatments in pediatric patients.[33] Other recent efforts with healthy participants and sophisticated study designs have examined the impact of repeated tDCS sessions. Findings suggest that multiple sessions of tDCS spaced with specific time intervals induce late long-term potentiation in the motor cortex.[34] Player and colleagues examined changes in PAS measures of plasticity during the course of tDCS treatment for depression in 18 adults. Findings demonstrated improvements in mood and cortical plasticity among participants, although these two improvements were not correlated.[35]

Recent EEG studies have suggested that various EEG indices may serve as predictors of response for patients with MDD treated with rTMS. In an open-label study of rTMS for MDD, P300 amplitude, anterior individual alpha peak frequency, frontocentral theta, prefrontal cordance in delta and beta bands all demonstrated differences between responders and non-responders. Combining these measures in a discriminant analysis yielded a fairly reliable means of identifying non-responders within the study sample.[36] Noda and colleagues investigated the lasting effects of rTMS on EEG by collecting resting and spectrum power EEG before and after 10 sessions of left prefrontal rTMS in 25 adults with treatment-resistant depression. Theta power in prefrontal electrodes increased with treatment. Clinical improvements correlated with F3 and F4 alpha power increased.[37] A subsequent study with 90 adult patients with MDD and 17 healthy controls examined nonlinear EEG methods with the goals of identifying differences in depression and potential utility to predict treatment response to rTMS treatment. There were no differences among healthy controls and patients with MDD. However, responders displayed an increase in Lempel–Ziv complexity measures from minute 1 to minute 2, while non-responders had a decrease, thereby suggesting this method may be used to predict treatment response.[38] Quantitative EEG cordance measures have also been used to probe changes with rTMS treatment for MDD with the hope of identifying predictive markers.[39,40]

Numerous previous studies have examined the impact of rTMS applied to the motor cortex on measures of motor cortical excitability and inhibition in adults. In general, low-frequency rTMS is thought to elicit a transient decrease in MEP amplitudes and has little effect on measures of cortical inhibition. High-frequency rTMS produces a long-lasting increase in MEP amplitudes and decreases in cortical inhibition. However, less is known regarding frequencies above 5 Hz.[27] Recent work with healthy participants suggests that potentiation of the CSP may serve as a useful biomarker for dose-finding in therapeutic rTMS protocols for MDD.[41] One prior study examined changes in cortical excitability as assessed with motor threshold measurements in adolescents with treatment-resistant depression undergoing 30 sessions of high-frequency rTMS applied to the left prefrontal cortex. Mean motor threshold measurements (percentage of total machine output) decreased over time and this reached statistical significance at week 5, suggesting that high-frequency rTMS sessions increased cortical excitability over time. Significant limitations of this initial study include a small sample size (eight adolescents) and visualization of movement assessments for motor threshold determination.[42]

NEUROIMAGING

Mechanistic studies of tDCS and rTMS frequently employ imaging modalities. To date, this is the most common approach for human mechanistic studies of NIBS for adult MDD.[43] Functional magnetic resonance imaging (fMRI) studies can assess functional or structural connectivity of brain regions with resting state protocols.[44] Other approaches examine variations in regional blood flow by assessing blood oxygen level dependence contrasts.[45] Near-infrared spectroscopy (NIRS) also examines cortical hemodynamic changes but with limited depth and spatial resolution. However, NIRS may have inherent utility in the study of pediatric depression and NIBS.[46] Early imaging studies of rTMS also often involved functional studies with positron emission tomography (PET)[47] and single-photon emission computerized tomography (SPECT).[48] These approaches are problematic for studies in depressed children and adolescents given the use of high-energy gamma radiation. Diffusion tensor imaging protocols have previously examined changes in white matter microstructure during the course of rTMS sessions.[49] Proton magnetic resonance spectroscopy (^{1}H-MRS) techniques examine regional neurochemistry albeit with limited temporal and spatial resolution (see Fig. 15.2). To date, ^{1}H-MRS has been the favored approach for studying depressed youth and young adults receiving rTMS.[50,51]

Current approaches with tDCS in adult MDD are often based on the supposition that depressed patients have decreased left hemispheric cortical activity and increased right hemispheric cortical activity. Anodal stimulation over the left prefrontal cortex thereby increases cortical activity

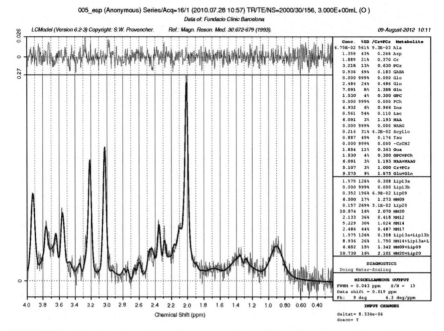

FIGURE 15.2   Sample proton magnetic resonance (¹H-MRS) from an anterior cingulate cortex voxel analyzed with LCModel. Cramer–Rao lower bounds and respective metabolite concentrations are on the right. Recent studies have demonstrated that ¹H-MRS has potential utility for the study of neurochemical changes during the course of brain stimulation treatments. Further, studies with concurrent ¹H-MRS and transcranial magnetic stimulation neurophysiology paradigms offer the opportunity for both receptor-mediated and neurochemical measures of inhibitory/excitatory neurotransmission. *Reproduced with permission from: Oritz AE, Ortiz AG, Falcon C, et al. ¹H-MRS of the anterior cingulate cortex in childhood and adolescent obsessive-compulsive disorder: A case–control study. Eur Neuropsychopharmacol. 2015;25:60–68.*

with improvements in mood and executive functioning. Conversely, right prefrontal cathodal stimulation modulates excessive right hemispheric activity. Emerging evidence further implies top–down modulatory effects for subcortical areas. Treatment with tDCS may also address deficits in neuroplasticity through NMDA-mediated neurotransmission.[52] Prior tDCS studies have examined these theories with clinical and neurocognitive approaches. Unfortunately, neuroimaging studies are lacking.[43] Stagg and colleagues have proposed that anodal tDCS applied to the left primary motor cortex decreases GABA concentrations in healthy controls as assessed with ¹H-MRS.[53] Furthermore, a recent multimodal imaging study demonstrated a negative relationship between GABA levels and the magnitude of resting motor network connectivity. In 10 healthy participants, anodal tDCS applied to the left primary motor cortex increased resting motor network connectivity immediately after the stimulation session.[54]

Another recent study with 11 healthy participants examined resting-state fMRI and ^{1}H-MRS immediately prior to and at the cessation of anodal tDCS applied to the right parietal cortex. Results demonstrated an increase in the combined (Glx) measure of glutamate and glutamine below the anode. Within-network increases were observed in the superior parietal, inferior parietal, left frontal-parietal, salience, and cerebellar intrinsic network. Conversely, anterior cingulate and basal ganglia connectivity decreased. Hunter and colleagues suggested that individual differences between glutamatergic metabolites and network connectivity may inform future tDCS work in psychiatric disease.[55] Unfortunately, imaging studies of tDCS for depression are lacking, particularly in pediatric patients.[43]

Collectively fMRI, PET, and SPECT studies of depressed adults suggest that rTMS changes cerebral blood flow at the site of stimulation and remotely.[56–59] However, it is difficult to reach definitive conclusions based on heterogeneous patient samples, different rTMS treatment protocols, and variable imaging techniques. Although not universal, some studies suggest that high-frequency rTMS applied to the left prefrontal cortex corrects hypoactivity in conjunction with antidepressant effects.[45,58] Some studies suggest more individual variability in effects with non-dominant hemisphere 1-Hz rTMS.[60] Most prior approaches have relied on regional activity or blood flow changes. Contemporary studies have begun to examine the potential network effects of rTMS in the context of treatment response. Liston and colleagues recently examined resting state fMRI functional connectivity measures of the frontoparietal central executive network (CEN) and the medial prefrontal-medial parietal default mode network (DMN) in 17 depressed adult participants prior to and after 5 weeks of high-frequency rTMS (see Fig. 15.3). In comparison to healthy controls, depressed patients had increased DMN functional connectivity and decreased CEN functional connectivity. Treatment with rTMS modulated increased subgenual connectivity in the DMN. Subgenual connectivity was also predictive of clinical response to treatments. The authors maintained that this work adds to growing literature implicating the DMN in depression and emotional regulation.[44,61] Such linking of cortical treatment targets with subcortical DBS targets, such as the subgenu, via connectivity imaging may represent a powerful tool to better understand modulation of functional networks in major depression.[62]

Peng and colleagues examined the impact of high-frequency rTMS on white matter integrity of young adults. Thirty participants with treatment-resistant MDD were randomized to 4 weeks of either high-frequency rTMS (15 Hz) or sham stimulation applied to the left dorsolateral prefrontal cortex. Diffusion tensor imaging scans were completed at baseline and after 4 weeks of rTMS treatment. Participants with treatment-resistant MDD demonstrated decreased fractional anisotropy in the left middle frontal gyrus in comparison to a sample of 25 healthy participants. Post-treatment

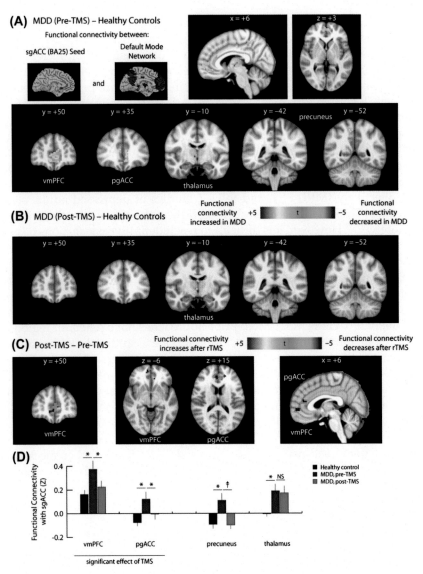

FIGURE 15.3 Transcranial magnetic stimulation modulates (TMS) increased default mode network (DMN) connectivity. Depressed patients demonstrated increased subgenual anterior cingulate connectivity within the DMN (A). Aside from the thalamus, all other areas of hyperconnectivity attenuated after TMS (B). The effects of TMS on hyperconnectivity normalization reached significant effects in the ventromedial prefrontal cortex and pregenual anterior cingulate cortex (C). Summary of data quantification from respective coordinates of peak t statistic (D). This work suggests that antidepressant effects of TMS involve the modulation of functional connectivity in the DMN and subgenual cingulate in particular. Further work in youth could assist with biomarker development and advance knowledge of neurodevelopment. *Reproduced with permission from: Liston C, Chen AC, Zebley BD, et al. Default mode network mechanisms of transcranial magnetic stimulation in depression. Biol Psychiatry. 2014;76(7):517–526.*

assessments revealed that participants receiving active stimulation had increased fractional anisotropy values and greater improvements in depressive symptoms compared to participants receiving sham stimulation. Of note, additional analyses also revealed inverse relationships between fractional anisotropy measures and depression severity measures at baseline and post-treatment in all participants. The authors postulated that the mechanism of action for therapeutic rTMS in this population may involve alterations in white matter properties between the prefrontal cortex and limbic structures.[63]

Two other studies with depressed adolescents and young adults receiving rTMS focused on ^{1}H-MRS measures.[50,51] Zheng and colleagues recruited 34 young adults with treatment-resistant depression for a 4-week, double-blind, sham-controlled trial of high-frequency (15 Hz) rTMS. At baseline, depressed participants and a sample of 28 healthy controls underwent multiple voxel ^{1}H-MRS. Metabolites were corrected to creatine and included myo-inositol (M-Ino), choline (Cho), creatine/phosphocreatine (Cr), N-acetyl aspartate (NAA), and glutamate/glutamine (Glx). The majority of participants (12 of 19) receiving active rTMS responded to treatment. At baseline M-Ino levels in the left prefrontal cortex were lower in depressed participants compared to healthy controls. There were no other statistically significant differences in metabolites among participants with treatment-resistant depression and healthy controls. Post ^{1}H-MRS scans revealed that responders had an increase in left prefrontal cortex M-Ino from baseline scans. Further M-Ino levels increased in the right prefrontal cortex of responders at a trend level. The authors contended that M-Ino deficits in depressed patients may be reflective of decreased glial density or functionality in young adults with MDD.[50] Yang and colleagues reported on a sample of six adolescents with treatment-resistant MDD who received high-frequency (10 Hz) rTMS applied to the left dorsolateral prefrontal cortex for 3 weeks. The authors collected ^{1}H-MRS measures of the left dorsolateral prefrontal cortex at baseline and post-treatment with the aim of examining changes in glutamate neurochemistry. Treatment responders demonstrated an increase in left dorsolateral prefrontal cortex glutamate levels. At baseline, patients not benefiting from rTMS had elevated glutamate levels which decreased further at post-treatment.[51] Modulation of glutamatergic neurotransmission may have a key role in the mechanism of rTMS in adolescent depression and ^{1}H-MRS measures of glutamate may have utility in predicting response to rTMS. However, this must be confirmed with further work. Given the safety profile and utility in studying inhibitory and excitatory neurochemistry, ^{1}H-MRS will likely play a key role in examining mechanisms of treatment. Multimodal imaging studies of rTMS in youth with testable, a priori hypotheses will inform the understanding of antidepressant mechanisms.

PERIPHERAL BIOMARKERS

Serum measures such as brain-derived neurotrophic factor, cortisol, thyroid hormones, reproductive hormones, and neurotransmitters have been considered as potential biomarkers for adults with MDD undergoing tDCS or rTMS.[43] These approaches are appealing as inexpensive, non-invasive tests to study mechanism and ultimately index patient responsivity at baseline. However, interpreting correlational brain effects with peripheral measures is challenging. Results have most often been inconsistent or negative when examined in a systematic fashion.[64] To date, there are no published studies of peripheral biomarkers in depressed youth undergoing NIBS.

NEUROCOGNITION

Neurocognitive testing and approaches may have promise in both optimizing NIBS treatment for depression and elucidating relevant mechanisms of action. While there are no relevant studies of tDCS in children and adolescents, early work has examined the neurocognitive impact of rTMS in adolescents and young adults. Wall and colleagues[65] reported on 18 adolescents with treatment-resistant MDD undergoing 30 treatments of high-frequency rTMS (10 Hz) applied to the left dorsolateral prefrontal cortex. Evaluations at baseline and post-treatment included the Children's Auditory Verbal Learning Test[66] and Delis–Kaplan Executive Function System Trail Making Test.[67] Results were reassuring as improvements were noted in immediate memory and delayed recall while other indices remained stable after rTMS.[65] Mayer and colleagues reported similar findings in six adolescents with MDD who underwent a course of rTMS.[68] Participants had follow-up neurocognitive testing with the Cambridge Neuropsychological Test Automated Battery[69] three years after treatment with rTMS. No decrements in neurocognitive functioning were noted in this small sample.[68] These studies demonstrate preliminary evidence of safety and suggest that neurocognitive approaches have importance in planning for future mechanistic studies of NIBS in depressed youth.[70]

CONCLUSION

There is a critical need for effective brain-based interventions in child and adolescent depression. Work with NIBS in child and adolescent depression is nascent but promising. A detailed, neurodevelopmental understanding of tDCS and rTMS mechanisms in youth depression is an ethical and pragmatic necessity for this body of research to mature. Extant evidence suggests that NIBS exerts antidepressant effects by modulating imbalances in excitatory and inhibitory neurotransmission. Definitive understanding in youth will be challenging as the roles of relevant neurotransmitter systems and cortical networks shift during development in health and disease. Preclinical, neurophysiological, neuroimaging, neurocognitive, and multimodal approaches will likely play key roles in developing this knowledge base to optimize the treatment of depression across development. State-dependent NIBS protocols with depressed youth would likely catalyze mechanistic understanding as well. These approaches could optimize therapeutic outcomes while providing improved functional resolution of NIBS in developing youth.[71]

References

1. Krishnan C, Santos L, Peterson MD, Ehinger M. Safety of noninvasive brain stimulation in children and adolescents. *Brain Stimul*. January–February 2015;8(1):76–87.
2. Donaldson AE, Gordon MS, Melvin GA, Barton DA, Fitzgerald PB. Addressing the needs of adolescents with treatment resistant depressive disorders: a systematic review of rTMS. *Brain Stimul*. January-February 2014;7(1):7–12.
3. Stagg CJ, Nitsche MA. Physiological basis of transcranial direct current stimulation. *Neuroscientist*. February 2011;17(1):37–53.
4. Reis J, Schambra HM, Cohen LG, et al. Noninvasive cortical stimulation enhances motor skill acquisition over multiple days through an effect on consolidation. *Proc Natl Acad Sci USA*. February 3, 2009;106(5):1590–1595.
5. Brunoni AR, Valiengo L, Baccaro A, et al. The sertraline vs. electrical current therapy for treating depression clinical study: results from a factorial, randomized, controlled trial. *JAMA Psychiatry*. April 2013;70(4):383–391.
6. Chung SW, Hoy KE, Fitzgerald PB. Theta-burst stimulation: a new form of Tms treatment for depression? *Depress Anxiety*. November 28, 2014;32(3):182–192.
7. Huang YZ, Edwards MJ, Rounis E, Bhatia KP, Rothwell JC. Theta burst stimulation of the human motor cortex. *Neuron*. January 20, 2005;45(2):201–206.
8. Bickford RG, Guidi M, Fortesque P, Swenson M. Magnetic stimulation of human peripheral nerve and brain: response enhancement by combined magnetoelectrical technique. *Neurosurgery*. January 1987;20(1):110–116.
9. George MS, Wassermann EM, Post RM. Transcranial magnetic stimulation: a neuropsychiatric tool for the 21st century. *J Neuropsychiatry Clin Neurosci*. Fall 1996;8(4): 373–382.
10. Parent A. Giovanni Aldini: from animal electricity to human brain stimulation. *Can J Neurol Sci*. November 2004;31(4):576–584.
11. Salvador R, Miranda PC. Transcranial magnetic stimulation of small animals: a modeling study of the influence of coil geometry, size and orientation. *Conf Proc IEEE Eng Med Biol Soc*. 2009;2009:674–677.
12. Bindman LJ, Lippold OC, Redfearn JW. The action of brief polarizing currents on the cerebral cortex of the rat (1) during current flow and (2) in the production of long-lasting after-effects. *J Physiol*. August 1964;172:369–382.
13. Creutzfeldt OD, Fromm GH, Kapp H. Influence of transcortical d-c currents on cortical neuronal activity. *Exp Neurol*. June 1962;5:436–452.
14. Purpura DP, McMurtry JG. Intracellular activities and evoked potential changes during polarization of motor cortex. *J Neurophysiol*. January 1965;28:166–185.
15. Pedron S, Monnin J, Haffen E, Sechter D, Van Waes V. Repeated transcranial direct current stimulation prevents abnormal behaviors associated with abstinence from chronic nicotine consumption. *Neuropsychopharmacology*. March 2014;39(4):981–988.

16. Keck ME, Welt T, Post A, et al. Neuroendocrine and behavioral effects of repetitive transcranial magnetic stimulation in a psychopathological animal model are suggestive of antidepressant-like effects. *Neuropsychopharmacology*. April 2001;24(4):337–349.
17. Fleischmann A, Prolov K, Abarbanel J, Belmaker RH. The effect of transcranial magnetic stimulation of rat brain on behavioral models of depression. *Brain Res*. November 13, 1995;699(1):130–132.
18. Post A, Muller MB, Engelmann M, Keck ME. Repetitive transcranial magnetic stimulation in rats: evidence for a neuroprotective effect in vitro and in vivo. *Eur J Neurosci*. September 1999;11(9):3247–3254.
19. Gersner R, Kravetz E, Feil J, Pell G, Zangen A. Long-term effects of repetitive transcranial magnetic stimulation on markers for neuroplasticity: differential outcomes in anesthetized and awake animals. *J Neurosci*. May 18, 2011;31(20):7521–7526.
20. Yue L, Xiao-lin H, Tao S. The effects of chronic repetitive transcranial magnetic stimulation on glutamate and gamma-aminobutyric acid in rat brain. *Brain Res*. March 13, 2009;1260:94–99.
21. Sun P, Wang F, Wang L, et al. Increase in cortical pyramidal cell excitability accompanies depression-like behavior in mice: a transcranial magnetic stimulation study. *J Neurosci*. November 9, 2011;31(45):16464–16472.
22. Muller PA, Dhamne SC, Vahabzadeh-Hagh AM, Pascual-Leone A, Jensen FE, Rotenberg A. Suppression of motor cortical excitability in anesthetized rats by low frequency repetitive transcranial magnetic stimulation. *PLoS One*. 2014;9(3):e91065.
23. Benali A, Trippe J, Weiler E, et al. Theta-burst transcranial magnetic stimulation alters cortical inhibition. *J Neurosci*. January 26, 2011;31(4):1193–1203.
24. Grehl S, Viola HM, Fuller-Carter PI, et al. Cellular and molecular changes to cortical neurons following low intensity repetitive magnetic stimulation at different frequencies. *Brain Stimul*. January-February 2015;8(1):114–123.
25. Baskaran A, Milev R, McIntyre RS. The neurobiology of the EEG biomarker as a predictor of treatment response in depression. *Neuropharmacology*. September 2012;63(4):507–513.
26. Jin Y, Phillips B. A pilot study of the use of EEG-based synchronized Transcranial Magnetic Stimulation (sTMS) for treatment of Major Depression. *BMC Psychiatry*. 2014;14:13.
27. Fitzgerald PB, Fountain S, Daskalakis ZJ. A comprehensive review of the effects of rTMS on motor cortical excitability and inhibition. *Clin Neurophysiol*. December 2006;117(12):2584–2596.
28. Croarkin PE, Nakonezny PA, Husain MM, et al. Evidence for increased glutamatergic cortical facilitation in children and adolescents with major depressive disorder. *JAMA Psychiatry*. March 2013;70(3):291–299.
29. Daskalakis ZJ, Farzan F, Barr MS, Maller JJ, Chen R, Fitzgerald PB. Long-interval cortical inhibition from the dorsolateral prefrontal cortex: a TMS-EEG study. *Neuropsychopharmacology*. November 2008;33(12):2860–2869.
30. Player MJ, Taylor JL, Weickert CS, et al. Neuroplasticity in depressed individuals compared with healthy controls. *Neuropsychopharmacology*. October 2013;38(11):2101–2108.
31. Powell TY, Boonstra TW, Martin DM, Loo CK, Breakspear M. Modulation of cortical activity by transcranial direct current stimulation in patients with affective disorder. *PLoS One*. 2014;9(6):e98503.
32. Tremblay S, Beaule V, Lepage JF, Theoret H. Anodal transcranial direct current stimulation modulates GABAB-related intracortical inhibition in the M1 of healthy individuals. *Neuroreport*. January 9, 2013;24(1):46–50.
33. Croarkin PE, Nakonezny PA, Husain MM, et al. Evidence for pretreatment LICI deficits among depressed children and adolescents with nonresponse to fluoxetine. *Brain Stimul*. March-April 2014;7(2):243–251.
34. Monte-Silva K, Kuo MF, Hessenthaler S, et al. Induction of late LTP-like plasticity in the human motor cortex by repeated non-invasive brain stimulation. *Brain Stimul*. May 2013;6(3):424–432.

35. Player MJ, Taylor JL, Weickert CS, et al. Increase in PAS-induced neuroplasticity after a treatment course of transcranial direct current stimulation for depression. *J Affect Disord.* October 2014;167:140–147.

36. Arns M, Drinkenburg WH, Fitzgerald PB, Kenemans JL. Neurophysiological predictors of non-response to rTMS in depression. *Brain Stimul.* October 2012;5(4):569–576.

37. Noda Y, Nakamura M, Saeki T, Inoue M, Iwanari H, Kasai K. Potentiation of quantitative electroencephalograms following prefrontal repetitive transcranial magnetic stimulation in patients with major depression. *Neurosci Res.* September–October 2013;77(1–2):70–77.

38. Arns M, Cerquera A, Gutierrez RM, Hasselman F, Freund JA. Non-linear EEG analyses predict non-response to rTMS treatment in major depressive disorder. *Clin Neurophysiol.* July 2014;125(7):1392–1399.

39. Erguzel TT, Ozekes S, Gultekin S, Tarhan N, Hizli Sayar G, Bayram A. Neural network based response prediction of rTMS in major depressive disorder using QEEG cordance. *Psychiatry Investig.* January 2015;12(1):61–65.

40. Ozekes S, Erguzel T, Sayar GH, Tarhan N. Analysis of brain functional changes in high-frequency repetitive transcranial magnetic stimulation in treatment-resistant depression. *Clin EEG Neurosci.* April 14, 2014;45.

41. de Jesus DR, Favalli GP, Hoppenbrouwers SS, et al. Determining optimal rTMS parameters through changes in cortical inhibition. *Clin Neurophysiol.* April 2014;125(4):755–762.

42. Croarkin PE, Wall CA, Nakonezny PA, et al. Increased cortical excitability with prefrontal high-frequency repetitive transcranial magnetic stimulation in adolescents with treatment-resistant major depressive disorder. *J Child Adolesc Psychopharmacol.* February 2012;22(1):56–64.

43. Fidalgo TM, Morales-Quezada JL, Muzy GS, et al. Biological markers in noninvasive brain stimulation trials in major depressive disorder: a systematic review. *J Ect.* March 2014;30(1):47–61.

44. Liston C, Chen AC, Zebley BD, et al. Default mode network mechanisms of transcranial magnetic stimulation in depression. *Biol Psychiatry.* October 1, 2014;76(7):517–526.

45. Nahas Z, Lomarev M, Roberts DR, et al. Unilateral left prefrontal transcranial magnetic stimulation (TMS) produces intensity-dependent bilateral effects as measured by interleaved BOLD fMRI. *Biol Psychiatry.* November 1, 2001;50(9):712–720.

46. Kozel FA, Tian F, Dhamne S, et al. Using simultaneous repetitive Transcranial Magnetic Stimulation/functional Near Infrared Spectroscopy (rTMS/fNIRS) to measure brain activation and connectivity. *Neuroimage.* October 1, 2009;47(4):1177–1184.

47. Speer AM, Kimbrell TA, Wassermann EM, et al. Opposite effects of high and low frequency rTMS on regional brain activity in depressed patients. *Biol Psychiatry.* December 15, 2000;48(12):1133–1141.

48. Teneback CC, Nahas Z, Speer AM, et al. Changes in prefrontal cortex and paralimbic activity in depression following two weeks of daily left prefrontal TMS. *J Neuropsychiatry Clin Neurosci.* Fall 1999;11(4):426–435.

49. Kozel FA, Johnson KA, Nahas Z, et al. Fractional anisotropy changes after several weeks of daily left high-frequency repetitive transcranial magnetic stimulation of the prefrontal cortex to treat major depression. *J Ect.* March 2011;27(1):5–10.

50. Zheng H, Zhang L, Li L, et al. High-frequency rTMS treatment increases left prefrontal myo-inositol in young patients with treatment-resistant depression. *Prog Neuropsychopharmacol Biol Psychiatry.* October 1, 2010;34(7):1189–1195.

51. Yang XR, Kirton A, Wilkes TC, et al. Glutamate alterations associated with transcranial magnetic stimulation in youth depression: a case series. *J Ect.* September 2014;30(3):242–247.

52. Brunoni AR, Ferrucci R, Fregni F, Boggio PS, Priori A. Transcranial direct current stimulation for the treatment of major depressive disorder: a summary of preclinical, clinical and translational findings. *Prog Neuropsychopharmacol Biol Psychiatry.* October 1, 2012;39(1):9–16.

53. Stagg CJ, Best JG, Stephenson MC, et al. Polarity-sensitive modulation of cortical neurotransmitters by transcranial stimulation. *J Neurosci.* April 22, 2009;29(16):5202–5206.

54. Stagg CJ, Bachtiar V, Amadi U, et al. Local GABA concentration is related to network-level resting functional connectivity. *Elife*. 2014;3:e01465.
55. Hunter MA, Coffman BA, Gasparovic C, Calhoun VD, Trumbo MC, Clark VP. Baseline effects of transcranial direct current stimulation on glutamatergic neurotransmission and large-scale network connectivity. *Brain Res*. January 12, 2015;1594:92–107.
56. Li X, Nahas Z, Kozel FA, Anderson B, Bohning DE, George MS. Acute left prefrontal transcranial magnetic stimulation in depressed patients is associated with immediately increased activity in prefrontal cortical as well as subcortical regions. *Biol Psychiatry*. May 1, 2004;55(9):882–890.
57. Barrett J, Della-Maggiore V, Chouinard PA, Paus T. Mechanisms of action underlying the effect of repetitive transcranial magnetic stimulation on mood: behavioral and brain imaging studies. *Neuropsychopharmacology*. June 2004;29(6):1172–1189.
58. Kito S, Fujita K, Koga Y. Changes in regional cerebral blood flow after repetitive transcranial magnetic stimulation of the left dorsolateral prefrontal cortex in treatment-resistant depression. *J Neuropsychiatry Clin Neurosci*. Winter 2008;20(1):74–80.
59. Li CT, Wang SJ, Hirvonen J, et al. Antidepressant mechanism of add-on repetitive tran-scranial magnetic stimulation in medication-resistant depression using cerebral glucose metabolism. *J Affect Disord*. December 2010;127(1–3):219–229.
60. Loo CK, Sachdev PS, Haindl W, et al. High (15 Hz) and low (1 Hz) frequency transcra-nial magnetic stimulation have different acute effects on regional cerebral blood flow in depressed patients. *Psychol Med*. August 2003;33(6):997–1006.
61. Sheline YI, Barch DM, Price JL, et al. The default mode network and self-referential pro-cesses in depression. *Proc Natl Acad Sci USA*. February 10, 2009;106(6):1942–1947.
62. Figee M, Luigjes J, Smolders R, et al. Deep brain stimulation restores frontostriatal net-work activity in obsessive-compulsive disorder. *Nat Neurosci*. 2013;16(4):386–387.
63. Peng H, Zheng H, Li L, et al. High-frequency rTMS treatment increases white matter FA in the left middle frontal gyrus in young patients with treatment-resistant depression. *J Affect Disord*. February 2012;136(3):249–257.
64. Brunoni AR, Baeken C, Machado-Vieira R, Gattaz WF, Vanderhasselt MA. BDNF blood levels after non-invasive brain stimulation interventions in major depres-sive disorder: a systematic review and meta-analysis. *World J Biol Psychiatry*. February 2015;16(2):114–122.
65. Wall CA, Croarkin PE, McClintock SM, et al. Neurocognitive effects of repetitive transcranial magnetic stimulation in adolescents with major depressive disorder. *Front Psychiatry*. 2013;4:165.
66. Talley JL. *Children's Auditory Verbal Learning Test*. Lutz, FL: Psychological Assessment Resources; 1993.
67. Delis DC, Kramer JH, Kaplan E, Holdnack J. Reliability and validity of the Delis-Kaplan Executive Function System: an update. *J Int Neuropsychol Soc*. March 2004;10(2):301–303.
68. Mayer G, Aviram S, Walter G, Levkovitz Y, Bloch Y. Long-term follow-up of adolescents with resistant depression treated with repetitive transcranial magnetic stimulation. *J ECT*. June 2012;28(2):84–86.
69. Luciana M. Practitioner review: computerized assessment of neuropsychological function in children: clinical and research applications of the Cambridge Neuropsychological Testing Automated Battery (CANTAB). *J Child Psychol Psychiatry*. July 2003;44(5): 649–663.
70. De Raedt R, Vanderhasselt MA, Baeken C. Neurostimulation as an intervention for treatment resistant depression: from research on mechanisms towards targeted neuro-cognitive strategies. *Clin Psychol Rev*. November 4, 2014;41:61–69.
71. Silvanto J, Pascual-Leone A. State-dependency of transcranial magnetic stimulation. *Brain Topogr*. September 2008;21(1):1–10.

Brain Stimulation in Childhood Mental Health: Therapeutic Applications

F.P. MacMaster, M. Sembo, K. Ma

University of Calgary, Calgary, AB, Canada

P. Croarkin

Mayo Clinic College of Medicine, Rochester, MN, United States

OUTLINE

INTRODUCTION

The treatment of pediatric mental illness and neurodevelopmental disorders is at a crossroads. Few novel medications are in development and existing treatments are often limited in effectiveness and tolerability in children and adolescents. Indeed, there is a growing consensus that the traditional strategy of treatment development has failed in these conditions.[1,2] In psychiatry, moving from pharmacology to pathophysiology has not yielded the progress we need (see Fig. 16.1). Past approaches are not personalized or well targeted to the disorder, and rely largely on trial and error. While most discoveries and innovations fail to earn a place in the clinic,[3,4] one very promising avenue for intervention in youth is brain stimulation. As discussed in this volume, the advantages of brain stimulation in children and adolescents are its safety profile, short duration to clinical effect, potential for durable effects, and direct mechanism of action. For example, a brain region relevant to the symptoms of the disorder is targeted for either inhibition or activation. In this chapter, we discuss the application of brain stimulation technology to the treatment of pediatric mental illness and neurodevelopmental disorders. We will address two non-invasive brain stimulation technologies—transcranial magnetic stimulation (TMS) and transcranial direct-current stimulation (tDCS). TMS was first introduced as a means to indirectly excite peripheral nerves.[5] Since then, TMS has evolved into a non-invasive intervention that stimulates targeted cortical areas via brief magnetic pulses.[6] tDCS utilizes a constant current applied to the head through electrodes on the scalp.[7] Here we will focus on clinical therapeutic applications in neuropsychiatry,

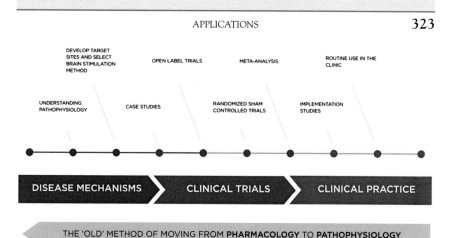

FIGURE 16.1 This schematic highlights the differences between the standard approach in mental illness (moving from pharmacology to pathophysiology) to the development path necessitated when applying brain stimulation. By its very nature, brain stimulation demands a specific target be developed. To do so, some understanding of the disorder is required.

while potential biological mechanisms are discussed elsewhere in this volume (see Chapter 15: Brain Stimulation in Pediatric Depression: Biological Mechanisms). Brain stimulation approaches have been applied in a variety of disorders in children and adolescents; namely, for patients experiencing attention-deficit hyperactivity disorder (ADHD), autism spectrum disorders (ASD), depression, schizophrenia, and Tourette syndrome.

APPLICATIONS

Depression

Major depressive disorder (or MDD) in adolescents is a major public health problem. MDD affects approximately 15% of adolescents[8]; it is associated with impairment in social, family, and academic functioning, and it is a major risk factor for suicide—a leading cause of death in adolescents.[9,10] MDD is characterized by depressed mood, markedly diminished interest or pleasure in activities, significant weight loss or weight gain, insomnia or hypersomnia, psychomotor agitation or retardation, fatigue or loss of energy, feelings of guilt or worthlessness, a diminished ability to concentrate, and recurrent thoughts of death or suicide. These symptoms can appear at any age, although the likelihood of experiencing symptoms increases with puberty.

Need for Novel Approaches and Biological Rationale for Targets

Unfortunately, there are few effectively targeted treatment options for this age group. Selective serotonin reuptake inhibitors (SSRIs) are the only class of medication approved for treating MDD in adolescents, but rates of response and remission following treatment with SSRIs are only 60 and 30%, respectively.[11] Furthermore, ongoing controversies persist regarding the safety of antidepressant use in youth. Cognitive behavior therapy is associated with similar remission rates[12] and access to therapy is limited. Based on the epidemiological data available, between 30% and 60% of people with MDD will have treatment-resistant MDD.[13] An optimal strategy for treating treatment-resistant MDD has not yet been identified.[14,15] This is especially problematic in youth, as a "lost" year (or more) during this critical period has an impact on education, occupational outcomes, and social development.

The left DLPFC has been the primary region of focus for studies of rTMS in adults with MDD[16,17] due to its proposed role in both the affective and cognitive symptoms of MDD. Indeed, there is growing evidence implicating the DLPFC in MDD in children and adolescents.[18–23] In the only neurobiological study to date on the effect of rTMS in youth with MDD, an increase in DLPFC glutamate concentration was associated with clinical responsiveness to rTMS,[21] similar to what has been noted in adults with MDD.[24]

Evidence

As a probe of function, TMS has offered insight into brain function in youth with MDD. First, it has been suggested that rTMS treatment alters cortical excitability.[25] A study of intracortical facilitation in youth with MDD suggested a possible excessive amount of glutamatergic activity.[26] Second, less inhibition and impaired GABA-B functioning (as measured with long-interval cortical inhibition) was suggested to be predictive of response to fluoxetine in youth with MDD.[27] Third, a significant negative relationship of age with motor threshold was observed in a sample of youth with MDD in both the right and left hemispheres, but not in healthy controls.[28] Finally, in youth with MDD, long-interval cortical inhibition (200-ms interval) was found to increase with age in both the right and left hemispheres.[28] This indicates an altered neurobiology and developmental trajectory for these measures in youth with depression.

In the adult literature, over 1300 subjects have been treated safely with rTMS (see Refs 16, 23, 29–32 as examples). rTMS is twice as likely to result in response (relative risk or RR: 2.35 (95% confidence interval or CI: 1.70–3.25)) and remission (RR: 2.24 (95% CI: 1.53–3.27)) than a sham procedure.[17] The evidence supporting the use of rTMS in adults is considered level A (definite efficacy).[33] Despite such potential, studies in children have been limited. This is surprising given the evidence suggesting younger adult subjects with MDD respond better to rTMS (56% response rate) compared to older subjects.[34,35]

Initial case reports applying rTMS in adolescents with depression were promising.[36,37] The first prospective, open, multicenter trial of adjunctive rTMS was conducted with eight adolescents with major depressive disorder that had not responded to two adequate antidepressant medication trials.[38] rTMS was applied to the left dorsolateral prefrontal cortex (120% of motor threshold; 10 Hz; 4-s trains; 26-s intertrain interval; 75 trains) for a total of 3000 stimulations per treatment session 5 days per week for 6–8 weeks. Seven of eight participants completed rTMS treatment and no safety issues were identified. In three participants with marked suicidal ideation, improvement was shown with treatment. Depression severity scores also improved and persisted to the 6-month follow-up.

In a large, double-blind study with a 2-week extended antidepressant phase, 60 first-episode younger adult (18–45 years) major depressive patients were randomly assigned to citalopram in combination with 2 weeks of either active or sham rTMS treatment.[39] There were a significantly greater number of early improvers observed in the active rTMS group compared to sham. The active rTMS group showed a significantly faster score reduction compared to the sham group at 2 weeks, which was maintained at 4 weeks. However, there was no significant difference observed in responder rates or in remission rates between the two groups (rTMS + citalopram or sham + citalopram) at 4 weeks. Participants improved in executive performance, as measured by the Wisconsin Card Sorting Test and completed Trail-Making Test. This study suggests that rTMS may accelerate the antidepressant response in first-episode younger depressive patients.

In a 3-year follow-up study, there was no evidence of deterioration in symptoms of depression or cognitive functioning compared to the last assessment after rTMS.[40] Indeed, a secondary post hoc analysis of neurocognitive outcome in adolescents who were treated with open-label rTMS in two separate studies found improvement in depressive symptoms and a statistically significant improvement in memory and delayed verbal recall.[41]

A recent case series of six adolescents with MDD found that along with a decrease in depressive symptoms, there was an increase in left dorsolateral prefrontal cortex glutamate concentration.[21] This effect was also noted in adults.[24] Interestingly, non-responders had higher glutamate concentrations at baseline, possibly indicating that glutamate may be a biomarker/predictor of response to rTMS in youth with MDD. More recently, we have found that thinner DLPFC, lower cerebral blood flow, and poor DLPFC to parietal connectivity predicts a better response to TMS in youth (unpublished data; see Fig. 16.2). This supports the targeting concept, as the level of dysfunction or impairment may allow for treatment response. Further study is needed.

There has been a single case of seizure associated with rTMS for depression in youth.[42] This participant was heavily intoxicated the night prior to

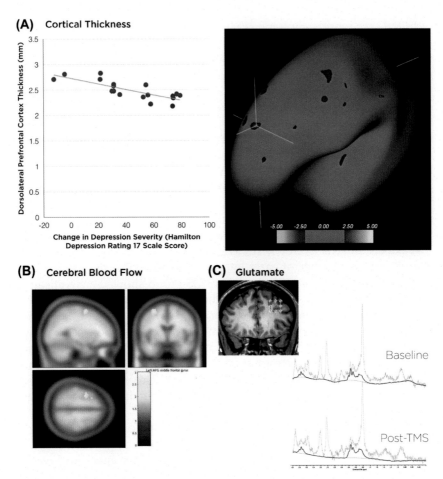

FIGURE 16.2 Unpublished data from our open-label trial of TMS (transcranial magnetic stimulation) in youth with major depressive disorder. (A) Association of left dorsolateral prefrontal cortex cortical thickness with change in depression severity with TMS indicating that thinner cortex predicts a better response to TMS. (B) Association of left dorsolateral prefrontal cortex cerebral blood flow with change in depression severity with TMS indicating that lower cerebral blood flow predicts a better response to TMS. (C) Magnetic resonance spectroscopy data showing the change in glutamate in the left dorsolateral prefrontal cortex in response to a positive clinical outcome with TMS.

the session; and this in combination with their psychotropic medications (olanzapine and sertraline) may have precipitated their seizure.[43] More commonly, affected youth and their parents found rTMS acceptable in terms of adverse effects and treatment experience.[44]

Summary for Depression

Of the disorders discussed here, brain stimulation to treat youth with MDD is the most mature application. This is not surprising given the weight of evidence in adults. It is also compelling that the clinical improvement can be maintained and it seems that some subjects may derive long-term benefit from the rTMS course.[40] Safety evidence is excellent, though alcohol and substance use should be considered.[42] Larger multisite randomized controlled trials are needed to earn FDA approval and the evidence strongly supports moving in that direction. The potential of predictive biomarkers of response to TMS also requires further exploration.

Attention-Deficit Hyperactivity Disorder

ADHD begins in childhood and occurs in about 5% of children. ADHD is characterized by a pattern of persistent inattention or hyperactivity-impulsivity that is present in multiple settings (home, school, or work) and interferes with functioning or development. Inattentive symptoms include difficulty in focusing on a task, lacking persistence, or disorganization. These symptoms are not due to lack of comprehension, hostility, or defiance. Hyperactivity refers to inappropriate and excessive motor activity, fidgeting, or talkativeness. Impulsive symptoms describe actions that are taken without forethought, with a high potential of harm to the child. To be diagnosed with ADHD, a child must display several developmentally inappropriate symptoms of inattention or hyperactivity-impulsivity in at least two settings, for at least 6 months before the age of 12 years. ADHD affects academic performance and social functioning. Children with ADHD are often rejected by their peers, and in turn are at higher risk for academic problems, school dropout, delinquency, substance abuse, and higher rates of psychopathology.[45]

Need for Novel Approaches and Biological Rationale for Targets

While stimulants are largely effective in ADHD in children, response is not universal. Current pharmacological interventions also have variable tolerability and parental discomfort is common. For instance, 32% of preschoolers discontinued the use of methylphenidate treatment for reasons including adverse effects or lack of efficacy.[46] This underscores the need for novel approaches.

The neurotransmitter dopamine has been implicated in ADHD. Dopaminergic cell burst firing is associated with reward stimuli,[47] and lack of motivation is a key symptom of ADHD patients. A meta-analysis of gene studies found that ADHD is highly associated with the D4 and D5 dopamine receptor genes.[48] D4 dopamine receptors are found in abundance in the prefrontal cortex, whereas D5 receptors are concentrated in deeper areas of the brain.[49] Given that tDCS and TMS target cortical areas of the brain, the prefrontal cortex stands out as a potential target to modulate the activity of the D4 dopamine receptor in ADHD.

Evidence

Children with ADHD have significantly reduced intracortical inhibition compared to healthy controls.[50] After methylphenidate treatment, a significant enhancement in TMS measures of intracortical inhibition was observed[50] along with intracortical facilitation.[51] A shorter duration of ipsilateral silent period in pediatric ADHD compared to controls was also noted.[52] Longer ipsilateral silent period latencies[53] may be the result of abnormal myelination of transcallosal fibers. As ipsilateral silent period latencies decreased with age in controls but not in pediatric ADHD, this may indicate a different neurodevelopmental trajectory.[53] Methylphenidate increases the ipsilateral silent period while reducing ADHD symptom severity.[54] Methylphenidate also reduced Connor's scores, which correlated with changes in motor-evoked potentials.[55] The effects of methylphenidate on motor system excitability may be interpreted in the sense of a "fine-tuning," with these mainly dopaminergic effects also depending on genetic parameters (DAT1 transporter).[56] Central motor conduction time is also elevated in pediatric ADHD,[57] indicating a delay in the maturation of the cortico-motor system. It appears that top–down control of motor cortical inhibition is reduced in children with ADHD,[58,59] similarly to adults.[60,61] When comorbid with Tourette syndrome, a reduced intracortical inhibition, as well as a shortened cortical silent period, is observed in pediatric ADHD.[62] After effective treatment with atomoxetine, children with ADHD and comorbid Tourette syndrome demonstrated reductions in ADHD symptoms that correlated with a reduction in short-interval cortical inhibition.[63] Reduced short-interval cortical inhibition was also observed in pediatric ADHD with[64] and without[65,66] comorbid Tourette syndrome. Interestingly, hyperactivity accounts for much of the variance in short-interval cortical inhibition.[67] The association between short-interval cortical inhibition and ADHD symptoms may, however, be more consistent or direct than the association between SICI and tics.[64]

In a double-blind crossover study of 12 children with ADHD using transcranial oscillating direct-current stimulation (toDCS) at 0.75 Hz and a sham condition, Prehn-Kristensen et al.[68] found that stimulation enhanced slow oscillation power and boosted memory performance to the same level as in healthy children. This suggests that increasing slow oscillation power during sleep by toDCS can alleviate declarative memory deficits in children with ADHD. In youth with ADHD, rTMS applied to the right prefrontal cortex (10 Hz, at 100% of the observed motor threshold, 2000 pulses per session, 10-session course, 2 weeks, sham-controlled crossover design) did not result in any significant changes in neuropsychological measures, and clinical measures improved in both active and sham groups.[69]

Summary for ADHD

While brain stimulation has been an effective probe of function and query of the effect of medication, there is a paucity of evidence to support its application in pediatric ADHD as a treatment. However, this absence of evidence speaks to a need for further study as opposed to a closure of the promise of brain stimulation techniques to reduce ADHD symptoms. As stated earlier, given that tDCS and TMS can target the prefrontal cortex—an area heavily implicated in pediatric ADHD—such studies are needed. For additional details, please see Chapter 12: Brain Stimulation in Children Born Preterm – Promises and Pitfalls.

Autism Spectrum Disorder and Developmental Disorders

ASD defines an array of conditions characterized by impairments in social interactions, communication, relationships, and by restricted and repetitive behaviors and interests. These impairments may manifest verbally and non-verbally, and can include intellectual and language impairments as well as motor deficits. A diagnosis of ASD requires that these symptoms be present from early childhood, and that they impair everyday functioning. It is important to note that some individuals with ASD may not be diagnosed until later in life, when social demands become greater than their capacity. The spectrum aspect of this disorder indicates that manifestation of symptoms and symptom severity varies greatly among individuals. ASD is reported in 1% of the population. Consequences of this disorder include difficulty coping with change, organization, or planning. Additionally, the social impairment can have negative effects on learning and academic achievement. In adulthood, many individuals with ASD struggle with independent living, social isolation, and poor communication.

Need for Novel Approaches and Biological Rationale for Targets

Pharmacological and behavioral therapies dominate ASD treatment regimens but, to date, these treatments are limited. Current medication options are limited in efficacy, as pharmacological therapy is less effective for ASD compared to ADHD or obsessive-compulsive disorder (OCD) patients. The side effect burden of pharmacotherapy in ASD patients is also problematic.[70] Furthermore, there exists insufficient literature of pharmacological interventions in adolescent populations.[71] Early intensive behavioral intervention (EIBI) for young children has been shown to be effective, but studies including robust randomized control trials are still lacking.[72] Cognitive behavioral therapy may be effective in reducing anxiety symptoms for this population.[73] Overall, current treatment approaches for ASD are suboptimal and do not target the relevant pathophysiology of ASD. Brain-based, neurostimulation approaches may reduce the disease burden for these patients.

A recent review by Minshew and Keller[74] suggested that the specialization of many cortical networks of the human brain has failed to develop fully in high-functioning individuals with autism. It is clear that children and youth with ASD follow a differing developmental trajectory than healthy children.[75,76] Possible cortical targets could include inferior frontal and parietal regions for social imitation, Broca's area for verbal communication, right inferior temporal gyrus for facial processing, and frontal cortex for stereotyped behaviors.[76]

Evidence

Brain stimulation has been applied as a probe of function in children and adults with ASD. In a study of 19 high-functioning males ages 9–18 with ASD, Oberman et al.[77] measured corticospinal excitability by applying single pulses of TMS to the primary motor cortex both before and following a 40-s train of continuous theta burst stimulation. They found a positive linear relationship between age and duration of modulation of rTMS aftereffects—ie, the older participants had a longer-lasting response. Additionally, while this protocol characteristically suppresses corticospinal excitability in adults, over a third of participants had a paradoxical facilitatory response to the stimulation. This could suggest aberrant plasticity and possible GABAergic dysfunction in this population.[77] Furthermore, cortical inhibition is significantly reduced in high-functioning autism compared with both Asperger syndrome and healthy controls.[78] This may suggest disruption of activity at GABA-A receptors. These deficits in cortical inhibition may underlie motor dysfunction in autism, and may relate to clinical symptoms (eg, repetitive behaviors). Jung et al.[79] found a significant impairment of long-term potentiation-like plasticity induced by paired associative stimulation in individuals with high-functioning autism and Asperger syndrome compared with healthy control participants. This suggests reduced excitatory synaptic connectivity and deficits in sensory–motor integration in ASD. In three cases of Rett syndrome (aged 4, 6, and 13 years), Nezu et al.[80] found that central motor conduction time was shorter in the pseudostationary stage (stage III) cases. Central motor conduction time shortening is thought to indicate hyperexcitability. Conversely, in 13 participants with Wilson disease, prolonged central motor conduction time was noted.[81]

ASD often presents with symptoms indicative of autonomic nervous system (ANS) dysfunction. Casanova et al.[82] sought to determine the effect of 18 sessions of low-frequency rTMS on ANS function in children with ASD by recording electrocardiogram and electrodermal activity pre, post and during each rTMS session. The dorsolateral prefrontal cortex was targeted bilaterally (six session over the left, six over the right, and six equally over both the left and right hemispheres). After 18 sessions of low-frequency rTMS, ASD participants demonstrated increased cardiac vagal control and reduced sympathetic arousal. It has also been shown that 12 sessions of low-frequency rTMS applied bilaterally to the dorsolateral prefrontal cortices in children and adolescents with ASD resulted in significant improvement in event-related potential indices of selective attention evoked at later (ie, 200–600 ms) stages of attentional processing and improvement in motor response error rate.[83] In addition, there were significant reductions in both repetitive behavior and irritability according to clinical behavioral questionnaires after administration of rTMS, though this was unblinded.

Sokhadze et al.[84] performed 18 sessions of low-frequency (1 Hz) rTMS applied over the dorsolateral prefrontal cortex in an attempt to improve executive functioning in 27 children and adolescents with ASD. Compared to a waitlist control group, event-related potential responses and behavioral reactions during executive function tests normalized, and clinical evaluations improved. Additional work by the same group combined prefrontal rTMS sessions with electroencephalographic (EEG) neurofeedback to prolong and reinforce TMS-induced EEG changes in 42 children with ASD (mean age 14.5 years). Combined rTMS and neurofeedback was superior to waitlist group performance on behavioral tasks as measured by event-related potentials.[85] Twelve weeks of low-frequency rTMS over bilateral dorsolateral prefrontal cortex in high-functioning children with ASD resulted in changes in error-related negativity and error-related positivity with an error-monitoring task.[86] Again by this group,[87] low-frequency rTMS application to dorsolateral prefrontal cortex resulted in improved selectivity in early cortical responses and better stimulus differentiation at later-stage responses in a novelty-processing task in children and adolescents with ASD. Finally, using tDCS targeting the dorsolateral prefrontal cortex (0.08 mA/cm² within the 30-min treatment period) to facilitate language acquisition in 10 children with immature syntax, Schneider and Hopp[88] found that treatment improved syntax acquisition as measured by a modified Bilingual Aphasia Test. A recent double-blind, randomized trial of deep rTMS enrolled 28 adults with ASD. Participants received 2 weeks of 15-min rTMS treatment sessions with 5-Hz stimulation (10-s trains with 20-s intertrain intervals) applied to the dorsomedial prefrontal cortex. Participants treated with active stimulation demonstrated improvements in social relatedness and social anxiety. These preliminary results in adults with ASD are encouraging and suggest that future studies with thoughtful methodology could involve younger participants.[89]

Summary for ASD and Developmental Disorders

Evidence to support the application of brain stimulation to treat the symptoms of ASD is emerging. However, a coherent approach to this complex set of disorders and dysfunctions has not yet emerged. The premise of combining brain stimulation with specific training or therapy[84] is compelling. Brain stimulation may allow for enhanced neuroplasticity so that the training/therapy can then focus on improved functioning. For additional details, please see Chapter 12: Brain Stimulation in Children Born Preterm – Promises and Pitfalls.

Childhood-Onset Schizophrenia

Childhood-onset schizophrenia is a chronic, severe form of schizophrenia, and is typically treatment-resistant. Schizophrenia is characterized by delusions, hallucinations, and/or disorganized speech, for a period of at least 1 month, accompanied by grossly disorganized or catatonic behavior, or diminished emotional expression or avolition. Individuals diagnosed with childhood-onset schizophrenia may experience depersonalization or derealization, as well as inappropriate affect, a dysphoric mood, somatic complaints, and sleep concerns. Lack of awareness of the symptoms of the disorder can also occur. Individuals with childhood-onset schizophrenia may experience persistent cognitive impairments and severe negative symptoms. Because earlier-onset schizophrenia is associated with poorer prognosis, those diagnosed are more likely to experience academic difficulties, problems maintaining employment, and limited social contact later in life.

Need for Novel Approaches and Biological Rationale for Targets

Even with optimal medication regimens, seven out of 10 pediatric patients have enduring psychotic symptoms and impaired cognition.[90] This extremely limited efficacy indicates a profound need for novel interventions in this patient population. Schizophrenia is associated with progressive brain changes—particularly gray matter loss—and an altered developmental trajectory.[91] This holds true for childhood-onset schizophrenia as well.[92] Brain regions thought to subserve auditory verbal hallucinations—like the Broca's area, auditory cortex, and cingulate gyrus—are considered viable targets for brain stimulation.[93,94] A possible aim would be to inhibit abnormal brain activity in these regions.

Evidence

There have been no studies of childhood-onset schizophrenia with rTMS to date aside from a single case report.[95] Ten sessions of low-frequency fMRI-guided rTMS over the left temporal-parietal cortex reduced auditory hallucinations in an 11-year-old male with drug-resistant schizophrenia. Despite the lack of evidence in youth, in adults rTMS shows promise for treating auditory hallucinations. A recent meta-analysis has shown that low-frequency (1 Hz) rTMS is an effective therapy for auditory verbal hallucinations in adults with schizophrenia.[96,97] Using theta burst TMS as compared to 1 Hz rTMS, Kindler et al.[94] targeted the auditory verbal hallucinations associated in adults with schizophrenia or schizoaffective disorder (ICD-10) in a single-blind randomized controlled study. It was found that theta burst TMS demonstrated equal clinical effects compared to the low-frequency TMS. In a second study by the same group,[93] it was observed that TMS-treated patients showed positive clinical effects coupled with decreases in cerebral blood flow in the primary auditory cortex, left Broca's area, and cingulate gyrus. This decrease in cerebral blood flow in the primary auditory cortex correlated with a decrease in auditory verbal hallucinations. For the symptoms associated with schizophrenia, alpha EEG-guided TMS (αTMS) may show promise.[98] With αTMS, the stimulus rate and threshold are determined by the individual's characteristic alpha frequency and motor threshold. Seventy-eight participants with schizophrenia were divided into four groups for a randomized double-blind sham-controlled study.[98] The four groups were frontal active and sham, and parietal active and sham conditions. General and positive symptoms improved with αTMS, while target site did not seem to matter. Combination with typical antipsychotics had greater efficacy than with atypical antipsychotics. Normalization of the subject's EEG also predicted a greater reduction in symptoms overall. Such personalized or precision medicine approaches may be more easily capitalized on with brain stimulation approaches.

Mattai et al.[99] investigated the tolerability of tDCS in 12 participants with childhood-onset schizophrenia. Participants were assigned to either (1) bilateral anodal dorsolateral prefrontal cortex stimulation ($n=8$) or (2) bilateral cathodal superior temporal gyrus stimulation ($n=5$). Participants (10–17 years of age) received either 2 mA of active tDCS treatment or sham for 20 min for 10 sessions over 2 weeks. tDCS was well-tolerated, with no serious adverse events but clear evidence of efficacy was not observed.

Summary for Childhood-Onset Schizophrenia

tDCS appears well-tolerated by participants with childhood-onset schizophrenia.[99] It has also been suggested that tDCS may be the favored option as TMS treatment might be too burdensome for this severely ill population.[90] In adults with schizophrenia, low-frequency (1 Hz) rTMS is an effective therapy for auditory verbal hallucinations[96,97] with theta burst TMS appearing comparable.[94] The paucity of studies in youth is a limitation and further studies are warranted.

Tourette Syndrome

Tics, the hallmark symptom of Tourette syndrome, are repetitive, sudden, semivoluntary movements and sounds. Tourette syndrome affects between 0.3% and 1% of people[100] and negatively impacts quality of life in children—at home, school, with friends and family.[101] Tic severity predicts poor outcomes across physical, psychological, and cognitive domains in youth.[102] For approximately 15% of children with Tourette syndrome, the tics can be especially disabling.

Need for Novel Approaches and Biological Rationale for Targets

Currently, there is a lack of knowledge regarding the underlying neurobiology of Tourette syndrome and how best to treat it. Antipsychotics are the most effective treatment for tics.[103] Unfortunately, antipsychotics have many risks, serious side effects, limited efficacy, and do not adequately address the underlying pathophysiology of this disease. Potential side effects include extrapyramidal symptoms, weight gain, increase in cholesterol, sedation, and changes in heart function.[104,105] Indeed, the recommendation to use antipsychotics in children is only made when the tics are so disabling that they make these risks acceptable.[103] Behavioral therapy is another option. Habit reversal therapy is a promising intervention.[106–109] In the Canadian Guidelines, habit reversal therapy earned a "strong recommendation" based on the high quality of evidence to support it.[108] Success of this modality is predicated on the patient's awareness of their tics. This may be a limiting factor in children where such awareness may be decreased. Others may be aware, but unable to suppress their tics.

Tics in Tourette syndrome are thought to arise from two mechanisms. First, dysregulation of excitation/inhibition in brain regions related to motor function. Second, inputs from the supplementary motor area cause an increase in excitability in the primary motor cortex. This results in the activation of striatal neurons, and hyperexcitation of the primary motor cortex, leading to the expression of tics. In Tourette syndrome, TMS measures of GABA transmission are reduced in the supplementary motor area and primary motor cortex.[110] Tics tend to increase over time, often peaking during adolescence.[111] At the same time, the ability to suppress tics increases. Tic suppression is associated with an increase in tonic inhibition in the supplementary motor area, and correlates with increased GABA neurotransmission and levels.[112] Consequently, the supplementary motor area is a potential target for therapeutic intervention.

Evidence

Brain stimulation has been applied as a probe of function in children and adults with Tourette syndrome. In 20 adults with Tourette syndrome as compared to 21 healthy control participants, the cortical silent period was shortened and intracortical inhibition was reduced.[113] Furthermore, children with Tourette syndrome also show a shortened cortical silent period.[62] This suggests that tics in Tourette syndrome are associated with (1) disinhibited afferent signals from subcortical regions affecting the motor cortex or (2) from impaired inhibition directly at the level of the motor cortex, or both. In addition, in a small study of adolescents, Heise et al.[114] used the difference between resting and active motor thresholds during the cortical silent period to determine voluntary motor drive. Participants with distal tics showed reduced voluntary motor drive compared to healthy controls, while patients without distal tics did not differ from controls. The findings support the idea that Tourette syndrome shows a reduction of voluntary motor drive that is associated with central motor threshold alterations confined to the motor networks related to the tics observed. Interestingly, a case report found that in an adolescent male with Tourette syndrome, delta 9-tetrahydrocannabinol (delta 9-THC) treatment improved tics significantly and intracortical inhibition increased—enhancing short-interval intracortical inhibition and the prolongation of the cortical silent period.[115]

Short-interval intracortical inhibition (SICI) correlates with ADHD hyperactivity scores in children and adults with Tourette syndrome.[64,67] Indeed, Tourette syndrome with comorbid ADHD is associated with more extensive changes in the excitability of motor cortex circuits than uncomplicated Tourette syndrome or Tourette syndrome + OCD.[116] Participants with Tourette syndrome + ADHD had greater intracortical facilitation than controls, Tourette syndrome participants or Tourette syndrome + OCD participants. Uncomplicated Tourette syndrome participants or Tourette syndrome + ADHD patients had less short-latency afferent inhibition than controls or Tourette syndrome + OCD participants. Using intermittent theta burst TMS in 10 adults with Tourette syndrome and 11 healthy control participants, Wu and Gilbert[117] found that motor-evoked potential amplitudes changed more in controls than in patients.

Many patients gain control over their tics during adolescence. This suggests that increased control results from the development of the ability to suppress corticospinal excitability ahead of volitional movements. Draper et al.[118] used single-pulse TMS in conjunction with a manual Go/NoGo task to investigate changes in corticospinal excitability ahead of volitional movements in 10 adolescents with Tourette syndrome. They found that corticospinal excitability was significantly reduced in adolescents with Tourette syndrome immediately preceding a finger movement. This finding was consistent with a previous report[119] that suggested the cortical hyperexcitability that may give rise to tics in Tourette syndrome is actively suppressed by cognitive control mechanisms. In the Draper et al.[118] study, this effect was negatively correlated with measures of tic severity, suggesting that patients with severe motor tics are the least able to regulate motor cortical excitability.

In studies of brain stimulation as an intervention to reduce tic severity in Tourette syndrome, a consensus around a single approach appears to be coalescing. In a case series of an adult and an adolescent with Tourette syndrome, low-frequency rTMS was used to target the supplementary motor area (10 sessions, 1200 stimuli/day, 110% of the individual RMT).[120] Participants had a 52% reduction in tic severity. In an open-label pilot study of 10 children with Tourette syndrome, all males, Kwon et al.[121] applied low-frequency rTMS over the supplementary motor area for 10 daily sessions (1 Hz, 100% of motor threshold, and 1200 stimuli/day). Participants did not report any side effects or worsening of comorbidity with rTMS, while tics improved significantly over the 12 weeks of the study based on Yale Global Tourette's Syndrome Severity Scale and Clinical Global Impression (CGI) scores. In 25 children with Tourette syndrome (all under 16 years) who received 20 daily sessions of 1-Hz rTMS to the supplementary motor area (110% of RMT), Le et al.[122] found that after 4 weeks of treatment there were statistically significant reductions on a number of tic severity and clinical rating scales. Furthermore, improvement in tics correlated with an increase in right and left RMT that was stable at 6-months follow-up. This suggests that 1-Hz rTMS to the supplementary motor area may improve clinical symptoms in children with Tourette syndrome for at least 6 months. Randomized sham-controlled studies are needed before any determination of efficacy can be considered as valid.

Summary for Tourette Syndrome

Tourette syndrome is associated with a shortened cortical silent period, reflecting decreased motor inhibition. It appears that low-frequency rTMS over the supplementary motor area demonstrates the most promise for treating the tics associated with Tourette syndrome. In the absence of a sham control, placebo response cannot be excluded. Additional studies using (1) sham, (2) blinded, balanced, and parallel designs, and (3) examining augmentation with behavior therapy such as habit reversal therapy are warranted.

DISCUSSION

Brain stimulation techniques are an emerging and promising approach for a plethora of mental health and neurodevelopmental disorders (see Fig. 16.3 for a summary of target sites). In particular, tDCS and rTMS are non-invasive techniques that allow for the modulation of brain activity, resulting in the alleviation of a variety of symptoms. While the evidence is largely scant, these approaches show promise and require further studies to refine and develop.

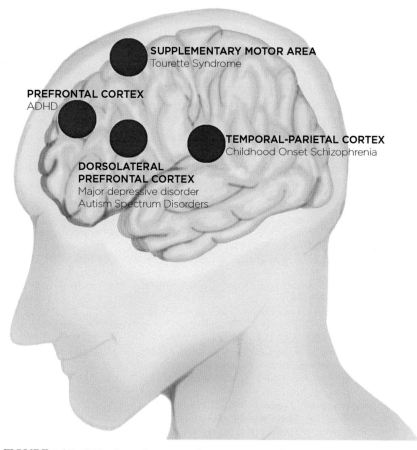

FIGURE 16.3 This schematic sums up the target sites used to date in neurodevelopmental and pediatric mental disorders. *ADHD*, attention-deficit hyperactivity disorder.

WHERE DO WE STAND AND WHERE DO WE NEED TO GO?

When it comes to developing novel approaches to treat mental illness and neurodevelopmental disorders in children and youth, they are by and large "research orphans."[123] Most studies described here are limited by small sample sizes and a lack of replication. Larger randomized sham-controlled trials are sorely needed for progress. This cannot occur until a certain level of consistency of approach is achieved (see Fig. 16.1). This appears to be the case in the use of rTMS in youth with MDD. Furthermore, the field also needs to develop the biological correlates that provide the rationale for targets and may be used to predict treatment response.

CONCLUSION

As stated at the outset, by its nature, brain stimulation requires the development of a biologically rational target for intervention. This is in line with the push for the development of rational therapeutics—based on the underlying neurobiology, rather than serendipity.[1,2] This is a tremendous advantage in the development of novel interventions. Evidence to support the application in youth with MDD is growing strong. The next logical step is a larger randomized control trial. The application of brain stimulation in ASD, ADHD, childhood onset schizophrenia, and Tourette syndrome is only developing.

References

1. Insel TR, Scolnick EM. Cure therapeutics and strategic prevention: raising the bar for mental health research. *Mol Psychiatry*. 2006;11(1):11–17.
2. Insel TR, Quirion R. Psychiatry as a clinical neuroscience discipline. *JAMA*. 2005;294(17):2221–2224.
3. Ioannidis JP. Evolution and translation of research findings: from bench to where? *PLoS Clin Trials*. 2006;1(7):e36.
4. CIHR. *Canada's Strategy for Patient-Oriented Research*; 2011.
5. Barker AT, Jalinous R, Freeston IL. Non-invasive magnetic stimulation of human motor cortex. *Lancet*. 1985;1(8437):1106–1107.
6. Croarkin PE, Wall CA, McClintock SM, Kozel FA, Husain MM, Sampson SM. The emerging role for repetitive transcranial magnetic stimulation in optimizing the treatment of adolescent depression. *J ECT*. 2010;26(4):323–329.
7. Tortella G, Casati R, Aparicio LV, et al. Transcranial direct current stimulation in psychiatric disorders. *World J Psychiatry*. 2015;5(1):88–102.
8. Lewinsohn PM, Hops H, Roberts RE, Seeley JR, Andrews JA. Adolescent psychopathology: I. Prevalence and incidence of depression and other DSM-III-R disorders in high school students. *J Abnorm Psychol*. 1993;102(1):133–144.
9. Birmaher B, Ryan ND, Williamson DE, et al. Childhood and adolescent depression: a review of the past 10 years. Part I. *J Am Acad Child Adolesc Psychiatry*. 1996;35(11): 1427–1439.
10. Birmaher B, Ryan ND, Williamson DE, Brent DA, Kaufman J. Childhood and adolescent depression: a review of the past 10 years. Part II. *J Am Acad Child Adolesc Psychiatry*. 1996;35(12):1575–1583.
11. Carvalho AF, Cavalcante JL, Castelo MS, Lima MC. Augmentation strategies for treatment-resistant depression: a literature review. *J Clin Pharm Ther*. 2007;32(5):415–428.
12. Emslie GJ, Mayes T, Porta G, et al. Treatment of Resistant Depression in Adolescents (TORDIA): week 24 outcomes. *Am J Psychiatry*. 2010;167(7):782–791.
13. Fava M. Diagnosis and definition of treatment-resistant depression. *Biol Psychiatry*. 2003;53(8):649–659.
14. Souery D, Serretti A, Calati R, et al. Switching antidepressant class does not improve response or remission in treatment-resistant depression. *J Clin Psychopharmacol*. 2011;31(4):512–516.
15. Souery D, Papakostas GI, Trivedi MH. Treatment-resistant depression. *J Clin Psychiatry*. 2006;67(suppl 6):16–22.
16. George MS. Transcranial magnetic stimulation for the treatment of depression. *Expert Rev Neurother*. 2010;10(11):1761–1772.

17. Unit HTA. *Repetitive Transcranial Magnetic Stimulation for Treatment Resistant Depression: A Health Technology Assessment.* University of Calgary; 2014.
18. Farchione TR, Moore GJ, Rosenberg DR. Proton magnetic resonance spectroscopic imaging in pediatric major depression. *Biol Psychiatry.* 2002;52(2):86–92.
19. Caetano SC, Fonseca M, Olvera RL, et al. Proton spectroscopy study of the left dorsolateral prefrontal cortex in pediatric depressed patients. *Neurosci Lett.* 2005;384(3):321–326.
20. Reynolds S, Carrey N, Jaworska N, Langevin LM, Yang XR, Macmaster FP. Cortical thickness in youth with major depressive disorder. *BMC Psychiatry.* 2014;14:83.
21. Yang XR, Kirton A, Wilkes TC, et al. Glutamate alterations associated with transcranial magnetic stimulation in youth depression: a case series. *J ECT.* 2014;30(3):242–247.
22. Mannie ZN, Taylor MJ, Harmer CJ, Cowen PJ, Norbury R. Frontolimbic responses to emotional faces in young people at familial risk of depression. *J Affect Disord.* 2011;130(1–2):127–132.
23. Fitzgerald PB, Brown TL, Marston NA, Daskalakis ZJ, De Castella A, Kulkarni J. Transcranial magnetic stimulation in the treatment of depression: a double-blind, placebo-controlled trial. *Arch Gen Psychiatry.* 2003;60(10):1002–1008.
24. Luborzewski A, Schubert F, Seifert F, et al. Metabolic alterations in the dorsolateral prefrontal cortex after treatment with high-frequency repetitive transcranial magnetic stimulation in patients with unipolar major depression. *J Psychiatr Res.* 2007;41(7):606–615.
25. Croarkin PE, Wall CA, Nakonezny PA, et al. Increased cortical excitability with prefrontal high-frequency repetitive transcranial magnetic stimulation in adolescents with treatment-resistant major depressive disorder. *J Child Adolesc Psychopharmacol.* 2012;22(1):56–64.
26. Croarkin PE, Nakonezny PA, Husain MM, et al. Evidence for increased glutamatergic cortical facilitation in children and adolescents with major depressive disorder. *JAMA Psychiatry.* 2013;70(3):291–299.
27. Croarkin PE, Nakonezny PA, Husain MM, et al. Evidence for pretreatment LICI deficits among depressed children and adolescents with nonresponse to fluoxetine. *Brain Stimul.* 2014;7(2):243–251.
28. Croarkin PE, Nakonezny PA, Lewis CP, et al. Developmental aspects of cortical excitability and inhibition in depressed and healthy youth: an exploratory study. *Front Hum Neurosci.* 2014;8:669.
29. Bretlau LG, Lunde M, Lindberg L, Unden M, Dissing S, Bech P. Repetitive transcranial magnetic stimulation (rTMS) in combination with escitalopram in patients with treatment-resistant major depression: a double-blind, randomised, sham-controlled trial. *Pharmacopsychiatry.* 2008;41(2):41–47.
30. Garcia KS, Flynn P, Pierce KJ, Caudle M. Repetitive transcranial magnetic stimulation treats postpartum depression. *Brain Stimul.* 2010;3(1):36–41.
31. Triggs WJ, Ricciuti N, Ward HE, et al. Right and left dorsolateral pre-frontal rTMS treatment of refractory depression: a randomized, sham-controlled trial. *Psychiatry Res.* 2010;178(3):467–474.
32. O'Reardon JP, Solvason HB, Janicak PG, et al. Efficacy and safety of transcranial magnetic stimulation in the acute treatment of major depression: a multisite randomized controlled trial. *Biol Psychiatry.* 2007;62(11):1208–1216.
33. Lefaucheur JP, Andre-Obadia N, Antal A, et al. Evidence-based guidelines on the therapeutic use of repetitive transcranial magnetic stimulation (rTMS). *Clin Neurophysiol.* 2014;125(11):2150–2206.
34. Fregni F, Marcolin MA, Myczkowski M, et al. Predictors of antidepressant response in clinical trials of transcranial magnetic stimulation. *Int J Neuropsychopharmacol.* 2006;9(6):641–654.
35. Figiel GS, Epstein C, McDonald WM, et al. The use of rapid-rate transcranial magnetic stimulation (rTMS) in refractory depressed patients. *J Neuropsychiatry Clin Neurosci.* 1998;10(1):20–25.

36. Loo C, McFarquhar T, Walter G. Transcranial magnetic stimulation in adolescent depression. *Australas Psychiatry*. 2006;14(1):81–85.
37. Bloch Y, Grisaru N, Harel EV, et al. Repetitive transcranial magnetic stimulation in the treatment of depression in adolescents: an open-label study. *J ECT*. 2008;24(2):156–159.
38. Wall CA, Croarkin PE, Sim LA, et al. Adjunctive use of repetitive transcranial magnetic stimulation in depressed adolescents: a prospective, open pilot study. *J Clin Psychiatry*. 2011;72(9):1263–1269.
39. Huang ML, Luo BY, Hu JB, et al. Repetitive transcranial magnetic stimulation in combination with citalopram in young patients with first-episode major depressive disorder: a double-blind, randomized, sham-controlled trial. *Aust N Z J Psychiatry*. 2012;46(3):257–264.
40. Mayer G, Aviram S, Walter G, Levkovitz Y, Bloch Y. Long-term follow-up of adolescents with resistant depression treated with repetitive transcranial magnetic stimulation. *J ECT*. 2012;28(2):84–86.
41. Wall CA, Croarkin PE, McClintock SM, et al. Neurocognitive effects of repetitive transcranial magnetic stimulation in adolescents with major depressive disorder. *Front Psychiatry*. 2013;4:165.
42. Chiramberro M, Lindberg N, Isometsa E, Kahkonen S, Appelberg B. Repetitive transcranial magnetic stimulation induced seizures in an adolescent patient with major depression: a case report. *Brain Stimul*. 2013;6(5):830–831.
43. Wall C, Croarkin P, Bandel L, Schaefer K. Response to repetitive transcranial magnetic stimulation induced seizures in an adolescent patient with major depression: a case report. *Brain Stimul*. 2014;7(2):337–338.
44. Mayer G, Faivel N, Aviram S, Walter G, Bloch Y. Repetitive transcranial magnetic stimulation in depressed adolescents: experience, knowledge, and attitudes of recipients and their parents. *J ECT*. 2012;28(2):104–107.
45. Staikova E, Gomes H, Tartter V, McCabe A, Halperin JM. Pragmatic deficits and social impairment in children with ADHD. *J Child Psychol Psychiatry*. 2013;54(12):1275–1283.
46. Vitiello B, Abikoff HB, Chuang SZ, et al. Effectiveness of methylphenidate in the 10-month continuation phase of the Preschoolers with Attention-Deficit/Hyperactivity Disorder Treatment Study (PATS). *J Child Adolesc Psychopharmacol*. 2007;17(5):593–604.
47. Mirenowicz J, Schultz W. Importance of unpredictability for reward responses in primate dopamine neurons. *J Neurophysiol*. 1994;72(2):1024–1027.
48. Bobb AJ, Castellanos FX, Addington AM, Rapoport JL. Molecular genetic studies of ADHD: 1991 to 2004. *Am J Med Genet B Neuropsychiatr Genet*. 2005;132B(1):109–125.
49. Meador-Woodruff JH, Damask SP, Wang J, Haroutunian V, Davis KL, Watson SJ. Dopamine receptor mRNA expression in human striatum and neocortex. *Neuropsychopharmacology*. 1996;15(1):17–29.
50. Moll GH, Heinrich H, Trott G, Wirth S, Rothenberger A. Deficient intracortical inhibition in drug-naive children with attention-deficit hyperactivity disorder is enhanced by methylphenidate. *Neurosci Lett*. 2000;284(1–2):121–125.
51. Moll GH, Heinrich H, Rothenberger A. Methylphenidate and intracortical excitability: opposite effects in healthy subjects and attention-deficit hyperactivity disorder. *Acta Psychiatr Scand*. 2003;107(1):69–72.
52. Buchmann J, Wolters A, Haessler F, Bohne S, Nordbeck R, Kunesch E. Disturbed transcallosally mediated motor inhibition in children with attention deficit hyperactivity disorder (ADHD). *Clin Neurophysiol*. 2003;114(11):2036–2042.
53. Garvey MA, Barker CA, Bartko JJ, et al. The ipsilateral silent period in boys with attention-deficit/hyperactivity disorder. *Clin Neurophysiol*. 2005;116(8):1889–1896.
54. Buchmann J, Gierow W, Weber S, et al. Modulation of transcallosally mediated motor inhibition in children with attention deficit hyperactivity disorder (ADHD) by medication with methylphenidate (MPH). *Neurosci Lett*. 2006;405(1–2):14–18.

55. Buchmann J, Gierow W, Weber S, et al. Restoration of disturbed intracortical motor inhibition and facilitation in attention deficit hyperactivity disorder children by methylphenidate. *Biol Psychiatry*. 2007;62(9):963–969.

56. Hoegl T, Bender S, Buchmann J, Kratz O, Moll GH, Heinrich H. Transcranial magnetic stimulation (TMS), inhibition processes and attention deficit/hyperactivity disorder (ADHD) - an overview. *Z Kinder Jugendpsychiatr Psychother*. 2014;42(6):415–428. quiz 428–429.

57. Ucles P, Serrano JL, Rosa F. Central conduction time of magnetic brain stimulation in attention-deficit hyperactivity disorder. *J Child Neurol*. 2000;15(11):723–728.

58. Bruckmann S, Hauk D, Roessner V, et al. Cortical inhibition in attention deficit hyperactivity disorder: new insights from the electroencephalographic response to transcranial magnetic stimulation. *Brain J Neurol*. 2012;135(Pt 7):2215–2230.

59. D'Agati E, Hoegl T, Dippel G, et al. Motor cortical inhibition in ADHD: modulation of the transcranial magnetic stimulation-evoked N100 in a response control task. *J Neural Transm*. 2014;121(3):315–325.

60. Schneider M, Retz W, Freitag C, et al. Impaired cortical inhibition in adult ADHD patients: a study with transcranial magnetic stimulation. *J Neural Transm*. 2007;(suppl 72): 303–309.

61. Richter MM, Ehlis AC, Jacob CP, Fallgatter AJ. Cortical excitability in adult patients with attention-deficit/hyperactivity disorder (ADHD). *Neurosci Lett*. 2007;419(2):137–141.

62. Moll GH, Heinrich H, Trott GE, Wirth S, Bock N, Rothenberger A. Children with comorbid attention-deficit-hyperactivity disorder and tic disorder: evidence for additive inhibitory deficits within the motor system. *Ann Neurol*. 2001;49(3):393–396.

63. Gilbert DL, Zhang J, Lipps TD, et al. Atomoxetine treatment of ADHD in Tourette syndrome: reduction in motor cortex inhibition correlates with clinical improvement. *Clin Neurophysiol*. 2007;118(8):1835–1841.

64. Gilbert DL, Bansal AS, Sethuraman G, et al. Association of cortical disinhibition with tic, ADHD, and OCD severity in Tourette syndrome. *Mov Disord*. 2004;19(4):416–425.

65. Gilbert DL, Isaacs KM, Augusta M, Macneil LK, Mostofsky SH. Motor cortex inhibition: a marker of ADHD behavior and motor development in children. *Neurology*. 2011;76(7):615–621.

66. Hoegl T, Heinrich H, Barth W, Losel F, Moll GH, Kratz O. Time course analysis of motor excitability in a response inhibition task according to the level of hyperactivity and impulsivity in children with ADHD. *PLoS One*. 2012;7(9):e46066.

67. Gilbert DL, Sallee FR, Zhang J, Lipps TD, Wassermann EM. Transcranial magnetic stimulation-evoked cortical inhibition: a consistent marker of attention-deficit/hyperactivity disorder scores in tourette syndrome. *Biol Psychiatry*. 2005;57(12):1597–1600.

68. Prehn-Kristensen A, Munz M, Goder R, et al. Transcranial oscillatory direct current stimulation during sleep improves declarative memory consolidation in children with attention-deficit/hyperactivity disorder to a level comparable to healthy controls. *Brain Stimul*. 2014;7(6):793–799.

69. Weaver L, Rostain AL, Mace W, Akhtar U, Moss E, O'Reardon JP. Transcranial magnetic stimulation (TMS) in the treatment of attention-deficit/hyperactivity disorder in adolescents and young adults: a pilot study. *J ECT*. 2012;28(2):98–103.

70. Politte LC, Henry CA, McDougle CJ. Psychopharmacological interventions in autism spectrum disorder. *Harv Rev Psychiatry*. 2014;22(2):76–92.

71. Dove D, Warren Z, McPheeters ML, Taylor JL, Sathe NA, Veenstra-VanderWeele J. Medications for adolescents and young adults with autism spectrum disorders: a systematic review. *Pediatrics*. 2012;130(4):717–726.

72. Reichow B, Barton EE, Boyd BA, Hume K. Early intensive behavioral intervention (EIBI) for young children with autism spectrum disorders (ASD). *Cochrane Database Syst Rev*. 2012;10:CD009260.

73. Ung D, Selles R, Small BJ, Storch EA. A systematic review and meta-analysis of cognitive-behavioral therapy for anxiety in youth with high-functioning autism spectrum disorders. *Child Psychiatry Hum Dev*. 2015;46(4):533–547.

74. Minshew NJ, Keller TA. The nature of brain dysfunction in autism: functional brain imaging studies. *Curr Opin Neurol*. 2010;23(2):124–130.

75. Vinette SA, Bray S. Variation in functional connectivity along anterior-to-posterior intraparietal sulcus, and relationship with age across late childhood and adolescence. *Dev Cogn Neurosci*. 2015;13:32–42.

76. Verhoeven JS, De Cock P, Lagae L, Sunaert S. Neuroimaging of autism. *Neuroradiology*. 2010;52(1):3–14.

77. Oberman LM, Pascual-Leone A, Rotenberg A. Modulation of corticospinal excitability by transcranial magnetic stimulation in children and adolescents with autism spectrum disorder. *Front Hum Neurosci*. 2014;8:627.

78. Enticott PG, Rinehart NJ, Tonge BJ, Bradshaw JL, Fitzgerald PB. A preliminary transcranial magnetic stimulation study of cortical inhibition and excitability in high-functioning autism and Asperger disorder. *Dev Med Child Neurol*. 2010;52(8):e179–e183.

79. Jung NH, Janzarik WG, Delvendahl I, et al. Impaired induction of long-term potentiation-like plasticity in patients with high-functioning autism and Asperger syndrome. *Dev Med Child Neurol*. 2013;55(1):83–89.

80. Nezu A, Kimura S, Takeshita S, Tanaka M. Characteristic response to transcranial magnetic stimulation in Rett syndrome. *Electroencephalogr Clin Neurophysiol*. 1998;109(2):100–103.

81. Jhunjhunwala K, Prashanth DK, Netravathi M, Nagaraju BC, Pal PK. Alterations of cortical excitability and central motor conduction time in Wilson's disease. *Neurosci Lett*. 2013;553:90–94.

82. Casanova MF, Hensley MK, Sokhadze EM, et al. Effects of weekly low-frequency rTMS on autonomic measures in children with autism spectrum disorder. *Front Hum Neurosci*. 2014;8:851.

83. Casanova MF, Baruth JM, El-Baz A, Tasman A, Sears L, Sokhadze E. Repetitive transcranial magnetic stimulation (rTMS) modulates event-related potential (ERP) indices of attention in autism. *Transl Neurosci*. 2012;3(2):170–180.

84. Sokhadze EM, El-Baz AS, Sears LL, Opris I, Casanova MF. rTMS neuromodulation improves electrocortical functional measures of information processing and behavioral responses in autism. *Front Syst Neurosci*. 2014;8:134.

85. Sokhadze EM, El-Baz AS, Tasman A, et al. Neuromodulation integrating rTMS and neurofeedback for the treatment of autism spectrum disorder: an exploratory study. *Appl Psychophysiol Biofeedback*. 2014;39(3–4):237–257.

86. Sokhadze EM, Baruth JM, Sears L, Sokhadze GE, El-Baz AS, Casanova MF. Prefrontal neuromodulation using rTMS improves error monitoring and correction function in autism. *Appl Psychophysiol Biofeedback*. 2012;37(2):91–102.

87. Sokhadze E, Baruth J, Tasman A, et al. Low-frequency repetitive transcranial magnetic stimulation (rTMS) affects event-related potential measures of novelty processing in autism. *Appl Psychophysiol Biofeedback*. 2010;35(2):147–161.

88. Schneider HD, Hopp JP. The use of the Bilingual Aphasia Test for assessment and transcranial direct current stimulation to modulate language acquisition in minimally verbal children with autism. *Clin Linguist Phon*. 2011;25(6–7):640–654.

89. Enticott PG, Fitzgibbon BM, Kennedy HA, et al. A double-blind, randomized trial of deep repetitive transcranial magnetic stimulation (rTMS) for autism spectrum disorder. *Brain Stimul*. 2014;7(2):206–211.

90. David CN, Rapoport JL, Gogtay N. Treatments in context: transcranial direct current brain stimulation as a potential treatment in pediatric psychosis. *Expert Rev Neurother*. 2013;13(4):447–458.

91. Olabi B, Ellison-Wright I, McIntosh AM, Wood SJ, Bullmore E, Lawrie SM. Are there progressive brain changes in schizophrenia? A meta-analysis of structural magnetic resonance imaging studies. *Biol Psychiatry*. 2011;70(1):88–96.

92. Rapoport JL, Gogtay N. Childhood onset schizophrenia: support for a progressive neu-rodevelopmental disorder. *Int J Dev Neurosci*. 2011;29(3):251–258.

93. Kindler J, Homan P, Jann K, et al. Reduced neuronal activity in language-related regions after transcranial magnetic stimulation therapy for auditory verbal hallucinations. *Biol Psychiatry*. 2013;73(6):518–524.

94. Kindler J, Homan P, Flury R, Strik W, Dierks T, Hubl D. Theta burst transcranial mag-netic stimulation for the treatment of auditory verbal hallucinations: results of a ran-domized controlled study. *Psychiatry Res*. 2013;209(1):114–117.

95. Jardri R, Lucas B, Delevoye-Turrell Y, et al. An 11-year-old boy with drug-resistant schizophrenia treated with temporo-parietal rTMS. *Mol Psychiatry*. 2007;12(4):320.

96. Aleman A, Sommer IE, Kahn RS. Efficacy of slow repetitive transcranial magnetic stim-ulation in the treatment of resistant auditory hallucinations in schizophrenia: a meta-analysis. *J Clin Psychiatry*. 2007;68(3):416–421.

97. Demeulemeester M, Amad A, Bubrovszky M, Pins D, Thomas P, Jardri R. What is the real effect of 1-Hz repetitive transcranial magnetic stimulation on hallucinations? Controlling for publication bias in neuromodulation trials. *Biol Psychiatry*. 2012;71(6): e15–e16.

98. Jin Y, Kemp AS, Huang Y, et al. Alpha EEG guided TMS in schizophrenia. *Brain Stimul*. 2012;5(4):560–568.

99. Mattai A, Miller R, Weisinger B, et al. Tolerability of transcranial direct current stimula-tion in childhood-onset schizophrenia. *Brain Stimul*. 2011;4(4):275–280.

100. Robertson MM, Eapen V, Cavanna AE. The international prevalence, epidemiology, and clinical phenomenology of Tourette syndrome: a cross-cultural perspective. *J Psy-chosom Res*. 2009;67(6):475–483.

101. Eddy CM, Rizzo R, Gulisano M, et al. Quality of life in young people with Tourette syndrome: a controlled study. *J Neurol*. 2011;258(2):291–301.

102. Cavanna AE, David K, Orth M, Robertson MM. Predictors during childhood of future health-related quality of life in adults with Gilles de la Tourette syndrome. *Eur J Paediatr Neurol*. 2012;16(6):605–612.

103. Pringsheim T, Doja A, Gorman D, et al. Canadian guidelines for the evidence-based treatment of tic disorders: pharmacotherapy. *Can J Psychiatry*. 2012;57(3):133–143.

104. Gao K, Kemp DE, Ganocy SJ, Gajwani P, Xia G, Calabrese JR. Antipsychotic-induced extrapyramidal side effects in bipolar disorder and schizophrenia: a systematic review. *J Clin Psychopharmacol*. 2008;28(2):203–209.

105. Pringsheim T, Lam D, Ching H, Patten S. Metabolic and neurological complications of second-generation antipsychotic use in children: a systematic review and meta-analysis of randomized controlled trials. *Drug Saf*. 2011;34(8):651–668.

106. Piacentini J, Woods DW, Scahill L, et al. Behavior therapy for children with Tourette disorder: a randomized controlled trial. *JAMA*. 2010;303(19):1929–1937.

107. McGuire JF, Piacentini J, Brennan EA, et al. A meta-analysis of behavior therapy for Tourette Syndrome. *J Psychiatr Res*. 2014;50:106–112.

108. Steeves T, McKinlay BD, Gorman D, et al. Canadian guidelines for the evidence-based treatment of tic disorders: behavioural therapy, deep brain stimulation, and transcranial magnetic stimulation. *Can J Psychiatry*. 2012;57(3):144–151.

109. Wile DJ, Pringsheim TM. Behavior therapy for tourette syndrome: a systematic review and meta-analysis. *Curr Treat Options Neurol*. 2013;15(4):385–395.

110. Franzkowiak S, Pollok B, Biermann-Ruben K, et al. Motor-cortical interaction in Gilles de la Tourette syndrome. *PLoS One*. 2012;7(1):e27850.

111. Burd L, Kerbeshian PJ, Barth A, Klug MG, Avery PK, Benz B. Long-term follow-up of an epidemiologically defined cohort of patients with Tourette syndrome. *J Child Neurology*. 2001;16(6):431–437.

112. Draper A, Stephenson MC, Jackson GM, et al. Increased GABA contributes to enhanced control over motor excitability in Tourette syndrome. *Curr Biol*. 2014;24(19): 2343–2347.

113. Ziemann U, Paulus W, Rothenberger A. Decreased motor inhibition in Tourette's disorder: evidence from transcranial magnetic stimulation. *Am J Psychiatry*. 1997;154(9):1277–1284.
114. Heise CA, Wanschura V, Albrecht B, et al. Voluntary motor drive: possible reduction in Tourette syndrome. *J Neural Transm*. 2008;115(6):857–861.
115. Hasan A, Rothenberger A, Munchau A, Wobrock T, Falkai P, Roessner V. Oral delta 9-tetrahydrocannabinol improved refractory Gilles de la Tourette syndrome in an adolescent by increasing intracortical inhibition: a case report. *J Clin Psychopharmacol*. 2010;30(2):190–192.
116. Orth M, Rothwell JC. Motor cortex excitability and comorbidity in Gilles de la Tourette syndrome. *J Neurol Neurosurg Psychiatry*. 2009;80(1):29–34.
117. Wu SW, Gilbert DL. Altered neurophysiologic response to intermittent theta burst stimulation in Tourette syndrome. *Brain Stimul*. 2012;5(3):315–319.
118. Draper A, Jude L, Jackson GM, Jackson SR. Motor excitability during movement preparation in Tourette syndrome. *J Neuropsychol*. 2015;9(1):33–44.
119. Jackson SR, Parkinson A, Manfredi V, Millon G, Hollis C, Jackson GM. Motor excitability is reduced prior to voluntary movements in children and adolescents with Tourette syndrome. *J Neuropsychol*. 2013;7(1):29–44.
120. Mantovani A, Leckman JF, Grantz H, King RA, Sporn AL, Lisanby SH. Repetitive transcranial magnetic stimulation of the supplementary motor area in the treatment of Tourette Syndrome: report of two cases. *Clin Neurophysiol*. 2007;118(10):2314–2315.
121. Kwon HJ, Lim WS, Lim MH, et al. 1-Hz low frequency repetitive transcranial magnetic stimulation in children with Tourette's syndrome. *Neurosci Lett*. 2011;492(1):1–4.
122. Le K, Liu L, Sun M, Hu L, Xiao N. Transcranial magnetic stimulation at 1 Hertz improves clinical symptoms in children with Tourette syndrome for at least 6 months. *J Clin Neurosci*. 2013;20(2):257–262.
123. Arnold LE, Zametkin AJ, Caravella L, Korbly N. Ethical issues in neuroimaging research with children. In: Ernst M, Rumsey JM, eds. *Functional Neuroimaging in Child Psychiatry*. Cambridge, UK: Cambridge University Press; 2000.

Transcranial Magnetic Stimulation Neurophysiology of Pediatric Traumatic Brain Injury

K.M. Barlow, T.A. Seeger

University of Calgary, Calgary, AB, Canada

Traumatic brain injury has been studied since antiquity. The Edwin Smith Papyrus outlines how academics in Ancient Egypt and Mesopotamia studied how traumatic brain injury affected the brain. With the evolution of the scientific method, and the advent of non-invasive stimulation techniques, we are uniquely positioned to investigate the electrophysiological and neuronal network changes that occur following traumatic brain injury (TBI) in ways that have previously been unattainable. This chapter will first introduce the reader to the clinical and physiological aspects of TBI and the unique challenges that are associated with pediatric TBI. It will review neuroplasticity and how the cortical networks change after TBI. The literature exploring the insights that TMS can provide in TBI is summarized, and finally some early results from the first study of cortical excitability in TBI in children.

EPIDEMIOLOGY AND CLINICAL SPECTRUM OF TRAUMATIC BRAIN INJURY

Traumatic brain injury (TBI) is common, especially mild TBI (mTBI), and occurs more frequently in childhood than at any other time of life. Traumatic brain injury is defined as an injury to the brain that occurs when biomechanical forces result in an alteration of brain function.[1] The incidence of TBI varies widely across studies, occurring at a frequency of between 341 and 1750 per 100,000 persons per year[2–5] and depends on the methodologies and populations investigated. Approximately 70–90% of all TBIs are considered mild.[5] As many as 25% of mTBIs do not receive medical attention, the true incidence is likely to be much higher.[6] Young children and older teens are at greater than double the risk of sustaining a TBI when compared to adults.[7] In addition, more than four adverse life events, such as parental divorce, abuse, or deaths in the family, can increase the risk of TBI by as much as three times.[8]

The causes of TBI are varied. Falls are the most common cause in young children, whereas motor vehicle accidents and assaults are more common in older teens and adults.[9] Sport-related injuries are an important cause of mild TBI and concussion.[3] One third of individuals who sustain a concussion will incur at least one additional TBI.[11] Concussion is more likely to occur in certain sports, ie, ice hockey, women's soccer, and men's football, which all have injury rates above 0.40 per 1000 athlete exposures.[1] In high school, football has the highest injury rate at 0.47 injuries per 1000 athlete exposures. As athletes are more likely to undergo repeated subconcussive events, ie, hits to the head without overt clinical symptoms, these figures are concerning.[11] With such high injury rates in these contact sports, the potential long-term consequences of repetitive injury, such as possible psychiatric illness and chronic traumatic encephalopathy,[10,11] become very

TABLE 17.1 American Academy of Rehabilitative Medicine Criteria for Traumatic Brain Injury Severity Classification[14-16]

Severity	Mild	Moderate	Severe
Glasgow coma scale	13–15	9–12	3–8
Loss of consciousness	<30 min	30 min to 24 h	>24 h
Posttraumatic amnesia	<24 h	24 h to 7 days	>7 days

important. Other sports such as mountain biking and sports involving motorized vehicles, are associated with more severe forms of TBI.

The severity of a TBI ranges from mild to severe and is assessed using clinical measures such as the presence or absence of a loss of consciousness, focal neurological signs, behavioral disturbances, and the length of amnesia following the injury (posttraumatic amnesia) (see Table 17.1). Whereas a severe TBI will be associated with a prolonged period of coma (lasting 7 days or more), an mTBI may or may not be associated with a loss of consciousness. Somewhat confusingly, these criteria may not be used in sports medicine.[5] Here, the term "concussion" is usually used to imply a milder form of mTBI[12,13] and may exclude patients with a Glasgow coma scale of 13 and/or a loss of consciousness for longer than 5 min. For these reasons, the inclusion criteria in studies examining concussion and/or mTBI should be explicit, and the reader should be aware of the patient population and their potential differences when drawing any conclusions.

OUTCOME

Outcome after TBI is predicted by preinjury demographic factors, injury-related factors, and postinjury factors. Preinjury demographic predictors include age, prior intellectual ability, and the presence of behavioral or psychiatric conditions, socioeconomic status, and family dynamics.[17] Outcomes strongly correlate with injury severity (eg, duration of unconsciousness). Post-traumatic injury predictors include success with rehabilitation programs and interventions. In children, TBI patients showed persistently lower quality of life than extracranial injury controls of a similar age.[18] The long-term impairments following pediatric TBI include behavioral problems, neuropsychological deficits (eg, problems with attention, executive function, memory, and speed of processing), motor deficits, and decreased social skills.[19] The pattern of interaction between child behavior problems and family functioning 6 months after injury is predictive of the burden on the family, parental stress, and child behavior problems at 1 year.[20] Recovery takes longer as the injury severity increases.[21]

Following an mTBI, there may be problems with fatigue, headache, drowsiness, difficulty sleeping, irritability, concentration difficulties, memory problems, and mood disturbance.[22] These symptoms have typically been considered to last between 7 and 10 days,[23] however, the median persistence time of symptoms in children has recently been shown to be 29 days (95% confidence intervals: 26.1, 31.9 days), and 11% of children remain symptomatic at 3 months post injury.[24,25] Postconcussion syndrome (PCS) is diagnosed when there is a constellation of three or more postconcussive symptoms lasting for over 3 months.[25]

PATHOPHYSIOLOGY OF TRAUMATIC BRAIN INJURY

TBI begins with an insult to the brain, and resultant complex pathological processes and cascades ensue. Consider the brain to be a high-density semisolid with defined borders suspended in a liquid, the cerebrospinal fluid (CSF), all of which are contained within a rigid skull. When a force is applied, it causes the whole structure to accelerate or decelerate. When the skull stops or reverses direction (eg, in a whiplash movement), the brain continues to move and may collide with the inside skull and its bony buttresses. The brain will deform when it impacts the skull (even with "subconcussive" forces). Injury occurs when such deformations exceed the biomechanical structural limits of any tissues undergoing strain (blood vessels, neurons, pial tissues, etc.). The brain is not a homogeneous mass but has gray and white matter, blood vessels, and CSF spaces. As gray and white matter have differing water content and densities (which change throughout brain development as myelination occurs), these structures move according to their inertia. This results in shear stresses on the axons, blood vessels, and oligodendrocytes as they cross the different densities. These forces cause the membranes to leak, known as mechanoporation, and results in a breakdown of the electrochemical gradient needed for effective signal transmission and disruption to cellular organelles.

An indiscriminate release of neurotransmitters occurs following TBI, activating enzyme pathways and other cellular processes that subsequently cause secondary damage to the brain. The pathophysiology of TBI has been reviewed by Giza et al.[26,27] This indiscriminate release of neurotransmitters begins a process called excitotoxicity. Excitotoxicity occurs when presynaptic neurons release large amounts of glutamate that binds to n-methyl-D-aspartate (NMDA) receptors on postsynaptic neurons and causes uncontrolled neuronal activation. These pathways, combined with mechanoporation, result in a shift in ionic gradients across membranes, an increase in cellular metabolism, and increases in reactive oxygen species. The net effect may lead to cytoskeletal and intracellular transport system degradation.

Calcium plays a key role in the activation of the many enzymatic cascades and intracellular disruptions associated with TBI. Calcium ion accumulation stems from excitotoxicity as NMDA receptors are permeable to calcium ions. In cells that have undergone mechanoporation, calcium will flow down its large concentration gradient into the cell. Increased intracellular calcium at the synaptic terminals can also lead to uncontrolled neurotransmitter release and further excitotoxicity. Calpain enzymes become activated, and may lead to breakdown of the cytoskeleton. These processes are activated to varying degrees and so can result in different cellular injury phenotypes even in adjacent neurons.

Indiscriminate activation of intracellular processes leads to the consumption of adenosine triphosphate (ATP), the primary unit of energy in the cell. Axons and cell bodies of injured neurons may swell due to ionic influx resulting in cerebral edema. Multiple secondary injury cascades and reduced blood flow to the brain exacerbate the energy crisis. Although many neurons will recover, others may be too severely damaged to recuperate, and will undergo apoptosis or necrosis. Damaged or dead neurons may not all be closely grouped, but may be spread throughout the brain in what is called diffuse axonal injury (DAI). This distribution (focal or diffuse) depends on the primary insult, the biomechanical properties of the brain, the size and propagation of force waves through the tissues, and the resultant location of stress points.

Recovery from TBI is less well understood than the acute response to injury. There is strong pathological evidence for a prolonged immune response in TBI, with microglial activation, astrocyte activation, and microvascular changes in the blood–brain barrier identifiable months and years after injury. This may contribute to poorer long-term outcomes.[28–32] The release of ATP from damaged neurons initiates the activation of the innate immune system, resulting in the release of inflammation-promoting mediators, such as cytokines, chemokines, and reactive oxygen and nitrogen species. Proinflammatory processes are intended to clear the central nervous system of potentially harmful substances. Anti-inflammatory processes follow this, performing reparative and regenerative functions considered to be beneficial to neuronal survival. An unbalanced or prolonged inflammatory response can be harmful.[33]

Changes in Cortical Excitability Following TBI

It is advisable to consider TBI as a dynamic process, and not just a static insult to the brain.[34] Cortical networks that remain intact enough to propagate signals may show short-term and long-term changes in excitability. The excitatory–inhibitory balance may be altered due to both the number and type of neurons that survive as well as changes in the number and type of receptor expression. The exact mechanisms and durations of neurotransmitter dysfunctions following TBI remain poorly elucidated, although these changes are more prominent in children, and have the potential to lead to longer-lasting deficits due to the concurrent developmental processes.[35]

Changes in neuronal excitation occur and can be immediate. This is reflected by the increased incidence of seizures in the acute postinjury period. Excitation tends to decrease over time. As glutamate and calcium levels normalize, longer-term changes in receptor expression occur. For example, in a juvenile mouse TBI model the relative expression of the NMDA receptor NR1, and NR2 subunit subtypes A and B differ after TBI. Subtype NR2B and NR1 remain unaffected, while NR2A subunit expression in the synapse is reduced.[36] Changes in receptor subunit or subtype expression can cause decreased ligand affinity, change protein transportation targets, and alter the functional mechanism (eg, ionotropic instead of metabotropic) of the receptor.

GABAergic interneurons (most commonly inhibitory) may be particularly vulnerable to injury.[37] GABA receptors may also show changes in their relative subtype expression following TBI. For example, some GABA A subtype receptor subunits are downregulated (eg, ϵ and θ subtypes in the thalamus and hypothalamus),[38] whereas other are upregulated (α4 subunit increase in the hippocampus).[39] However, there are multiple subtypes of GABA receptors, and though the GABAa receptor is more commonly studied, GABA B subtype receptors are also noteworthy. Although less studied, GABAb receptor alterations may have an important role in network changes as they are metabotropic and so have longer-lasting effects. The resultant inhibitory alteration depends on the type of GABAergic interneurons affected, receptor subtype expression, their location in the brain, and the time after injury.

Neuroplasticity Following TBI

One of the most interesting aspects of recovery following a TBI that can be investigated using TMS is neuroplasticity (see also Chapter: Neuroplasticity Protocols: Inducing and Measuring Change). Neuroplasticity is the ability of the neurons to adapt to changes in their environment. It encompasses the formation of new neuronal connections, uncovering latent connections, and strengthening (or lessening) the modulatory influence of an existing connection.[40] Connection formation and recruitment of latent connections are costly in terms of cellular resources and also takes longer to achieve than strengthening an already active connection. Therefore, strengthening an established connection is more efficient, especially in a mild injury.[35] This can be achieved by adding more presynaptic neurotransmitter vesicles or postsynaptic receptors. In contrast, as severe injuries have a greater number of cell deaths, new connections will be required. Axonal growth and the formation of new synapses are related to the action of growth factors such as nerve growth factor induced gene A, homer, activity regulated cytoskeletal-associated protein, and brain-derived neurotrophic factor (BDNF).[41] BDNF, for example, increases the number of dendritic spines, and aids in the formation of new synapses.

Long-term potentiation (LTP) is a building block of plasticity. Synapses that have undergone LTP tend to have stronger electrical responses to stimuli than other synapses and is associated with increases in NMDA receptor concentration at the postsynaptic membrane. This can be artificially induced with high-frequency stimulation in vitro.[42,43] The opposing form of neuroplasticity to LTP is long-term depression (LTD). Decreasing the NMDA receptor concentration at the postsynaptic neuron is the crux of LTD. LTP-like and LTD-like effects have been found non-invasively in humans using high- and low-frequency stimulation in repetitive TMS (rTMS), respectively.[44,45]

Plasticity is not solely a function of the relationship between two cells but involves groups of cells. These microcircuits include many different cell types, protein expression patterns, and functions. Interneurons are thought to play a key role in network adaptation. Interneurons are short-ranging neurons that modulate the activation of other cell types. They may receive inputs from different neurons or brain regions than that of the "target" neuron. If an inhibitory interneuron synapses just above the axon terminal on the presynaptic neuron, this will decrease the likelihood of neurotransmitter release (see Fig. 17.1). By this mechanism and similar mechanisms, a presynaptic/postsynaptic neuron synapse will be inhibited, and may produce weaker than normal signals. If these neurons had undergone LTP, this inhibition will counteract the LTP. This GABA-mediated inhibition, therefore, filters neuroplasticity at excitatory inputs.[43] Autoreceptors on the interneuron moderate this activity. At the interneuron–presynaptic neuron synaptic cleft, GABAb autoreceptors on the interneuron bind GABA released from the interneuron. When this happens, the interneuron becomes hyperpolarized and is therefore less likely to release neurotransmitter. As a result, the presynaptic neuron does not become hyperpolarized, and is now more likely to generate an action potential.

Changes occurring after TBI are of course more complex than the changes occurring in individual cells or small groups of synapsing cells. Collections of microcircuits form networks, sometimes ranging long distances across the brain. Information is conveyed in these networks using a careful balance between excitation and inhibition through successive levels of processing. Dysfunction in single or groups of cells after TBI may affect the excitability of neuronal circuits, and such disruption will affect the larger network. Therefore, when measuring cortical network activity after TBI, the TMS investigator must be cognizant that the reparative

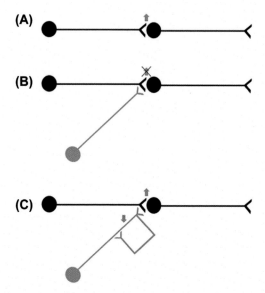

FIGURE 17.1   GABAergic filtering of synaptic plasticity. (A) A presynaptic/postsynaptic neuron pair (black), which will propagate signals from left to right, with a small *blue arrow* indicating long-term potentiation (LTP). (B) The same circuit expanded to include an inhibitory interneuron (blue) synapsing onto the presynaptic terminal which negates the LTP effects. However, (C) shows autoinhibition of the same interneuron, which restores the effects of LTP.

mechanisms of the network may alter the network's response to stimulation. This is true in single- and paired-pulse TMS, where the investigator is attempting to evaluate the current state of the network, and modulate its plasticity, such as in rTMS. rTMS will be attempting to modulate plasticity in a system that is already undergoing an increased amount of neuroplasticity compared to the normal brain, and outputs will reflect the interaction between the ongoing neuroplasticity and that evoked by rTMS.

INSIGHTS ABOUT TBI USING TMS: A LITERATURE REVIEW

TMS has been used to examine various networks after TBI, including changes in the excitability of corticospinal, intracortical, and interhemispheric motor networks. The existing TMS studies have examined mild, moderate, and severe TBI, in both the athlete and general population at different time points post injury. As most studies do not include children, it is not known whether their findings can be generalized to the pediatric age range. TMS findings in adults after TBI primarily fall into three categories: acute mTBI, chronic mTBI, and severe TBI. As the populations vary considerably, as does the time of study after the injury, it is hard to draw definite conclusions about exactly how cortical excitability and cortical inhibition change during recovery. Overall, the literature suggests that recovery is associated with an increase in cortical inhibition. The evidence is summarized in the following sections and in Tables 17.2 and 17.3.

MILD TRAUMATIC BRAIN INJURY

Changes in the excitation–inhibition balance can be measured using the rest motor threshold (measuring the most excitable neurons in the area) and the stimulus response curves (SRCs).[46] Studies that report changes in RMT after mTBI, even in the acute stages after injury, are contradictory. About half of the small number of studies describe no difference in thresholds, while the other half describe an elevation of the RMT. For example, increased RMTs were found in mTBI patients with or without sleep disturbances[47,48] and those whose injury was over 3 years prior.[49] Conversely, other studies have failed to show RMT differences in either mild or moderate TBI in both acute and chronic phases.[50–53] This is highlighted in the study by Miller et al.,[52] which examined patients repeatedly between 72 h and 2 months after injury. Although some studies suggest a slight shift to increased inhibition, RMT is highly variable between individuals and also may not be sensitive enough to detect the often-subtle changes associated with a mild injury.

Excitation can be measured using MEP amplitude, which is the ability of corticospinal tract neurons to effectively summate, and create a unified descending volley of action potentials. This measure is a function of the RMT and is affected by the skull thickness. Some small studies found MEP amplitude was increased within the same subject over the first 10 days following an mTBI in athletes.[50,54,55] However, this was not substantiated in larger studies,[56,57] or in a repeated measures study examining changes over the first 2 months postinjury.[52] Stimulus response curves (SRC) or input/output (IO) curves expand these investigations to measure recruitment of higher-threshold corticospinal tract neurons and the synchrony of descending action potential volleys.[58] SRCs are normal following mTBI,[53,55] although they show increased variability.[59] Together, these two measures, MEP amplitude and SRC, indicate that the neuron population that excites the target muscle pathways is still fully capable of eliciting a normal output following stimulation.

The white matter tract injuries often associated with TBI may be detected using techniques exploring conduction times, eg, MEP latency and corticomotor conduction times (CMCT). In the acute stages after mTBI, subtle conduction abnormalities have not been identified using TMS in the setting of normal RMT and MEP amplitudes.[50,55] In a larger

TABLE 17.2 Summary of the Literature Using Single- and Paired-Pulse Transcranial Magnetic Stimulation Paradigms After Mild and Moderate Traumatic Brain Injury (TBI)

					TMS Paradigms																	
					Single-Pulse TMS								Silent Period							Paired Pulse		
						MEP Amplitude			SRC					CSP			ISP					
First Author	Year(s)	Time After Injury	TBI Severity	Notes	RMT	Peak to Peak	MEP/M Wave	Variability	Between Groups	V50 Curve	Slope	MEP Latency	CMCT	Duration	SPT	V50 Curve and Slope	TCI	TCI Onset Latency	TCT	SICI	ICF	LICI
Powers	2014	Acute	Mild	Asymptomatic	x									x						x	↑	x
Livingston	2010, 2012	Acute	Mild	Day 1	x	x						x	x									
				Day 3	x	x						x	x									
				Day 5	x	x						x	x									
				Day 10	x	x						x	x									
Chistyakov	1998	Acute	Mild		↑			↑														
Miller	2014	Acute	Mild		x	x						x		↑								
Pearce	2014	Chronic	Mild	Retrospective						x	x			↓		x						↓
De Beaumont	2012	Chronic	Mild	Asymptomatic	x	x								x								↑
De Beaumont	2011	Chronic	Mild	Asymptomatic										↑								↓
Tremblay	2011	Chronic	Mild	Concussion	0									↑						x		↑
Tremblay	2014	Chronic	Mild											x								x
Tallus	2012, 2013	Chronic	Mild	Symptomatic	↑																	
				Asymptomatic	↑																	

Study	Year			Subgroup										
De Beaumont	2007	Chronic	Mild	Single	↑			x					x	x
				Multiple	↑			x			↑		x	x
Chistyakov	2001	Acute	Mild–Mod	Minor	x		x			x	x		x	x
				Mild	↑		↓			↑	↑	↑	↑	↑
				Moderate	↑		↓			↑	↑	↑	↑	↑
				Concussion	x		x			x	x	x	x	x
				Focal lesion	↑		↓			↑	↑	↑	↑	↑
				Diffuse injury	↑		↓			↑	↑	↑	↑	↑
				Combined	↑		↓			↑	↑	↑	↑	↑
Nardone	2011	Acute	Mild–Mod	Fatigue	x					x	x		x	x
				Subjective EDS	x					x	x		x	x
				Objective EDS	↑					x			↑	↑
				Hypersomnia	x					x			x	x
Chistyakov	1999	Acute	Mild	Diffuse injury	↑			↑		x	↑			
				Focal injury	↑			↑		x	x			
			Mod	Diffuse injury	↑			↑		↑	↑			
				Focal injury	↑			↑		x	↑			

↑ indicates a significant increase in the paradigm after TBI, while ↓ indicates a decrease. "x" indicates that no significant differences were found.

TMS, transcranial magnetic stimulation; MEP, motor-evoked potential; SRC, stimulus response curve; CMCT, corticomotor conduction time; CSP, cortical silent period; SPT, silent period threshold; SP, ipsilateral silent period; TCI, transcallosal inhibition; TCT, transcallosal conduction time; SICI, short-interval intracortical inhibition; ICF, intracortical facilitation; LICI, long-interval intracortical inhibition; EDS, excessive daytime sleepiness.

TABLE 17.3 Summary of the Literature Using Single- and Paired-Pulse Transcranial Magnetic Stimulation Paradigms After Severe Traumatic Brain Injury (TBI)

First Author	Year(s)	Acuteness	TBI Severity	Notes	RMT	Peak to Peak	MEP/M Wave	Variability	Between Groups	V50 Curve	Slope	MEP latency	CMCT	Duration	SPT	V50 Curve and Slope	TCI	TCI Onset Latency	TCT	SICI	ICF	LICI
						MEP Amplitude				**SRC**				**CSP**				**ISP**		**Paired Pulse**		
														Silent Period								
Mazzini	1999	Acute	Severe	Upper								x	↑									
				Lower								↑	x									
Jang	2005	Chronic	Severe		x	x																
Bagnato	2012	Chronic	Severe	VS	x																↓	
Fujiki	2006	Chronic	Severe	DAI	x															↓	↓	
Takeuchi	2006	Chronic	Severe		x	x							x				↓	x	x			
Bernabeu	2009	Chronic	Severe	Non-Paretic	↑	x			x					x								
				Paretic	↑	↑			↑					x								
				Mild DAI	↑	x			x					x								
				Severe DAI	↑	x			↑					x								
				Overall	↑	↑			x					↑								

TMS, transcranial magnetic stimulation; *RMT*, rest motor threshold; *MEP*, motor-evoked potential; *SRC*, stimulus response curve; *CMCT*, corticomotor conduction time; *CSP*, cortical silent period; *SPT*, silent period threshold; *ISP*, ipsilateral silent period; *TCI*, transcallosal inhibition; *TCT*, transcallosal conduction time; *SICI*, short-interval intracortical inhibition; *ICF*, intracortical facilitation; *LICI*, long-interval intracortical inhibition; *VS*, vegetative state; *DAI*, diffuse axonal injury. Upper and lower (Mazzini et al.[80]) refer to the upper or lower limbs, respectively.

study of individuals 2 weeks after injury, Chistyakov et al.[60] found an increase in MEP latency associated with increased RMT and MEP amplitude variability. This conduction delay could be due to axonal dysfunction in the motor tracts, as longer axons are more vulnerable to stretch and torsion forces during the insult. However, this was not substantiated by Miller et al.,[52] who did not find any differences in MEP amplitudes or latencies between controls and mTBI participants at 72 h, and 1, 2, 4, and 8 weeks postinjury. Ultimately, the interpretation of such studies is difficult due to the heterogeneity of the populations studied (Miller et al. studied younger people with sport-related concussion) and the severities of injury even in mild TBI. This is corroborated by CMCT studies. CMCT mathematically eliminates the peripheral motor conduction times from the MEP latency. A study by Chistyakov et al.[59] divided the participants into four groups: concussion, focal injury, diffuse injury, and combined focal and diffuse injury. The focal, diffuse, and combined injury groups consisted of both mild and moderate TBI participants. This study found that only the more severe injuries (diffuse injuries) had increased CMCT compared to controls.

The role of inhibition following TBI has also been explored. Inhibition can be measured using single- or paired-pulse paradigms, such as cortical and ipsilateral silent period, and short- and long-interval intracortical inhibition, respectively. GABAergic inhibitory mechanisms can be confirmed by increasing GABA in the synapse, eg, Tiagabine inhibits GABA reuptake into the neurons and astrocytes and by using GABAb receptor agonists such as baclofen. Both tiagabine and baclofen increase the duration of the cortical silent period.[61,62] Miller et al. explored inhibition in the acute stages of TBI and provided strong evidence for increased inhibition by demonstrating a prolonged CSP in mTBI. CSP was prolonged at 72 h postinjury and remained increased until at least 8 weeks.[52] This is in keeping with other adult TBI studies.[57,59,63,64]

Postconcussive symptoms have usually resolved by 3 months in 80–85% of children with mTBI.[24] It is therefore reasonable to consider any changes in cortical excitability persisting for 3 months or longer as reflecting a chronic change. Increases in CSP have been found between 3 months and 30 months post injury.[53,57,63,65] This implies a change in cortical inhibition, and further, that these changes may be persistent years after the injury. They may also correlate with the number of previous concussions.[63] In contrast, 22 years after a concussion, Pearce et al.[56] found a shortened CSP (during an "active contraction"). These findings implicate a further change in cortical inhibition over time. The biological advantage and neurophysiological substrates underpinning this however are unknown. There may be a slow evolution in the regulation of the GABAb receptor protein expression over time, and/or a decrease in GABA release, which may eventually "overcompensate" for the injury.

As inhibitory interneurons are vulnerable in TBI however,[37] a prolonged CSP seems counterintuitive. These changes could represent alterations in GABAb receptor expression (similar to that found in murine models of TBI), although it is also possible that the individuals with concussion studied by Miller et al. were different even before the injury, ie, cortical inhibition was increased *before* the acute injury.[52] These increases in inhibition could be due to previous concussive or subconcussive events. Further validation and exploration of these findings together with ipsilateral silent period interrogation are needed.

Intracortical inhibition has been studied using the paired-pulse paradigms short-interval and long-interval intracortical inhibition (SICI and LICI, respectively). SICI is thought to be the effect of interneuronal GABAa receptor-mediated inhibition of the pyramidal tract neurons. Changes in SICI are again variable after TBI, and methodologies vary significantly across studies. A wide range of populations as well as time points postinjury were studied making it difficult to draw conclusions about changes in intracortical inhibition over time.[48,51,56,57,64,66,67] For example, Pearce et al. found chronically decreased SICI following mTBI.[56] Conversely, SICI was increased in individuals with excessive daytime sleepiness early after mild and moderate TBI.[66] The mechanisms underlying the modulation of SICI after TBI are poorly understood and warrant further investigation.

LICI is thought to be due to the GABAb receptor-mediated inhibition of the pyramidal tract neurons by the conditioning TMS stimulus. Tremblay et al. found that LICI was increased in the first 12 months post mTBI and that this inhibition may decrease in the long term.[68] Although LICI and CSP share a common pathway (via GABAb receptors), they do examine slightly different mechanisms of intracortical inhibition.[46,69,70] Nevertheless, increases in LICI are consistent with increases in CSP following adult TBI supporting a significant role for GABAb receptor-mediated intracortical inhibition in the recovery from TBI.[52,55] Further studies exploring the role of intracortical inhibition in the recovery from a TBI over time may well provide useful insights into potential therapeutic interventions in TBI in the future.

Chistyakov et al.[60] report an interesting finding of fatigability which was present at 2 weeks postinjury, and resolved by 3 months. To do this, they used suprathreshold rTMS trains of 50 stimuli and found that MEP amplitude progressively decreases within trains of stimulation, with a marked irregularity in the shape of the MEP waveform. Although this study does suffer from attrition, these abnormalities improved at 3 months and may correlate with symptom improvement.[60]

Combining TMS with functional tests can be more informative in mTBI, as cognitive deficits may only be found in the most difficult tasks, and may persist after symptom resolution.[71-73] The similar persistence of TMS abnormalities (specifically cSP) may provide a correlate to these findings after mTBI. One study did find a differential recovery in symptoms, neuropsychological test scores and TMS parameters after sport-related concussion within the first 10 days after injury, but these results need corroboration.[74]

SEVERE TRAUMATIC BRAIN INJURY

Studies using TMS to investigate severe TBI in both the acute and chronic states are rare. Most studies report RMT values in severe TBI (including minimally conscious states following TBI) to be similar to control populations[75–78] except in one study. This study by Bernabeu et al.[79] classified the participants based on (i) motor function (paretic vs. non-paretic) and (ii) radiological findings (mild and moderate DAI, severe DAI, and combined DAI and focal lesions). Increased RMT and decreased area under the MEP curve was present in severe TBI, especially in the presence of paresis and/or severe diffuse axonal injury. This is consistent with the clinical picture and the neuronal loss following severe diffuse axonal injury. The concurrent dysregulation of the SRCs demonstrates that neuronal recruitment was lower in those recovering from TBI who have the greatest functional motor impairment.

Conduction times may be more sensitive to the consequences of severe brain injury than RMT and MEP amplitude. MEP latency was increased following severe TBI.[76,80] It is not clear whether CMCT or MEP latencies are better to document these abnormalities as the current studies are small and conflicting.[78,80]

Bernabeu et al.[79] also investigated the chronic effects of diffuse axonal injury on intracortical excitability and inhibition. CSP prolongation, indicating an increase in GABAb receptor-mediated inhibition, also occurs in severe TBI. Such increased inhibition may be a neuroprotective mechanism by blunting the effects of indiscriminate excitatory glutamate release.[81]

Corpus callosal injury is often found in severe TBI.[82–84] In a normal motor cortex, the transcallosal fibers originating from one motor cortex will synapse onto inhibitory interneurons in the contralateral cortex, thereby inhibiting contralateral motor output. Evidence of decreased transcallosal cortical inhibition has been found 2 years following adult severe TBI and correlated with injury severity. Takeuchi et al. found that there is a decreased magnitude of inhibition in the contracting hand (ipsilateral to stimulation) after severe TBI, but that the duration of the inhibition elicited via these transcallosal tracts (ipsilateral silent period, iSP) was not different.[77] This suggests a decrease in the *number* of inhibitory interneurons stimulated in the contralateral hemisphere, which is in keeping with transcallosal tracts damage and might also explain the presence of the mirror movements that can be seen after severe TBI.

TREATMENT STUDIES USING rTMS

The safety of performing rTMS in patients with TBI is still being investigated, but early reports suggest that it is probably safe. The literature contains one report of a seizure following TMS as well as an increase in postconcussion symptoms. These seem to be uncommon occurrences but further study is needed.[85,86] The studies using rTMS as a treatment modality in TBI are few and have explored its use in postconcussion syndrome following mTBI, and in minimally conscious states and dysphagia following severe TBI.

rTMS over the dorsolateral prefrontal cortex (DLPFC) has been used to successfully treat patients with depression[87,88] (see also Chapters 15 and 16: Brain Stimulation in Pediatric Depression: Biological Mechanisms and Brain Stimulation in Childhood Mental Health: Therapeutic Applications). As many patients with persistent postconcussion symptoms have quite marked mood disturbance, it is logical to explore its similar use in PCS. In a pilot study, Koski et al. examined the safety and use of rTMS stimulating the DLPFC to treat persistent PCS which had been present for 6 months or longer.[86] At the end of 20 sessions over a 4-week period, although one patient reported an increase in symptoms, participants reported a significant decrease in symptoms overall. Given these favorable results and good safety profile, larger studies exploring rTMS in mTBI studies are ongoing.[89,90]

Minimally conscious states (including vegetative state) have an extremely poor prognosis following TBI. There are case series of rTMS being used to try to improve the conscious level in these patients following severe TBI but the results are disappointing.[91] One patient was reported to show minimal improvement (changing from a vegetative to minimally conscious state), but the other five patients showed no response. Another study investigated whether rTMS to the mylohyoid motor cortex improves swallowing postinjury in severe TBI. The protocol was similar to those used in the treatment of stroke except that the target differed (ie, contralesional low-frequency and ipsilesional high-frequency stimulations for 10 s, repeated every minute for 20 min for 10 days). Gains were made in two of the three dysphagia measures when compared to sham rTMS treatment.[92] More research is needed but rTMS may be a useful adjunct therapy in the rehabilitation of people with TBI in the future, especially for specific deficits.

TMS IN PEDIATRIC MILD TRAUMATIC BRAIN INJURY

There is no literature exploring cortical excitability using TMS in children. We report our pilot study investigating cortical excitability in children during their recovery from a mild TBI. This is a controlled longitudinal cohort study. Children with mTBI are eligible for study if they are between 8 and 18 years of age, and are "symptomatic" at 1 month postinjury. They are compared to two control groups: (1) children who have recovered by 1 month postinjury and so are "asymptomatic" and (2) normal children without a history of mTBI. Exclusion criteria include an injury of greater severity, the use of any medications that could affect TMS, and/or a history of serious medical or psychiatric illness.

TMS methodology: Biphasic stimuli were given using a Magstim BiStim 200 Transcranial Magnetic Stimulator (The Magstim Company Limited, Carmarthenshire) with a figure-of-eight coil (70 mm inner diameter, each circle) to the hotspot for the first dorsal interosseous muscle. Electromyography (EMG) was recorded bilaterally from the first dorsal interosseous muscle using Ag/AgCl EMG electrodes (Amplified 1000 times, band pass filtered 20–2000 Hz, 5000 Hz sampling rate). Motor thresholds were adjusted to the minimum stimulation intensity required to elicit 5/10 consecutive MEPs of the required amplitude. Active contractions were measured using an EMG oscilloscope (GwINSTEK GDS-1022, 25 MHz, 250 M Sa/s, Good Will Instrument Co., New Taipei City, Taiwan), and all contractions were 20% of the maximal voluntary contraction, unless stated. If a participant's 50-µV RMT was above 66% of the maximal stimulator output, they could not participate in the rest of the SRC trial. If their 1000-µV RMT was above the maximal stimulator output, they could only perform the active SRC. Data were processed using Matlab (MATLAB and Statistics Toolbox Release 2012b, The MathWorks, Inc., Natick, Massachusetts, USA). Graphs were created in Sigmaplot 13.0 (Systat Software, Inc., San Jose, California, USA, www.sigmaplot.com). Line graphs show the means with standard deviations. Boxplots show the group median as a black horizontal line inside the box. The top edge of the box is the 75th percentile, and the bottom of the box is the 25th percentile, with the group mean in the middle of the box. Outside the box are the "whiskers" which denote the outer fence (1.5*interquartile range). Outliers are shown as points.

TMS outcome measures included single TMS and paired-pulse TMS paradigms. The primary outcome measure was CSP. Other measures included RMT (50 µV and 1 mV), AMT, active and resting SRC, short-interval intracortical facilitation (SICF), active and resting SICI, active and resting ICF, LICI, and iSP. Secondary outcome measures included the tolerability of the TMS session using a questionnaire designed for TMS in children.[101] Active and resting SRC were 10% intervals from 100% to 150% of the resting and active motor threshold. For active SRC, we also measured the silent period after each stimulus to create a silent period recruitment curve. CSP was evoked in the active condition using a stimulus at the 1000-µV RMT.[63,65] CSP duration was from the onset of disrupted waveform after the MEP to the resumption of normal EMG. iSP duration was found similarly, with contraction at 50% of the maximal voluntary effort in the hand ipsilateral to the stimulation.[78] SICI and ICF stimuli were separated by ISIs of 2 ms

TABLE 17.4   Demographic Information of Pediatric Mild Traumatic Brain Injury Pilot Study

		Healthy	Asymptomatic	Symptomatic
Sample size		27	15	15
Age at assessment	Mean (SD)	14.28 (3.10)	14.04 (2.27)	14.99 (2.04)
	Mean rank	27.88	24.88	31.40
Hemisphere stimulated	Left:right	23:3	15:0	11:4
Gender	Male:female	12:14	6:9	6:9
Active SRC only (1000 μV > 100% MSO)		3	2	3
No rest SRC		3	0	0
Full TMS		22	14	13

SRC, stimulus response curves; *TMS*, transcranial magnetic stimulation; *MSO*, maximal stimulator output.

for SICI and 10 ms for ICF.[93] The active state SICI/ICF used a conditioning stimulus at 70% AMT, then a test stimulus at 1000-μV RMT. Resting state SICI/ICF used a conditioning stimulus applied at 90% RMT, and a test stimulus at 1000-μV RMT. LICI used conditioning and test stimuli at 1000-μV RMT, separated by 100 ms. SICF used a conditioning stimulus at the 1000-μV RMT and a test stimulus at 90% of 50-μV RMT, with ISIs at 1.5, 2.6, and 4.3 ms.[94] For paired-pulse techniques, all ISIs were pseudorandomized with a test stimulus alone, which was compared offline.

Preliminary analysis has been completed on 27 symptomatic and 15 asymptomatic mTBI participants, and 15 healthy controls. Groups are comparable in age and gender (see Table 17.4). Two asymptomatic controls had 50-μV RMTs that exceeded the maximal stimulator output (2 T), preventing further study. RMT thresholds exceeded 66% of maximal stimulator output in one healthy and two symptomatic participants, who therefore could not participate in resting SRC. In addition, two participants from each group had 1000-μV RMTs greater than the maximal stimulator output, and so could only perform active SRC.

SAFETY AND TOLERABILITY

No serious adverse events were reported. Tolerability scores for ranking TMS out of eight activities were similar between groups ($P=0.866$): mean symptomatic scores were 3.75 (SD±1.49); asymptomatic scores were 3.86 (SD±1.88); and normal controls 4.08 (SD±2.02).[95] Healthy controls reported mild headache (one participant), tingling sensations (one participant), and nausea (one participant). Asymptomatic participants reported one mild instance of neck pain, and one instance of mild tingling. Symptomatic participants reported one instance each of mild neck pain, light-headedness, and nausea, and four instances of mild tingling in the fingers or at the stimulation site. The frequency of reported minor adverse side effects was not different between participants.

RESULTS

All three motor thresholds showed no differences between groups. CSP times did not differ between groups: 86.87 ± 37.85 ms in the symptomatic group, 115.82 ± 33.74 ms in the asymptomatic group, and 113.99 ± 46.37 ms in healthy control ($F(2, 47) = 2.35$, $P = 0.108$) (see Fig. 17.2).

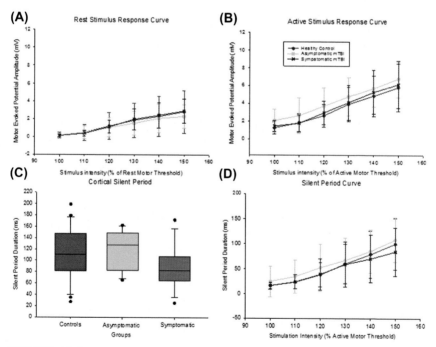

FIGURE 17.2 Stimulus response curve measures and cortical silent periods. (A) The rest stimulus response curve mean (±standard deviation) of the amplitude of motor-evoked potentials expressed as percentages of the rest motor threshold. (B) The active stimulus response curve mean (±standard deviation) of the amplitude of motor-evoked potentials expressed as percentages of the active motor threshold at a contraction of 20% of the participant's maximum voluntary effort. (C) A boxplot for the distribution of the cortical silent period for the controls, asymptomatic and symptomatic mTBI groups. (D) The mean (±standard deviation) silent period curve calculated from the active stimulus response curve.

TABLE 17.5   Transcranial Magnetic Stimulation Results 1 month After Pediatric Mild Traumatic Brain Injury

Measure	Healthy	Asymptomatic	Symptomatic	Stat	P
CSP	113.99 (46.37)	115.82 (33.74)	86.87 (37.85)	$F(2, 47) = 2.35$	0.107
ISP 1 mV	15.09 (7.78)	11.97 (5.69)	9.83 (6.00)	$F(2, 45) = 0.86$	0.430
ISP 75% MSO	20.09 (12.18)	14.76 (10.33)	10.28 (6.16)	$F(2, 42) = 2.36$	0.108
Rest ICF	1.01 (0.62)	1.68 (0.99)	1.34 (1.02)	$F(2, 35) = 1.56$	0.224
Active ICF	0.40 (0.41)	0.96 (0.09)	1.04 (0.25)	$F(2, 47) = 0.63$	0.536
Rest SICI	0.61 (0.49)	0.78 (0.36)	1.17 (0.79)	$F(2, 35) = 1.07$	0.356
Active SICI	0.93 (0.23)	0.88 (0.13)	0.99 (0.24)	$F(2, 47) = 1.29$	0.284
LICI	0.32 (0.38)	0.41 (0.74)	0.99 (0.80)	$\chi^2 = 7.62$	0.022

Single-pulse TMS paradigms include: silent periods, with values in milliseconds; and stimulus response curve values area under the curve, slope, and max, which are all functions of the motor-evoked potential amplitude. Paired-pulse paradigms include intracortical facilitation (ICF) and long- and short-interval intracortical inhibition (LICI and SICI, respectively). Paired-pulse paradigms are expressed as a ratio of the conditioned stimulus to the test stimulus alone (unconditioned), where values less than 1 indicates inhibition, and values greater than 1 indicate facilitation.

CSP, cortical silent period; ISP, ipsilateral silent period; ICF, intracortical facilitation; SICI, short-interval intracortical inhibition; LICI, long-interval intracortical inhibition.

These values are similar to healthy children in other studies[96,97] and our lab (Kirton, personal communication, June 2, 2015). iSP at 1000-µV RMT stimulation was similar between groups, $F(2, 45) = 0.86$, $P = 0.430$. The details of each of the TMS outcome measures are given in Tables 17.5, 17.6 and 17.7.

The SRC at rest showed a significant effect of stimulation intensity $(F(1.47, 68.99) = 71.97, P < 0.01)$, but was not different between groups $(F(2, 47) = 0.93, P = 0.911)$. Similarly, active stimulus response curves showed an effect for stimulus intensity $(F(1.66, 82.74) = 143.01, P < 0.01)$ but no effect of group, or interaction (shown in Table 17.6). Silent period recruitment curves during active SRC are shown in Fig. 17.2. The duration of the silent period response curve during active stimulus response curve increased with increasing stimulation $(F(1.43, 68.84) = 101.66, P < 0.01)$. There was no main effect of group $(F(2, 48) = 0.56, P = 0.578)$.

SICF has not previously been explored in TBI. Groups were similar $(F(2, 22) = 0.97, p = 0.395)$, but there was an effect of ISI $(F(1.49, 32.85) = 4.59, p = 0.026)$ (see Fig. 17.3). ICF and SICI for active or resting conditions was also similar between groups. However, the variability in resting SICI was higher than expected. As expected, during the

TABLE 17.6 Mean (Standard Deviations) Amplitudes in Millivolts of the Motor-Evoked Potential (MEP) by the Stimulation Used for Stimulus Response Curves

TMS Paradigm			Amplitude (mV)			Interaction		Between Subjects		Within Subjects	
TMS	Stimulation		Healthy	Asymptomatic	Symptomatic	Stat	P	Stat	P	Stat	P
Rest stimulus response curve	100% RMT		0.17 (0.31)	0.15 (0.15)	0.15 (0.14)	$F_{(2.96, 68.99)}=0.24$	0.867	$F_{(2, 47)}=0.09$	0.911	$F_{(1.47, 68.99)}=71.97$	<0.001
	110% RMT		0.47 (0.90)	0.47 (0.68)	0.40 (0.31)						
	120% RMT		1.19 (1.52)	1.06 (1.00)	0.98 (0.67)						
	130% RMT		1.85 (1.94)	1.57 (1.22)	1.77 (1.24)						
	140% RMT		2.30 (2.22)	2.12 (1.50)	2.36 (1.44)						
	150% RMT		2.82 (2.37)	2.40 (1.27)	2.67 (1.43)						
Active stimulus response curve	100% AMT		1.26 (0.68)	2.03 (1.23)	1.72 (0.51)	$F_{(3.31, 82.74)}=0.32$	0.833	$F_{(2, 50)}=2.22$	0.120	$F_{(1.66, 82.74)}=143.01$	<0.001
	110% AMT		1.84 (1.03)	2.58 (1.91)	2.12 (0.87)						
	120% AMT		1.64 (1.29)	3.77 (1.98)	3.43 (1.13)						
	130% AMT		3.94 (1.90)	5.05 (2.27)	4.63 (1.62)						
	140% AMT		4.88 (2.82)	5.99 (2.25)	5.79 (1.95)						
	150% AMT		5.83 (2.67)	7.05 (2.19)	6.75 (2.01)						

Statistical comparisons from mixed models analysis of variance are shown by interaction, between subjects and within subjects.
RMT, rest motor threshold; AMT, active motor threshold, performed with 20% maximal voluntary contraction.

TABLE 17.7 Mean (Standard Deviations) Silent Periods in Milliseconds During 20% Maximal Voluntary Contraction by the Stimulation Used

TMS Paradigm		Silent Period (ms)			Interaction		Between Subjects		Within Subjects	
TMS	Stimulation	Healthy	Asymptomatic	Symptomatic	Stat	P	Stat	P	Stat	P
Active stimulus response curve silent period	100% AMT	16.46 (5.66)	16.36 (43.41)	8.84 (30.68)	$F_{(2.87, 68.84)}$ $=0.63$	0.594	$F_{(2, 48)}$ $=0.56$	0.578	$F_{(1.43, 68.84)}$ $=101.66$	<0.001
	110% AMT	23.21 (12.40)	25.41 (46.88)	16.41 (34.68)						
	120% AMT	37.89 (23.89)	42.21 (60.15)	31.09 (47.28)						
	130% AMT	59.84 (39.91)	55.75 (60.84)	50.95 (60.51)						
	140% AMT	78.80 (47.16)	72.74 (62.95)	60.42 (64.62)						
	150% AMT	99.56 (51.82)	94.07 (68.45)	71.81 (67.32)						

Statistical comparisons from mixed models analysis of variance are shown by interaction, between subjects, and within subjects.
AMT, active motor threshold.

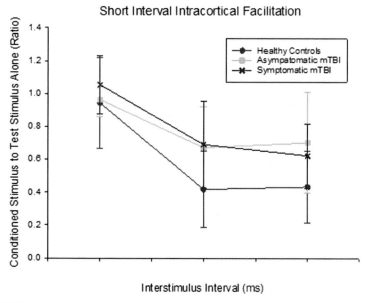

FIGURE 17.3 Short-interval intracortical facilitation conditioned stimulus to test stimulus ratios for each group. The lines represent different groups, while each point on the x-axis represents the interstimulus interval at which the stimuli were given. All data are shown as the ratio of the amplitudes of conditioned stimuli to the test stimulus alone (unconditioned). The red line indicates 1, or the dividing point between inhibition (<1) and facilitation (>1).

active SICI/ICF paradigm, inhibition and facilitatory effects of these paired-pulse paradigms is negated[46] (see Table 17.6). All SICI and ICF paradigms are shown in Fig. 17.4. LICI differed significantly ($\chi^2 = 7.62$, p = 0.022) in that symptomatic participants had a decreased inhibitory response ($P = 0.029$) (Fig. 17.5).

These preliminary data show that it is feasible and safe to examine cortical excitability in children following mild TBI. Although most measures did not differ in mTBI at 1 month post injury, symptomatic children were found to have a decreased long-interval intracortical inhibitory response (ie, decreased LICI). This finding is consistent with chronic findings in the literature.[56,63] These findings may suggest that GABAb receptor-mediated inhibition is decreased in symptomatic participants and that this returns to normal as symptoms decrease. Larger sample sizes are required to confirm this finding given the amount of between subject variability of TMS measures.

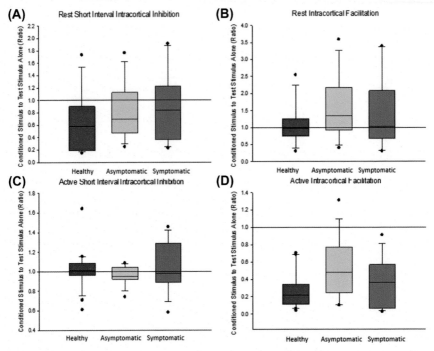

FIGURE 17.4 Intracortical facilitation and short-interval intracortical inhibition during rest and active states. The resting short-interval intracortical inhibition (SICI; A) and resting intracortical facilitation (ICF; B) are on the top row, while the same measures in the active state are on the bottom row (C&D). The red line on each graph indicates 1, or the dividing point between inhibition (<1) and facilitation (>1).

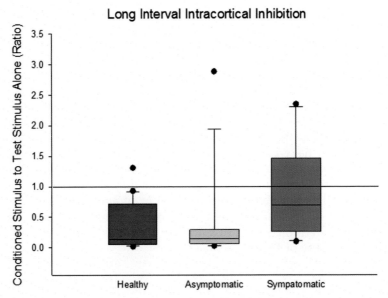

FIGURE 17.5 Long-interval intracortical inhibition ratios for each group. All data are shown as the ratio of the amplitudes of conditioned stimuli to the test stimulus alone (unconditioned). The red line on each graph indicates 1, or the dividing point between inhibition (<1) and facilitation (>1).

SUMMARY

The study of cortical excitability in TBI is uniquely challenging in that it is a highly heterogeneous condition that is lacking in accurate quantification of its severity. TMS is also highly variable between individuals in both adults and children. The current literature is insufficient to draw conclusions about what changes in cortical excitability occur after TBI and their neurophysiological drivers. Although most studies suggest an increase in cortical inhibition, the studies are small, populations are heterogeneous, and TMS methodologies used vary considerably. Larger studies are required, especially in the setting of the developing brain.[74]

We report our findings in a well-controlled group of children with and without persistent symptoms 1 month following an mTBI. Our provisional results suggest that PCS symptoms are associated with a lack of cortical inhibition, although the sample is small and preinjury values are unknown. The study of young athletic populations may offer a way to study changes in cortical excitability over time due to the concussion. However, these studies should control factors that change over time including age, pubertal development, exercise training, and the use of prescribed and non-prescribed medications. Future studies should be carefully controlled for severity of the injury (using sensitive measures), multiple injuries, functional outcome, cognition, and the presence or absence of medical conditions that may alter cortical excitability (eg, attention deficits[98]; migraine[99,100]).

In summary, TBI may exhibit various changes in cortical excitability, even years after the injury, in sports or the general population, potentially showing a shift toward inhibition. Changes in cortical excitability may help to better explain the persistence of symptoms after TBI and poorer outcomes associated with pediatric TBI.[18,101]

References

1. Prins M, Greco T, Alexander D, Giza CC. The pathophysiology of traumatic brain injury at a glance. *Dis Model Mech.* 2013;6:1307–1315.
2. Ryu WHA, Feinstein A, Colantonio A, Streiner DL, Dawson DR. Early identification and incidence of mild TBI in Ontario. *Can J Neurol Sci.* 2009;36:429–435.
3. Langlois JA, Rutland-Brown W, Wald MM. The epidemiology and impact of traumatic brain injury: a brief overview. *J Head Trauma Rehabil.* 2006;21:375–378.
4. Macpherson A, Fridman L, Scolnik M, Corallo A, Guttmann A. A population-based study of paediatric emergency department and office visits for concussions from 2003 to 2010. *Paediatr Child Heal.* 2014;19:543–546.
5. McKinlay A, et al. Prevalence of traumatic brain injury among children, adolescents and young adults: prospective evidence from a birth cohort. *Brain Inj.* 2008;22:175–181.
6. Sosin DM, Sniezek JE, Thurman DJ. Incidence of mild and moderate brain injury in the United States, 1991. *Brain Inj.* 1996;10:47–54.
7. Faul M, Xu L, Wald MM, Coronado VG, Ward MM. Traumatic Brain Injury in the United States. *US Dep Heal Hum Serv.* 2010;74.

8. McKinlay A, Grace RC, Horwood LJ, Fergusson DM, MacFarlane MR. Long-term behavioural outcomes of pre-school mild traumatic brain injury. *Child Care Health Dev.* 2010;36:22–30.

9. Bruns J, Hauser WA. The epidemiology of traumatic brain injury: a review. *Epilepsia.* 2003;44(suppl 1):2–10.

10. Maroon JC, et al. Chronic traumatic encephalopathy in contact sports: a systematic review of all reported pathological cases. *PLoS One.* 2015;10:e0117338.

11. Gardner RC, Yaffe K. Epidemiology of mild traumatic brain injury and neurodegenerative disease. *Mol Cell Neurosci.* 2015;66:75–80.

12. Keith Owen Y, Taylor HG. Neurobehavioural outcomes of mild head injury in children and adolescents. *Dev Neurorehabil.* 2005;8:5–16.

13. Bigler ED. Neuropsychology and clinical neuroscience of persistent post-concussive syndrome introduction: brief history. *J Int Neuropsychol Soc.* 2008;14:1–22.

14. Kay T, et al. Definition of mild traumatic brain injury. *J Head Trauma Rehabil.* 1993;8:86–87.

15. Malec JF, et al. The mayo classification system for traumatic brain injury severity. *J Neurotrauma.* 2007;24:1417–1424.

16. Pearl Chung FK. Traumatic brain injury (TBI): overview of diagnosis and treatment. *J Neurol Neurophysiol.* 2013;05:1–10.

17. Catroppa C, Godfrey C, Rosenfeld J, Hearps S, Anderson V. Functional recovery ten years after pediatric traumatic brain injury: outcomes and predictors. *J Neurotrauma.* 2012;29:2539–2547.

18. Moran LM, et al. Quality of life in pediatric mild traumatic brain injury and its relationship to postconcussive symptoms. *J Pediatr Psychol.* 2012;37:736–744.

19. O'Flaherty SJ, et al. The Westmead Pediatric TBI Multidisciplinary Outcome Study: use of functional outcomes data to determine resource prioritization. *Arch Phys Med Rehabil.* 2000;81:723–729.

20. Yeates K, Taylor H, Walz N, Stancin T, Wade S. The family environment as a moderator of psychosocial outcomes following traumatic brain injury in young children. *Neuropsychology.* 2010;24:345–356.

21. Babikian T, et al. The UCLA longitudinal study of neurocognitive outcomes following mild pediatric traumatic brain injury. *J Int Neuropsychol Soc.* 2011;17:886–895.

22. Barlow KM. Postconcussion syndrome: a review. *J Child Neurol.* 2014. http://dx.doi.org/10.1177/0883073814543305.

23. Eisenberg MA, Andrea J, Meehan W, Mannix R. Time interval between concussions and symptom duration. *Pediatrics.* 2013;132:8–17.

24. Barlow KM, et al. Epidemiology of postconcussion syndrome in pediatric mild traumatic brain injury. *Pediatrics.* 2010;126:e374–e381.

25. Barlow KM, Crawford S, Brooks BL, Turley B, Mikrogianakis A. The incidence of Post Concussion Syndrome remains stable following mild Traumatic Brain Injury in children. *Pediatr Neurol.* 2015. http://dx.doi.org/10.1016/j.pediatrneurol.2015.04.011.

26. Giza CC, Hovda DA. The neurometabolic cascade of concussion. *J Athl Train.* 2001; 36:228–235.

27. Barkhoudarian G, Hovda DA, Giza CC. Concussion in sports the molecular pathophysiology of concussive brain injury. *Clin Sports Med.* 2011;30:33–48.

28. Shultz SR, Bao F, Weaver LC, Cain DP, Brown A. Treatment with an anti-CD11d integrin antibody reduces neuroinflammation and improves outcome in a rat model of repeated concussion. *J Neuroinflammation.* 2013;10:26.

29. Ramlackhansingh AF, et al. Inflammation after trauma: microglial activation and traumatic brain injury. *Ann Neurol.* 2011;70:374–383.

30. Ziebell JM, Morganti-Kossmann MC. Involvement of pro- and anti-inflammatory cytokines and chemokines in the pathophysiology of traumatic brain injury. *Neurotherapeutics.* 2010;7:22–30.

31. Johnson VE, Stewart W, Smith DH. Axonal pathology in traumatic brain injury. *Exp Neurol.* 2013;246:35–43.
32. Korn A, Golan H, Melamed I, Pascual-Marqui R, Friedman A. Focal cortical dysfunction and Blood-Brain barrier disruption in patients with postconcussion syndrome. *J Clin Neurophysiol.* 2005;22:1–9.
33. Mayer CL, Huber BR, Peskind E. Traumatic brain injury, neuroinflammation, and post-traumatic headaches. *Headache.* 2013;53:1523–1530.
34. Menon DK, Schwab K, Wright DW, Maas AI. Position statement: definition of traumatic brain injury. *Arch Phys Med Rehabil.* 2010;91:1637–1640.
35. Giza CC, Mink RB, Madikians A. Pediatric traumatic brain injury: not just little adults. *Curr Opin Crit Care.* 2007;13:143–152.
36. Giza CC, Santa Maria NS, Hovda DA. N-Methyl-D-Aspartate receptor subunit changes after traumatic injury to the developing brain. *J Neurotrauma.* 2008;23:950–961.
37. Cantu D, et al. Traumatic brain injury alters cortical glutamate network function by compromising gabaergic inhibition. *Epilepsy Curr.* 2014;14:472.
38. Huusko N, Pitkänen A. Parvalbumin immunoreactivity and expression of GABAA receptor subunits in the thalamus after experimental TBI. *Neuroscience.* 2014;267:30–45.
39. Raible DJ, Frey LC, Cruz Del Angel Y, Russek SJ, Brooks-Kayal AR. GABA(A) receptor regulation after experimental traumatic brain injury. *J Neurotrauma.* 2012;29(16):2548–2554.
40. Giza CC, Prins ML. Is being plastic fantastic? Mechanisms of altered plasticity after developmental traumatic brain injury. *Dev Neurosci.* 2006;28:364–379.
41. Wieloch T, Nikolich K. Mechanisms of neural plasticity following brain injury. *Curr Opin Neurobiol.* 2006;16:258–264.
42. Obrenovitch TP, Urenjak J, Zilkha E. Effects of increased extracellular glutamate levels on the local® eld potential in the brain of anaesthetized rats. *Br J Pharmacol.* 1997;122:372–378.
43. Davies C, Starkey S, Pozza M, Collingridge G. GABAb autoreceptors regulate the induction of LTP. *Nature.* 1991;349:609611.
44. Reis J, et al. Contribution of transcranial magnetic stimulation to the understanding of cortical mechanisms involved in motor control. *J Physiol.* 2008;586:325–351.
45. Dayan E, Censor N, Buch ER, Sandrini M, Cohen LG. Noninvasive brain stimulation: from physiology to network dynamics and back. *Nat Neurosci.* 2013;16:838–844.
46. Chen R, et al. The clinical diagnostic utility of transcranial magnetic stimulation: report of an IFCN committee. *Clin Neurophysiol.* 2008;119:504–532.
47. Nardone R, et al. The role of the ipsilateral primary motor cortex in movement control after spinal cord injury: a TMS study. *Neurosci Lett.* 2013;552:21–24.
48. Golaszewski SM, et al. Modulation of motor cortex excitability by different levels of whole-hand afferent electrical stimulation. *Clin Neurophysiol.* 2012;123:193–199.
49. Tallus J, et al. Transcranial magnetic stimulation-electroencephalography responses in recovered and symptomatic mild traumatic brain injury. *J Neurotrauma.* 2013;30:1270–1277.
50. Livingston SC, et al. A preliminary investigation of motor evoked potential abnormalities following sport-related concussion. *Brain Inj.* 2010;24:904–913.
51. Powers KC, Cinelli ME, Kalmar JM. Cortical hypoexcitability persists beyond the symptomatic phase of a concussion. *Brain Inj.* 2014;28:465–471.
52. Miller NR, et al. Acute and longitudinal changes in motor cortex function following mild traumatic brain injury. *Brain Inj.* 2014;28:1270–1276.
53. De Beaumont L, Lassonde M, Leclerc S, Théoret H. Long-term and cumulative effects of sports concussion on motor cortex inhibition. *Neurosurgery.* 2007;61:329–336. discussion 336–7.

54. Livingston SC, et al. Motor-evoked potential amplitudes are correlated with prior number of concussions in the acutely concussed collegiate athlete. *J Head Trauma Rehabil.* 2009;24:400–401.

55. Livingston SC. Electrophysiologic evidence for the effects of acute cerebral concussion in a collegiatewomen's soccer player. *J Head Trauma Rehabil.* 2012;27:E12–E13.

56. Pearce AJ, et al. The long-term effects of sports concussion on retired Australian football players: a study using transcranial magnetic stimulation. *J Neurotrauma.* 2014;31:1139–1145.

57. De Beaumont L, Tremblay S, Poirier J, Lassonde M, Théoret H. Altered bidirectional plasticity and reduced implicit motor learning in concussed athletes. *Cereb Cortex.* 2012;22:112–121.

58. Carroll TJ, Riek S, Carson RG. Reliability of the input – output properties of the corticospinal pathway obtained from transcranial magnetic and electrical stimulation. *J Neurosci Methods.* 2001;112:193–202.

59. Chistyakov AV, et al. Excitatory and inhibitory corticospinal responses to transcranial magnetic stimulation in patients with minor to moderate head injury. *J Neurol Neurosurg Psychiatry.* 2001;70:580–587.

60. Chistyakov AV, Soustiel JF, Hafner H, Elron M, Feinsod M. Altered excitability of the motor cortex after minor head injury revealed by transcranial magnetic stimulation. *Acta Neurochir (Wien).* 1998;140:467–472.

61. Werhahn KJ, Kunesch E, Noachtar S, Benecke R, Classen J. Differential effects on motorcortical inhibition induced by blockade of GABA uptake in humans. *J Physiol.* 1999;517:591–597.

62. Siebner HR, Dressnandt J, Auer C, Conrad B. Continuous intrathecal baclofen infusions induced a marked increase of the transcranially evoked silent period in a patient with generalized dystonia. *Muscle Nerve.* 1998;21:1209–1212.

63. De Beaumont L, et al. Persistent motor system abnormalities in formerly concussed athletes. *J Athl Train.* 2011;46:234–240.

64. Tremblay SS, et al. Multimodal assessment of primary motor cortex integrity following sport concussion in asymptomatic athletes. *Clin Neurophysiol.* 2014;125:1371–1379.

65. Tremblay S, de Beaumont L, Lassonde M, Théoret H, Theoret H. Evidence for the specificity of intracortical inhibitory dysfunction in asymptomatic concussed athletes. *J Neurotrauma.* 2011;28:493–502.

66. Nardone R, et al. Cortical excitability changes in patients with sleep-wake disturbances after traumatic brain injury. *J Neurotrauma.* 2011;28:1165–1171.

67. Kawasaki Y, et al. Short latency afferent inhibition associated with cortical compression and memory impairment in patients with chronic subdural hematoma. *Clin Neurol Neurosurg.* 2012;114:976–980.

68. Tremblay S, De Beaumont L, Lassonde M, Theoret H. Specificity of neurophysiologic dysfunctions in asymptomatic concussed athletes. *Clin Neurophysiol.* 2011;122: S185.

69. McDonnell MN, Orekhov Y, Ziemann U. The role of GABA(B) receptors in intracortical inhibition in the human motor cortex. *Exp Brain Res.* 2006;173:86–93.

70. Groppa S, et al. A practical guide to diagnostic transcranial magnetic stimulation: report of an IFCN committee. *Clin Neurophysiol.* 2012;123:858–882.

71. Nordström A, Edin BB, Linstrom S, Nordstrom P. Cognitive function and other risk factors for mild traumatic brain injury in young men : nationwide cohort. *Br Med J.* 2013;346:1–9.

72. Davis Ga, Iverson GL, Guskiewicz KM, Ptito a, Johnston KM. Contributions of neuroimaging, balance testing, electrophysiology and blood markers to the assessment of sport-related concussion. *Br J Sports Med.* 2009;43(suppl 1):i36–i45.

73. Gagnon I, Swaine B, Friedman D, Forget R. Visuomotor response time in children with a mild traumatic brain injury. *J Head Trauma Rehabil.* 2004;19:391–404.

74. Livingston SC, et al. Differential rates of recovery after acute sport-related concussion: electrophysiologic, symptomatic, and neurocognitive indices. *J Clin Neurophysiol.* 2012;29:23–32.

75. Fujiki M, et al. Navigated brain stimulation for preoperative anatomic and functional identification of impaired motor cortex in a patient with meningioma. *Neurosurg Q.* 2007;17:33–39.

76. Jang SH, et al. Motor recovery mechanism of diffuse axonal injury: a combined study of transcranial magnetic stimulation and functional MRI. *Restor Neurol Neurosci.* 2005;23: 51–56.

77. Bagnato S, et al. Patients in a vegetative state following traumatic brain injury display a reduced intracortical modulation. *Clin Neurophysiol.* 2012;123:1937–1941.

78. Takeuchi N, Ikoma K, Chuma T, Matsuo Y. Measurement of transcallosal inhibition in traumatic brain injury by transcranial magnetic stimulation. *Brain Inj.* 2006;20: 991–996.

79. Bernabeu M, et al. Abnormal corticospinal excitability in traumatic diffuse axonal brain injury. *J Neurotrauma.* 2009;26:2185–2193.

80. Mazzini L, et al. Somatosensory and motor evoked potentials at different stages of recovery from severe traumatic brain injury. *Arch Phys Med Rehabil.* 1999;80:33–39.

81. Pangilinan PH, Giacoletti-Argento A, Shellhaas R, Hurvitz Ea, Hornyak JE. Neuropharmacology in pediatric brain injury: a review. *PM R.* 2010;2:1127–1140.

82. Dennis EL, et al. White matter disruption in moderate/severe pediatric traumatic brain injury: advanced tract-based analyses. *Neuroimage (Amst).* 2015;7:493–505.

83. Dennis EL, et al. Callosal function in pediatric traumatic brain injury linked to disrupted white matter integrity. *J Neurosci.* 2015;35:10202–10211.

84. Moen KG, et al. Traumatic axonal injury: the prognostic value of lesion load in corpus callosum, brain stem, and thalamus in different magnetic resonance imaging sequences. *J Neurotrauma.* 2014;11:1–11.

85. Cavinato M, Iaia V, Piccione F. Repeated sessions of sub-threshold 20-Hz rTMS. Potential cumulative effects in a brain-injured patient. *Clin Neurophysiol.* 2012;123: 1893–1895.

86. Koski L, et al. Noninvasive brain stimulation for persistent postconcussion symptoms in mild traumatic brain injury. *J Neurotrauma.* 2014;44:38–44.

87. Kedzior KK, Reitz SK. Short-term efficacy of repetitive transcranial magnetic stimulation (rTMS) in depression- reanalysis of data from meta-analyses up to 2010. *BMC Psychol.* 2014;2:1–19.

88. Baeken C, et al. The impact of accelerated HF-rTMS on the subgenual anterior cingulate cortex in refractory unipolar major depression: insights from 18FDG PET brain imaging. *Brain Stimul.* 2015;8:808–815.

89. Louise-Bender Pape T, Rosenow J, Lewis G. Transcranial magnetic stimulation: a possible treatment for TBI. *J. Head Trauma Rehabil.* 2006;21(5):437–451. http://dx.doi.org/ovidsp. ovid.com/ovidweb.cgi?T=JS&PAGE=reference&D=med5&NEWS=N&AN=16983227.

90. University of Manitoba. Treatment and recovery monitoring of post TBI symptoms. *Clin.* [Internet]. *Bethesda Natl. Libr. Med. (US).* 2000–2016. Available from http://dx.doi.org/ clinicaltrials.gov/ct2/show/NCT02426749 NLM Identifier NCT02426749.

91. Manganotti P, et al. Effect of high-frequency repetitive transcranial magnetic stimulation on brain excitability in severely brain-injured patients in minimally conscious or vegetative state. *Brain Stimul.* 2013;6:913–921.

92. Kim L, Chun MH, Kim BR, Lee SJ. The effect of repetitive transcranial magnetic stimulation on brain injured patients with dysphagia. *Neurorehabil Neural Repair.* 2011;35:407.

93. Kujirai T, et al. Corticocortical inhibition in human motor cortex. *J Physiol.* 1993;471: 501–519.

94. Ni Z, Bahl N, Gunraj CA, Mazzella F, Chen R. Increased motor cortical facilitation and decreased inhibition in Parkinson disease. *Neurology.* 2013;80:1746–1753.

95. Seeger TA, Kirton CA, Barlow KM. Cortical excitability changes after paediatric mild traumatic brain injury: preliminary data. In: *Alberta Child. Hosptial Res. Institute/Can. Investig. Heal. Res. Grad. Student Symp. Calgary, Canada. April 15–16, 2015*; 2015.
96. Garvey M, Ziemann U, Becker D, Barker C, Bartko J. New graphical method to measure silent periods evoked by transcranial magnetic stimulation. *Clin Neurophysiol.* 2001;112:1451–1460.
97. Garvey MA, et al. Cortical correlates of neuromotor development in healthy children. *Clin Neurophysiol.* 2003;114:1662–1670.
98. Gilbert DL, Isaacs KM, Augusta M, Macneil LK, Mostofsky SH. Motor cortex inhibition: a marker of ADHD behavior and motor development in children. *Neurology.* 2011;76:615–621.
99. Brigo F, Storti M, Tezzon F, Manganotti P, Nardone R. Primary visual cortex excitability in migraine: a systematic review with meta-analysis. *Neurol Sci.* 2013;34:819–830.
100. Brigo F, et al. Transcranial magnetic stimulation of visual cortex in migraine patients: a systematic review with meta-analysis. *J Headache Pain.* 2012;13:339–349.
101. Yeates KO, et al. Longitudinal trajectories of postconcussive symptoms in children with mild traumatic brain injuries and their relationship to acute clinical status. *Pediatrics.* 2009;123:735–743.

Brain Stimulation Applications in Pediatric Headache and Pain Disorders

T. Rajapakse

Alberta Children's Hospital, Calgary, AB, Canada

A. Kirton

University of Calgary, Calgary, AB, Canada

OUTLINE

HISTORY OF NEUROSTIMULATION IN HEADACHE AND PAIN

Scribonius Largus was the court physician to Roman emperor Claudius. In 46 AD his compendium of medical treatments known as *Compositiones* described a novel treatment for headache. In this medical text, he prescribed that "to immediately remove and permanently cure a headache, however long-lasting and intolerable, a live black torpedo is put on the place which is in pain, until the pain ceases and the part grows numb." The black torpedo referred to was a bioelectric fish. This finding largely fell into the background of headache treatment until the 19th century[1] when, with electrical current production discovered from mechanical sources, individual generators and transcutaneous current were used to treat various forms of pain. In the 20th century, the availability of batteries allowed forms of neurostimulation to become portable and invasive, and modalities such as the vagal nerve stimulator were developed. The first use of vagus nerve stimulation (VNS) for primary headache disorders began in 1999 following observational data and retrospective analysis suggesting possible clinical effectiveness.[1] This was preceded by deep brain stimulation (DBS) with temporary electrodes for pain syndromes in the 1950s and occipital nerve stimulation (ONS) for head pain in the 1970s.[1]

In the pediatric population, while transcranial magnetic stimulation (TMS) is reported as a method used in the study of pediatric pain and headache neurophysiology,[2] there are currently no reports of invasive (DBS, ONS) or non-invasive brain stimulation (repetitive transcranial magnetic stimulation (rTMS), transcranial direct-current stimulation (tDCS)) methods being used as treatment for pediatric headache or pain.

PEDIATRIC HEADACHE

The World Health Organization's 2010 Global Burden of Disease study ranked migraine as the eighth leading cause of years lived with disability.[3] By the age of 15 years, >75% of children have experienced a significant headache,[4] and the impact of migraine on affected children and adolescents, their families, and society is considerable. The US National Health Interview Survey determined that 975,000 children had experienced a migraine within a 2-week period resulting in 164,454 missed school days.[5]

Pediatric migraine has specific proposed criteria as designated by the International Headache Society,[6] with other headache disorders in children defined via extrapolation from current adult headache definitions. Table 18.1 discusses current definitions, epidemiology, diagnostic criteria, and treatment options available to children and adolescents with headache disorders.

Although all of the above headache disorders cause pain and impairment in functioning, chronic migraine is especially disabling for children and their families and has deleterious effects on quality of life including school, extracurricular activities, home life, and work.[13]

Unfortunately, there is little high-quality research evidence to guide prophylactic treatment choices for chronic migraine in children and adolescents. In addition, currently used migraine preventative therapies consist largely of oral pharmacologic treatments that are not always effective and burdened with side effects and low adherence rates in patients.[14]

This has led to a search for a better understanding of the neurobiology of child and adolescent migraine to inform the development of safe, effective, and well-tolerated treatments for childhood headache. Central and peripheral nervous system neurostimulation, both non-invasive and invasive, promises to become an exciting new avenue in the understanding of the pathophysiology and treatment of childhood headache and pain disorders.

SAFETY AND TOLERABILITY OF NEUROSTIMULATION IN CHILDREN

It is now well established that non-invasive brain stimulation techniques, such as TMS, are safe in children.[15,16] While still early in its development for use in children, tDCS also promises to have a similar safety profile to TMS. Unique issues are being identified as researchers develop rational approaches to dose customization in children such as computer modelling studies.[17]

More invasive brain stimulation devices, such as DBS and ONS, have yet to be used in children and adolescents for the treatment of refractory headache or chronic pain. While DBS has FDA approval for the treatment of drug-refractory primary dystonia in children over the age of 7 years,[18] it has not been reported in the literature as used in anyone under the age of 18 years for the treatment of a refractory pain disorder.

TABLE 18.1 Current Headache Definitions, Epidemiology, Diagnostic Criteria, and Treatment Options Available to Children and Adolescents

Headache Disorder	Prevalence	International Headache Society Diagnostic Criteria[6]	Treatment Options

Migraine	7.7%[7] Global	*Modified Pediatric criteria*
		A. At least five attacks fulfilling criteria B–D
		B. Headache attacks lasting 1–72 h
		C. Headache has at least two of the following characteristics:
		  1. Bilateral or unilateral location
		  2. Pulsating quality
		  3. Moderate or severe pain intensity
		  A. Aggravation by or causing avoidance of routine physical activity
		  B. During headache at least one of the following:
		    1. Nausea and/or vomiting
		    2. Photophobia and phonophobia
		Not attributed to another disorder. Episodic: <15 days per month
		Chronic: >15 days per month
			Acute: Non-steroidal anti-inflammatory (NSAID) triptan
			Prophylactic:
			• Lifestyle modifications (diet, exercise, sleep)
			• Behavioral therapies (biofeedback, cognitive behavioral therapy (CBT)
			• Tricyclic antidepressants (amitriptyline/nortriptyline)
			• Antiepileptics (topiramate, valproic acid, gabapentin)
			• B-blockers (propranolol)
			• Calcium channel blockers (flunarizine)
			• Nutraceutical: Riboflavin, magnesium, coenzyme Q10
			• Interventional: Botulinum toxin injections, occipital nerve block

Tension-type headache[8]	9.8%[9] (strict IHS criteria) *Sweden*
	23%[9] (modified IHS criteria) *Sweden*
		A. At least 10 episodes fulfilling criteria B–D
		B. Headache lasting from 30 min to 7 days
		C. Headache has at least two of the following characteristics:
		  1. Bilateral location
		  2. Pressing/tightening (non-pulsating) quality
		  3. Mild or moderate intensity
		  4. Not aggravated by routine physical activity such as walking or climbing stairs
		D. Both of the following:
		  1. No nausea or vomiting (anorexia may occur)
		  2. No more than one of photophobia or phonophobia
		Not attributed to another disorder
		Infrequent: <1 day per month
		Frequent: 1–5 days Per month
		Chronic: >15 days per month
		Modified IHS:
		Removal of (A) number of episodes and (B) duration criteria
			Lifestyle modifications (see above)
			Behavioral therapy (biofeedback/CBT)
			Melatonin
			Amitriptyline
			Valproic acid, topiramate, gabapentin

Chronic migraine with medication overuse	1.75%[10] *USA*	**A.** Headache occurring on _15days per month in a patient with a pre-existing headache disorder (chronic migraine) **B.** Regular overuse for >3months of one or more drugs that can be taken for acute and/or symptomatic treatment of headache **C.** Not better accounted for by another ICHD-3 diagnosis Acetaminophen, ASA, other NSAID Use >15days per month for >3months Triptans, ergotamine, opioids, combination analgesics: Use >10days per month for >3months	Discontinuation of overuse medication Long-acting triptan during detoxification period Topamax
New daily persistent headache (NDPH)[11]	Pediatric: Unknown, rare Adult 0.3%[12]	**A.** Headache that, within 3 days of onset, fulfills criteria B–D **B.** Headache is present daily, and is unremitting, for >3months **C.** At least two of the following pain characteristics: **1.** Bilateral location **2.** Pressing/tightening (non-pulsating) quality **3.** Mild or moderate intensity **4.** Not aggravated by routine physical activity such as walking or climbing **D.** Both of the following: **1.** No more than one of photophobia, phonophobia, or mild nausea **2.** Neither moderate or severe nausea nor vomiting Not attributed to another disorder	Lifestyle modification (sleep, diet, exercise) Behavioral therapy (biofeedback) If obese; weight loss Antiepileptics (topiramate, valproic acid, gabapentin) Tricyclic antidepressants (amitriptyline)

Generally, repetitive TMS (rTMS) is well tolerated in children; however some subjects in a study at our institution have reported a brief, mild, self-resolving headache located near the scalp region of stimulation rTMS. The study is an open-label trial of rTMS for the treatment of medically refractory depression of adolescents. Stimulation parameters include 10-Hz rTMS stimulation of the left dorsolateral prefrontal cortex for a total of 3000 pulses per day over 15 days (3 weeks). An interim safety and tolerability analysis showed that while during week 1, 25% of participants reported a mild, self-limited headache over the site of stimulation that did not require treatment, during weeks 2–3 this headache was not reported. No subject withdrew from the study due to headache. These preliminary results suggest pediatric subjects are able to habituate to the mild scalp discomfort of rTMS over time[19] and children also appear to tolerate a single session of theta-burst TMS,[20] despite mild headache in some.

Cutaneous allodynia or pain resulting from application of a non-noxious stimulus to normal skin is a known complication of chronic pain and migraine[21] and can make subjects exquisitely sensitive to sensory stimuli perceived as neutral by subjects without chronic head pain. Thus, while the initial mild scalp discomfort of rTMS appears to be habituate over time, care should be taken to ensure brain stimulation techniques to treat headache do not worsen existing head pain.

PATHOPHYSIOLOGY OF HEAD PAIN AND MIGRAINE

The creation of head pain is a complex process with neuroanatomic relays between peripheral afferent nerves of the head, neck and brainstem, subcortical relay centers, and higher order processing cortices of the brain[1] with a detailed review found here.[22] The trigeminocervical complex describes the interplay of head sensation existing between the trigeminal nerves supplying sensation to the anterior head and upper cervical nerves supplying sensation to the posterior head. While the brain itself is largely insensate, trigeminovascular afferents can create what is perceived as headache in humans through stimulation of blood vessels in the dura mater. The trigeminovascular system consists of the dura mater and large intracranial vessel innervated by the ophthalmic division of the trigeminal nerve and the trigeminal nerve's afferent projection to the central trigeminal nucleus caudalis (TNC). This critical relay center transmits nociceptive inputs from the cranial vasculature to the brainstem and higher-order processing centers. Associated with activation of this system are neurovascular changes involving multiple vasoactive peptides such as calcitonin gene-related peptide (CGRP), substance P, vasoactive intestinal peptide (VIP), neurokinin A, and nitric oxide synthase (NOS). Extensive basic science work to explain this complex system is summarized in a review by Akerman et al.[22]

Animal evidence supports functional links between the trigeminocervical complex of the brainstem and peripheral pathways including the greater occipital nerves. In cats, Goadsby et al. found that stimulation of the greater occipital nerve has been shown to increase metabolic activity in the trigeminal nucleus caudalis and cervical dorsal horn. This important finding suggested that the well-recognized clinical phenomenon of pain at the front and back of the head and in the upper neck was likely a consequence of overlap of processing of nociceptive information at the level of second-order neurons.[23]

Stimulation of vascular structures including veins (superior sagittal sinus)[24] and arteries (middle meningeal artery)[25] also demonstrated increased metabolic activity in the trigeminocervical nucleus. These seminal studies were able to demonstrate peripheral and central anatomical and functional connections and identified attractive peripheral targets for potential alteration of central functioning.[1] Unfortunately, animal models exploring the neurobiology of headache neurostimulation have been limited.

Non-Invasive Brain Stimulation to Understand Headache and Pain Neurophysiology

With the clinical observation that many migraine patients with aura describe visual symptoms, the occipital cortex has long been a focus of neurophysiology research. TMS can study occipital cortical excitability in migraine using phosphine thresholds. When single-pulse TMS is applied to the primary visual cortex (V1), subjects report perceiving an artificial flash-like visual percept, or "phosphene." A paired-pulse paradigm, where an additional, equipotent conditioning stimulus is added at an interstimulus interval of 50 ms, may enhance the effect.[26] This allows for more efficient provocation of phosphenes with less stimulus intensity and therefore, reduced scalp discomfort.

TMS-induced phosphenes can be used to explore the neural dynamics underlying visual perception and they show several unique properties.[27] Subjects are typically seated and blindfolded in a dark room. A TMS coil is placed anatomically over the dominant primary visual cortex with landmarks measured 2 cm rostrally and 2 cm laterally to the right of the inion/occipital protuberance.[28] Phosphenes typically increase in frequency and intensity with increased stimulation intensity, allowing measurement of an individualized "phosphene threshold" for each subject. The percentage of maximal stimulator output that achieves perceptible phosphenes with >50% of stimulation trials defines the phosphene threshold (PT). Phosphene thresholds can be reduced (i.e. more easily elicited) with a conditioning TMS stimulus to associated areas such as the posterior parietal cortex or the frontal eye fields where TMS can facilitate perception of visual stimuli.[29]

TMS induction of artificial phosphenes and PT is an informative method to assess the excitability of the occipital cortex in subjects with and without migraine. A recent systematic review of adult patients[30] described the prevalence and characteristics of phosphenes and PT values across 15 single-pulse TMS studies of 369 migraine patients and 269 controls. Patients with migraine with aura had statistically lower PT compared to controls when a circular coil was used. This difference, however, was not seen when a figure-of-eight coil was used. Higher phosphene prevalence was seen in subjects with migraine with aura but no significant difference was seen within migraineurs without aura and controls. PT values with figure-of-eight coils did not show a significant difference in those with migraine without aura compared to controls. In general, the authors concluded that the results *slightly* support the hypothesis of primary visual cortex hyperexcitability in migraine with aura but that there is not enough evidence for migraine without aura. Significant methodological heterogeneity continues to limit our understanding of occipital lobe excitability in migraine patients. Until an organized, uniformly followed protocol for measuring phosphenes and PT is established, it will be difficult to understand the true nature of occipital lobe excitability in adult and pediatric patients with migraine. However, brain stimulation methodology may still provide the most essential tool currently missing from migraine research: an objective, easily applied neurophysiological biomarker of disease biology.

PEDIATRIC DATA

The literature on phosphene thresholds and cortical excitability in children with migraine is sparse but holds the same great potential for study. Siniatchkin et al. have studied regional excitability of the occipital (round coil) and motor (figure-of-eight coil) cortices in 10 children with migraine without aura and 10 healthy age-matched controls without migraine.[2] Patients were studied 1–2 days before and after a migraine attack as well as during the inter-migraine interval. Motion after-effect (MAE) was also studied as an index of cortical reactivity to moving visual stimuli. MAE refers to the illusory impression of motion that is seen after viewing moving displays, with its presence thought to be enhanced in migraine and localized to the extrastriate V5/MT region.[31] Motor cortex excitability, similar to that seen in adult headache patients, was not altered in patients and did not change during the migraine cycle. Subjects with migraine had lower phosphene thresholds compared to the healthy participants at each time point, suggesting increased occipital cortex excitability. This increase however, was attenuated 1–2 days before a migraine attack as indicated by a relative increase in PT. The increase in PTs prior to the next migraine attack was associated with a stronger TMS-induced suppression of visual perception and a prolongation of MAE. This was thought to reflect spontaneous fluctuations of occipital cortical excitability linked to the migraine cycle. The neural suppression underlying MAE has been attributed in part to cellular hyperpolarization and in part to a decrease in mutual excitation in the network of cells responsible for responding to the adapting visual display.[32,33] Although there were no significant group differences, the immediate MAE increased significantly before a migraine compared to at intermigraine timepoints. This suggests that changes to the rate of cellular recovery following activity (prolonged hyperpolarization) rather than synaptic changes alone may account for the changes in MAE across the migraine cycle. It was also proposed that cellular recovery rate is slower before a migraine attack because of this phenomenon. Slower cellular recovery was thought to imply a reduction in cortical excitability, which was seen with the corresponding change in PTs and increased in immediate MAE duration before a migraine attack. Overall the authors concluded that their evidence along with that of others[32,33] suggested that PT, the disruption of visual perception and the MAE are caused by activity in different neuronal substrates in the visual cortex. Further, it was considered most likely that premigraine decreases in occipital cortex excitability and increases in occipital inhibition co-occurred with increases in intracortical inhibition in various visually related networks, although a distinct causal relationship still remains unclear.[2]

An ongoing study at our institution has captured some additional, unique pediatric headache brain stimulation data. Using paired-pulse TMS protocols, children with migraine with and without aura are studied for phosphene thresholds pre and post the initiation of common migraine prophylactic agents including topiramate, flunarizine, and riboflavin. Children undergo a standardized phosphene protocol with good tolerability. They are asked to draw their perceived phosphenes using a variety of shapes and colors on a preset graph of their visual field. The subject-illustrated results in Fig. 18.1 are all preprophylaxis and very interesting; despite precisely landmarked stimulation of the non-dominant (right) primary visual cortex with a figure-of-eight coil, three of five children illustrated bilateral phosphene perception in a variety of shapes and colors, suggesting radiation to extrastriate visual cortices (see Fig. 18.1). Probing cortical physiology in occipital and possibly other locations appears to harbor potential to advance the understanding of migraine pathophysiology in children.

THERAPEUTICS

Transcranial Magnetic Stimulation

TMS has not yet been studied as an acute or prophylactic treatment for migraine in the pediatric population. However, numerous approaches have been tested in adult migraine and chronic pain. Low-frequency (1 Hz) rTMS is considered inhibitory, while high frequency (>10 Hz) is

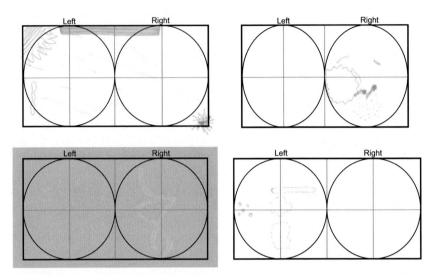

FIGURE 18.1 **Phosphene drawings.** Preprophylaxis subject illustrations of perceived phosphenes after paired-pulse TMS stimulation of non-dominant primary visual cortex. *With permission, from the authors.*

considered excitatory. In a recent systematic review of therapeutic rTMS applications,[34] there was Level A evidence (of definite efficacy) found to support the use of high-frequency rTMS of M1 (primary motor cortex) contralateral to the pain side in neuropathic pain. A possible analgesic effect (Level C) of high-frequency rTMS over M1 contralateral to pain was found in patients with complex regional pain syndrome type I. Low-frequency rTMS of M1 contralateral to pain was thought to be probably ineffective in managing neuropathic pain (Level B evidence). Unfortunately, in the absence of replicated, large controlled studies, no recommendations were made for the application of any rTMS protocol in migraineurs. Due to heterogeneous study protocols and conflicting results, similarly no recommendations were made regarding appropriate stimulation targets or indications for rTMS treatment in patients with fibromyalgia.

Despite the above lack of evidence for use of rTMS in adult migraine, single-pulse TMS (sTMS) was studied in a large, randomized clinical trial of 201 adults with migraine with aura.[35] A portable TMS device (Neuralieve, Sunnyvale, CA, USA) was developed for patient self-administration on demand with an identical sham stimulator also developed by the company. Subjects were randomly allocated either sham stimulation ($n=99$) or sTMS ($n=102$). They were instructed to treat up to three attacks over 3 months with the sTMS device while experiencing aura. Patients applied the machine to the occiput, just below the occipital bone and administered two pulses approximately 30 s apart. In the "treat" mode a single magnetic field pulse of 0.9 T peak is delivered. Patients heard a brief sound when the magnetic pulse (or sham pulse) occurred. The patient then pressed the charge and treat buttons for a second time for the device to deliver the second pulse. The primary outcome was pain-free response 2h after the first attack, and co-primary outcomes were non-inferiority at 2h for nausea, photophobia, and phonophobia. Pain-free response rates after 2h were significantly higher with sTMS (32/82 (39%)) than with sham stimulation (18/82 (22%)), for a therapeutic gain of 17% (95% CI:3–31%; $P=0.0179$). Sustained pain-free response rates significantly favored sTMS at 24 and 48h post-treatment. Non-inferiority was shown for nausea, photophobia, and phonophobia. No device-related serious adverse events were recorded, and incidence and severity of adverse events were similar between sTMS and sham groups. While the authors concluded sTMS could be a promising acute treatment for migraine with aura, the study and use of sTMS in this specific population of migraineurs has its limitations. True sham blinding is difficult to achieve with no scalp stimulation, as a traditional TMS pulse will cause a sensation in the scalp as the magnetic pulse is delivered to the cortex. The therapeutic gain seen in this trial of 17% is less effective than other currently used more cost-effective and easily accessible acute treatments such as triptans. Finally, the focus on migraneurs with aura excludes the majority (70–90%) of migraine patients who do not have aura, leaving the utility of sTMS in this population still unknown.[36] Additional trials of sTMS are likely required before extrapolation into pediatric populations.

The use of rTMS in prophylaxis of chronic migraine has been investigated in adults with mixed results. In one study, 27 migraineurs treated with 1-Hz rTMS with a round coil over the vertex (two trains of 500 monophasic pulses over five consecutive days) was compared to sham stimulation with a figure-of-eight coil. The rTMS treatment was well tolerated but demonstrated no evidence in favor of efficacy for migraine prophylaxis compared to sham.[37] Another study of 11 patients randomized to rTMS ($n=6$) or sham rTMS ($n=5$) demonstrated 20-Hz high-frequency rTMS over the left dorsolateral prefrontal cortex (DLPFC) suggesting significantly reduced migraine frequency and use of abortive treatments compared to sham.[38] These existing trials suffer from multiple limitations including small sample sizes and masking issues. Lack of consensus regarding target brain regions for stimulation, stimulation parameters (including frequency, intensity, coil type, and cumbersome equipment issues) are additional challenges to advancing rTMS treatments for headache (Fig. 18.2).[36]

Transcranial Direct-Current Stimulation

tDCS has not been used in children with headache or chronic pain as a treatment modality. However, use in adults is growing despite some early study design limitations and concerns. tDCS, including both cathodal (so-called "inhibitory") and anodal ("excitatory") applications have

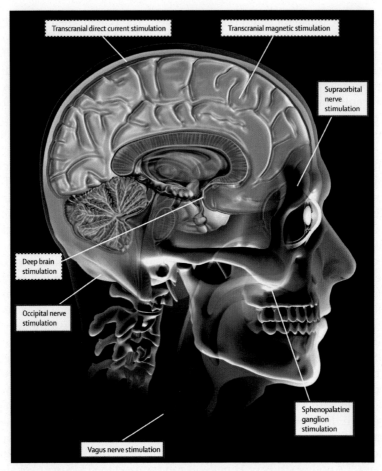

FIGURE 18.2 **Neurostimulation sites in headache treatment.** *Reprinted from Magis D, Schoenen J. Advances and challenges in neurostimulation for headache.* Lancet Neurol. *2012;11(8):708–719. With permission from Elsevier.*

been reported in small pilot studies of adult chronic pain subjects to show some efficacy including traumatic spinal cord injury central pain[39] and fibromyalgia.[40] A small study of 26 adults with chronic migraine equally randomized subjects to 6 weeks of either cathodal or sham stimulation over V1 visual cortex three times per week.[41] Patients treated by cathodal tDCS showed a significant reduction in the duration of attacks, the intensity of pain and the number of migraine-related days post-treatment as compared to the baseline period, but not in the frequency of the attacks. However, compared to the sham group, only the intensity of the pain (and not headache frequency) was significantly reduced post-stimulation. Tolerability was favorable and there were no severe adverse events. The authors concluded that the application of cathodal stimulation over the V1 might be an effective prophylactic therapy in migraine, at least with regard to pain control. The lack of effect on reducing migraine frequency, at least in this small study, currently limits its potential for clinical effectiveness.

A recent meta-analysis of adult studies however reported that current clinical evidence does not support the use of tDCS in the treatment of pain disorders[42] but perhaps with better study design, commonly accepted stimulation parameters and locations for stimulation, this non-invasive, well-tolerated neurostimulation modality may become of greater use in the pediatric pain population.

Peripheral Stimulation

Occipital Nerve Stimulation

Wall and Melzack's "gate control" theory of pain[43] proposed that afferent sensory Aβ fibers block segmental transmission of nociceptive input from Aδ and C pain fibers. Direct stimulation of peripheral nerves causes reduced excitability, increased electrical threshold, and a transient slowing in conduction velocity.[44] The paresthesia felt during peripheral nerve stimulation is mediated by Aβ fibers, and consistent with the gate theory of pain, it is possible the peripheral stimulation is similar to spinal cord stimulation in that these fibers travel in the dorsal column of the spinal cord.[45] Low-frequency stimulation of Aδ fibers in a rat model demonstrated potentiation of long-term depression of monosynaptic and polysynaptic excitatory post-synaptic potentials at the substantia gelatinosa with an effect that was seen to last several hours.[46]

The first percutaneous nerve stimulator for occipital neuralgia was inserted in 1999 by Weiner and Reed in a series of 15 patients. Interestingly, eight of the patients were diagnosed with chronic migraine but, despite this, they continued to have successful treatment of their symptoms and as a result, ONS was recommended as a treatment for chronic migraine (Fig. 18.3).[47]

The mechanism of action of ONS is based on the convergence of the trigeminal, dural, and cervical afferents in the brainstem. Activation of afferents from the caudal trigeminal nucleus at C2 can induce pain in both the trigeminal and cervical distributions. It is thought that electrical stimulation to neuromodulate the occipital nerves can affect pain in areas innervated by the cervical and trigeminal nerves. With the greater occipital nerves being branches of the C2 root, they represent a peripherally accessible target that can affect the functioning of central structures.[45]

A positron emission tomography (PET) study of eight patients with chronic migraines treated with ONS showed changes in regional blood flow correlated to pain relief in the dorsal rostral pons, anterior cingulate

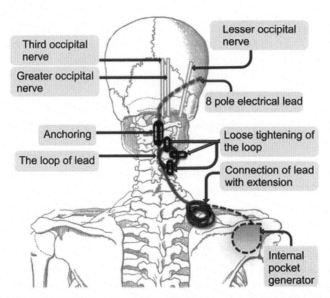

FIGURE 18.3   **Occipital nerve stimulation (ONS).** This particular stimulator was inserted at level C1–C2 and an eight-pole electrical lead provided coverage of greater, lesser, and third occipital nerves. After a 1-week trial period, permanent implantation of ONS occurred with internal pulse generator inserted in the right infraclavicular region. The patient experienced significant improvement in headache after stimulator implantation. *Reprinted from Shin JH, Kim YC, Jang IK, Kim J-H, Park SY, Lee SC. Occipital nerve stimulation in a patient with an intractable chronic headache - A case report.* Korean J Anesthesiol. *2011;60(4):298–301.* http://dx.doi.org/10.4097/kjae.2011.60.4.298. *Open access license:* http://creativecommons.org/licenses/by-nc/3.0/.

cortex, and cuneus.[48] Based on this favorable evidence, three large randomized controlled clinical trials were organized which will be discussed below.

The Occipital Nerve Stimulation for the Treatment of Intractable Migraine (ONSTIM) feasibility study randomized 66 patients with medically refractory migraine to bilateral ONS and randomized them to adjustable or preset stimulation vs. medical management. As a primary outcome, this study identified responders as a rate of >50% reduction of headache days per month or a reduction in pain intensity of at least three points on the visual analogue scale (VAS) at 3-month follow-up. In the stimulation group, 39% responded to adjustable stimulation and 6% responded with preset stimulation. There was no response in the medical management arm. No unanticipated device events were seen, although lead migration occurred in 12/51 (24%) of subjects. The data from this feasibility study suggested that ONS for medically intractable chronic migraine can be carried out relatively safely and was worthy of further study. They found that, although not powered for efficacy evaluation, the 39% responder rate was comparable with response rates achieved with widely used medications such as topiramate.[49]

The Precision Implantable Stimulator for Migraine (PRISM) study randomized 132 patients to an active versus sham stimulation paradigm with 12 weeks of stimulation and a primary outcome defined as changes in headache days per month. The active group had a reduction of 5.5 headache days per month versus the sham group with 3.9 days per month reduction. Mean reduction of migraine days was 27% in the active group and 20% in the sham group. The results were not statistically significant.[50]

The most recent randomized controlled trial studied 157 patients with refractory migraine and randomized them 2:1 to active or sham stimulation. The primary outcome was a reduction by 50% in the VAS scale and no increase in headache frequency or duration. At 12 weeks there was no significant difference demonstrated based on the 50% reduction benchmark. However, statistically significant differences were seen in the percentage of patients who achieved a 30% reduction in VAS. Compared to sham-treated patients, there were also significant differences in reduction of number of headache days (-3.1 per month, $P = 0.008$) and a reduction in migraine-related disability ($P = 0.001$), and direct reports of pain ($P = 0.001$). The most common adverse event reported was persistent implant site pain. Although the study failed to meet its primary endpoint, it was deemed the first large-scale study of peripheral neurostimulation of the occipital nerves in chronic migraine patients and demonstrated significant reductions in pain, headache days, and migraine-related disability. The authors concluded that additional controlled studies using endpoints that have recently been identified and accepted as clinically meaningful are warranted in this highly disabled population.[51]

Retrospective analysis of an open-label study of ONS in chronic refractory headache suggested that while ONS may be effective in some patients, the majority of the 15 studied (60%) required lead revision within 1 year with one patient requiring generator revision. The commonly required surgical revisions prompted the authors to push for more studies including RCTs to assess safety and efficacy of ONS in headache disorders.[52]

There are currently no published reports of occipital nerve stimulators implanted in people under the age of 18 years. While this treatment is quite invasive and true efficacy remains to be seen, it may in time develop as a safe and effective method of treating intractable headache in adolescents.

Clinical Data in Supraorbital and Combined ONS/Supraorbital Nerve Stimulation

An adult patient with chronic cluster headache responding to unilateral supraorbital nerve stimulation has been reported.[53] A combination of continuous baseline stimulation was given in addition to superimposed increased stimulation to be used as a "rescue" method. The postimplantation attack frequency decreased from two to three daily attacks to one attack every other day. After 2months the attacks resolved completely for the 14months of follow-up, aside from one attack occurring after discontinuation of constant stimulation. Resumption of constant stimulation returned the patient to an attack-free state.

Reed et al. described a case series of seven adult patients with chronic migraine responding to combination occipital and supraorbital nerve stimulation.[54] The initial group of eight patients was evaluated with trial stimulation and seven of the eight patients responded. Patient characteristics described showed none of the patients had an occipital focus of pain or medication overuse headache, some factors described in previous studies as potential inclusion and exclusion criteria, respectively. One patient was refractory to occipital nerve block in the past; another had a history of hemiplegic migraine. Permanent combined occipital nerve–supraorbital nerve neurostimulation systems were implanted and the relative responses to two stimulation programs were evaluated: one that stimulated only the occipital leads and one that stimulated both the occipital and supraorbital leads together. Stimulation was bilateral and patient-controlled. Follow-up ranged from 1 to 35months and all patients reported a full therapeutic response but only to combined supraorbital-occipital neurostimulation. Five out of seven patients no longer required medication. ONS alone provided a markedly inferior and inadequate response. One patient had a lead migration and another patient had an infection around the lead. The authors concluded that combined occipital nerve–supraorbital nerve neurostimulation systems may provide effective treatment for patients with chronic migraine and refractory chronic migraine headaches. They proposed that for patients with chronic migraine headaches the response to combined systems appeared to be substantially better than ONS alone.

Vagus Nerve Stimulation in Migraine

The vagus nerve innervates many organs in the body and relays sensory information regarding the state of viscera to the central nervous system via its 80% afferent and 20% efferent fibers.[55] Patients receiving VNS for epilepsy treatment reported a reduction in headache frequency and intensity,[56] along with a headache prophylactic effect of VNS.[57]

The mechanisms of such effects on migraine are still unclear. A recent reverse translational animal model examined non-invasive vagus nerve stimulation (nVNS) for the treatment of trigeminal allodynia in rats. Rats were infused with inflammatory mediators directly onto the dura which led to chronic trigeminal allodynia. Administration of nVNS for 2min decreased periorbital sensitivity in rats with trigeminal allodynia for up to 3.5h after stimulation. Using microdialysis, extracellular neurotransmitters in the trigeminal nucleus caudalis in allodynic rats showed a 7.7-fold increase in extracellular glutamate after chemical headache trigger, with only a 2.3-fold increase seen in allodynic rats who received nVNS. Even when nVNS treatment was delayed by 120 min after chemical headache trigger, the high levels of glutamate in the trigeminal nucleus caudalis were reversed after nVNS. No significant changes to heart rate or blood pressure were seen with the nVNS parameters used in the study. The authors concluded nVNS may be of use in the treatment of trigeminal allodynia.[58]

Recent reports have examined non-invasive VNS as a novel potential therapy for the treatment of acute migraine.[59] An open label trial of nVNS in 30 enrolled patients (25 females, five males, median age 39 years) analyzed a total of 27 patients treating 80 attacks with pain. The pain-free rate at 2 h was 4/19 (21%) for the first treated attack with a moderate or severe headache at baseline. For all moderate or severe attacks at baseline, the pain-free rate was 12/54 (22%). An adverse event was reported in 13 patients, including neck twitching ($n=1$), raspy voice ($n=1$), and redness at the device site ($n=1$). No unanticipated, serious, or severe adverse events were reported. The authors of the study concluded nVNS may be an effective and well-tolerated treatment for acute migraine in certain patients.

While the non-invasive aspects of potential nVNS treatment in children are attractive, further research must be done to determine whether this is a safe and efficacious potential therapy for children with episodic and chronic migraine.

Auriculotemporal Nerve Stimulation

A report of an adult successfully treated with stimulation of bilateral auriculotemporal nerves has identified a potential novel target for peripheral neurostimulation in headache therapy.[60] The patient was a 52-year-old woman with refractory pain in the bilateral temporal distribution and marked phonophobia as a result of chronic migraine. After a successful 3-week trial period, the patient underwent implantation of bilateral peripheral nerve stimulators targeting the auriculotemporal nerves. At 16 months of follow up, her average pain intensity declined from 8 to 9/10 on the numeric rating scale to 5/10. Her function improved as assessed by the Migraine Disability Assessment, from total disability (grade IV) to mild disability (grade II). Her phonophobia also became far less debilitating. The authors of the study concluded that auriculotemporal nerve stimulation may be useful tool in the treatment of refractory pain in the temporal distribution due to chronic migraine.

This case highlights what may indeed be a useful future target in adolescents with refractory chronic migraine and pain in the temporal head regions but no similar studies have been done to date in the pediatric population.

Other Headache Disorders

Cluster Headache

While neuromodulation of refractory cluster headache was initially aimed at central targets in the hypothalamus, the significant morbidity associated with this approach led to search for less invasive targets.[61] ONS was shown in a PET study in adults to reverse the hypermetabolism seen in the hypothalamus, midbrain, and pons in cluster headache.[62]

A pilot open-label trial of ONS in eight patients with medication-refractory cluster headache implanted a suboccipital neurostimulator on the side of the headache and measured outcomes of attack frequency, intensity, and symptomatic treatment before and after continuous ONS. To detect changes in cephalic and extracephalic pain-processing authors measured electrical and pressure pain thresholds and the nociceptive blink reflex. Two patients were pain-free after a follow-up of 16 and 22 months; one of them still had occasional autonomic attacks. Three patients had around a 90% reduction in attack frequency. Two patients, one of whom had had the implant for only 3 months, had improvement of around 40%. Mean follow-up was 15.1 months (SD 9.5, range 3–22). Intensity of attacks rather than frequency tended to decrease earlier during ONS and was on average improved by 50% in remaining attacks. All but one patient were able to substantially reduce preventive drug treatment during ONS. Interestingly, interruption of ONS by switching off the stimulator or because of battery failure was followed within days by recurrence and increase of attacks in all improved patients. ONS did not significantly modify pain thresholds and the amplitude of the nociceptive blink reflex increased with longer durations of ONS. There were no serious adverse events seen. The authors of the study concluded ONS could be an efficient treatment for drug-resistant chronic cluster headache and could be a safer alternative to deep hypothalamic stimulation. The observed delay of 2 months or more between implantation and significant clinical improvement suggested that ONS acts via slow neuromodulatory processes at the level of the upper brain stem or diencephalic centers.[63]

In a later open label study, 14 patients with intractable cluster headache were implanted with bilateral suboccipital electrodes for ONS and retrospective assessment of their clinical outcome was obtained. At a median follow-up of 17.5 months (range 4–35 months), 10 of 14 patients reported improvement and nine of these recommended ONS. Three patients noticed a marked improvement of 90% or better, three a moderate improvement of 40% or better, and four a mild improvement of 20–30%. Improvement occurred within days to weeks for those who responded most and, similarly to previous studies, patients consistently reported their attacks returned within hours to days when the device was off. One patient found that ONS helped abort acute attacks. Adverse events of concern were lead migrations and battery depletion. The study authors concluded that ONS was a safe and effective, non-invasive and non-neurally destructive option for some patients with chronic cluster headache.[64]

Study design for future studies and current evidence has been limited by the difficulty in blinding as patients in the active arm can feel paresthesia and are thus able to determine whether they are in the active or sham treatment categories.[45]

However, a new RCT design in the ICON study is comparing low- and high-amplitude ONS in each group as a method of blinding in adults with chronic cluster headache. Final results are pending but this study design is promising for the development of further RCTs involving ONS.

As primary cluster headache is a very rare entity in children with a history toward medical and spontaneous remission,[65] ONS is unlikely to be useful as a treatment modality in this population. However, it may be kept in mind as an option for medically refractory cases.

FUTURE AVENUES IN PEDIATRIC BRAIN STIMULATION FOR PAIN

Recently, a group of European experts presented guidelines citing Level A (definite efficacy) evidence to support the use of high-frequency rTMS over the primary motor cortex (M1) contralateral to the site of pain.[34] The panel found in review of all available studies of a total of 511 adult patients with chronic neuropathic pain that high-frequency rTMS over the contralateral M1 was found to produce significant pain-relieving effects.

A Level C recommendation was made regarding high-frequency rTMS of M1 in patients with chronic regional pain syndrome (CRPS) type 1 based on two class II–III studies involving a total of 32 patients with a conclusion of a "possible analgesic effect" seen in these patients and no data to assess rTMS efficacy in CRPS type 2. Similarly, the panel was unable to offer guidelines regarding the role of rTMS in patients with fibromyalgia, largely due to study design variability and small sample sizes.

Fibromyalgia has been estimated to occur in as many as 6% of school-aged children[66] with high disability and health resource utilization.[67] While the epidemiology of pediatric complex regional pain syndrome is not yet clearly defined, it also represents a challenging and disabling disorder for children, families, and caregivers.[68] The above guidelines however suggest there is compelling evidence to expand non-invasive brain stimulation as a possible treatment modality into the realm of pediatric pain disorders, such as fibromyalgia and CRPS. With rigorous development of uniform parameters regarding stimulation location, frequency and large enough samples sizes, non-invasive brain stimulation methods, such as TMS, could revolutionize pediatric pain research and treatment.

References

1. Jenkins B, Tepper SJ. Neurostimulation for primary headache disorders, part 1: pathophysiology and anatomy, history of neuromodulation in headache treatment, and review of peripheral neuromodulation in primary headaches. *Headache*. 2011;51(8):1254–1266. http://dx.doi.org/10.1111/j.1526-4610.2011.01966.x.
2. Siniatchkin M, Reich A-L, Shepherd AJ, van Baalen A, Siebner HR, Stephani U. Peri-ictal changes of cortical excitability in children suffering from migraine without aura. *Pain*. 2009;147(1–3):132–140. http://dx.doi.org/10.1016/j.pain.2009.08.028.
3. Vos T, Flaxman AD, Naghavi M, et al. Years lived with disability (YLDs) for 1160 sequelae of 289 diseases and injuries 1990–2010: a systematic analysis for the Global Burden of Disease Study 2010. *Lancet*. 2012;380(9859):2163–2196. http://dx.doi.org/10.1016/S0140-6736(12)61729-2.
4. Bille BS. Migraine in school children. A study of the incidence and short-term prognosis, and a clinical, psychological and electroencephalographic comparison between children with migraine and matched controls. *Acta Paediatr Suppl*. 1962;136:1–151.
5. Stang PE, Osterhaus JT. Impact of migraine in the United States: data from the National Health Interview Survey. *Headache*. 1993;33(1):29–35.
6. Headache Classification Committee of the International Headache Society (IHS). The international classification of headache disorders, 3rd edition (beta version). *Cephalalgia Int J Headache*. 2013;33(9):629–808. http://dx.doi.org/10.1177/0333102413485658.
7. Abu-Arafeh I, Razak S, Sivaraman B, Graham C. Prevalence of headache and migraine in children and adolescents: a systematic review of population-based studies. *Dev Med Child Neurol*. 2010;52(12):1088–1097. http://dx.doi.org/10.1111/j.1469-8749.2010.03793.x.
8. Bonfert M, Straube A, Schroeder AS, Reilich P, Ebinger F, Heinen F. Primary headache in children and adolescents: update on pharmacotherapy of migraine and tension-type headache. *Neuropediatrics*. 2013;44(1):3–19. http://dx.doi.org/10.1055/s-0032-1330856.
9. Laurell K, Larsson B, Eeg-Olofsson O. Prevalence of headache in Swedish schoolchildren, with a focus on tension-type headache. *Cephalalgia Int J Headache*. 2004;24(5):380–388. http://dx.doi.org/10.1111/j.1468-2982.2004.00681.x.
10. Lipton RB, Manack A, Ricci JA, Chee E, Turkel CC, Winner P. Prevalence and burden of chronic migraine in adolescents: results of the chronic daily headache in adolescents study (C-dAS). *Headache*. 2011;51(5):693–706. http://dx.doi.org/10.1111/j.1526-4610.2011.01885.x.
11. Baron EP, Rothner AD. New daily persistent headache in children and adolescents. *Curr Neurol Neurosci Rep*. 2010;10(2):127–132. http://dx.doi.org/10.1007/s11910-010-0097-3.
12. Grande RB, Aaseth K, Lundqvist C, Russell MB. Prevalence of new daily persistent headache in the general population. The akershus study of chronic headache. *Cephalalgia*. 2009;29(11):1149–1155. http://dx.doi.org/10.1111/j.1468-2982.2009.01842.x.
13. Powers SW, Patton SR, Hommel KA, Hershey AD. Quality of life in childhood migraines: clinical impact and comparison to other chronic illnesses. *Pediatrics*. 2003;112(1 Pt 1):e1–e5.
14. Ramsey RR, Ryan JL, Hershey AD, Powers SW, Aylward BS, Hommel KA. Treatment adherence in patients with headache: a systematic review. *Headache*. 2014;54(5):795–816. http://dx.doi.org/10.1111/head.12353.

15. Rajapakse T, Kirton A. Non-invasive brain stimulation in children: applications and future directions. *Transl Neurosci.* 2013;4(2). http://dx.doi.org/10.2478/s13380-013-0116-3.
16. Gilbert DL, Garvey MA, Bansal AS, Lipps T, Zhang J, Wassermann EM. Should transcranial magnetic stimulation research in children be considered minimal risk? *Clin Neurophysiol Off J Int Fed Clin Neurophysiol.* 2004;115(8):1730–1739. http://dx.doi.org/10.1016/j.clinph.2003.10.037.
17. Gillick BT, Kirton A, Carmel JB, Minhas P, Bikson M. Pediatric stroke and transcranial direct current stimulation: methods for rational individualized dose optimization. *Front Hum Neurosci.* 2014;8:739. http://dx.doi.org/10.3389/fnhum.2014.00739.
18. Peña C, Bowsher K, Samuels-Reid J. FDA-approved neurologic devices intended for use in infants, children, and adolescents. *Neurology.* 2004;63(7):1163–1167.
19. Kirton A. *Safety and Tolerability of TMS Neurophysiology and Interventional rTMS in Children.* Montreal: Canadian Neurological Sciences Federation; 2013.
20. Wu SW, Shahana N, Huddleston DA, Lewis AN, Gilbert DL. Safety and tolerability of theta-burst transcranial magnetic stimulation in children. *Dev Med Child Neurol.* 2012;54(7):636–639. http://dx.doi.org/10.1111/j.1469-8749.2012.04300.x.
21. Landy S, Rice K, Lobo B. Central sensitisation and cutaneous allodynia in migraine: implications for treatment. *CNS Drugs.* 2004;18(6):337–342.
22. Akerman S, Holland PR, Goadsby PJ. Diencephalic and brainstem mechanisms in migraine. *Nat Rev Neurosci.* 2011;12(10):570–584. http://dx.doi.org/10.1038/nrn3057.
23. Goadsby PJ, Knight YE, Hoskin KL. Stimulation of the greater occipital nerve increases metabolic activity in the trigeminal nucleus caudalis and cervical dorsal horn of the cat. *Pain.* 1997;73(1):23–28.
24. Goadsby PJ, Zagami AS. Stimulation of the superior sagittal sinus increases metabolic activity and blood flow in certain regions of the brainstem and upper cervical spinal cord of the cat. *Brain J Neurol.* 1991;114(Pt 2):1001–1011.
25. Hoskin KL, Zagami AS, Goadsby PJ. Stimulation of the middle meningeal artery leads to Fos expression in the trigeminocervical nucleus: a comparative study of monkey and cat. *J Anat.* 1999;194(Pt 4):579–588.
26. Gerwig M, Niehaus L, Kastrup O, Stude P, Diener HC. Visual cortex excitability in migraine evaluated by single and paired magnetic stimuli. *Headache.* 2005;45(10):1394–1399. http://dx.doi.org/10.1111/j.1526-4610.2005.00272.x.
27. Taylor PCJ, Walsh V, Eimer M. The neural signature of phosphene perception. *Hum Brain Mapp.* 2010;31(9):1408–1417. http://dx.doi.org/10.1002/hbm.20941.
28. Elkin-Frankston S, Fried PJ, Pascual-Leone A, Rushmore RJ, Valero-Cabr A. A novel approach for documenting phosphenes induced by transcranial magnetic stimulation. *JoVE.* 2010;(38). http://dx.doi.org/10.3791/1762.
29. Silvanto J, Muggleton N, Lavie N, Walsh V. The perceptual and functional consequences of parietal top-down modulation on the visual cortex. *Cereb Cortex (New York, NY: 1991).* 2009;19(2):327–330. http://dx.doi.org/10.1093/cercor/bhn091.
30. Brigo F, Storti M, Tezzon F, Manganotti P, Nardone R. Primary visual cortex excitability in migraine: a systematic review with meta-analysis. *Neurol Sci Off J Ital Neurol Soc Ital Soc Clin Neurophysiol.* 2013;34(6):819–830. http://dx.doi.org/10.1007/s10072-012-1274-8.
31. Shepherd AJ. Local and global motion after-effects are both enhanced in migraine, and the underlying mechanisms differ across cortical areas. *Brain J Neurol.* 2006;129(Pt 7): 1833–1843. http://dx.doi.org/10.1093/brain/awl124.
32. Carandini M, Movshon JA, Ferster D. Pattern adaptation and cross-orientation interactions in the primary visual cortex. *Neuropharmacology.* 1998;37(4–5):501–511.
33. Carandini M. Visual cortex: fatigue and adaptation. *CB.* 2000;10(16):R605–R607.
34. Lefaucheur J-P, André-Obadia N, Antal A, et al. Evidence-based guidelines on the therapeutic use of repetitive transcranial magnetic stimulation (rTMS). *Clin Neurophysiol Off J Int Fed Clin Neurophysiol.* 2014;125(11):2150–2206. http://dx.doi.org/10.1016/j.clinph.2014.05.021.

35. Lipton RB, Dodick DW, Silberstein SD, et al. Single-pulse transcranial magnetic stimulation for acute treatment of migraine with aura: a randomised, double-blind, parallel-group, sham-controlled trial. *Lancet Neurol.* 2010;9(4):373–380. http://dx.doi.org/10.1016/S1474-4422(10)70054-5.

36. Jürgens TP, Leone M. Pearls and pitfalls: neurostimulation in headache. *Cephalalgia Int J Headache.* 2013;33(8):512–525. http://dx.doi.org/10.1177/0333102413483933.

37. Teepker M, Hötzel J, Timmesfeld N, et al. Low-frequency rTMS of the vertex in the prophylactic treatment of migraine. *Cephalalgia Int J Headache.* 2010;30(2):137–144. http://dx.doi.org/10.1111/j.1468-2982.2009.01911.x.

38. Brighina F, Piazza A, Vitello G, et al. rTMS of the prefrontal cortex in the treatment of chronic migraine: a pilot study. *J Neurol Sci.* 2004;227(1):67–71. http://dx.doi.org/10.1016/j.jns.2004.08.008.

39. Fregni F, Boggio PS, Lima MC, et al. A sham-controlled, phase II trial of transcranial direct current stimulation for the treatment of central pain in traumatic spinal cord injury. *Pain.* 2006;122(1–2):197–209. http://dx.doi.org/10.1016/j.pain.2006.02.023.

40. Fregni F, Gimenes R, Valle AC, et al. A randomized, sham-controlled, proof of principle study of transcranial direct current stimulation for the treatment of pain in fibromyalgia. *Arthritis Rheum.* 2006;54(12):3988–3998. http://dx.doi.org/10.1002/art.22195.

41. Antal A, Kriener N, Lang N, Boros K, Paulus W. Cathodal transcranial direct current stimulation of the visual cortex in the prophylactic treatment of migraine. *Cephalalgia Int J Headache.* 2011;31(7):820–828. http://dx.doi.org/10.1177/0333102411399349.

42. Luedtke K, Rushton A, Wright C, Geiss B, Juergens TP, May A. Transcranial direct current stimulation for the reduction of clinical and experimentally induced pain: a systematic review and meta-analysis. *Clin J Pain.* 2012;28(5):452–461. http://dx.doi.org/10.1097/AJP.0b013e31823853e3.

43. Melzack R, Wall PD. Pain mechanisms: a new theory. *Science.* 1965;150(3699):971–979.

44. Ignelzi RJ, Nyquist JK. Direct effect of electrical stimulation on peripheral nerve evoked activity: implications in pain relief. *J Neurosurg.* 1976;45(2):159–165. http://dx.doi.org/10.3171/jns.1976.45.2.0159.

45. Goroszeniuk T, Pang D. Peripheral neuromodulation: a review. *Curr Pain Headache Rep.* 2014;18(5):412. http://dx.doi.org/10.1007/s11916-014-0412-9.

46. Sandkühler J, Chen JG, Cheng G, Randić M. Low-frequency stimulation of afferent Adelta-fibers induces long-term depression at primary afferent synapses with substantia gelatinosa neurons in the rat. *J Neurosci Off J Soc Neurosci.* 1997;17(16):6483–6491.

47. Weiner RL, Reed KL. Peripheral neurostimulation for control of intractable occipital neuralgia. *Neuromodulation J Int Neuromodulation Soc.* 1999;2(3):217–221. http://dx.doi.org/10.1046/j.1525-1403.1999.00217.x.

48. Matharu MS, Bartsch T, Ward N, Frackowiak RSJ, Weiner R, Goadsby PJ. Central neuromodulation in chronic migraine patients with suboccipital stimulators: a PET study. *Brain J Neurol.* 2004;127(Pt 1):220–230. http://dx.doi.org/10.1093/brain/awh022.

49. Saper JR, Dodick DW, Silberstein SD, et al. Occipital nerve stimulation for the treatment of intractable chronic migraine headache: ONSTIM feasibility study. *Cephalalgia Int J Headache.* 2011;31(3):271–285. http://dx.doi.org/10.1177/0333102410381142.

50. Lipton R, Goadsby P, Cady R. PRISM study: occipital nerve stimulation for treatment-refractory migraine. *Cephalalgia Int J Headache.* 2009;29(Suppl 1):30.

51. Silberstein SD, Dodick DW, Saper J, et al. Safety and efficacy of peripheral nerve stimulation of the occipital nerves for the management of chronic migraine: results from a randomized, multicenter, double-blinded, controlled study. *Cephalalgia Int J Headache.* 2012;32(16):1165–1179. http://dx.doi.org/10.1177/0333102412462642.

52. Schwedt TJ, Dodick DW, Hentz J, Trentman TL, Zimmerman RS. Occipital nerve stimulation for chronic headache—long-term safety and efficacy. *Cephalalgia Int J Headache.* 2007;27(2):153–157. http://dx.doi.org/10.1111/j.1468-2982.2007.01272.x.

53. Narouze SN, Kapural L. Supraorbital nerve electric stimulation for the treatment of intractable chronic cluster headache: a case report. *Headache*. 2007;47(7):1100–1102. http://dx.doi.org/10.1111/j.1526-4610.2007.00869.x.

54. Reed KL, Black SB, Banta CJ, Will KR. Combined occipital and supraorbital neurostimulation for the treatment of chronic migraine headaches: initial experience. *Cephalalgia Int J Headache*. 2010;30(3):260–271. http://dx.doi.org/10.1111/j.1468-2982.2009.01996.x.

55. Agostoni E, Chinnock JE, De Daly MB, Murray JG. Functional and histological studies of the vagus nerve and its branches to the heart, lungs and abdominal viscera in the cat. *J Physiol*. 1957;135(1):182–205.

56. Hord ED, Evans MS, Mueed S, Adamolekun B, Naritoku DK. The effect of vagus nerve stimulation on migraines. *J Pain Off J Am Pain Soc*. 2003;4(9):530–534.

57. Lenaerts ME, Oommen KJ, Couch JR, Skaggs V. Can vagus nerve stimulation help migraine? *Cephalalgia Int J Headache*. 2008;28(4):392–395. http://dx.doi.org/10.1111/j.1468-2982.2008.01538.x.

58. Oshinsky ML, Murphy AL, Hekierski H, Cooper M, Simon BJ. Noninvasive vagus nerve stimulation as treatment for trigeminal allodynia. *Pain*. 2014;155(5):1037–1042. http://dx.doi.org/10.1016/j.pain.2014.02.009.

59. Goadsby PJ, Grosberg BM, Mauskop A, Cady R, Simmons KA. Effect of noninvasive vagus nerve stimulation on acute migraine: an open-label pilot study. *Cephalalgia Int J Headache*. 2014;34(12):986–993. http://dx.doi.org/10.1177/0333102414524494.

60. Simopoulos T, Bajwa Z, Lantz G, Lee S, Burstein R. Implanted auriculotemporal nerve stimulator for the treatment of refractory chronic migraine. *Headache*. 2010;50(6):1064–1069. http://dx.doi.org/10.1111/j.1526-4610.2010.01694.x.

61. Leone M, Franzini A, Proietti Cecchini A, Mea E, Broggi G, Bussone G. Deep brain stimulation in trigeminal autonomic cephalalgias. *Neurother J Am Soc Exp Neurother*. 2010;7(2):220–228. http://dx.doi.org/10.1016/j.nurt.2010.02.001.

62. Magis D, Bruno M-A, Fumal A, et al. Central modulation in cluster headache patients treated with occipital nerve stimulation: an FDG-PET study. *BMC Neurol*. 2011;11:25. http://dx.doi.org/10.1186/1471-2377-11-25.

63. Magis D, Allena M, Bolla M, De Pasqua V, Remacle J-M, Schoenen J. Occipital nerve stimulation for drug-resistant chronic cluster headache: a prospective pilot study. *Lancet Neurol*. 2007;6(4):314–321. http://dx.doi.org/10.1016/S1474-4422(07)70058-3.

64. Burns B, Watkins L, Goadsby PJ. Treatment of intractable chronic cluster headache by occipital nerve stimulation in 14 patients. *Neurology*. 2009;72(4):341–345. http://dx.doi.org/10.1212/01.wnl.0000341279.17344.c9.

65. Arruda MA, Bonamico L, Stella C, Bordini CA, Bigal ME. Cluster headache in children and adolescents: ten years of follow-up in three pediatric cases. *Cephalalgia Int J Headache*. 2011;31(13):1409–1414. http://dx.doi.org/10.1177/0333102411418015.

66. Buskila D, Press J, Gedalia A, et al. Assessment of nonarticular tenderness and prevalence of fibromyalgia in children. *J Rheumatol*. 1993;20(2):368–370.

67. Anthony KK, Schanberg LE. Juvenile primary fibromyalgia syndrome. *Curr Rheumatol Rep*. 2001;3(2):165–171.

68. Borucki AN, Greco CD. An update on complex regional pain syndromes in children and adolescents. *Curr Opin Pediatr*. June 2015. http://dx.doi.org/10.1097/MOP.0000000000000250.

69. Magis D, Schoenen J. Advances and challenges in neurostimulation for headaches. *Lancet Neurol*. 2012;11(8):708–719. http://dx.doi.org/10.1016/S1474-4422(12)70139-4.

70. Shin JH, Kim YC, Jang IK, Kim J-H, Park SY, Lee SC. Occipital nerve stimulation in a patient with an intractable chronic headache - A case report. *Korean J Anesthesiol*. 2011;60(4):298–301. http://dx.doi.org/10.4097/kjae.2011.60.4.298. Open access license: http://creativecommons.org/licenses/by-nc/3.0/.

SECTION III

INVASIVE BRAIN STIMULATION IN CHILDREN

Deep Brain Stimulation in Children: Clinical Considerations

J.-P. Lin

Guy's and St. Thomas' NHS Foundation Trust, London, United Kingdom

HISTORICAL BACKGROUND

Deep brain stimulation (DBS) represents one of the most remarkable developments in neuroscience of the mid to late 20th century. When successful, the clinical results today would surely surpass the expectations of the early pioneers of DBS neurophysiology and clinical applications. For perhaps the first time, chronic disabling movement disorders such as parkinsonian tremor, essential tremor, and dystonia became amenable to long-term neuromodulation capable of abolishing disabling symptoms for an indefinite period, subject to the design and capacity of a succession of pulse-generator improvements. The history of DBS offers a fascinating interaction between intractable clinical need, advances in neurosurgical technique, the development of clinical outcome scales, engineering design and developments, and the pacemaker industry.[1]

As a general rule, the effects of DBS are more successful when directed against rapid movement disorders such as tremors, rapid dystonias, and myoclonus. In adult parkinsonism, DBS is most successful in cases with a good response to levodopa or dopa-agonists. There are two reasons for this: the first is that a response to levodopa or analogues indicates relative preservation of the neuronal composition of the striatum; the second is that DBS in effect manages the levodopa-induced dyskinesia. So the DBS is dealing with the rapid or phasic element of a movement disorder.

By contrast parkinsonian rigidity, bradykinesia, and akinesia are less likely to respond or may indeed worsen or be exacerbated by DBS.

The lessons from DBS use for parkinsonism, a neurodegenerative disorder of man, from the 1980s and 1990s onwards have helped shape our perspectives of the scope of clinical usefulness and limitations of DBS and also the stimulation parameters likely to be beneficial in this and kindred conditions. During these times, pallidotomy was the favored stereotactic approach to the management of severe dystonia.[2] Applications of DBS in dystonia were first generally introduced into clinical practice by Philippe Coubes in Montpellier in the mid-1990s, when a critically ill child in intensive care with a genetic dystonia caused by the DYT-1 gene mutation in Torsin A had bilateral DBS to the globus pallidus internus (GPi): leading to the first case series of successful management of dystonia by DBS.[3,4] Since these fundamental beginnings, the spectrum of DBS interventions has continued to expand, including emerging applications in the developing brains of children.

PRIMARY DYSTONIAS AND DBS

Idiopathic torsion dystonia associated with the DYT-1 mutation in torsin A is a rare movement dystonia and, when generalized, is most likely to have an onset before the age of 10 years.[5] Typical presentations first affect one foot, then the other, followed by spreading up to involve the upper limbs, trunk, neck, and face.

Other reports followed[6] and the field began to explore the clinical characteristics of the best response to DBS in dystonia.[7] This resulted in a root and branch clinical-genetic and neuroimaging analysis of the dystonias, including the important question of whether there would be differences in response to DBS between *DYT-1 positive* and *DYT-1 negative* primary dystonias. The first randomized and blinded evaluations of DBS for primary dystonias[8] were quickly followed by studies of DBS efficacy on *generalized* as opposed to *segmental* body distribution of primary dystonias.[9]

PHYSIOLOGY OF DBS EFFECT IN PRIMARY DYSTONIAS

Exploration of the physiological effects of DBS[10] indicated that globus pallidus internus (GPi) DBS effects were of slower onset than those seen in parkinsonism, often taking weeks or months and exerting both long-term and short-term effects on cortical and subcortical circuits. These in turn may result in a progressive normalization of spinal brainstem excitability, correlating with clinical improvement.[10] Tisch et al. concluded that

GPi DBS modified the basal ganglia outputs to brainstem, thalamus, and cortex leading to "neuronal reorganization," which explained the slow, progressive onset of DBS efficacy in the dystonia.

EFFICACY OF DBS ON DYT-1-POSITIVE AND DYT-1-NEGATIVE DYSTONIAS

Genetic diagnosis (*DYT-1 positive* and *DYT-1 negative*) and temporal characteristics such as age at onset, duration of dystonia at the time of GPi DBS, dystonia severity, and associated comorbidities such as fixed deformity/contracture indicated that shorter duration of dystonia improved the chances of DBS success.[11] This was further confirmed by the first meta-regression study of GPi DBS in dystonia.[12]

SEGMENTAL VERSUS GENERALIZED DYSTONIA AND DBS

Further evaluation of 3- and 5-year benefits of DBS[13] were obtained in 38 cases using sham versus GPi DBS in segmental and generalized dystonias leading to a 20-point reduction in BFMDRS-M severity. This improvement continued from the 6-month time-point resulting in a further absolute BFMDRS-M reduction of 5.7 (SD 8.4) or 34% reduction in dystonia compared to baseline which was sustained at 3 and 5 years. *The interpretation of absolute as opposed to percentage changes in dystonia rating scales is an important methodological issue in this field and will be addressed later.*

EFFECTS OF SWITCHING DBS "OFF"

Would GPi DBS effects wane over time? Long-term follow-up studies provided evidence that DBS control of primary DYT-1 dystonia were maintained provided that, with some exceptions, stimulation was maintained.[14,15]

However some long-term stability after DBS has been evaluated using "on" and "off" studies in primary dystonias.[16] In essence, very little change was noted 2 days after switching "off," indicating some long-term neural reorganization of the nervous system. But patients with the highest cerebral plasticity when DBS was "on" exhibited the greatest retention of benefit when the DBS was switched "off." The opposite was also observed. One working definition of "higher plasticity" was a patient achieving clinical DBS benefit with relatively low current. Conversely, the higher the current delivery required to achieve a clinical response, the lower the plasticity of the subject.

A further interesting study demonstrated persistent abnormal neuro-physiological parameters in patients receiving DBS for severe or life-threatening genetic dystonias.[17] When clinically stable, DBS patients were studied in the "DBS-off" condition and persistently abnormal physiological variables were recorded.[17] This was interpreted as representing a method of determining whether a DBS patient was at risk of a severe dystonia relapse in the event of DBS being switched off or DBS failing.[17] Another way of viewing these findings is to say that these subjects are "DBS-dependent."

DBS FOR SECONDARY DYSTONIAS

A question which needed to be answered was whether DBS might have more universal applications for more common, acquired, secondary dystonias? If so, this would surely influence clinical-diagnostic practice and service provision for these disorders, often managed in under-resourced community envirnoments.

DBS FOR NEURODEGENERATION WITH BRAIN IRON ACCUMULATION (NBIA)

Already in 2005, Pierre Castelnau with the Montpellier group demonstrated benefits of GPi DBS in the extremely severely affected Pantothenic Kinase Associated Neuro degeneration (PKAN) caused by a PANK2 mutation, a form of neurodegeneration with brain iron accumulation (NBIA).[18] This autosomal recessive condition is severe in its infantile form and in very early-onset cases, may be mistaken for a particularly severe form of cerebral palsy until clues are followed up leading to a correct diagnosis with neuroimaging, which shows the MRI features of the "eye of the tiger" sign in the anterior GPi and subsequent genetic confirmation.

PRIMARY AND SECONDARY DYSTONIA RESPONSES TO DBS

A prospective, blinded evaluation comparing DBS in nine secondary dystonia and four patients with cranial dystonia/spasmodic torticollis[19] reported notable objective improvements in 11/13 cases using a range of dystonia impairment scales such as the *Unified Dystonia Rating Scale (UDRS)*, *Toronto Western Spasmodic Torticollis Rating Scale (TWSTRS)*, *Burke–Fahn–Marsden Dystonia Rating Scale (BFMDRS)*, or *Abnormal Involuntary Movement Scale (AIMS)*. Quality of life scoring was assessed using a standardized 7-point Global Rating Scale. DBS programming was completed over 6.5 visits over 5.9 months.

Pretto et al. began to elaborate a fundamental clinical question relating to who could benefit from GPi DBS as well as illustrating the utility of available impairment scales (UDRS, TWSTRS, BFMDRS, AIMS) and the need for quality-of-life measurements.[19] The authors concluded: "Although the benefits were variable and not fully predictable, they were of sufficient magnitude to justify offering the procedure when medications and botulinum toxin injections have failed."[19]

SECONDARY DYSTONIAS AND FUNCTIONAL IMAGING BEFORE AND AFTER DBS

Katsakiori et al.[20] reported eight cases of secondary dystonia undergoing DBS including cerebral palsy ($n=2$), drug-induced ($n=1$), postencephalitis ($n=2$) and postanoxic dystonia ($n=3$). The target for DBS was the GPi in seven patients and in the one patient with postanoxic pallidal injury, the *ventralis oralis anterior* (Voa) nucleus of the thalamus was chosen.[20] Across the treatment group, the percentage change in BFM-DRS motor and disability scores was lower (improved) after DBS: 41.4% (0–94.3) and 29.5% (0–84.2), respectively.[20] Furthermore, before and after SPECT scans showed reduced regional cerebral blood flow (rCBF) during the "on DBS" compared with the "off DBS" condition.[20] This case series advanced the study of a much wider clinical group of dystonias, covering a range of common pathologies including cerebral palsy, encephalitis, and postanoxic disorders.

PATIENT SELECTION FOR DBS

A dozen or so years after the first GPi DBS for dystonia reports, it was possible to discuss patient selection in routine clinical practice.[21] A study reviewing GPi DBS in 341 primary and 109 secondary dystonias reported that certain secondary dystonias, such as *tardive dystonia, myoclonus*, and NBIA (principally associated with the PANK2 mutation) responded very well to GPi DBS.

Aside from neurosurgical technical issues, worse outcomes were related to comorbidities including long duration of disease, joint contractures, scoliosis, and additional motor disorders including spasticity and cerebellar dysfunction.[21]

POOLED COHORTS OF DBS DATA

Pooling of cases has helped highlight benefits of DBS in rare disorders, such as PANK2 NBIA,[22] revealing more modest, but sustained benefits of DBS in this severe neurodegenrative disorder. Mean percentage improvement in dystonia severity was 28.5% at 2–6 months and 25.7% at 9–15 months. At 9–15 months postoperatively, two-thirds of cases showed an improvement of 20% or more in severity of dystonia, and 31.3% showed an improvement of 20% or more in disability. Fourteen of these NBIA cases had onset under the age of 10 years, in eight cases onset was under 4 years of age, ie, infantile-onset presentations. The other nine cases became symptomatic in the second decade: mean age at onset 7.8 (SD ± 4.8; range 1–15) years. The mean age at onset was 7.8 (SD ± 4.8) years, but the mean age at GPi DBS surgery was 18.8 (SD ± 8.8) years, resulting in a mean duration of dystonia symptoms prior to surgery of 10.2 (SD ± 6.4) years, respectively.[22] Once again the impact of early life living with dystonia in PANK2 disease prior to GPi DBS could influence both benefit from and disease-modifying consequences of DBS.

CHILDREN WITH DYSTONIA AND DBS

Do children benefit from DBS? It is clear from the preceding sections that many primary dystonias have their onset in childhood so why not bring forward the time of DBS surgery into childhood itself? This was comparatively rare, even as late as 2011 when Air et al. published their review of 815 consecutive DBS cases of whom 31/815 (3.8% of the cohort) had been children implanted over a 12-year period.[23] Mean age at DBS surgery was 13.2 years (4–17 years old) and again, like their adult counterparts, the primary dystonias responded best, while the secondary dystonias exhibited only small improvements.

Baseline Severity, Skeletal Maturity, and DBS Outcome in Cerebral Palsy

Inevitably, the focus would soon turn to cerebral palsy (CP), the leading cause of neurological disability in children. GPi DBS case-series include Marks et al.[24] who reported on 14 cases, eight of whom had GPi DBS before completion of skeletal maturity and did comparatively better than the six cases who received GPi DBS after skeletal maturity.

Pooled CP cases contributed to a meta-analysis comprising 20 reports involving 68 cases undergoing GPi DBS.[25] The mean baseline BFMDRS-Movement score of 69.94 (SD ± 25.4) dropped to a mean score of 50.5 (SD ± 26.77) also expressed as a 23.6% ($P < 0.001$) change at a median follow-up of 12 months.[25] The corresponding BFMDRS-Disability score also diminished from 18.54 (SD ± 6.15) to 16.83 (SD ± 6.42) or a percentage mean reduction of 9.2% ($P < 0.001$) after GPi DBS.[25] A *significant negative correlation* was also found between dystonia severity at baseline and response to DBS.[25]

Skeletal maturity and dystonia severity at baseline therefore seem to adversely influence the success of GPi DBS in CP patients.

Proportion of Life Lived With Dystonia (PLD) and Response to DBS

The comparative outcomes of GPI DBS in children with primary, secondary (including CP), and neurodegenerative dystonias has been explored.[26] Overall, the comparisons suggest that the PLD is significantly negatively correlated with percentage BFMDRS-M dystonia improvement after DBS (Fig. 19.1).

Percentage changes in BFMDRS-M after GPi DBS are plotted in Fig. 19.1 adapted from Lumsden et al.[26,46], with additional data from Marks et al.[24],

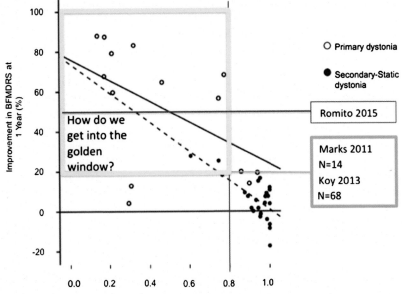

FIGURE 19.1 **Percentage changes in the BFMDRS-M score at 1 year after DBS (vertical axis) against proportion of life lived with dystonia (PLD) (horizontal axis).** Primary dystonias (white circles), with shorter PLD experience a much higher percentage change in BFMDRS-M than the secondary dystonias (black circles) who tend to have a higher PLD. However, a few secondary dystonia cases do seem to respond better to GPi DBS: particularly those with a PLD of less than 0.8 of life. Data for reported percentage BFMDRS-M changes in dystonic and dyskinetic CP following bilateral GPi DBS reported by Marks et al.[24]: 23% change; Koy et al.[25]: 24% change and Romito et al.[25a]: 49.5% change are also superimposed on the data from Lumsden et al.[26,46] together with what can be considered the 'ideal' or desirable 'golden box' of dystonia change. According to this data, reducing the PLD seems to correlate with better dystonia reduction. *After Lumsden DE, Kaminska M, Gimeno H, et al. Proportion of life lived with dystonia inversely correlates with response to pallidal deep brain stimulation in both primary and secondary childhood dystonia. Dev Med Child Neurol. 2013;55(6):567–574; Lumsden DE, Ashmore J, Charles-Edwards G, Lin JP, Ashkan K, Selway R. Accuracy of stimulating electrode placement in paediatric pallidal deep brain stimulation for primary and secondary dystonia. Acta Neurochir (Wien). 2013;155(5):823–836.*

Koy et al.[25] and Romito 2015, the latter showing a 49.5% change in BFM-DRS-M at 2 years after GPi DBS surgery in 15 adults with CP compared to the 1-year follow-up in the Marks, Koy, and Lumsden studies respectively. One reason why the Romito data are interesting is that it reflects an up to 5-year follow-up, indicating continued benefits from DBS with each successive year.

Limitations in BFMDRS-M Percentage Change Outcome

Methodologically, the BFMDRS-M as a measure of change in dystonia has several limitations.[27] The most obvious limitation is that when a percentage change in baseline BFMDRS is measured, more severe cases appear to respond less well than mild cases. For instance a 10-point change in BFMDRS-M corresponds to a 50% response if the baseline BFMDRS-M score is 20, but only a 10% change if the BFMDRS-M score is 100 at baseline. This leads to the question: "To whom does a 10-point change in BFM-DRS-M score matter most? The milder or more severe dystonia case?" This question is answered in part by a new comparative study correlating the BFMDRS-M with the GMFCS, MACS, and CFCS (see below).[36]

What Matters Most to Children With Dystonia and Their Caregivers

Dystonia severity reduction may not adequately capture what matters most to children or their family and caregivers. One study used the Canadian Occupational Performance Measure (COPM), a subjective scale of performance and satisfaction in personalized goals, in 57 children with dystonia of mean age 11.2 years (3.5–18.1), including 25% primary dystonias and 75% secondary dystonias. All responders expressed at least one concern around self-care and difficulties with routine functions such as *access to assistive technology, pain, dressing activities, use of tools, and social participation*. The nature and presence of concerns did not statistically differ according to etiology or the severity of the gross motor function classification system (GMFCS) or manual function classification system (MACS) impairments.[28]

Although it is possible to demonstrate improvement after bilateral GPI DBS in primary and secondary dystonias using the Melbourne Assessment of Unilateral Upper Limb Function (MAUULF),[29] goal-setting using the COPM before administering bilateral GPi DBS shows that primary, secondary static, and secondary progressive dystonias successfully achieve the preoperative goals in both domains of "performance" and "satisfaction," even though the BFMDRS-M scores change very little.[30] This indicates that a small reduction in dystonia may release a measurable degree of "activity" and "participation" after DBS in children. Such subjective scales reflecting patient and family goals must be included in future pediatric DBS clinical trials.

The Impact of Dystonia in Children

A large UK cohort of 279 children with dystonia[31] was referred for consideration of intrathecal baclofen (ITB) pump implantation or DBS. It was possible to determine at the time of the first visit that despite access to the full range of available medical and physical therapies available at local and regional level, in approximately two-thirds of cases the dystonia had worsened and in a further third, the dystonia had remained as severe despite all conventional medical and allied therapy inputs. Only 8% of cases experienced some sort of improvement.[31] Not surprisingly, the secondary and neurodegenerative dystonia groups were dominated by severe motor impairments as measured by the GMFCS. Whereas about two-thirds of cases with primary, primary-plus, and neurodegenerative dystonias had experienced a period of normal development, only one-fifth of secondary dystonias had done so.[31] This factor may influence outcome after DBS, since secondary dystonias lack a pattern of previous normal motor function prior to the onset of dystonia which could re-emerge, as in primary dystonias, after DBS. This period of normal early motor development can be considered as "money in the bank," functional capacity available to draw upon when the dystonia is moderated by DBS.

Interventional Studies for Dystonia in Childhood and Outcome Measures

Given the pessimistic outlook for childhood-onset dystonias, it was found that in 70 interventional studies, 46 neurosurgical and 24 pharmacological, the majority of neurosurgical studies (34/46) used impairment scales to measure change and pharmacological studies typically reported more subjective changes in motor symptoms.[32] This survey highlights the relative poverty of clinical work in the field of dystonia in children. Perhaps this lack of appropriate measures represents a major factor when considering "DBS efficacy for managing dystonia in children" beyond the impairment measures alone.

Improving Patient Selection for DBS in Children

As can be determined by published studies, primary dystonias, with and without known genetic causes respond best to DBS. This response may correlate with pallidal volume which is largest in DYT-1-positive dystonias.[33]

Thus, good clinical phenotyping plays an important role in patient selection. However, in children, there is still some confusion about diagnosing dystonia as a symptom or a sign and it is likely that there is relative overdiagnosis of "spasticity" rather than a broader disorder of movement and posture.[34,35]

A recent aid to interpretation of clinical difference and shared characteristics of children with "primary" and "secondary" dystonias is a "Rosetta Stone" correlation[36] of the Burke–Fahn–Marsden Dystonia Rating Scale (BFMDRS-M) with the Gross Motor Function Classification System (GMFCS), the Manual Abilities Function Classification System, and the Communication Function Classification System (CFCS), which represent the main motor severity measures used within the cerebral palsy literature.[36] The Rosetta Stone being an ancient tablet with the same text written in the three classical languages: ancient Egyptian hieroglyphs, Greek, and Latin. One aim of this study was to provide a common language of dystonia motor severity across all etiologies.

In addition, the phenotype–genotype characterizations have become increasingly complicated, due to genetic pleiotropy in which a gene mutation may code for many phenotypes. Modern methodologies are required, including either very broad genetic panels or next-generation whole-exome sequencing and bioinformatic support to fully explore the genetic factors underpinning movement disorders in children[37] and perhaps better define candidates and success predictors for DBS in children.

Neuroimaging and Neurophysiology Biomarkers?

Neuroimaging may not always correlate with physiological findings. For instance McClelland et al.[38] found that 80% of children with dystonia and abnormal MRI brain scans had normal central motor conduction times (CMCTs), measured using transcranial magnetic stimulation (TMS) (Fig. 19.2).

A follow-up of this study in 49 children (mean age 9 years, range 3–19 years) undergoing assessment for possible DBS found no relationship between CMCT values and the fractional anisotropy (FA) values of the corticospinal tracts using Tract Based Spatial Statistics (TBSS).[39]

Currently, neurophysiological and tractography imaging represent independent means of assessing the burden of functional disturbance in children with dystonia, bearing in mind that the CMCT values may take up to 6 years to mature in healthy children and this value may be further prolonged in the dystonic child.

Cerebral Plasticity in Children

Paired afferent stimulation may be used to evaluate cortical plasticity in children[40] and this is important since cortical plasticity is increased in dystonia[41] and the greater the cerebral plasticity, the better the response to DBS.[16,17] It is therefore important to measure these neurophysiological

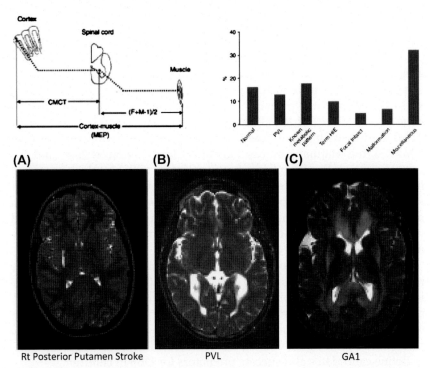

FIGURE 19.2   Transcranial magnetic stimulation (TMS) for central motor conduction time (CMCT) evaluation. Of 60 children with dystonia 50/62 had normal CMCT and 42/50 with *normal CMCT* had *abnormal neuroimaging indicating that children with abnormal scans may have normal physiology of the corticospinal pathways. After McClelland V, Mills K, Siddiqui A, Selway R, Lin JP. Central motor conduction studies and diagnostic magnetic resonance imaging in children with severe primary and secondary dystonia.* Dev Med Child Neurol. 2011;53(8):757–763.

biomarkers in children who may be candidates for DBS and to determine if responsiveness to DBS can be more accurately predicted.

Fluorodeoxyglucose-Positron Emission Tomography CT (^{18}FDDPET-CT) Brain Scanning

Functional imaging may give a clue to underlying brain dysfunction in dystonia, particularly in secondary dystonias where function may be severely altered. Katsakiori et al.[20] used single photon emission computed tomography (SPECT) to analyze overall cerebral activity in eight cases with secondary dystonias, in whom all showed a reduction in brain activity in the "on DBS" condition as compared with the "off DBS" condition.

We have used ^{18}FDGPET-CT as a baseline in all children undergoing DBS, partly since MRI brain follow-up was not possible in our center.[42] We were interested to find that the PANK2 disease children showed very

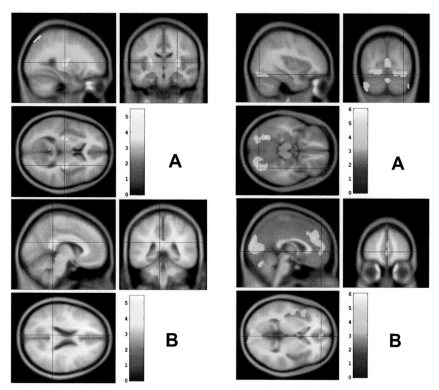

FIGURE 19.3 **Comparative 18-fluorodeoxy glucose positron emission tomography-CT study uptake in primary dystonias and PANK2 disease children at baseline before DBS. Increased uptake (red–yellow):** Statistical parametric mapping analysis showing relative increases in normalized ^{18}F-fluorodeoxyglucose uptake in neurodegeneration with brain iron accumulation versus primary dystonia. Increased signal is evident in the bilateral posterior putamen (A – appearing at the uncorrected cluster size threshold; however, they are reported given the strong hypothesis of disruption of basal ganglial metabolism) and in the bilateral retrosplenial posterior cingulate cortex (B – appearing at corrected cluster size thresholds). **Reduced FDG uptake (blue-turquoise):** Statistical parametric mapping analysis showing relative decreases in normalized ^{18}F-fluorodeoxyglucose uptake in neurodegeneration with brain iron accumulation versus primary dystonia. Decreased signal is evident in the bilateral inferior occipital cortex and cerebellum (A) and in the bilateral primary visual cortex (midline occipital cortex), anterior cingulate cortex and left-sided anterior and posterior insula cortex (B). Corrected cluster size thresholds. *Adapted from Szyszko TA, Dunn JT, O'Doherty MJ, Reed L, Lin JP. Role of ^{18}F-FDG PET imaging in paediatric primary dystonia and dystonia arising from neurodegeneration with brain iron accumulation. Nucl Med Commun. 2015;36(5):469–476.*

little, if any, individual differences in regional metabolism, but when compared with primary dystonias, the PANK2 cases showed relative hypermetabolism in the posterior putamen (striatum), when compared with primary dystonias (Fig. 19.3).

The posterior putamen is packed with dopaminergic neurons and tends to be activated when changing from a stable to an unstable motor pattern.

By contrast, there was relative hypometabolism in the anterior cingulate (which is usually activated when faced with conflicting choices) as well as lingual and fusiform gyri (visual processing areas required for naming objects, recognizing faces and understanding visual story lines) as well as the cerebellar vermis.

These findings may explain how PANK2 cases differ from primary dystonias and why certain activities, such as changing from familar to unfamiliar routines; making choices, speaking; dealing with compex visual information or attempting to stand up and walk: all of which may provoke stress and hence worsen dystonia.

TECHNICAL ISSUES INFLUENCING DBS EFFICACY

Neuromodulation with DBS is a major technological and neurosurgical intervention.

Unfortunately, as the old non-rechargeable neurostimulators failed, so DBS efficacy waned,[43] but the comparatively recent and smaller rechargeable neurostimulators available since 2008 in the United Kingdom, have been well accepted by children and their carers.[44]

"Electrode failure" due to wire fatigue or fracture leads to gradual or sudden DBS efficacy failure, with return of severe dystonia, often experienced bilaterally. Fractured wires must be replaced promptly to restore DBS efficacy. Gradually increasing resistance (impedence) makes it more difficult to judge the timing of DBS electrode replacement, but this should also be considered promptly if DBS efficacy is waning. Rarely, DBS wires can literally migrate out of the skull, usually following some particularly boisterous activity such as a somersault (or rolling over the back of the sofa!). Children may present with very severe worsening dystonia, sometimes also in frank status dystonicus.

Twiddling with the neurostimulator under the skin may result in flipping over the neurostimulator which may not charge well. Fortunately, it is possible to "flip" the neurostimulator back.[45]

Head growth may influence electrode depth in the growing child.

Accuracy of stereotactic DBS intracranial leads in children of various shapes and sizes is essential.[46] Surprisingly, although there is a projected increase of 5–10 mm in DBS electrode depth between ages 4–18 years, there was no significant variation in intracerebral DBS electrode depth between cases of all ages 4–18 years in a study observing and modeling DBS electrode depth in the GPi target in the developing brain.[47]

DBS and Cognitive Function in Children

A recent evaluation of the cognitive performance of children with primary dystonia undergoing GPi DBS shows no deterioration in cognitive performance at 1 year.[48] Similar findings have been found in children with PANK2 disease[49] and in those with secondary dystonia undergoing GPi DBS (*Owen T personal communication*).

DBS and Compulsive and Impulsive Behaviors

Certain dystonia-plus syndromes such as myoclonus-dystonia (DYT-11) due to the epsilon sarcoglycan mutation are commonly associated with obsessive-compulsive disorder (OCD). DBS for dystonia myoclonus is aimed at relieving the dystonia and myoclonus, typically by targeting bilateral GPi.[50,51] However the effects of DBS on OCD symptoms are less well understood.

Lesch–Nyhan disease, a disorder of urate metabolism, represents one of the most distressing and severe childhood-onset motor disorders resulting in repeated compulsive self-mutilation, including biting of lips and fingers and stabbing of the eyes. Small case-series of DBS in this extremely deserving group of children indicate that moderation of dystonia and compulsive behaviors may be achieved with pallidal DBS but further work is required to establish this.[52]

Complications and Safety

A common complication of DBS in children with dystonia is the risk of disappointment with the results.

Our survey of 279 children with dystonia referred for possible entry into the neurosurgical program for ITB or DBS aged 5–18 years demonstrated that 92% had been refractory to all medical and allied-health therapy interventions including botulinum toxin administration by the time of referral.[31] Since over 60% had worsening dystonia from onset of dystonia to first visit in the supraregional ITB/DBS service, the natural history of dystonias in childhood is therefore to worsen or remain severe. So, it is no surprise that some children who do not respond to DBS actually worsen as illustrated in Fig. 19.1. This is disappointing and some of the mitigating factors have been discussed at length in the foregoing sections. The surgical risks are small with functional neurosurgery such as DBS, chief among which being the risk of DBS implant infection. Our current overall risk of infection is approximately 8% of more than 130 children and young people implanted at our center and about 9.5% (2/21) for a first DBS implant under the age of 7 years (unpublished results). Presurgical disinfectant washes 5 days before surgery and perioperative antibiotics vary from center to center. Shorter surgery duration is thought to reduce infection risks.

It seems that surgical and hardware complications might be reduced if the DBS extension cables and battery could be implanted on the skull rather than in the usual infraclavicular or abdominal positions, which result in more movement of the implant and a greater risk of erosions.

The cochlear implant technologies might provide a guide to what is possible in terms of future child-friendly DBS designs.

Challenges for DBS in Children

Common descriptors of motor severity and dystonia are required to understand outcomes in dystonia management.[36]

The challenge for neuromodulation in children with dystonia and other movement disorders is to unlock the full potential of this remarkable technique and deliver it at the optimal time in the child's development before dystonia has resulted in fixed deformities and before the window of opportunity for developmental neuroplasticity has been lost. Better understanding of the neurophysiology of DBS and neuroplasticity of the brain is required. Techniques for establishing the degree of neuroplasticity in healthy children have been reported and it now remains to assess neuroplasticity in children with dystonia and other movement disorders.

Ethics and DBS in Children: "The Broad Clinical Gaze"

A recent sociological study of DBS in children has shown that far from turning children into machines, DBS has an ultimately humanizing effect, allowing children to realize more of their human potential.[53]

Acknowledgments

I would like to thank all the children, parents, and carers and their referring clinicians.

Also thanks to my colleagues within the Complex Motor Disorders Service and colleagues within the Evelina London Children's Hospital, particularly nursing and allied health professional staff on Savannah Ward and on the neurosurgical Lion ward at King's College Hospital and my neurosurgical colleagues, Mr Richard Selway and Professor Keyoumars Ashkan. Particular thanks to Professor Philippe Coubes and Dr Laura Cif, Montpellier, who helped us set up our service in June 2005, and to the numerous colleagues who have given advice and answered questions and provided support over the years.

I would like to thank the following medical charities for supportive grants:

Guy's and St Thomas Charity *New Services and Innovation Grant*: G060708.

The Dystonia Society (UK): Grants 01/2011 and 07/2013.

Action Medical Research: AMR – GN2097.

References

1. Gardner J. A history of deep brain stimulation: technological innovation and the role of clinical assessment tools. *Soc Stud Sci*. 2013;43(5):707–728.
2. Lozano AM, Kumar R, Gross RE, et al. Globus pallidus internus pallidotomy for generalized dystonia. *Mov Disord*. 1997;12:865–870.
3. Coubes P, Roubertie A, Vayssiere N, Hemm S, Echenne B. Treatment of DYT1-generalised dystonia by stimulation of the internal globus pallidus. *Lancet*. June 24, 2000;355(9222):2220–2221.

4. Coubes P, Cif L, El Fertit H, et al. Electrical stimulation of the globus pallidus internus in patients with primary generalized dystonia: long-term results. *J Neurosurg.* 2004;101(2):189–194.

5. Fahn S, Bressman S, Marsden CD. Classification of dystonia. In: Fahn S, Marsden CD, DeLong M, eds. *Classification of Dystonia.* Philadelphia: Lipponcot-Raven; 1998:1–10.

6. Krause M, Fogel W, Kloss M, et al. Pallidal stimulation for dystonia. *Neurosurgery.* 2004;55:1361–1368.

7. Eltahawy HA, Saint-Cyr J, Giladi N, et al. Primary dystonia is more responsive than secondary dystonia to pallidal interventions: outcome after pallidotomy or pallidal deep brain stimulation. *Neurosurgery.* 2004;54:613–619.

8. Vidailhet M, Vercueil L, Houeto JL, et al. Bilateral deep-brain stimulation of the globus pallidus in primary generalized dystonia. *N Eng J Med.* 2005;352:459–467.

9. Kupsch A, Benecke R, Muller J, et al. Pallidal deep-brain stimulation in primary generalized or segmental dystonia. *N Engl J Med.* 2006;355:1978–1990.

10. Tisch S, Rothwell JC, Limousin P, et al. The physiological effects of pallidal deep brain stimulation in dystonia. *IEEE Trans Neural Syst Rehabil Eng.* 2007;15:166–172.

11. Isaias IU, Alterman RL, Tagliati M. Deep brain stimulation for primary generalized dystonia: long-term outcomes. *ArchNeurol.* 2009;66:465–470.

12. Andrews C, Aviles-Olmos I, Hariz M, Foltynie T. Which patients with dystonia benefit from deep brain stimulation? A metaregression of individual patient outcomes. *J Neurol Neurosurg Psychiatry.* 2010;81:1383–1389.

13. Volkmann J, Wolters A, Kupsch A, et al. DBS study group for dystonia. Pallidal deep brain stimulation in patients with primary generalised or segmental dystonia: 5-year follow-up of a randomised trial. *Lancet Neurol.* 2012;11(12):1029–1038.

14. Vidailhet M, Yelnik J, Lagrange C, et al. Bilateral pallidal deep brain stimulation for the treatment of patients with dystonia-choreoathetosis cerebral palsy: a prospective pilot study. *Lancet Neurol.* 2009;8:709–717.

15. Cif L, Vasques X, Gonzalez V, et al. Long-term follow-up of DYT1 dystonia patients treated by deep brain stimulation: an open-label study. *Mov Disord.* 2010;25(3):289–299.

16. Ruge D, Laura Cif L, Limousin P, et al. Shaping reversibility? Long-term deep brain stimulation in dystonia: the relationship between effects on electrophysiology and clinical symptoms. *Brain.* 2011;134(7):2106–2115.

17. Ruge D, Laura Cif L, Limousin P, et al. Longterm deep brain stimulation withdrawal: clinical stability despite electrophysiological instability. *J Neurol Sci.* July 15, 2014;342(1–2):197–199.

18. Castelnau P, Cif L, Valente EM, et al. Pallidal stimulation improves pantothenate kinase-associated neurodegeneration. *Ann Neurol.* 2005;57(5):738–741.

19. Pretto TE, Dalvi A, Kang UJ, Penn RD. A prospective blinded evaluation of deep brain stimulation for the treatment of secondary dystonia and primary torticollis syndromes. *J Neurosurg.* 2008;109:405–409.

20. Katsakiori PF, Kefalopoulou Z, Markaki E, et al. Deep brain stimulation for secondary dystonia: results in 8 patients. *Acta Neurochir (Wien).* 2009;151:473–478.

21. Speelman JD, Contarino MF, Schuurman PR, et al. Deep brain stimulation for dystonia: patient selection and outcomes. *Eur J Neurol.* 2010;17(suppl 1):102–106.

22. Timmermann L, Pauls KAM, Wieland K. Dystonia in neurodegeneration with brain iron accumulation: outcome of bilateral pallidal stimulation. *Brain.* 2010;133(3):701–712.

23. Air EL, Ostrem JL, Sanger TD, Starr PA. Deep brain stimulation in children: experience and technical pearls. *J Neurosurg Pediatr.* 2011;8:566–574.

24. Marks WA, Honeycutt J, Acosta Jr F, et al. Dystonia due to cerebral palsy responds to deep brain stimulation of the globus pallidus internus. *Mov Disord.* 2011;26:1748–1751.

25. Koy A, Hellmich M, Pauls KA, et al. Effects of deep brain stimulation in dyskinetic cerebral palsy: a meta-analysis. *Mov Disord.* 2013;28(5):647–654.

25a. Romito LM, Zorzi G, Marras CE, Franzini A, Nardocci N, Albanese A. Pallidal stimulation for acquired dystonia due to cerebral palsy: beyond 5 years. *Eur J Neurol.* 2015;22(3):426–e32.

III. INVASIVE BRAIN STIMULATION IN CHILDREN

26. Lumsden DE, Kaminska M, Gimeno H, et al. Proportion of life lived with dystonia inversely correlates with response to pallidal deep brain stimulation in both primary and secondary childhood dystonia. *Dev Med Child Neurol.* 2013;55(6):567–574.
27. Gimeno H, Tustin K, Selway R, Lin JP. Beyond the Burke-Fahn-Marsden Dystonia Rating Scale: deep brain stimulation in childhood secondary dystonia. *Eur J Paediatr Neurol.* 2012;16(5):501–508.
28. Gimeno H, Gordon A, Tustin K, Lin JP. Functional priorities in daily life for children and young people with dystonic movement disorders and their families. *Eur J Paediatr Neurol.* 2013;17(2):161–168.
29. Gimeno H, Lumsden D, Gordon A, et al. Improvement in upper limb function in children with dystonia following deep brain stimulation. *Eur J Paediatr Neurol.* 2013;17(4):353–360.
30. Gimeno H, Tustin K, Lumsden D, et al. Evaluation of functional goal outcomes using the Canadian Occupational Performance Measure (COPM) following Deep Brain Stimulation (DBS) in childhood dystonia. *Eur J Paediatr Neurol.* 2014;18(3):308–316.
31. Lin JP, Lumsden DE, Gimeno H, Kaminska M. The impact and prognosis for dystonia in childhood including dystonic cerebral palsy: a clinical and demographic tertiary cohort study. *J Neurol Neurosurg Psychiatry.* 2014;85(11):1239–1244.
32. Lumsden DE, Gimeno H, Tustin K, et al. Interventional studies in childhood dystonia do not address the concerns of children and their carers. *Eur J Paediatr Neurol.* 2015;19(3):327–336.
33. Vasques X, Cif L, Hess O, et al. Prognostic value of globus pallidus internus volume in primary dystonia treated by deep brain stimulation. *J Neurosurg.* 2009;110(2):220–228.
34. Lin JP. The cerebral palsies: a physiological approach. *J Neurol Neurosurg Psychiatry.* 2003;74(suppl 1):i23–i29.
35. Lin JP. The contribution of spasticity to the movement disorder of cerebral palsy using pathway analysis: does spasticity matter? *Dev Med Child Neurol.* 2011;53:7–9.
36. Elze CM, Gimeno H, Tustin K, et al. Burke-Fahn-Marsden dystonia severity, Gross-Motor, Manual-Ability and Communication-Function Classification scales in childhood hyperkinetic movement disorders including CP: a Rosetta Stone study. *Dev Med Child Neurol.* 2016. [accepted September 2015].
37. Silveira-Moriyama L, Lin JP. A field guide to current advances in paediatric movement disorders. *Curr Opin Neurol.* 2015;28(4):437–446.
38. McClelland V, Mills K, Siddiqui A, Selway R, Lin JP. Central motor conduction studies and diagnostic magnetic resonance imaging in children with severe primary and secondary dystonia. *Dev Med Child Neurol.* 2011;53(8):757–763.
39. Lumsden DE, McClelland V, Ashmore J, et al. Central Motor Conduction Time and diffusion tensor imaging metrics in children with complex motor disorders. *Clin Neurophysiol.* 2015;126(1):140–146.
40. Damji O, Keess J, Kirton A. Evaluating developmental motor plasticity with paired afferent stimulation. *Dev Med Child Neurol.* 2015;57(6):548–555.
41. Quartarone A, Bagnato S, Rizzo V, et al. Abnormal associative plasticity of the human motor cortex in writer's cramp. *Brain.* 2003;126:2586–2596.
42. Szyszko TA, Dunn JT, O'Doherty MJ, Reed L, Lin JP. Role of ^{18}F-FDG PET imaging in paediatric primary dystonia and dystonia arising from neurodegeneration with brain iron accumulation. *Nucl Med Commun.* 2015;36(5):469–476.
43. Lumsden DE, Kaminska M, Tustin K, et al. Battery life following pallidal deep brain stimulation (DBS) in children and young people with severe primary and secondary dystonia. *Childs Nerv Syst.* 2012;28(7):1091–1097.
44. Kaminska M, Lumsden DE, Ashkan K, et al. Rechargeable deep brain stimulators in the management of paediatric dystonia: well tolerated with a low complication rate. *Stereotact Funct Neurosurg.* 2012;90(4):233–239.
45. Chelvarajah R, Lumsden D, Kaminska M, et al. Shielded battery syndrome: a new hardware complication of deep brain stimulation. *Stereotact Funct Neurosurg.* 2012;90(2):113–117.

46. Lumsden DE, Ashmore J, Charles-Edwards G, Lin JP, Ashkan K, Selway R. Accuracy of stimulating electrode placement in paediatric pallidal deep brain stimulation for primary and secondary dystonia. *Acta Neurochir (Wien)*. 2013;155(5):823–836.

47. Lumsden DE, Ashmore J, Charles-Edwards G, Selway R, Lin JP, Ashkan K. Observation and modeling of deep brain stimulation electrode depth in the pallidal target of the developing brain. *World Neurosurg*. 2015;83(4):438–446.

48. Owen T, Gimeno H, Selway R, Lin JP. Cognitive function in children with primary dystonia before and after deep brain stimulation. *Eur J Paediatr Neurol*. 2015;19(1):48–55.

49. Mahoney R, Selway R, Lin JP. Cognitive functioning in children with pantothenate-kinase-associated neurodegeneration undergoing deep brain stimulation. *Dev Med Child Neurol*. 2011;53(3):275–279.

50. Gruber D, Kühn AA, Schoenecker T, et al. Pallidal and thalamic deep brain stimulation in myoclonus-dystonia. *Mov Disord*. 2010;25(11):1733–1743.

51. Azoulay-Zyss J, Roze E, Welter ML, et al. Bilateral deep brain stimulation of the pallidum for myoclonus-dystonia due to ε-sarcoglycan mutations: a pilot study. *Arch Neurol*. 2011;68(1):94–98.

52. Visser JE, Schretlen DJ, Cif L, et al. Long-term follow-up of DBS in Lesch-Nyhan disease: patient reported outcome. *Mov Disord*. 2014;29(suppl 1):1270. [abstract].

53. Gardner J. *The Broad Clinical Gaze in Paediatric Deep Brain Stimulation Science, Medicine, and Anthropology*; 2014. http://somatosphere.net/2014/03/the-broad-clinical-gaze-in-paediatric-deep-brain-stimulation.html.

20

Deep Brain Stimulation Children – Surgical Considerations

R.D. Bhardwaj

Sanford Children's Hospital, Sioux Falls, SD, United States

BACKGROUND

The application of deep brain stimulation (DBS) as a therapeutic intervention in children suffering from neurological problems has been gaining favor and interest over the past decades. The field has been learning much over this span of time regarding optimization of patient selection, discovering new medical indications, as well as improving surgical techniques. We are currently at a very interesting time in DBS and neuromodulation, as we are beginning to ask questions of how the delivery of chronic electrical stimulation into a growing brain changes brain function. Answers to such questions will undoubtedly lead us to better understand neuroplasticity.

In certain movement disorders, such as primary dystonia, the effects of DBS can sometimes be instantaneous and demonstrate efficacy bordering on the miraculous. Some children starting from an extremely disabled baseline are able to regain the ability to mobilize postoperatively, and then normalize their lives within months. Interestingly, these improvements can be durable with long-term follow-up over 10 years showing similar efficacy.[1] Two of the goals within the practice of medicine are to prevent suffering and to promote healing. These two goals are sometimes met in the surgical treatment of movement disorders, within the setting of DBS therapy. The treatment of pediatric dystonia with the use of DBS is one of the rare opportunities in medicine where one is able to definitively reverse a clinically striking, apparently irreversible, and progressively debilitating medical condition with dramatic effect following a definitive surgical operation. With dramatic improvements such as this, researchers within the field are now attempting to better understand how electrical current can elicit such improvement.

The clinical indications for DBS have been expanding quite broadly over the past decades. Whereas initial DBS indications were mainly for tremor and dyskinesias, new trials have been performed tackling maladies including obesity, depression, dementia, and other diseases. This has also transferred to pediatric indications in some cases, although the barrier to treat a child with a surgically invasive procedure always holds a high ethical level of standard.[2] Neuromodulation is definitely an expanding therapeutic option, as many novel brain circuitry disorders will begin to be addressed in coming years.

From a purely pediatric vantage, the beauty of DBS in children is that the device is providing neuromodulatory effect into a developing organ. The child's brain is dynamic, it is plastic, and it is still growing. These fundamental properties that we often take for granted in pediatrics afford a very exciting intersection with traditional neuroscience, and especially neurophysiology.

As will be discussed, it is clear to see that DBS will offer many new and diverse opportunities to better treat children with brain disorders, and also provide an excellent perspective into the neuroplasticity of a growing brain.

TECHNICAL ASPECTS OF DBS

The implantation of DBS systems has been occurring for many decades now and over 100,000 devices have been implanted globally. Although the majority of systems have been implanted in adults suffering from

Parkinson disease and tremor, it is estimated that about 1% of systems have been placed into children. Historically, electrical stimulation to the brain has been used from the time of Roman emperors (using electrical torpedo fish to treat headache) and evolved to the initial DBS prototype of Professor Delgado's famous experiments with brain stimulation in a charging bull. This lead to the first human DBS implantation in 1987.[3]

The DBS device (see Fig. 20.1A) is composed of three parts: the implanted intracranial electrodes shown bilaterally entering the skull at the implant site, the subcutaneously tunneled leads, and the connection to the subcutaneously implanted pacemaker-like device, also known as the internal pulse generator (IPG). Typically, the entire surgical procedure is staged over two procedures, spaced 3–10 days apart. In stage 1, the lead (comprised of four insulated wires with four electrode contacts seen at the electrode tip, see Fig. 20.1B) is carefully and precisely placed at the target site, with the leads then being connected to the extension wires. In stage 2 of the operation, the extension wires are tunneled subcutaneously under the skin and then connected to the IPG, which houses the battery that supplies the neurostimulator current. After 2 weeks, the DBS system is usually activated within the setting of the clinic by a radio device that is in contact with the skin near the implanted IPG.

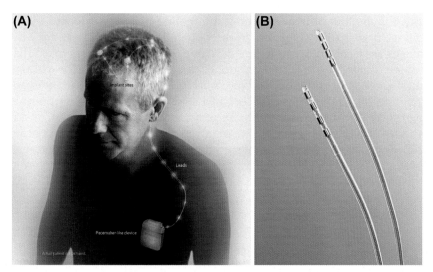

FIGURE 20.1 (A and B) The DBS device (see A) is composed of three components: the implanted intracranial electrodes shown bilaterally entering the skull at the implant site which connect to the subcutaneously tunneled leads, and then join into the subcutaneously implanted pacemaker-like device, also known as the internal pulse generator (IPG). (B) A magnified view of the end of the implanted electrodes, one can see the four metal contacts, the source of the therapeutic electrical current. *Images courtesy of Medtronic Inc.*

Great care and thought is focused towards accurate placement of electrodes to the intended neuroanatomical target, in combination with consideration of maximal safety. The anatomic target depends on the clinical indication; for example, in dystonia the commonly selected target would be the globus pallidus interna. There are many different manners in which one can plan and surgically access the desired target, and these are comprehensively discussed in this source.[4] I will briefly outline safe and reasonable approach taken by the author in placing DBS systems in children.

First, a high-resolution (3T) magnetic resonance imaging (MRI) of the brain is taken and the relevant anatomy is studied. Children with cerebral palsy may have significantly altered basal ganglia anatomy, and the perinatal strokes often cause lateral asymmetry to arise. In adults, many centers perform DBS surgery while the patient is awake, and brain recording is also done. The benefit of this strategy is that one can obtain direct neurophysiological recordings, and have greater certainty that the correct neuroanatomic target has been found. However, doing awake surgery on children has obvious disadvantages mainly due to patient safety, comfort, and logistical practicalities of children not able to tolerate this procedure under local anesthesia. Another key point to factor into this strategy is that multiple passes of the electrodes are often performed with monitoring, and each additional electrode passage carries an accumulated and additional amount of risk of hemorrhage and damage.

In place of doing awake recording to verify electrode placement accuracy, the method of neuroanatomical targeting is a very safe and viable option in pediatric surgery.[5] In order to follow this method, the MRI images are then uploaded to a neuronavigation planning system, so that a three-dimensional (3D) brain model can be constructed. The volume can then be seen in various coronal, axial, or sagittal planes with relevant anatomy highlighted (as depicted in Fig. 20.2A–E). The target point is then chosen and a safe electrode trajectory is then created in 3D. Considerations to avoid the ventricle are made, as this is seen to elicit less brain shift. Trajectories avoiding brain sulci are advantageous, as cortical transit through a gyrus will have less probability of disrupting blood vessels and causing hemorrhage. The avoidance of passage through eloquent cortical regions is also prioritized.

Once the above planning is complete, there are a few options to precisely place leads. The utility of placing a stereotactic frame is an excellent method of safely and accurately targeting the correct brain regions (shown in Fig. 20.3). On the day of surgery, a stereotactic frame is placed on the head of the patient while under general anesthetic by way of pin placement into the skull. An image (computed tomography [CT] or MRI) is then taken with the head frame on, which one is then able to merge on the computer platform to the pre-existing MRI 3D volume that was previously done. The planning software then translates 3D computer

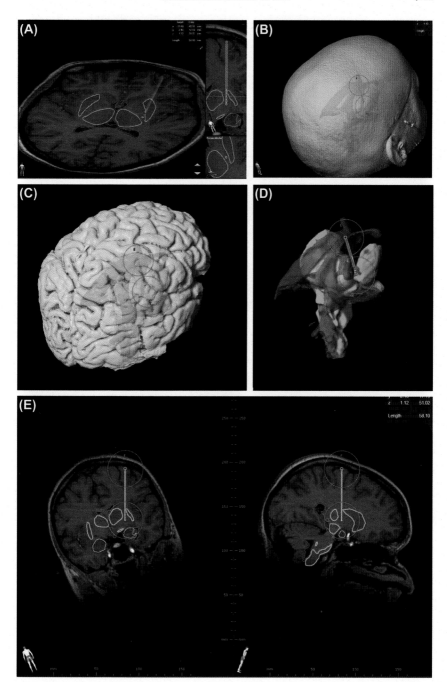

FIGURE 20.2 (A–E) These images represent the planning that occurs once 3D MRI is obtained and loaded onto the neuronavigation platform. These images are based on prototype software, but the key point is to be aware about the planning capacity that is and will be available in DBS planning. (A) An axial brain slice that demonstrates the ability to delineate various subcortical brain structures. One can visualize the deeper structures, through the head (B) and the brain (C). (D) These brain structures in 3D, in relation to the trajectory path. (E) The trajectory in sagittal and coronal planes. *Images courtesy of Brainlab Inc.*

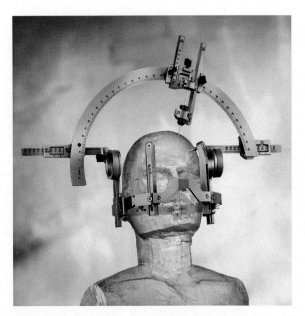

FIGURE 20.3   This figure depicts how a stereotactic frame is placed and fixed on a phantom head model. One can see the sets of x, y, and z millimeter marking settings, as well as arc and ring co-ordinates that enable one to achieve great precision and accuracy of electrode placement to the desired target within the operative setting. *Image courtesy of Elekta Inc.*

MRI coordinates into stereotactic coordinates, which are then used in the operating room. The stereotactic frame is then placed on the patient in the operating room, and the coordinates allow for very accurate and precise placement of the electrodes to millimeter precision values. Other methods that rely on neuronavigation also using interventional, intraoperative MRI in pediatric cases have been described.[6]

It is always prudent to avoid radiation delivery to children, and one should strive to use CT as judiciously as possible. One should utilize low-dose CT whenever possible, as in the case of the image obtained after the frame is placed.

When using neuroanatomic targeting, it is prudent to verify lead placement postoperatively. One also can rule out hemorrhage or other postoperative complications with imaging after surgery. Fig. 20.4 depicts postoperative CT merged with the planning MRI. One can observe that the white lines of the electrodes seen on CT imaging almost perfectly correlate with the blue lines that were created on the planning computer platform. This is a desirable outcome, as all clinical efficacy is directly related to the precise and accurate placement of the electrodes.

As mentioned above, the second stage of the procedure typically occurs 3–10 days later. Our practice is to wait about 2 weeks after the second stage

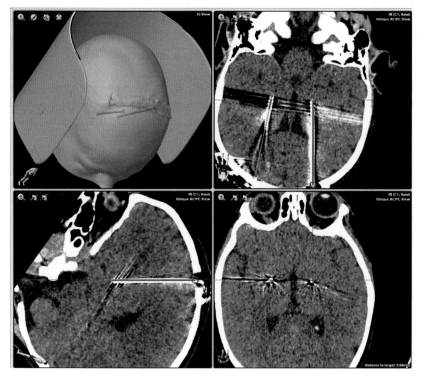

FIGURE 20.4 This image is a merged postoperative CT image overlaid on the MRI plan, as seen in the neuronavigation workstation. The blue lines represent the planned electrode trajectories as seen: on the 3D head model in the top left, followed by various differing planes. The white lines are the actual implanted electrodes seen on CT. One can see that there is very close approximation of the optimal theoretical electrode placement as compared to actual physical placement within the child's brain; with the ideal blue trajectory superimposed by the actual white trajectory.

before turning the device on, so that issues related to potential brain edema after electrode placement have resolved fully. This is done in the clinical setting, and the device is activated by radiofrequency control from a programmer held over the skin of the internal pulse generator. The importance of programming is high and lies in the ability to program the DBS device, by clinically selecting the best contacts (each electrode has four metal contacts) to provide the electrical current at the optimal site. One can change which contact provides current, the shape of the current (eg, bipolar or unipolar), at what voltage of current is administered, and also how much current is administered. It is also exceedingly important to note that the greater the volume of current administered, the greater is the chance of causing unintended stimulation of adjacent brain circuits. These secondary effects may then lead to clinical side effects, commonly seen as flashes of light, tingling sensation, discomfort, etc. depending on which anatomic regions are being stimulated. Thus, the actual optimization of programming is an "art" which requires a seasoned programmer to administer therapy correctly.

This constant modification of settings over months and years is necessary for the ongoing benefit to the patient. Typically, in our experience, the stimulation parameters are: amplitude 1.5–4.0 V, pulse width 60–120 μs, frequency 60–180 Hz, and most often a single electrode monopolar array.

Complications with pediatric implantation of DBS are mainly related to infection, with reports indicating that higher risks occur with patients less than 10 years of age.[7] Great effort must be taken to decrease infection risk in children, with meticulous attention to surgical detail, operative room vigilance, and antibiotic usage all being important factors to decrease potential risk. Lead fracture, lead extender fracture, inaccurate placement, infarction, and cerebrospinal fluid accumulation are all related complications that have also been reported.

Successful DBS surgery in the pediatric population is not a trivial task, but with correct care and attention, excellent results can be attained. As mentioned above, one first must have accurate and precise lead placement so that the programmer has a good opportunity for delivery of electrical stimulation in a therapeutic manner.

CLINICAL INDICATIONS

General indications for DBS have been mainly linked to movement disorders historically. Dystonia has been the major indication for DBS implantation in children. Adult DBS has been mainly focused upon the movement disorders, tremor, and Parkinson disease, but many new indications have arisen over the past decade.[3] Neuropsychiatric disorders (such as depression, obsessive-compulsive disorder, etc.), metabolic disorders such as obesity, decreased levels of consciousness states, and cognitive enhancement for dementia are all some of the new and exciting targets for adult DBS therapy over the past years. Surgical indications for children always carry a much higher barrier of entry, as ethical and moral issues must be clearly addressed in the setting of informed consent within the treatment of minors.[2] That being said, the pediatric indications for DBS are certainly expanding.

Dystonia is a neurological disorder with the hallmark of sustained, patterned, repetitive muscle contractions producing involuntary twisting or squeezing movements and abnormal postures. Dystonia is often divided into the categories of primary dystonia (often caused by a genetic mutation, eg, DYT1 mutations) and secondary dystonia; which was caused by an earlier insult (ischemia, metabolic, trauma, etc.). Over 300,000 people in the US are diagnosed with dystonia, making it the third most common movement disorder, after essential tremor and Parkinson disease. Treatment typically may include systemic pharmacological administration (oral or intrathecal administration), or locally administered agents to muscles. Children suffering from severe and chronic forms of dystonia are often misdiagnosed and suboptimally treated. The children with the diagnosis of primary dystonia initially enjoy early normal motor development with onset of symptoms typically not occurring until the middle of the first decade of life, and then gradually begin to lose that function over time. It is a common story to see a child that seems normal until kindergarten, and then gradually requires crutches, splints, walkers, and then wheelchairs for mobility over the following years. Treatment with neurologists and physiatrists often occurs, however the treatment with medications are of limited use and disease progression occurs steadily in the majority of cases.

DBS therapy has been performed in children suffering from both primary and secondary dystonia. Whereas primary dystonia is typically caused by genetic mutations, secondary dystonia is typically acquired. Ischemic insults, trauma, or metabolic (ie, Wilson disease) causes will all alter the anatomic structure of the brain. Thus, patients with secondary dystonia present with a far more varied constellation of symptoms and comorbidities. These factors make the treatment of secondary dystonia with DBS more challenging; as there is a wide spectrum of neuroanatomy, degree of motor dysfunction, and fixed abnormalities within the population of children who have been given the diagnosis of cerebral palsy.

DBS for the indications of other movement disorders in children; such as tremor, postinfarct hemidystonia,[8] and status dystonicus[9] have been reported. Lesch–Nyhan disease is a rare X-linked recessive genetic disorder that carries a very characteristic set of symptoms. Children with this neurobehavioral disorder exhibit dystonia, have delayed motor skills, and exhibit aggressive behavior in combination with self-mutilatory tendencies. DBS has been reported to be quite helpful in alleviating both the motor as well as behavioral components of this rare disease.[10]

Neuropsychiatric disorders such as Tourette syndrome[11] and obsessive-compulsive disorder[12] have also had attempts at amelioration through DBS therapy, with some pediatric cases reported. Trials investigating the efficacy of DBS therapy in adult refractory epilepsy have been conducted, and some improvements were noted.[13] This work has not been duplicated in children, but it would be interesting to see whether the continuing brain development within the pediatric populations would help or hinder the efficacy of DBS therapy within the context of aberrant circuitry seen in epilepsy. Adult DBS pilot studies for obesity in adults have been performed,[14] and this may be a potential pediatric indication in the future as well. Important ethical caveats are inherent when considering potential

DBS therapy in children for psychiatric and/or novel indications.[2] There is definitely potential for expanding clinical indications, however, great care and wisdom must be exercised in the contemplation of such proposed endeavors, especially within the neuropsychiatric disorders. These reviews provide an in-depth look at current and future pediatric clinical indications.[15–17]

TREATMENT EFFICACY

Since the first human implantation, the field of DBS therapy has evolved and matured with the publication of many trials demonstrating benefit. DBS has undergone large, randomized, blinded trials to show efficacy over the prior gold standard medical therapy in the treatment of certain populations of patients suffering from Parkinson disease. In the realm of pediatric medicine, the majority of reporting has been done in the treatment of dystonia.

The choice of optimal neuroanatomical target is profoundly important in the alleviation of symptoms within a given pathology. Constant fine-tuning of existing targets within patient subpopulations (eg, subthalamic nucleus, globus pallidus interna, and pedunculopontine nucleus as in treatment for Parkinson disease) is ongoing as well as the discovery of novel ones as indications expand, ie, addiction, obesity, cognition enhancement.

Once the ideal patient is selected with the right clinical indication, and the optimal surgical target has been appropriately accessed, then one is in a good position to achieve therapeutic improvement with the correct programming ability. The important steps postsurgery then lie in the programmer's ability to clinically select the best contacts (each electrode has four contacts) to provide the electrical current at the optimal dose. One can change which contact provides current, the shape of the current (eg, bipolar or unipolar), what voltage the current is administered at, and also how much current is administered. It is also exceedingly important to note that the greater the volume of current administered, the greater the chance of causing unintended stimulation of adjacent brain circuits. This will then lead to side effects, commonly seen as flashes of light, tingling sensation, discomfort, etc., depending on the brain structures that have been unintentionally stimulated. This constant modification of settings over months and years is necessary for the ongoing benefit to the patient. Typically, in our experience for pediatric dystonia, the stimulation parameters are: amplitude 1.5–4.0 V, pulse width 60–120 μs, frequency 60–180 Hz, and most often a single electrode monopolar array.

Many trials have been published over the past years that clearly demonstrate the efficacy of DBS in ameliorating the motor symptoms in children suffering from primary dystonia.[18–20] Factors that predict better response to DBS therapy in children suffering from primary dystonia have been reported to be younger age of DBS placement and shorter duration of disease.[21] Given this valuable information, it is the correct diagnosis of primary dystonia in symptomatic children and then the prompt referral to a pediatric DBS center, which represent key factors to helping these children in need.

In particular cases of known DYT1 primary dystonia, great efficacy has been reported with DBS therapy. This study shows robust and long-term benefits from DBS treatment in children over a 10-year period, with a known diagnosis of the genetic mutation of the DYT1 gene.[1] Again, it seems that a shorter duration of symptomatology is correlated with improved DBS efficacy, in this subset of children suffering from dystonia caused by the DYT1 mutation.[22]

Predicting and understanding DBS efficacy in children suffering from secondary dystonia is much more complicated. As stated in the last section, most children with secondary dystonia have the diagnosis of cerebral palsy. This patient population is very heterogeneous in terms of symptomatology, degree of movement disorder, cognitive disability, and magnitude of abnormal brain anatomy; so, obtaining relevant data from pilots studies published in various sites makes generalization of DBS efficacy challenging. An important meta-analysis looking at DBS therapy in children suffering from cerebral palsy seemed to suggest that DBS therapy may have helped in quality of life, showed a moderate but significant improvement in motor scales, and indicated that response to DBS seemed to be less dramatic than in children with the diagnosis of primary dystonia.[23]

PHYSIOLOGY OF NEUROMODULATION, NEUROPLASTICITY, AND PEDIATRIC CONSIDERATIONS

The intersection of many fields comes into prominence after one clearly sees the clinical efficacy of DBS on children in certain circumstances. How does the delivery of electrical stimulation affect brain function in the acute setting, as well as chronically? What is the mode of action of DBS? How does the stimulation affect plasticity of a growing brain? Age seems to matter with respect to treatment efficacy as seen in the last section. Many of these questions have no firm answers, but the field is making steady progress.

The precise working understanding of how DBS is able to exert its physiological activity is not fully known at this point in time. The electrical current that is delivered from the electrode contact will invariably affect the local cells within the vicinity, at either a subcellular, cellular, or white matter axonal bundle level. It is not clear whether the key phenomenon with DBS is elicited by the blockade of unwanted neuronal activity through inhibition, or whether it is the net result of hyperactivation of neuronal circuitry. Hypotheses regarding subcellular effects on dendrites or axonal bundles, with various perturbations in firing of neurons or with neurotransmitter release, have also been put forward. DBS effects on long-term potentiation, and the implied ramifications on plasticity, are also implicated in mechanistic understanding.

Even the cell of action has been put into question – are all effects based on neurons only or can this electrical stimulation also affect glia, with downstream glial–neuron perturbations occurring? Some hypotheses see DBS affecting primarily axons, so that DBS is mainly affecting white matter tracts. With this rationale, DBS can be seen as a treatment modality that is affecting brain circuits, not just individual cells. Insight in neural oscillations has increased tremendously, and the effects of DBS on brain activity coupled to the study of oscillatory patterns have led to novel research. The timescale is also important, since in some clinical DBS indications such as tremor, the effects occur almost instantaneously, as compared to much slower changes over weeks to months, as seen in depression. Thus, mechanisms ranging from neurotransmitter release and signal jamming that take in the order of milliseconds all the way to changes in neurogenesis over the scale of weeks to months, may be relevant within the appropriate clinical context. This excellent review article[24] goes into greater depth and provides a comprehensive overview.

With constant stimulation as seen in DBS, one could question why tolerance does not occur over time. As mentioned before, children suffering from DYT1-related dystonia often show benefit over years, and many have shown constant benefit to therapy in the range of decades. Children have been known to suffer rapid clinical deterioration after many years with DBS battery depletion, and then quickly regain efficacy postoperatively after battery replacement. However, a small series has shown that in two pediatric primary DYT1 dystonia cases, efficacy was lost after great improvements occurred initially within the first 6 months.[25] This is an interesting finding since one can observe how little we actually know about the mechanisms of DBS on growing and plastic brains. However, the vast majority of published, primary dystonic children show prolonged and sustained improvement with DBS treatment over years.

Increased plasticity and the pediatric populations have always been linked, given the well-known Kennard principle,[26] which argues for the presence of greater neuroplastic changes in younger versus older patients. The ability to study direct neuroplastic changes in children is difficult, but advances in imaging have recently improved these prospects. An ambitious study looked at children with hemidystonia, from the perspective of functional MRI.[27] In motor tasks, various changes in functional MRI activity were seen in thalamic, pallidal, and temporal cortical brain regions in comparison to ipsilateral and contralateral brain. It will be interesting to also study white matter changes longitudinally in cohorts of children with DBS implantation for the long term, in order to determine if one could possibly see neuroanatomical changes in white matter bundles. This could be very interesting as one could make a direct argument for physical manifestations of neuroplasticity.

If younger DBS implantation is better, are there physical limits to DBS placement from an age standpoint? Practically speaking, there is concern for implanting a DBS system into a child whose brain and skull will grow so much, that the electrodes would have relative movement to the desired brain target over time. A study that modeled this based on skull and brain growth concluded that after the age of 7 years, growth would be unlikely to cause the electrode to migrate significantly.[28] Thus, in children older than 7 years old, relative brain and skull growth should not adversely impact target changes over time. In children younger than this, clinical vigilance with imaging should be maintained through the early years in order to achieve optimal therapeutic performance.

FUTURE DEVELOPMENTS AND INNOVATION

There is considerable interest in the field of neuromodulation from within the field of clinical neurosciences and through the industrial partnership intrinsic to device manufacturing. DBS had ambitious yet humble beginnings within the bullring of Professor Delgado, with solid progress occurring up to the first human implantation in the last century. This new millennium now is applying the therapeutic stimulation from DBS into applications of the brain and mind; as opposed to just being relegated to modulating basal ganglia structures for specific movement disorders only. There is interest now in tackling disorders of the mind, and current trials now are focused upon improving the mind's cognitive powers as well as deep thought going into mood and complex disease states, such as schizophrenia and addiction recovery.[29–31]

Keeping up with the clinical momentum, device innovation has also been rapidly improving. DBS batteries that power the IPG now have the capacity to be rechargeable. The delivery of therapeutic current originating from the four contacts near the tip of the electrode is being improved. Thoughts of steering current into neuroanatomical regions that are desired and avoiding spread to undesirable regions are now coming into reality with next-generation devices. This will certainly add to improved therapeutic efficacy, while decreasing unintended stimulatory actions (ie, side effects); thus improving the focus and precision of the existing DBS devices. The notion of having four contacts has been standard over the past decades, but this will likely change and be improved upon as well.

Next-generation DBS devices will also have the ability to sense and record brain activity, in the range of local field potentials.[3,29] Initially, this type of data will be exceedingly important within the field of neurophysiology. This will allow us to directly observe the electrical firing patterns of different brain regions akin to the opportunity currently only seen within the setting of intraoperative neurophysiological recording sessions. However, this new ability will afford chronic recordings and data analysis over the long term, and researchers will be able to determine what a diseased-state brain's firing patterns look like, before and after applying therapeutic stimulation. We will be able to directly search for and possibly discover electrical biosignatures of diseased states (eg, Parkinson disease, dystonia, etc.) and then observe, in real-time, if and how the abnormal firing pattern is altered with the addition of therapeutic stimulation. This may be exceedingly helpful in the fine-tuning of DBS programming in the future, and thus add some science to the current "art."

With advanced forms of diffusion-tensor imaging MRI, one will likely be able to better focus precision of electrode placement neuroanatomically, with the notion that one is directly influencing white matter pathways and circuits via DBS. As imaging technologies improve over the coming years, field such as connectomics, functional MRI, and resting state MRI may prove to add valuable wealth to the capacity of neuromodulation. As has been mentioned, if DBS is able to exert its influence primarily on white matter tracts and circuits; then these novel imaging modalities will gain prominence in seeing changes with therapeutic effect and modulation of brain circuitry. This will elevate imaging and anatomy of the brain from the current static state, with major importance in planning and target selection, into the dynamic world of brain function and real-time modulation.

The application of the above improvements in technology and imaging will undoubtedly have great benefits in the field of pediatrics. Plasticity is a dynamic phenomenon, and these innovations will enable us to better see these subtle but exceedingly important changes. As resolution of imaging improves, we may be able to visualize plasticity directly; perhaps in the modifications of white matter thickness or the changes in gray matter density. This natural evolution of imaging coupled to neuromodulation will afford a very exciting time ahead in the field of pediatric plasticity, in the setting of DBS.

One should not neglect to reflect upon the fact that the human brain is the most complex, complicated (and beautiful) structure in the known universe, and that the opportunity to directly influence its functions through DBS is nothing short of miraculous, in certain circumstances. The opportunity offered through the technology of DBS allows one to rudimentarily speak (via electric current) to and with the existing circuits of the brain, and very soon, one will also be able to listen to the operating circuits in a functional, living, human brain. This will inevitably allow for exciting times ahead in both the DBS as well as neuroplasticity fields.

References

1. Cif L, Vasques X, Gonzalez V, et al. Long-term follow-up of DYT1 dystonia patients treated by deep brain stimulation: an open-label study. *Mov Disord*. 2010;25:289–299.
2. Woopen C, Pauls KA, Koy A, Moro E, Timmermann L. Early application of deep brain stimulation: clinical and ethical aspects. *Prog Neurobiol*. 2013;110:74–88.
3. Okun MS. Deep-brain stimulation–entering the era of human neural-network modulation. *N Engl J Med*. 2014;371:1369–1373.
4. Starr PA, Barbaro NM. *Functional Neurosurgery*. Thieme; 2011.
5. Vayssiere N, Hemm S, Zanca M, et al. Magnetic resonance imaging stereotactic target localization for deep brain stimulation in dystonic children. *J Neurosurg*. 2000;93:784–790.
6. Starr PA, Markun LC, Larson PS, Volz MM, Martin AJ, Ostrem JL. Interventional MRI-guided deep brain stimulation in pediatric dystonia: first experience with the ClearPoint system. *J Neurosurg Pediatr*. 2014;14:400–408.
7. Air EL, Ostrem JL, Sanger TD, Starr PA. Deep brain stimulation in children: experience and technical pearls. *J Neurosurg Pediatr*. 2011;8:566–574.
8. Witt J, Starr PA, Ostrem JL. Use of pallidal deep brain stimulation in postinfarct hemidystonia. *Stereotact Funct Neurosurg*. 2013;91:243–247.
9. Walcott BP, Nahed BV, Kahle KT, Duhaime AC, Sharma N, Eskandar EN. Deep brain stimulation for medically refractory life-threatening status dystonicus in children. *J Neurosurg Pediatr*. 2012;9:99–102.
10. Cif L, Biolsi B, Gavarini S, et al. Antero-ventral internal pallidum stimulation improves behavioral disorders in Lesch-Nyhan disease. *Mov Disord*. 2007;22:2126–2129.
11. Schrock LE, Mink JW, Woods DW, et al. Tourette syndrome deep brain stimulation: a review and updated recommendations. *Mov Disord*. 2015;30:448–471.
12. Garnaat SL, Greenberg BD, Sibrava NJ, et al. Who qualifies for deep brain stimulation for OCD? Data from a naturalistic clinical sample. *J Neuropsychiatry Clin Neurosci*. 2014;26:81–86.
13. Lega BC, Halpern CH, Jaggi JL, Baltuch GH. Deep brain stimulation in the treatment of refractory epilepsy: update on current data and future directions. *Neurobiol Dis*. 2010;38:354–360.
14. Dupre DA, Tomycz N, Oh MY, Whiting D. Deep brain stimulation for obesity: past, present, and future targets. *Neurosurg Focus*. 2015;38:E7.
15. Lipsman N, Ellis M, Lozano AM. Current and future indications for deep brain stimulation in pediatric populations. *Neurosurg Focus*. 2010;29:E2.
16. DiFrancesco MF, Halpern CH, Hurtig HH, Baltuch GH, Heuer GG. Pediatric indications for deep brain stimulation. *Childs Nerv Syst*. 2012;28:1701–1714.
17. Marks WA, Honeycutt J, Acosta F, Reed M. Deep brain stimulation for pediatric movement disorders. *Semin Pediatr Neurol*. 2009;16:90–98.
18. Haridas A, Tagliati M, Osborn I, et al. Pallidal deep brain stimulation for primary dystonia in children. *Neurosurgery*. 2011;68:738–743. [discussion 743].
19. Ghosh PS, Machado AG, Deogaonkar M, Ghosh D. Deep brain stimulation in children with dystonia: experience from a tertiary care center. *Pediatr Neurosurg*. 2012;48:146–151.
20. Vidailhet M, Vercueil L, Houeto JL, et al. Bilateral, pallidal, deep-brain stimulation in primary generalised dystonia: a prospective 3 year follow-up study. *Lancet Neurol*. 2007;6:223–229.
21. Isaias IU, Volkmann J, Kupsch A, et al. Factors predicting protracted improvement after pallidal DBS for primary dystonia: the role of age and disease duration. *J Neurol*. 2011;258:1469–1476.
22. Markun LC, Starr PA, Air EL, Marks Jr WJ, Volz MM, Ostrem JL. Shorter disease duration correlates with improved long-term deep brain stimulation outcomes in young-onset DYT1 dystonia. *Neurosurgery*. 2012;71:325–330.

23. Koy A, Hellmich M, Pauls KA, et al. Effects of deep brain stimulation in dyskinetic cerebral palsy: a meta-analysis. *Mov Disord*. 2013;28:647–654.

24. Lozano AM, Lipsman N. Probing and regulating dysfunctional circuits using deep brain stimulation. *Neuron*. 2013;77:406–424.

25. Miyagi Y, Koike Y. Tolerance of early pallidal stimulation in pediatric generalized dystonia. *J Neurosurg Pediatr*. 2013;12:476–482.

26. Finger S. Margaret Kennard on sparing and recovery of function: a tribute on the 100th anniversary of her birth. *J Hist Neurosci*. 1999;8:269–285.

27. Gonzalez V, Le Bars E, Cif L, et al. The reorganization of motor network in hemidystonia from the perspective of deep brain stimulation. *Brain Imaging Behav*. 2015;9:223–235.

28. Lumsden DE, Ashmore J, Charles-Edwards G, Selway R, Lin JP, Ashkan K. Observation and modeling of deep brain stimulation electrode depth in the pallidal target of the developing brain. *World Neurosurg*. 2015;83:438–446.

29. Hariz M, Blomstedt P, Zrinzo L. Future of brain stimulation: new targets, new indications, new technology. *Mov Disord*. 2013;28:1784–1792.

30. Beste C, Muckschel M, Elben S, et al. Behavioral and neurophysiological evidence for the enhancement of cognitive control under dorsal pallidal deep brain stimulation in Huntington's disease. *Brain Struct Funct*. 2014;220(4):2441–2448.

31. Clark VP, Parasuraman R. Neuroenhancement: enhancing brain and mind in health and in disease. *Neuroimage*. 2014;85(Pt 3):889–894.

Invasive Neuromodulation in Pediatric Epilepsy: VNS and Emerging Technologies

M. Ranjan, W.J. Hader

University of Calgary, Calgary, AB, Canada

OUTLINE

INTRODUCTION

Modulation of nervous system function by implantable surgical devices has revolutionized the treatment of neurological conditions, through the direct delivery of medications or the application of stimulation to both peripheral and central nervous system sites. Unlike traditional surgical treatment of epilepsy, which involves the removal of the epileptogenic focus, neuromodulation therapy is capable of delivering modifiable therapy in a safe and completely reversible manner to a variety of potential nervous system targets. Vagus nerve stimulation (VNS) was the first neuromodulatory therapeutic option approved by Food and Drug Administration (FDA) for intractable epilepsy in 1997 and still remains the most frequently utilized therapeutic neuromodulation option in clinical practice today with an experience of over two decades. Application of stimulation in patients with intractable epilepsy to additional peripheral nervous system targets including the trigeminal nerve and various central nervous system targets, including the cortex, thalamus, and hippocampus has emerged in the last decade. Two of these, deep brain stimulation (DBS) to the anterior nucleus of the thalamus and responsive cortical stimulation, have demonstrated reliable reduction of seizure frequency in recent clinical trials, generating hope for continued and better neurostimulation therapy for children with epilepsy.

EPILEPSY IN CHILDREN

Epilepsy is the fourth most common neurological disease and is second only to stroke in causing neurological morbidity.[1] There are an estimated 50 million people afflicted with epilepsy worldwide.[2] Age-specific incidences of epilepsy have demonstrated that a bimodal distribution exists with the greatest peak in childhood.[1] More than half of children with epilepsy eventually will achieve complete seizure remission,[3] though the natural course of childhood epilepsy is highly variable and the response to therapy is unpredictable.[4] Despite advances in the number of pharmacological options, 30–40% of patients will continue to have seizures and be considered medically refractory.[5] The impact of ongoing seizures during stages of rapid brain maturation in infants and children may lead to frank developmental arrest, progressive disturbances in cognitive function and behavior, and psychiatric comorbidity.[6] Cognitive disturbances when present, in particular verbal memory deficits, and emotional and behavioral difficulty significantly increase the risk of low health-related quality of life (HRQoL) in children with severe epilepsy.[7] Without remission and when considering progression to adulthood, the likelihood of permanent adverse psychosocial outcomes including incomplete education, unemployment, poverty, social isolation, and a lifetime of dependency increases. Overall people with intractable epilepsy are at risk from injury and status epilepticus, have a two- to threefold increased mortality,[8] and are 24 times more likely to die of sudden death compared with the general population.[9]

SURGICAL TREATMENTS FOR EPILEPSY

Consideration for epilepsy surgery evaluation is contemplated for pediatric patients with medically refractory epilepsy; those who have failed at least two antiepileptic drugs, have serious side effects, or have a readily identifiable lesion on MRI.[6] Investigations identifying a surgical focus amenable to resection offer the best possible chance to achieve seizure freedom and improved quality of life. In appropriately selected cases, long-term seizure freedom may be achieved in up to 80–90% of patients with lesionectomies,[10] 60–70% of patients with temporal resections,[11,12] and 30–40% of patients with extratemporal resections with normal MR imaging.[12] Permanent complications may occur in up to 4% of epilepsy surgical patients after resective surgery, the incidence of which may be higher in the pediatric population due to a greater proportion of surgery in extratemporal areas.[13] Postoperative deficits remain one of the most important predictors of dissatisfaction after epilepsy surgery.[14] In addition, a large proportion of patients may not be deemed surgical candidates either due to seizures originating from an eloquent brain region or being multifocal, diffuse, or bihemispheric in onset. Palliative surgical procedures may be offered including multiple subpial transections (MSTs), corpus callosotomy, posterior quadrant disconnections, or combination of disconnection/resective procedures. In an effort to provide surgical therapeutic options for these patients, various less invasive alternatives have emerged. These include focused stereotactic radiosurgery,[15] interstitial laser ablation therapy,[16] and neuromodulation options to multiple nervous system targets, both in the peripheral and central nervous systems.

VAGUS NERVE STIMULATION (VNS)

Rationale of Neuromodulation in Epilepsy

Epileptic seizures are clinical reflections of pathological neuronal hypersynchrony which has been well documented, both in human and animal models.[17,18] There is a wide experience and study of neuromodulation in Parkinson disease and similar to epilepsy, Parkinson disease is also the result of abnormal synchrony indicative of regional brain dysfunctions with a signature electrophysiological hallmark of beta oscillations. The pathological synchrony can be modulated with neurostimulation of STN with amelioration of the clinical symptoms.[19]

The potential importance of the vagus nerve for epilepsy to modulate abnormal cortical activity of an epileptic seizure dates back to 1938, when Bailey and Bremer demonstrated that high-frequency stimulation of the vagal nerve synchronized EEG fast activity in the orbitofrontal cortex.[20] Differential effects of VNS on the EEG were identified by Magnes,[21] who upon stimulation of the nucleus of the tractus solitarius (NTS), noted that synchronization of the EEG would occur at low frequency and desynchronization of the EEG at higher frequencies of stimulation. Further investigations demonstrated that VNS was capable of producing a profound desynchronization effect on the EEG,[22,23] an effect later confirmed to be able to eliminate chemically induced epileptiform EEG activity in cats.[24] Clear evidence of the potential therapeutic benefit of VNS emerged when repetitive electrical stimulation was proven to be capable of termination of a motor seizure in a canine model of epilepsy.[25] Possibly more important than the immediate antiseizure actions, studies showed that the inhibitory effects persisted long after the actual stimulation was terminated, up to four times as long as the stimulation duration itself. The optimal frequency of stimulation was determined and the maximal antiseizure effects appeared to be obtained at 20 Hz with a diminishing effect at frequencies greater than 60 Hz. Antiseizure effects plateaued at a stimulus amplitude of 20 V, suggesting that inhibitory fibers were likely small unmyelinated (C) fibers producing the effect, which was consistent with the findings of Woodbury and Woodbury[26] No effect of stimulus duration was detected and therefore the shortest pulse, 0.2 ms, to minimize current flow was to be utilized. With clarification of the efficacy of VNS and stimulation parameters determined, the first clinical experience of VNS followed soon in 1990.[27] Of the first four patients who received VNS, two experienced complete seizure control. Five clinical trials of VNS in human in the early 1990s led to approval of VNS as a neuromodulatory device for drug-resistant epilepsy by the FDA in 1997.

Anatomy and Surgical Details

Anatomically, the vagus nerve is a mixed nerve, however the majority is afferent information from the viscera projects diffusely to various brain structures via the NTS. This relays sensory afferents to the parabrachial nucleus, amygdala, hippocampus, thalamus, hypothalamus, and medullary reticular formation. VNS therapy (Fig. 21.1) consists of a

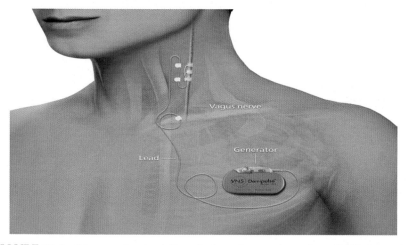

FIGURE 21.1   Diagram showing left cervical vagal nerve electrode and VNS demipulse system. *Courtesy of Cyberonics Inc, TX, USA, reprinted with permission.*

pacemaker-like device or generator implanted in the subclavicular area and thin thread-like wire, or a lead, which is anchored around the vagus nerve in the neck on the left side, to avoid adverse cardiac side effects due to innervation of the sinoatrial node by the right vagal nerve. The procedure is relatively quick and safe and usually performed as an outpatient procedure or day surgery. Mild, intermittently pulsed constant current is delivered at fixed intervals 24h per day in standard mode to the vagus nerve, which then through its connections with the NTS is capable of activating widespread various areas of the brain. The programming of the generator is done by an external dose adjustment system by a trained physician who is able to adjust accordingly, the stimulation duration, frequency, and intensity. In addition, patients or their caregivers, at onset of an aura or seizure, can use a pocket magnet, that when held over the generator, manually activates a single train of preprogrammed pulses which may shorten, stop, or reduce the recovery period from a seizure.

Mechanism of Action

Though there has been an accumulated experience of almost two decades with VNS, the exact mechanism of action remains unclear. Studies of EEG after VNS implantation in humans have been unable to detect any electrographic alterations of EEG activity similar to that seen in experimental models, despite stimulation at a wide range of frequencies.[28] PET studies following acute VNS have documented widespread changes in cerebral blood flow (CBF) in the rostral, dorsal-central medulla in the presumed vagal complex; right postcentral gyrus consistent with sensory activation, and various bilateral locations consistent with activation of the NTS including the hypothalami, thalami, insular cortex, cerebellar hemispheres, hippocampus, amygdala, and posterior cingulate gyri.[29] The initial conclusion was that synaptic changes likely occur in widespread bilateral areas as part of the polysynaptic connections of the vagus nerve. However, after VNS therapy during which seizure control improves, chronic VNS-induced CBF changes persisted only in the right post central gyrus, thalami, and hypothalamic regions yet were no longer identified in the cortical regions. Altered synaptic activities in subcortical sites of persisting VNS-induced CBF changes possibly reflect antiseizure actions and ongoing modulation of brain.[30] SPECT activation studies after left VNS have revealed decreased perfusion in the ipsilateral thalamus,[31] however, there was no correlation of VNS-related changes in the thalamus to a reduction in seizure.

Clinical Efficacy

The approval of VNS as a neuromodulatory device is primarily based on two randomized controlled trials, E03[32] and E05,[33] in which 114 and 199 patients were enrolled, respectively. Subjects included were at least 12 years old and had a minimum of six seizures per month. Patients were randomized to either a high-stimulation (therapeutic) or low-stimulation (subtherapeutic) group. The high-stimulation group received intermittent stimulation with output currents up to 3.5 mA, 30 Hz, 500 µs pulse width, with 30 s on time and 5 min of off time. The low-stimulation arm had similar output currents, but a frequency of 1 Hz, pulse width of 130 µs, on time of 30 s, and off time of 180 min. Patients in the low-stimulation group were adjusted to a minimum level so that a sensation of stimulation was still present. The assessment period was 12 weeks and the primary outcome measure was percent change in seizure frequency. A significant reduction of seizures was seen in the high-stimulation group in both the E03 study (24.5% vs. 6.1% $P=0.01$) and in the E05 study (27.9% vs. 15.2%, $P=0.04$) compared to the low-stimulation groups. In the E03 study the high-stimulation group had a 27% responder rate (patients with >50% reduction in seizures) compared to 13% of patients in the low-stimulation group. Several patients experienced a >75% reduction in seizures while no patients were seizure-free at the end of the 12 weeks of stimulation. Evaluation of the efficacy of VNS in percent reduction by the individual seizure type, either partial, complex partial, or secondarily generalized, revealed no significant difference between the three groups.

Long-Term Results

DeGiorgio et al.[34] prospectively evaluated the long-term efficacy of VNS after completion of the acute phase of the E05 study. They included patients who had originally received low-frequency stimulation who then transitioned to high-frequency stimulation. A progressive improvement in the median reduction in seizure frequency from 34% to 45% occurred between 3 and 12 months after implantation. In 1999, Morris and Mueller reported on 454 patients that had been enrolled in the five clinical trials of VNS between 1988 and 1995.[35] The progressive improvement in the median seizure frequency reductions was 35% at 1 year, 44.3% at 2 years, and 44.1% at 3 years compared with the baseline. An additional long-term study of VNS (both adult and pediatric patients) reported a mean reduction of 55.8% in seizure frequency at almost 5 years of follow up.[36] In the largest meta-analysis of VNS efficacy in epilepsy patients consisting of 74 clinical studies with 3321 patients suffering from intractable epilepsy, including three blinded, randomized controlled trials (class I evidence); two non-blinded, randomized controlled trials (class II evidence); 10 prospective studies (class III evidence); and numerous retrospective studies, seizure frequency was reduced by an average of 45%, with a 36% reduction in seizures at 3–12 months after surgery and a 51% reduction after >1 year of VNS therapy.[37] Using a meta-analysis of randomized controlled trials of VNS, Forbes et al. [38] estimated that six people required implantation in order for one person to experience a 50% reduction in seizure frequency. The cumulative clinical efficacy of VNS therapy reported supports the notion of a true neuromodulatory response in which the antiseizure effects are not limited to the period of stimulation alone and a progressive improvement in efficacy may be seen over time.

VNS Studies in Children

Although not the primary population of patients in the initial randomized trials, pediatric patients with intractable epilepsy would be confirmed to experience similar efficacy to the adult trials. Murphy[39] reported on 60 children aged 3–18 years, a 31% and 34% seizure reduction after 6 and 12 months, respectively. Similar to adults after VNS, the number of pediatric patients that achieve significant reductions (>50%) in baseline seizure frequency increases over time. Orosz et al.[40] studied the long-term outcome (24 months) in 347 pediatric patient and reported significant reductions of the predominant seizure type in 32.5%, 37.6%, and 43.8% of cases at 6, 12, and 24 months respectively consistent with large studies in which a 50% reduction in seizures was seen in 45% of cases with an overall 44% reduction in seizures.[41,42] Seizure types that were responsive to VNS were variable. In an uncontrolled study of 38 children, Patwardhan et al.[43] reported reductions in seizure types including atonic (80%), absence (65%), complex partial (48%), and generalized (45%) seizures.

In 2013, the Guideline Development Subcommittee of the American Academy of Neurology reviewed the relevant literature on VNS in epilepsy patients with special reference to patients younger than 12 years of age and published evidence-based guidelines for VNS for the treatment of epilepsy.[44] The committee noted that VNS is associated with a >50% seizure reduction in 55% (95% confidence interval [CI] 50–59%) of 470 children with partial or generalized epilepsy (13 class III studies) and >50% seizure reduction in 55% (95% CI 46–64%) of 113 patients with Lennox–Gastaut syndrome (LGS) (four class III studies). An increase (7%) in the proportion of patients achieving >50% seizure frequency reduction rates occurs from 1 to 5 years postimplantation (two class III studies). Overall, the committee concluded that VNS may have improved efficacy over time and is possibly effective for seizures (both partial and generalized) in children and for LGS-associated seizures.

Adverse Effects

Adverse effects of VNS can be separated into complications of the procedure itself or unpleasant side effects associated with intermittent stimulation. The surgical procedure and application of VNS is considered relatively safe. Serious implantation complications including asystole at the time of implantation with vagal nerve manipulation and injury to the vagus nerve resulting in vocal cord paresis are rare (1/1000). Infection at the VNS implantation site in children is increased relative to that in adults and additional care must be taken.[44] Stimulation effects were reported in 5% of patients in the large randomized trials and mainly consisted of pain (29%), coughing (14%), voice alteration (13%), chest pain (12%), and nausea (10%). The side effects tend to improve with time and are dose-dependent, which can be minimized with reductions in stimulation parameters. Rare events resulting in device malfunction can cause painful stimulation or direct-current stimulation due to an uninterrupted discharging generator or lead fracture. Nine deaths in 440 patients were reported in the EO1–EO5 studies, none related to complications of the implantation or stimulation itself. Annegers et al.[45] reviewed all deaths in 1819 patients with VNS. The patients were followed up for 3176.3 person-years from implantation and 25 deaths were reported. Rates of SUDEP were 4.1 per 1000 in patients treated with VNS and 4.5 per 1000 for a control population with refractory epilepsy. After stratification for duration of VNS use, the prevalence of SUDEP was 5.5 per 1000 for the first 2 years and then dropped to 1.7 per 1000 for the subsequent years, suggesting lower SUDEP rates after VNS therapy. A clear trend towards a reduction in standardized mortality ratios (SMRs) 2 year postimplantation has been reported, however, SUDEP rates did not change.[46] The conclusion was that SMRs and SUDEP rates after VNS were comparable to other cohorts of patients with intractable epilepsy and VNS likely did not lower the risk of premature death overall.

QOL and Other Health-Related Indices

In contrast to standard antiepileptic drugs, VNS poses no risk to cognition in children and additional potential beneficial effects have been reported. Follow-up in patients after VNS therapy is reported to be associated with a significant reduction in hospitalizations, emergency room visits, episodes of status epilepticus, and overall health care expenses compared to a cohort of patients with intractable epilepsy.[47,48] Reports on the impact of VNS on indices of quality of life appear to be variable from no changes or slight improvement in QOL[49,50] to significant improvements during the first year of treatment.[51] In a recent open prospective randomized long-term effectiveness trial (PULSE), the authors concluded that VNS therapy as a treatment adjunct to the best medical practice in patients with medically resistant focal seizures was associated with a significant improvement in HRQoL compared with best medical practice alone.[52] Improvements in seizure severity, QOL,[53] behavior, mood and depressive feelings,[54] and sleep quality[55] in addition to positive antiseizure effects, have also been reported after VNS. However, the improvement in these parameters may not be related to the antiseizure effect of VNS, which was suggested by a Canadian prospective, case-controlled study.[50] They noticed that QOL did not differ between medical patients and VNS patients and QOL improvements after VNS were not related to the control of seizures.

Recent Advances in VNS

In an effort to provide alternatives to implantable VNS stimulator systems, transcutaneous stimulation of branches of the vagus nerve has been proposed, though initially not for epilepsy.[56] Subsequently, this method was found to be feasible, less invasive with reduced side effects, yet provided similar efficacy[57,58] and areas of brain activation,[59,60] including in pediatric population.[61] The most recent modification to traditional VNS uses an on-demand method of stimulation in an attempt to influence the seizure severity or duration. Aspire SR (Cyberonics) incorporates a cardiac-based seizure detection algorithm capable of detecting ictal tachycardia, a common accompaniment of seizures, which when detected will trigger a predetermined stimulation. Early reports of the device have suggested apart from the cosmetic differences of implanting the new generator, short-term complications are similar to traditional VNS.[62] The efficacy of the device for the intended purpose of aborting seizures based on ictal tachycardia has yet to be proven in clinical trials.

TRIGEMINAL NERVE STIMULATION

Like the vagus nerve, the trigeminal nerve has widespread projections to the brainstem and remote supratentorial areas which are distinct from vagal afferents. The first report of a role for trigeminal nerve stimulation in epilepsy therapy was in 1976, when Maksimow reported interruption of ongoing GTCS with manual stimulation of the infraorbital nerve.[63] Subsequent animal studies confirmed clinical suspicion of effects on epilepsy when Fanselow et al.[64] showed reduced seizure activity with trigeminal stimulation in PTZ-treated rats. The first human clinical experience was reported by DeGiorgio et al.[65] in 2003 from a pilot clinical trial of transcutaneous infraorbital nerve stimulation in two patients who reported 39% and 76% reduction in seizure frequency without major side effects. A larger study by the same group on additional patients with infraorbital and supraorbital nerve stimulation reported a mean seizure frequency reduction of 66% at 3 months and 59% at 12 months.[66,67] One of the patients would achieve 90% reduction in seizure frequency at 12 months. Studies suggested that bilateral stimulation may be more effective than unilateral stimulation alone. A subsequent randomized controlled trial of trigeminal nerve stimulation in 50 patients with refractory partial seizures however, failed to demonstrate a significant difference in the overall responder rate (>50% reduction in seizure frequency) of treated patients compared to controls.[68] Stimulation of the trigeminal nerve was reported to be associated with significant improvements in mood.[68] Long-term follow-up of the prospective cohort from the previous trial reported the median seizure frequency for the original treatment group decreased by 2.39 seizures per month at 6 months (27.4%) and 3.03 seizures per month at 12 months (34.8%).[69] The 50% responder rates at 3, 6, and 12 months remained at 36.8% for the treatment group. However, the efficacy of TNS in refractory epilepsy has been questioned and some feel that TNS may not be effective in refractory partial seizures.[70] Trigeminal nerve stimulation is currently approved for treatment of epilepsy in adults and children 9 years and older in Europe and Canada. Though TNS is still investigational in the United States, approval from the FDA of its Investigational Device Exemption (IDE) has been given to perform a large, multicenter pivotal phase III trial for partial seizures with or without secondary generalization in people with epilepsy ages 12–65.

INTRACRANIAL NEUROSTIMULATION FOR EPILEPSY

Stimulation of specific areas of brain to modulate networks responsible for a variety of neurological conditions, in particular movement disorders, has a long clinical and research history. The earliest target of intracranial stimulation for epilepsy was the cerebellum where it had been shown that modulation of experimentally induced epileptogenic cortical activity could be inhibited by paleocerebellum stimulation.[71] A technique for successful chronic stimulation of the anterior lobe of the cerebellum in humans, in an attempt to treat disabling spasticity and rigidity, was described by Cooper in 1973. Cooper discovered that convulsive seizures present in four patients being treated were completely suppressed with stimulation over a short follow-up period.[72] While a subsequent blinded trial of cerebellar stimulation in patients with refractory epilepsy failed to reproduce the reduction in seizure frequency that he had seen, the groundwork was laid for the future investigation of alternative targets of neurostimulation for epilepsy.[73,74] While no formal assessment of patient-reported outcomes was completed, 11 of 12 patients felt that the stimulation had helped them despite no reduction in seizure frequency. Investigational targets of DBS for the treatment of refractory epilepsy have included the centromedian thalamic nucleus, caudate nucleus, subthalamic nucleus, mammillary bodies, amygdalohippocampal complex, locus ceruleus, and mammillothalamic tract.[75] Recent successful trials of anterior nucleus of thalamus stimulation and responsive neurostimulation (RNS) for patients with intractable epilepsy have been reported.

Anterior Nucleus of Thalamus Stimulation

The participation of cortical-subcortical networks in the propagation of focal epileptic seizures is well established.[76] The anterior nucleus of the thalamus (ANT) is an integral part of the Papez circuit originally, which was felt to connect the hypothalamus to the limbic lobe. The nucleus receives information from the mesial temporal lobe via the fornix and mammillothalamic tract and sends projections to the cingulum in the frontal lobe and back to the temporal lobe. In terms of electrophysiology, it was noted that the cortical EEG is highly coherent with anterior thalamic activity.[77,78] In fact in 1967, Mullan et al.[79] reported that the stereotactic lesioning of the anterior nucleus of thalamus in humans improved seizure. Furthermore, high-frequency stimulation (100 Hz) or lesion of AN or interruption of the mammillothalamic tract were noted to have an anticonvulsant action.[80,81] Interestingly, as with VNS, varying effects occur with altering the frequency of stimulation, low-frequency stimulation (8 Hz) exhibited proconvulsant effects, the opposite effect of that seen with high-frequency stimulation. Subsequently in 1984, Cooper et al.[82] reported the first clinical human experience of anterior nucleus of thalamus stimulation and showed significant (>50%) reduction in seizure frequency in five of their six patients. A number of uncontrolled studies validated the anterior nucleus as a potential target for intractable epilepsy (Fig. 21.2) as an important therapeutic option for limbic, frontal, and temporal seizures and led the way to a definitive clinical trial.

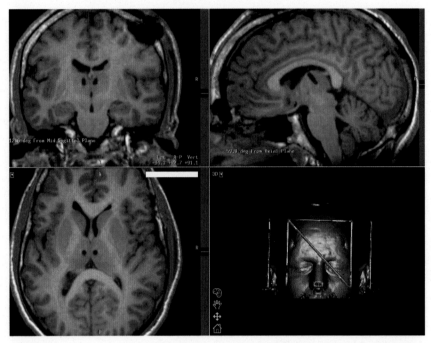

FIGURE 21.2   Postoperative T1W axial, coronal, and sagittal MR image showing anterior nucleus of thalamus DBS electrode position, fused on preoperative MR image. Original target (red dot). *Courtesy of Dr. Andres Lozano, University of Toronto.*

SANTE or Stimulation of the Anterior Nucleus of Thalamus for Epilepsy study was a large multicenter randomized, and blinded trial of bilateral stimulation of the anterior nucleus completed in 110 patients with medically refractory partial or secondarily generalized seizures.[83] During the blinded "on/off" evaluation, there was a significant reduction in seizure frequency among patients who received stimulation compared to the control group (40.4% vs. 14.5%). The effectiveness persisted or/and even improved at long-term follow-up. At 2 years of follow-up, 54% of patients had >50% seizure reduction and 13 patients had >90% reduction in seizure frequency including 13% of patients that became seizure-free for at least 6 months. Stimulation appeared to benefit patients with temporal lobe epilepsy compared to other locations. Reasons for this may be the close association of the mesial temporal lobe to thalamic pathways of the anterior nucleus via the limbic circuit of Papez. Questions remain about the true mechanism of action of DBS for epilepsy including why seizures remote from the thalamic site of stimulation are reduced. Stimulation of remote areas from the actual seizure focus, in an attempt to modulate epilepsy circuits indirectly through functional interconnections between the two locations, most often proven ineffectual. Even when areas have been shown to be effective in treating seizures, as the anterior nucleus of the thalamus, reasons for the reduction remains unclear.

Adverse Effects

Serious complications of DBS can occur including infection, hemorrhage, status epilepticus, and rarely death. Infection rates in the randomized trial was 12.7% which is comparable with rates reported for DBS in Parkinson disease patients. Asymptomatic hemorrhage, discovered only upon postoperative neuroimaging, was reported in up to 5% of patients and while no clinically significant hemorrhages were reported, hemiparesis may be seen in up to 3% of patients after DBS surgery for Parkinson disease.[84] In a collective analysis of complication of DBS surgery for various indications and various targets, including thalamic stimulation, Beric et al.[85], in 2001, reported an overall 6% of persistent neurological sequelae, however, there were no permanent severe disability or any mortality. In recent surgical series, the incidence of symptomatic intracranial hemorrhage was 1.1%, neurological complication was 1.7% and incidence of infection was 1.7%.[86] Unlike VNS, direct DBS for epilepsy is completed at much higher frequencies than peripheral stimulation and concerns for initiating an epileptic event are real. Induction of status epilepticus in the acute period after stimulator implantation prior to the onset of stimulation has been documented in up to 1% of patients. In addition, seizures associated with the onset of stimulation may be seen, the risk of which may be reduced by lowering stimulation voltages. No clear objective negative effects of anterior nucleus of thalamus DBS were seen on neuropsychological testing or alterations in mood during the prospective trial, however, a subjective decline in memory was reported by more participants in the stimulated group.

Hippocampus Stimulation

Stimulation at the site of the seizure onset directly has become the focus of attempts to define new targets for intracranial neurostimulation. Low-frequency electrical stimulation of the amygdala (which does not disrupt ongoing behavior) in animal models can have profound and long-lasting effects on seizure development, expression, and thresholds.[87] Mesial temporal lobe epilepsy (MTLE) remains the most common surgically remediable syndrome in patients with intractable epilepsy. Despite the success of resections in the mesial temporal lobe, many patients with clear MTLE may not be surgical candidates due to independent involvement of both temporal lobes in the generation of clinical seizures or concerns of memory decline after hippocampal resection.

For those patients and others who have failed previous temporal resections, reversible stimulation of the hippocampus (Fig. 21.3) provides an attractive option capable of reducing the seizure frequency without the risk of permanent neuropsychological sequelae.

Subacute high-frequency external stimulation of the hippocampus was first demonstrated to block both seizures and paroxysmal interictal discharges in 10 patients at the time of depth electrode implantation for the investigation of temporal lobe epilepsy.[88] Early chronic clinical studies of high-frequency (130 Hz) unilateral or bilateral hippocampal stimulation reported that the majority of patients would experience >50% reduction in seizures with a follow-up of 5–18 months.[89,90] Patients with normal MR imaging appeared to have a faster and more significant reduction in seizures compared to those with mesial temporal sclerosis. Patients were randomized and blinded to stimulation or no stimulation, however the period off stimulation lasted only 1 month. Assessment of unilateral hippocampal stimulation in a double-blind, multiple crossover, randomized controlled study, however, showed only a modest decrease of seizure frequencies of only 15% which was not significant.[91] A similar trial of bilateral hippocampal stimulation by the same group reported a significant reduction in seizures by around one-third was possible, and the results would persist during the wash out period after stimulation suggesting a modulating effect was occurring.[92] Although no adverse effects in objective or subjective memory were reported during the study, the overall seizure effect was felt to be modest as no patient had a clinically meaningful reduction in seizures during the study. The authors have stated that hippocampal stimulation remains an experimental therapy for epilepsy, and patients considered for this intervention should do so in the context

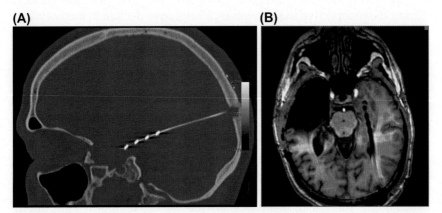

FIGURE 21.3   (A) Postoperative sagittal reconstruction of CT scan of brain showing left hippocampal electrode. (B) Postoperative T1W axial MRI with implanted left hippocampal electrode (postoperative status – right temporal lobectomy).

of a well-designed randomized controlled trial. A follow-up Canadian study "A Multicenter Study of Hippocampal Electrical Stimulation [HS, in Mesial Temporal Lobe Epilepsy] (METTLE)" was initiated; however, it was terminated because of poor patient enrolment.

Responsive Neurostimulation

Direct stimulation of the anatomic target responsible for ongoing seizures in an on-demand fashion is particularly appealing in the field of epilepsy, since the seizure, unlike persistently dysfunctioning targets as in movement disorders, represents an episodic and intermittent "electroneurological dysfunction." Responsive neurostimulation (Fig. 21.4) was felt capable of providing the similar benefits of continuous intermittent stimulation with avoidance of unpleasant side effects and added advantage of prolonging battery life. Penfield and Jasper observed that direct cortical stimulation resulted in flattening of the EEG at the time of electrocorticography in humans.[93] Cortical stimulation at the time of implanted subdural electrodes in humans has demonstrated that stimulation is capable of suppressing fast activity and interictal spikes[94] and terminating after discharges.[95] Arrest of electrographic seizure was possible in animal model[96] and thought to result from inhibitory hyperpolarization caused by the application of transmembrane current.[97] Experimental observations suggested that not only was direct cortical stimulation in response to epileptogenic activity possible, it was effective. This led to development of a closed loop system, consisting of an afferent loop for seizure detection and prediction device and an efferent loop consisting of electrical delivery system in response to seizure detection.

(A)   **(B)**

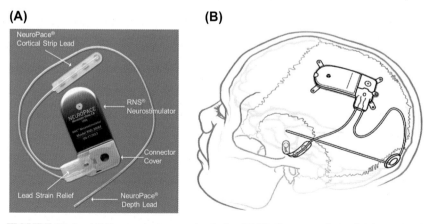

FIGURE 21.4   (A) Responsive neurostimulation (RNS). One cortical strip lead, one depth lead, and RNS device. (B) Schematic diagram of implanted RNS system. *Courtesy of Neuropace Inc., CA, USA, reprinted with permission.*

The most important role of the device was to detect seizures predictably and accurately. Proof of concept for an implantable system capable of reliably detecting defined EEG parameters indicative of an ictal onset followed by the delivery of predetermined stimulation was developed by Peters et al.[98] They successfully demonstrated precise detection of seizures prior to clinical onset in patients undergoing subdural electrode evaluation. An open trial of external responsive stimulation conducted on four patients, who had intracranial electrodes implanted for surgical investigations, demonstrated suppression of electrographic seizures.[99] Though the study was not designed to assess the efficacy, a reduction in seizure frequency was seen in patients with both temporal and extratemporal seizures.

The RNS System Trial (Responsive Cortical Stimulation for Treatment of Medically Intractable Partial Epilepsy) trial was a large multicenter, double-blind, randomized controlled trial conducted to assess the safety and effectiveness of responsive cortical stimulation as an adjunctive therapy for partial-onset seizures in adults with medically refractory epilepsy.[100] A statistically significant reduction in seizure frequency occurred in patients with RNS as compared to the control group (37.9% vs. 17.3%, $P = 0.012$) in the 12-week blinded period. During the open-label period, when stimulation was on for both groups, the seizure reduction was sustained in the treatment group, while reduction in seizure frequency was also noted in control group. No significant adverse effects related to stimulation or change in neuropsychological functioning occurred. Based on this class I evidence, the NEUROPACE RNS system was given premarket approval in 2013, as an adjunctive therapy in reducing the frequency of seizures in individuals 18 years of age or older with refractory and disabling partial-onset seizures with no more than two epileptogenic foci.

Other Neuromodulatory Devices

Application of cold saline to the cerebral cortex is a well-known phenomenon used intraoperatively to arrest after-discharges and seizures during electrocorticography and brain mapping. A pilot trial of a closed loop system capable of detecting a seizure followed by the application of focal cooling through an implanted device to the cortical epileptogenic area resulted in the termination of focal seizure.[101] The preliminary results are encouraging, however further studies for feasibility and validation for clinical application will be needed.

CONCLUSION

VNS is an effective and safe surgical option in pediatric patients with medically refractory epilepsy that may provide benefits in addition to improvements in seizure control. Alternative proven invasive methods of neuromodulation now exist, including DBS of the anterior nucleus of the thalamus and responsive focal cortical neurostimulation, which provide new hope for pediatric patients with intractable epilepsy.

References

1. Kotsopoulos IA, van Merode T, Kessels FG, de Krom MC, Knottnerus JA. Systematic review and meta-analysis of incidence studies of epilepsy and unprovoked seizures. *Epilepsia*. 2002;43(11):1402–1409.
2. Banerjee PN, Filippi D, Allen Hauser W. The descriptive epidemiology of epilepsy-a review. *Epilepsy Res*. 2009;85(1):31–45.
3. Camfield PR, Camfield CS. What happens to children with epilepsy when they become adults? Some facts and opinions. *Pediatr Neurol*. 2014;51(1):17–23.
4. Berg AT, Rychlik K. The course of childhood-onset epilepsy over the first two decades: a prospective, longitudinal study. *Epilepsia*. 2015;56(1):40–48.
5. Kwan P, Sander JW. The natural history of epilepsy: an epidemiological view. *J Neurol Neurosurg Psychiatry*. 2004;75(10):1376–1381.
6. Cross JH, Jayakar P, Nordli D, et al. International League against Epilepsy SfPES, Commissions of N, Paediatrics. Proposed criteria for referral and evaluation of children for epilepsy surgery: recommendations of the Subcommission for Pediatric Epilepsy Surgery. *Epilepsia*. 2006;47(6):952–959.
7. Hrabok M, Brooks BL, Fay-McClymont TB, Sherman EM. Wechsler Intelligence Scale for Children-fourth edition (WISC-IV) short-form validity: a comparison study in pediatric epilepsy. *Child Neuropsychol*. 2014;20(1):49–59.
8. Hauser WA, Annegers JF, Elveback LR. Mortality in patients with epilepsy. *Epilepsia*. 1980;21(4):399–412.
9. Ficker DM, So EL, Shen WK, et al. Population-based study of the incidence of sudden unexplained death in epilepsy. *Neurology*. 1998;51(5):1270–1274.
10. Fay-McClymont TB, Hrabok M, Sherman EM, et al. Systematic review and case series of neuropsychological functioning after epilepsy surgery in children with dysembryoplastic neuroepithelial tumors (DNET). *Epilepsy Behav*. 2012;23(4):481–486.
11. Wiebe S, Blume WT, Girvin JP, Eliasziw M. Effectiveness, Efficiency of Surgery for Temporal Lobe Epilepsy Study G. A randomized, controlled trial of surgery for temporal-lobe epilepsy. *N Engl J Med*. 2001;345(5):311–318.
12. Tellez-Zenteno JF, Dhar R, Wiebe S. Long-term seizure outcomes following epilepsy surgery: a systematic review and meta-analysis. *Brain*. 2005;128(Pt 5):1188–1198.
13. Hader WJ, Tellez-Zenteno J, Metcalfe A, et al. Complications of epilepsy surgery: a systematic review of focal surgical resections and invasive EEG monitoring. *Epilepsia*. 2013;54(5):840–847.
14. Macrodimitris S, Sherman EM, Williams TS, Bigras C, Wiebe S. Measuring patient satisfaction following epilepsy surgery. *Epilepsia*. 2011;52(8):1409–1417.
15. Barbaro NM, Quigg M, Broshek DK, et al. A multicenter, prospective pilot study of gamma knife radiosurgery for mesial temporal lobe epilepsy: seizure response, adverse events, and verbal memory. *Ann Neurol*. 2009;65(2):167–175.
16. Tovar-Spinoza Z, Carter D, Ferrone D, Eksioglu Y, Huckins S. The use of MRI-guided laser-induced thermal ablation for epilepsy. *Childs Nerv Syst*. 2013;29(11):2089–2094.
17. Garcia Dominguez L, Wennberg RA, Gaetz W, Cheyne D, Snead 3rd OC, Perez Velazquez JL. Enhanced synchrony in epileptiform activity? Local versus distant phase synchronization in generalized seizures. *J Neurosci*. 2005;25(35):8077–8084.
18. Truccolo W, Ahmed OJ, Harrison MT, et al. Neuronal ensemble synchrony during human focal seizures. *J Neurosci*. 2014;34(30):9927–9944.

19. Ranjan M, Honey CR. Subthalamic nucleus deep brain stimulation for Parkinson's disease. In: Eljamel S, Slavin K, eds. *Clinical Neurostimulation: Principles and Practice*. 1st ed. Oxford, UK: Wiley-Blackwell; 2013.

20. Bailey P, Bremer F. A sensory cortical representation of the vagus nerve. *J Neurophysiol*. 1938;1:405–412.

21. Magnes J, Moruzzi G, Pompeiano O. Synchronization of the EEG produced by low frequency electrical stimulation of the region of the solitary tract. *Arch Ital Biol*. 1961;99:33–67.

22. Zanchetti A, Wang SC, Moruzzi G. The effect of vagal afferent stimulation on the EEG pattern of the cat. *Electroencephalogr Clin Neurophysiol*. 1952;4(3):357–361.

23. Chase MH, Nakamura Y, Clemente CD, Sterman MB. Afferent vagal stimulation: neurographic correlates of induced EEG synchronization and desynchronization. *Brain Res*. 1967;5(2):236–249.

24. Stoica I, Tudor I. Effects of vagus afferents on strychninic focus of coronal gyrus. *Rev Roum Neurol*. 1967;4:287–295.

25. Zabara J. Inhibition of experimental seizures in canines by repetitive vagal stimulation. *Epilepsia*. 1992;33(6):1005–1012.

26. Woodbury DM, Woodbury JW. Effects of vagal stimulation on experimentally induced seizures in rats. *Epilepsia*. 1990;31(suppl 2):S7–S19.

27. Penry JK, Dean JC. Prevention of intractable partial seizures by intermittent vagal stimulation in humans: preliminary results. *Epilepsia*. 1990;31(suppl 2):S40–S43.

28. Hammond EJ, Uthman BM, Reid SA, Wilder BJ. Electrophysiological studies of cervical vagus nerve stimulation in humans: I. EEG effects. *Epilepsia*. 1992;33(6):1013–1020.

29. Henry TR, Bakay RA, Votaw JR, et al. Brain blood flow alterations induced by therapeutic vagus nerve stimulation in partial epilepsy: I. Acute effects at high and low levels of stimulation. *Epilepsia*. 1998;39(9):983–990.

30. Henry TR, Bakay RA, Pennell PB, Epstein CM, Votaw JR. Brain blood-flow alterations induced by therapeutic vagus nerve stimulation in partial epilepsy: II. prolonged effects at high and low levels of stimulation. *Epilepsia*. 2004;45(9):1064–1070.

31. Vonck K, Boon P, Van Laere K, et al. Acute single photon emission computed tomographic study of vagus nerve stimulation in refractory epilepsy. *Epilepsia*. 2000;41(5):601–609.

32. A randomized controlled trial of chronic vagus nerve stimulation for treatment of medically intractable seizures. The Vagus Nerve Stimulation Study Group. *Neurology*. 1995;45(2):224–230.

33. Handforth A, DeGiorgio CM, Schachter SC, et al. Vagus nerve stimulation therapy for partial-onset seizures: a randomized active-control trial. *Neurology*. 1998;51(1):48–55.

34. DeGiorgio CM, Schachter SC, Handforth A, et al. Prospective long-term study of vagus nerve stimulation for the treatment of refractory seizures. *Epilepsia*. 2000;41(9):1195–1200.

35. Morris 3rd GL, Mueller WM. Long-term treatment with vagus nerve stimulation in patients with refractory epilepsy. The Vagus Nerve Stimulation Study Group E01-E05. *Neurology*. 1999;53(8):1731–1735.

36. Elliott RE, Morsi A, Tanweer O, et al. Efficacy of vagus nerve stimulation over time: review of 65 consecutive patients with treatment-resistant epilepsy treated with VNS > 10 years. *Epilepsy Behav*. 2011;20(3):478–483.

37. Englot DJ, Chang EF, Auguste KI. Vagus nerve stimulation for epilepsy: a meta-analysis of efficacy and predictors of response. *J Neurosurg*. 2011;115(6):1248–1255.

38. Forbes RB, Macdonald S, Eljamel S, Roberts RC. Cost-utility analysis of vagus nerve stimulators for adults with medically refractory epilepsy. *Seizure*. 2003;12(5):249–256.

39. Murphy JV. Left vagal nerve stimulation in children with medically refractory epilepsy. The Pediatric VNS Study Group. *J Pediatr*. 1999;134(5):563–566.

40. Orosz I, McCormick D, Zamponi N, et al. Vagus nerve stimulation for drug-resistant epilepsy: a European long-term study up to 24 months in 347 children. *Epilepsia*. 2014;55(10):1576–1584.

41. Murphy JV, Torkelson R, Dowler I, Simon S, Hudson S. Vagal nerve stimulation in refractory epilepsy: the first 100 patients receiving vagal nerve stimulation at a pediatric epilepsy center. *Arch Pediatr Adolesc Med.* 2003;157(6):560–564.

42. Helmers SL, Wheless JW, Frost M, et al. Vagus nerve stimulation therapy in pediatric patients with refractory epilepsy: retrospective study. *J Child Neurol.* 2001;16(11):843–848.

43. Patwardhan RV, Stong B, Bebin EM, Mathisen J, Grabb PA. Efficacy of vagal nerve stimulation in children with medically refractory epilepsy. *Neurosurgery.* 2000;47(6):1353–1357. [discussion 1357–1358].

44. Morris 3rd GL, Gloss D, Buchhalter J, Mack KJ, Nickels K, Harden C. Evidence-based guideline update: vagus nerve stimulation for the treatment of epilepsy: report of the guideline development subcommittee of the american academy of neurology. *Epilepsy Curr.* 2013;13(6):297–303.

45. Annegers JF, Coan SP, Hauser WA, Leestma J. Epilepsy, vagal nerve stimulation by the NCP system, all-cause mortality, and sudden, unexpected, unexplained death. *Epilepsia.* 2000;41(5):549–553.

46. Granbichler CA, Nashef L, Selway R, Polkey CE. Mortality and SUDEP in epilepsy patients treated with vagus nerve stimulation. *Epilepsia.* 2015;56(2).

47. Helmers SL, Duh MS, Guerin A, et al. Clinical outcomes, quality of life, and costs associated with implantation of vagus nerve stimulation therapy in pediatric patients with drug-resistant epilepsy. *Eur J Paediatr Neurol.* 2012;16(5):449–458.

48. Helmers SL, Duh MS, Guerin A, et al. Clinical and economic impact of vagus nerve stimulation therapy in patients with drug-resistant epilepsy. *Epilepsy Behav.* 2011;22(2):370–375.

49. Dodrill CB, Morris GL. Effects of vagal nerve stimulation on cognition and quality of life in epilepsy. *Epilepsy Behav.* 2001;2(1):46–53.

50. McGlone J, Valdivia I, Penner M, Williams J, Sadler RM, Clarke DB. Quality of life and memory after vagus nerve stimulator implantation for epilepsy. *Can J Neurol Sci.* 2008;35(3):287–296.

51. Sirven JI, Sperling M, Naritoku D, et al. Vagus nerve stimulation therapy for epilepsy in older adults. *Neurology.* 2000;54(5):1179–1182.

52. Ryvlin P, Gilliam FG, Nguyen DK, et al. The long-term effect of vagus nerve stimulation on quality of life in patients with pharmacoresistant focal epilepsy: the PuLsE (Open Prospective Randomized Long-term Effectiveness) trial. *Epilepsia.* 2014;55(6):893–900.

53. Hallbook T, Lundgren J, Stjernqvist K, Blennow G, Stromblad LG, Rosen I. Vagus nerve stimulation in 15 children with therapy resistant epilepsy; its impact on cognition, quality of life, behaviour and mood. *Seizure.* 2005;14(7):504–513.

54. Klinkenberg S, van den Bosch CN, Majoie HJ, et al. Behavioural and cognitive effects during vagus nerve stimulation in children with intractable epilepsy – a randomized controlled trial. *Eur J Paediatr Neurol.* 2013;17(1):82–90.

55. Hallbook T, Lundgren J, Kohler S, Blennow G, Stromblad LG, Rosen I. Beneficial effects on sleep of vagus nerve stimulation in children with therapy resistant epilepsy. *Eur J Paediatr Neurol.* 2005;9(6):399–407.

56. Sanders I, Aviv J, Biller HF. Transcutaneous electrical stimulation of the recurrent laryngeal nerve: a method of controlling vocal cord position. *Otolaryngol Head Neck Surg.* 1986;95(2):152–157.

57. Aihua L, Lu S, Liping L, Xiuru W, Hua L, Yuping W. A controlled trial of transcutaneous vagus nerve stimulation for the treatment of pharmacoresistant epilepsy. *Epilepsy Behav.* 2014;39:105–110.

58. Rong P, Liu A, Zhang J, et al. Transcutaneous vagus nerve stimulation for refractory epilepsy: a randomized controlled trial. *Clin Sci (Lond).* 2014 Apr 1. [Epub ahead of print].

59. Fallgatter AJ, Neuhauser B, Herrmann MJ, et al. Far field potentials from the brain stem after transcutaneous vagus nerve stimulation. *J Neural Transm.* 2003;110(12):1437–1443.

60. Kraus T, Hosl K, Kiess O, Schanze A, Kornhuber J, Forster C. BOLD fMRI deactivation of limbic and temporal brain structures and mood enhancing effect by transcutaneous vagus nerve stimulation. *J Neural Transm.* 2007;114(11):1485–1493.

61. He W, Jing X, Wang X, et al. Transcutaneous auricular vagus nerve stimulation as a complementary therapy for pediatric epilepsy: a pilot trial. *Epilepsy Behav.* 2013;28(3):343–346.

62. Schneider UC, Bohlmann K, Vajkoczy P, Straub HB. Implantation of a new Vagus Nerve Stimulation (VNS) Therapy(R) generator, AspireSR(R): considerations and recommendations during implantation and replacement surgery-comparison to a traditional system. *Acta Neurochir (Wien).* 2015;157(4):721–728.

63. Maksimow K. Interruption of grand mal epileptic seizures by the trigeminal nerve stimulation. *Neurol Neurochir Pol.* 1976;10(2):205–208.

64. Fanselow EE, Reid AP, Nicolelis MA. Reduction of pentylenetetrazole-induced seizure activity in awake rats by seizure-triggered trigeminal nerve stimulation. *J Neurosci.* 2000;20(21):8160–8168.

65. DeGiorgio CM, Shewmon DA, Whitehurst T. Trigeminal nerve stimulation for epilepsy. *Neurology.* 2003;61(3):421–422.

66. DeGiorgio CM, Shewmon A, Murray D, Whitehurst T. Pilot study of trigeminal nerve stimulation (TNS) for epilepsy: a proof-of-concept trial. *Epilepsia.* 2006;47(7):1213–1215.

67. DeGiorgio CM, Murray D, Markovic D, Whitehurst T. Trigeminal nerve stimulation for epilepsy: long-term feasibility and efficacy. *Neurology.* 2009;72(10):936–938.

68. DeGiorgio CM, Soss J, Cook IA, et al. Randomized controlled trial of trigeminal nerve stimulation for drug-resistant epilepsy. *Neurology.* 2013;80(9):786–791.

69. Soss J, Heck C, Murray D, et al. A prospective long-term study of external trigeminal nerve stimulation for drug-resistant epilepsy. *Epilepsy Behav.* 2015;42:44–47.

70. Pack AM. Trigeminal nerve stimulation may not be effective for the treatment of refractory partial seizures. *Epilepsy Curr.* 2013;13(4):164–165.

71. Dow RS, Fernández-Guardiola A, Manni E. The influence of the cerebellum on experimental epilepsy. *Electroencephalogr Clin Neurophysiol.* 1962;14(3):383–398.

72. Cooper IS, Amin I, Gilman S. The effect of chronic cerebellar stimulation upon epilepsy in man. *Trans Am Neurol Assoc.* 1973;98:192–196.

73. Krauss GL, Fisher RS. Cerebellar and thalamic stimulation for epilepsy. *Adv Neurol.* 1993;63:231–245.

74. Wright GD, McLellan DL, Brice JG. A double-blind trial of chronic cerebellar stimulation in twelve patients with severe epilepsy. *J Neurol Neurosurg Psychiatry.* 1984;47(8):769–774.

75. Al-Otaibi FA, Hamani C, Lozano AM. Neuromodulation in epilepsy. *Neurosurgery.* 2011;69(4):957–979. [discussion 979].

76. Norden AD, Blumenfeld H. The role of subcortical structures in human epilepsy. *Epilepsy Behav.* 2002;3(3):219–231.

77. Mirski MA, Tsai YC, Rossell LA, Thakor NV, Sherman DL. Anterior thalamic mediation of experimental seizures: selective EEG spectral coherence. *Epilepsia.* 2003;44(3):355–365.

78. Wennberg RA, Lozano AM. Intracranial volume conduction of cortical spikes and sleep potentials recorded with deep brain stimulating electrodes. *Clin Neurophysiol.* 2003;114(8):1403–1418.

79. Mullan S, Vailati G, Karasick J, Mailis M. Thalamic lesions for the control of epilepsy. A study of nine cases. *Arch Neurol.* 1967;16(3):277–285.

80. Mirski MA, Ferrendelli JA. Anterior thalamic mediation of generalized pentylenetetrazol seizures. *Brain Res.* 1986;399(2):212–223.

81. Mirski MA, Ferrendelli JA. Interruption of the mammillothalamic tract prevents seizures in guinea pigs. *Science.* 1984;226(4670):72–74.

82. Cooper IS, Upton AR, Amin I, Garnett S, Brown GM, Springman M. Evoked metabolic responses in the limbic-striate system produced by stimulation of anterior thalamic nucleus in man. *Int J Neurol.* 1984;18:179–187.

83. Fisher R, Salanova V, Witt T, et al. Electrical stimulation of the anterior nucleus of thalamus for treatment of refractory epilepsy. *Epilepsia*. 2010;51(5):899–908.
84. Deep-Brain Stimulation for Parkinson's Disease Study Group. Deep-brain stimulation of the subthalamic nucleus or the pars interna of the globus pallidus in Parkinson's disease. *N Engl J Med*. 2001;345(13):956–963.
85. Beric A, Kelly PJ, Rezai A, et al. Complications of deep brain stimulation surgery. *Stereotact Funct Neurosurg*. 2001;77(1–4):73–78.
86. Fenoy AJ, Simpson Jr RK. Risks of common complications in deep brain stimulation surgery: management and avoidance. *J Neurosurg*. 2014;120(1):132–139.
87. Weiss SR, Li XL, Rosen JB, Li H, Heynen T, Post RM. Quenching: inhibition of development and expression of amygdala kindled seizures with low frequency stimulation. *Neuroreport*. 1995;6(16):2171–2176.
88. Velasco M, Velasco F, Velasco AL, et al. Subacute electrical stimulation of the hippocampus blocks intractable temporal lobe seizures and paroxysmal EEG activities. *Epilepsia*. 2000;41(2):158–169.
89. Vonck K, Boon P, Achten E, De Reuck J, Caemaert J. Long-term amygdalohippocampal stimulation for refractory temporal lobe epilepsy. *Ann Neurol*. 2002;52(5):556–565.
90. Velasco AL, Velasco F, Velasco M, Trejo D, Castro G, Carrillo-Ruiz JD. Electrical stimulation of the hippocampal epileptic foci for seizure control: a double-blind, long-term follow-up study. *Epilepsia*. 2007;48(10):1895–1903.
91. Tellez-Zenteno JF, McLachlan RS, Parrent A, Kubu CS, Wiebe S. Hippocampal electrical stimulation in mesial temporal lobe epilepsy. *Neurology*. 2006;66(10):1490–1494.
92. McLachlan RS, Pigott S, Tellez-Zenteno JF, Wiebe S, Parrent A. Bilateral hippocampal stimulation for intractable temporal lobe epilepsy: impact on seizures and memory. *Epilepsia*. 2010;51(2):304–307.
93. Penfield W, Jasper HH. *Epilepsy and the Functional Anatomy of the Human Brain*. 1st ed. Boston, MA: Little, Brown; 1954.
94. Kinoshita M, Ikeda A, Matsumoto R, et al. Electric stimulation on human cortex suppresses fast cortical activity and epileptic spikes. *Epilepsia*. 2004;45(7):787–791.
95. Lesser RP, Kim SH, Beyderman L, et al. Brief bursts of pulse stimulation terminate afterdischarges caused by cortical stimulation. *Neurology*. 1999;53(9):2073–2081.
96. Psatta DM. Control of chronic experimental focal epilepsy by feedback caudatum stimulations. *Epilepsia*. 1983;24(4):444–454.
97. Nakagawa M, Durand D. Suppression of spontaneous epileptiform activity with applied currents. *Brain Res*. 1991;567(2):241–247.
98. Peters TE, Bhavaraju NC, Frei MG, Osorio I. Network system for automated seizure detection and contingent delivery of therapy. *J Clin Neurophysiol*. 2001;18(6):545–549.
99. Kossoff EH, Ritzl EK, Politsky JM, et al. Effect of an external responsive neurostimulator on seizures and electrographic discharges during subdural electrode monitoring. *Epilepsia*. 2004;45(12):1560–1567.
100. Morrell MJ. RNS System in Epilepsy Study Group. Responsive cortical stimulation for the treatment of medically intractable partial epilepsy. *Neurology*. 2011;77(13):1295–1304.
101. Yang XF, Duffy DW, Morley RE, Rothman SM. Neocortical seizure termination by focal cooling: temperature dependence and automated seizure detection. *Epilepsia*. 2002;43(3):240–245.

22

Emerging Applications and Future Directions in Pediatric Neurostimulation

D.L. Gilbert

Cincinnati Children's Hospital Medical Center, Cincinnati, OH, United States

A. Kirton

University of Calgary, Calgary, AB, Canada

The developing brain represents an exciting, unique, and relatively unexplored window of opportunity for advances in brain stimulation. Rather than simply following along behind the terrific progress being made in the mature brain, the knowledge base and experience summarized in preceding chapters suggests those researching the developing brains of children have distinct opportunities to lead advances across multiple areas of non-invasive brain stimulation. With this in mind, the following sections summarize some promising and exciting avenues for future development of brain stimulation applications in children.

TRANSCRANIAL MRI-GUIDED FOCUSED ULTRASOUND

Systems have recently been developed to apply focused ultrasound to discrete brain targets within an MRI scanner.[1,2] Such systems have demonstrated early evidence of feasibility and efficacy of ablative procedures in adult patients with a variety of neurological disorders, the most prominent to date being essential tremor, neuropathic pain, and brain tumors.[3–6] The technological aspects of this method are beyond the scope of this chapter and are reviewed elsewhere.[1]

Such technologies have also been posited to afford new opportunities for non-invasive neuromodulation.[1] "Gentler," subablative approaches may be able to modulate neuronal activity in discrete brain areas. This may involve modifications of existing systems used in ablative procedures. Current focused ultrasound systems consist of a 30-cm diameter hemispheric 1024-element phased-array transducer operating at 650 kHz, but lower frequency systems (eg, 220 kHz) are being evaluated.

Animal studies have confirmed the ability of focused ultrasound to modulate functional brain activity at intensities well below those used for ablations. For example, studies in both rats and rabbits have demonstrated modulation of motor cortex including generation of motor-evoked potentials (see Fig. 22.1).[7,8] The same studies demonstrated the ability to assess modulatory effects using neurophysiology and functional imaging, while histological examination excluded evidence of tissue damage.

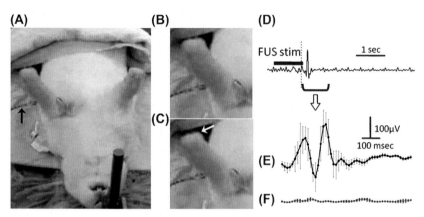

FIGURE 22.1 (A) Experimental setup of a rabbit with subdermal electrodes positioned in the left forepaw for motion readings (see *arrow*). Upon sonication, forepaw movement was observed ((B) and (C); *arrow* indicates movement). (D) An example of forepaw motion recording from an animal. The signal averaged across four repeated excitations (with standard error) recorded upon the completion of 1-s long sonication in the motor cortex (E) and away from the motor cortex (F) ($n = 4$).

Human studies of transcranial ultrasound neuromodulation are emerging. Focused ultrasound of somatosensory cortex in healthy controls elicits discrete subjective sensory changes with accompanying EEG changes and favorable initial safety.[9] Major potential advantages of focused ultrasound include an ability to non-invasively move to modulation of deeper structures, overcoming the limitation of TMS of cortical surface stimulation only. Real-time image guidance with improved spatial resolution may also be enhanced with such methods. Seminal work by Tufail and colleagues collectively highlights many potential advantages of transcranial focused ultrasound. They provide compelling preclinical evidence that focused ultrasound can modulate functional cortical areas such as motor cortex, affected deep structures such as hippocampus, and even modulate collective brain circuits with high spatial resolution (scale of a few millimeters) and no histological evidence of tissue injury.[10,11]

ROBOTICS

Intersubject variability in responses to TMS as well as methodological factors affecting accuracy and consistency of administration of TMS pose obstacles to advanced applications of TMS. Some of these precision and reproducibility issues may be addressed with emerging TMS robotic systems. Technologies such as those shown in Fig. 22.2 include a robotic arm that holds the TMS coil to allow precise placement including three-dimensional orientation. By sensing head motion, the robot can adjust to maintain this exact positioning in near-real time which may be particularly advantageous in younger or developmentally delayed children with decreased ability to remain entirely still. Integration with existing neuronavigation systems allows accurate co-registration with subject MR imaging, allowing individualized mapping and precise repositioning with serial sessions including treatment trials.

Experience to date has been limited but evidence is emerging. For example, using the Axilum TMS robot depicted Fig. 22.2, pilot studies have suggested increased accuracy of coil positioning.[12] Procedures are safe and well-tolerated, possibly with improved subject comfort. An example of the unique capacities afforded by TMS robotics was recently reported, demonstrating the potential for conscious brain-to-brain communication between human subjects.[13] One subject trained a brain computer interface (see in the following) to detect a simple motor imagery signal (think of "hand" or "foot"). The system then transmitted this message electronically to a second subject thousands of kilometers away. The second subject was positioned in the TMS robot where their occipital hotspot of phosphene generation was mapped. Upon receiving the "message" from the first subject, the robot would instantly orient the coil to either generate a phosphene (= "hand") or not (= "foot"). The second subject could then literally "read the mind" of the first, accurately determining which of the two thoughts they had.

Experience with TMS robotics has not yet been reported in the pediatric population. However, unique advantages are suggested in addition to the positioning and reproducibility issues mentioned earlier. For example, motor mapping is a well-established TMS methodology of particular utility in understanding motor system organization in stroke survivors and children with hemiplegic cerebral palsy.[14–16] Major limitations of current motor mapping methods are extensive time requirements and inconsistent targeting of grid mapping sites. The possibility of automated feedback of MEP

FIGURE 22.2   TMS robotics. The Axilum TMS robot facilitates precise and consistent coil positioning by adjusting for subject head motion and removing human error factors. Systems are integrated with established neuronaviagtion systems as shown. *With permission from Michel Berg, axilumrobotics.com.*

data to inform the very quick and accurate targeting available with a robotic TMS system creates a new opportunity to perform deeper, more accurate, and much more efficient motor mapping studies. This in turn enables new questions to be asked such as the colocalization of ipsilateral projections in hemiparetic cerebral palsy and the neurophysiological effects of interventions on motor cortex maps. Numerous other methodological improvements are anticipated as TMS robotic systems continue to develop.

BRAIN–COMPUTER INTERFACES (BCI)

The fundamental concepts of BCI are now well established in humans. As summarized in Fig. 22.3 and excellent review articles,[17] most BCI systems require the following three primary components.

First, functional messages or signal must be extracted from the brain, typically via electrical potentials. Originally this involved direct implantation of sensors within the brain parenchyma using electrocorticography or similar systems. The location would be logically paired with types of messages sought such as implantation in the primary motor cortex in order to detect motor plans or "commands." More recently, non-invasive systems, such as surface electroencephalography (EEG), have been developed to capture cortical signals with enough accuracy to create interpretable messages.

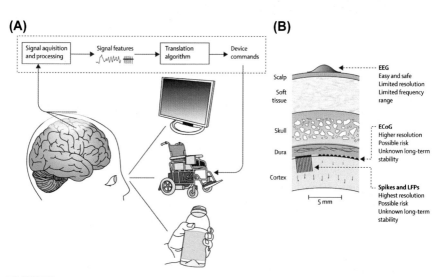

FIGURE 22.3 Brain computer interfaces. Cortical signal is measured and sent to computer software that applies algorithms to extract meaning following focused training. This in turn can then connect subject thoughts to execute specific actions in various effectors as shown (A). Different detection systems ranging from invasive implants to surface EEG can be used to detect cortical signals (B).

The second component is a computer-based learning system capable of receiving the above signals and extracting meaning from them. There are many complexities to this process but many systems use a simple "on/off" training paradigm whereby subjects will perform specific thought tasks, often only for seconds. Measureable changes in cortical activity such as event-related desynchronization (ERD) during this epoch will then be compared to baseline measures, the difference between them is specific enough to deliver a message to the learning system in essentially real time.

The third component is an effector capable of executing the desired task being trained. Remarkably, the majority of the most important daily functional tasks missed by those with severe physical disabilities require a limited number (five or less) of relatively simple command signals. Common examples include the control of a mouse on a computer screen where most people interact with the world via simple combined commands of up/down, left/right, and "click." Operation of a wheelchair also requires only a few simple signals for start/stop and direction. With the relative "strength of thought" BCI can also provide relative gain to such systems (eg, faster, slower). Robotic exoskeleton systems are another avenue for those with severe unilateral disabilities, such as amputees and stroke survivors.

BCI applications are relatively unexplored in the developing brain where additional complexities are encountered. An adult with high spinal cord injury or amyotrophic lateral sclerosis (ALS), for which BCI are commonly applied, already has a completely developed and anatomically normal motor cortex and network. In contrast, the motor network of children is constantly changing during its "installation" over decades. Add to this a disease such as cerebral palsy, where the cortex may be injured, and the complexity is amplified exponentially.

However, despite these and other limitations, progress is being made. Studies from children undergoing electrocorticography suggest that motor signals can be obtained and interpreted in a fashion similar to adults and compatible with BCI development.[18] Studies in cerebral palsy have been limited but suggest feasibility and also distinct challenges in extracting motor messages for BCI applications.[19,20] Our preliminary experience suggests children as young as 6 years with perinatal brain injuries can train commercially available BCI systems to perform multiple commands in less than 30 min.

BCIs are of course not really neurostimulation in the typical sense of the word. However, advancing their success and overcoming some of the hurdles mentioned earlier may be facilitated by non-invasive brain stimulation. For example, several recent studies have suggested the ability of neuromodulation to enhance BCI training. In healthy subjects, tDCS may be able to enhance the detection of event-related desynchronization[21] and impact neural coupling between different brain regions to modulate phase synchronization during motor imagery tasks typically used in BCI systems.[22] Capturing event-related desynchronization from the lesioned hemisphere of hemiparetic adult stroke patients appears to be enhanced by concurrent anodal tDCS.[23] Collectively, these results suggest that non-invasive neuromodulation may play an important role in the development of BCI systems, particularly in children with abnormal neurodevelopment where enhancement of signal generation may be particularly advantageous.

FUTURE DIRECTIONS FOR SINGLE- AND PAIRED-PULSE TMS STUDIES

To date, the most widespread use of TMS in pediatric publications is its use in motor cortex as a biomarker of function and dysfunction in normal development and in disorders and diseases. In most cases, the readout comes from surface EMG tracings. The "readout" is often the motor-evoked potential (MEP), usually the peak-to-peak amplitude. Sometimes the area under the MEP is used. Studies of evoked potential latencies are also performed, or durations of silent periods. In nearly all cases, the study participants are "at rest." Most laboratories test hypotheses in small samples of convenience, and most utilize healthy young adults.

In pediatrics, there still remain substantial opportunities for expansion of brain stimulation research within and outside of motor cortex. Validating observations from small studies through multisite collaborations enrolling large, representative, and well-characterized groups of pediatric patients is feasible and important.[24–26]

BRAIN STIMULATION BIOMARKERS

Biomarkers in Heterogeneous Neurobehavioral Syndromes

The use of TMS in complex, heterogeneous syndromes such as ADHD and Tourette syndrome is quite widespread, as discussed in Chapter 8,"TMS Applications in ADHD and Developmental Disorders." By contrast, there is relatively less exploration in children with autism spectrum disorder (ASD). Given the feasibility in using the technique to probe physiology in adults and adolescents with high-functioning autism,[27,28] moving TMS ASD studies into broader populations of children with ASD and extending it to younger ages may be feasible and valuable. Challenges with non-verbal children are likely and for this research to be successful the team will need to know how to work with and distract autistic children who may have more anxiety and sensory hypersensitivity than typically developing peers. Techniques that are short and involve lower-intensity TMS pulses may possibly be more tolerable. The use of systematic methods to collect adverse event data[24,29,30] as well as subjective impressions[31] is to be encouraged.

Biomarkers in Single-Gene Disorders

The use of TMS to probe networks involved in single-gene disorders affecting the CNS is relatively less well explored to date. However, there is tremendous translational potential. The generation of increasingly precise rodent models of human genetic disease often involves physiological studies to characterize central nervous system effects. A pathway-mechanism-based medication may then be used to treat the symptoms or "rescue the phenotype," with both clinical/behavioral outcomes and repeated neurophysiological measurements to understand those effects.

There are several examples to date where TMS measures have been used to probe non-invasively the analogous changes in humans at the macro level. A recent example of this is found in the NF1 literature. NF1 knockout mice were shown with cellular, hippocampal recordings to have increased GABA-mediated inhibition and deficient long-term potentiation.[32] In young adults, Mainberger and colleagues used TMS to evaluate analogous measures. They found that, compared to controls, NF1 patients had increased short-interval cortical inhibition (SICI) and decreased LTP-like plasticity, evaluated with paired associative stimulation (PAS), findings which were concordant with physiological results in mice. Lovastatin, an inhibitor of 3-hydroxy-3-methylglutaryl coenzyme A (HMG-CoA) that also inhibits RAS/mitogen-activated protein kinase (MAPK) activity, and similar drugs are under investigation to see whether downregulation of this overactive pathway in NF1 improves cognitive function. In a 4-day, randomized, placebo-controlled study, treated NF1 patients had a behavioral response, improved phasic alertness, and a physiologic "normalization," increased PAS neuroplasticity, and decreased SICI.[33]

Children with NF1 often have ADHD-type symptoms and cognitive delays or impairments. Meaningful improvement in intellectual function might be expected to take months or years, so one role of a biomarker might to be determine whether a medication "normalizes" physiology, with the idea being that improvements in plasticity, for example, might precede the actual gaining of skills or improvement in function. This presupposes that there is adequate test–retest reliability of these measures, which has been estimated for SICI[34] and for PAS.[35]

A similar rationale underlies using theta burst TMS (TBS) in motor cortex to evaluate neuroplasticity in boys with fragile X syndrome.[36] One could evaluate drug effects, eg, in studies of arbaclofen in ASD,[37] using a GABA-B sensitive TMS measure like the cortical silent period (CSP) or long-interval cortical inhibition (LICI) as well as TBS or PAS to study neuroplasticity. In a much rarer disease, succinic semialdehyde dehydrogenase deficiency, animal models suggest use-dependent downregulation of postsynaptic GABA(B) receptors plays a role in pathophysiology.[38] To translate this understanding into human disease, TMS CSP and LICI measures were employed in affected children. The study demonstrated shorter CSP and less LICI.[39] There remains tremendous potential to combine predictions made by an animal model with results from interventional studies to probe disease-related pathophysiology and treatment effects.

FUNCTIONAL TMS

The past 10 years have witnessed a growing body of literature in healthy individuals where TMS is used to probe motor cortex during tasks, as opposed to simply at rest. Examples include using TMS to probe the neural correlates of response inhibition, or inhibition of prepared actions, in Go/NoGo or Stop Signal Reaction Time tasks.[40–42] While to date this has mostly been performed in adults, there has been application of such evaluations of cortical inhibition during response inhibition to a pertinent population; children with ADHD.[43] Another important example probing theories about empathy and mirror neurons has been to probe motor cortex in persons with ASD observing hand movements and gestures.[44,45]

Similarly, the relationships between multiple aspects of reward and motor cortex physiology have been evaluated. This includes studies of perceptions of a desired reward, the probability of reward, and action to obtain reward, using single- and paired-pulse designs during real-time engagement in reward-related processes.[46–50] In healthy adults, motor cortex excitability appears to be affected by the probability, size, and subjective desirability of rewarding stimuli. As this may be related to dopaminergic tone, it has been tested in individuals with low dopamine, eg, those with Parkinson disease,[51] and could be extended to various populations of children, such as those with Tourette syndrome or ADHD, where dopamine dysregulation is suspected.

TMS FOR BIOFEEDBACK

TMS has also recently been used in a biofeedback training model where healthy young adults participating in TMS were shown the amplitude of their TMS-evoked MEPs. Through several methods, participants then used strategies to try to reduce their own cortical excitability, as demonstrated by seeing smaller MEP amplitudes.[52] Participants randomized to receive accurate information or "feedback" were able to suppress their cortical excitability. This raises the possibility that TMS could be used to develop greater motor control through selective suppression of motor cortex controlling muscles used for a task, or neighboring muscles not involved in the task but which might interfere with optimal performance. So, for example, hyperkinetic movement or behavioral disorders such as dystonia, Tourette syndrome, or even ADHD might be evaluated with such a paradigm. Similar to other forms of neurofeedback, training sessions for controlling motor cortex excitability could be tested as a means for improving motor or behavioral control outside of the laboratory setting.

NON-INVASIVE LOW-CURRENT ELECTRICAL STIMULATION

Transcranial electrical stimulation (TES) techniques have emerged as an important modality for understanding physiological disturbances in various neurological conditions and possibly modulating them to improve symptoms. Unlike TMS, the small size and relatively low cost of TES devices have resulted in a "democratization" of this technology, as individuals can make or buy their own devices. On the negative side, this of course means the likelihood of profit-driven misuse for neuro-quackery is emerging and likely to increase. However, on the positive side, if beneficial clinical applications are identified through rigorous research, disseminating these applications to patients may be much easier and less expensive than doing so with TMS/rTMS.

SHIFTING MEMBRANE POTENTIALS WITH TRANSCRANIAL DIRECT-CURRENT STIMULATION

The most widely used TES technique, tDCS, usually involves application of large anode and cathode electrodes to the scalp (see Chapter 5: Transcranial Direct-Current Stimulation (tDCS): Principles and Emerging Applications in Children). Both electrodes generate an extracellular electric field. The anode excites underlying cortex and the cathode inhibits it. In terms of understanding the mechanism of action, there are to date a number of pharmacophysiological studies in healthy adults showing effects of tDCS can be modified through concurrent treatment with NMDA antagonists and calcium channel blockers.[53,54]

With a large potential number of possible stimulation parameters, optimization in various clinical groups of children may require extensive studies. It will of course be important to carefully model and predict differences, and possibly individualize pediatric dosing, based on pediatric brain size or maturation.[55,56] Advances in tDCS technology, such as HD-tDCS, offer intriguing advantages that require exploration in children. Finally, cautious studies to evaluate tolerability are important, and these are ongoing.[57]

TABLE 22.1 Comparison of Different Transcranial Electrical Current Methods

Method	Detail	Parameter Example
Transcranial direct-current stimulation tDCS[67–69]	Uses weak electrical currents to shift resting membrane potentials, depending on polarity (anode excites cortex, cathode inhibits)	Current dose: 1–2 mA; densities 0.28–0.80 A/m², duration 20–40 min
Transcranial alternating-current stimulation (tACS)[64,70,71]	Uses alternating sinusoidal or rectangular wave currents restricted to single frequency within a physiologically relevant range to drive brain oscillations at that frequency	Current density 0.14 A/m² 40 Hz, duration 90 s (Feurra tuning 2011); 0.80 A/m², 15 Hz, 20 min; ripple range 140 Hz,[72] 1–5 kHz[73]
Transcranial random current stimulation[74,75]	Uses an electrical white-noise signal – a random level of current is generated from 0.1 to 640 Hz in a bell-shaped probability density function	101–640 Hz (random), 1 mA, 62.5 μA/cm², 10 min

ENTRAINING OR DISRUPTING INTRINSIC BRAIN OSCILLATIONS WITH TRANSCRANIAL ALTERNATING-CURRENT STIMULATION

The distinct oscillatory patterns underlying human brain function have been the focus of much physiological study, using a variety of invasive and non-invasive methodologies.[58–62] Transcranial magnetic and electrical brain stimulation techniques allow possibilities to better characterize the function of these oscillatory phenomena through disrupting them or entraining them briefly. Most such research has involved EEG recording before and after repetitive TMS, but as artifact-removing protocols have been developed, recording EEG during TMS and TES has allowed for demonstration of direct causal entrainment and plasticity influencing intrinsic brain oscillations. Pioneering studies have demonstrated that rhythmic, 10 Hz rTMS[63] and 10 Hz tACS[64] can synchronize and enhance alpha activity in parietal/occipital cortex. This has some potential clinical applications, as 10-Hz tACS may improve visual motion sensitivity[65] and motor learning.[66] More common is research probing interactions between electrical current stimulation and overall cortical excitability. Mechanisms are less clear, but both t-ACS and transcranial alternating current stimulation can exert excitatory effects in cortex (see details in Table 22.1), without the issues of polarity in tDCS.

CONCLUSION

The rapid growth of non-invasive brain stimulation technologies and applications witnessed over the last two decades will continue, likely at an accelerated pace. Novel neurotechnologies and methodologies promise to expand the armamentarium of both neurophysiological and modulatory therapeutic opportunities. As discussed in this chapter, areas of opportunity may include neurophysiological biomarkers, novel stimulation techniques, such as transcranial MRI-guided focused ultrasound and alternating-current stimulation, preclinical models, neurofeedback, functional TMS, robotics, and brain–computer interfaces.

References

1. Ghanouni P, Pauly KB, Elias WJ, et al. Transcranial MRI-guided focused ultrasound: a review of the technologic and neurologic applications. *AJR Am J Roentgenol.* 2015;205(1):150–159.
2. Lipsman N, Mainprize TG, Schwartz ML, Hynynen K, Lozano AM. Intracranial applications of magnetic resonance-guided focused ultrasound. *Neurotherapeutics.* 2014;11(3):593–605.
3. Martin E, Jeanmonod D, Morel A, Zadicario E, Werner B. High-intensity focused ultrasound for noninvasive functional neurosurgery. *Ann Neurol.* 2009;66(6):858–861.
4. Jeanmonod D, Werner B, Morel A, et al. Transcranial magnetic resonance imaging-guided focused ultrasound: noninvasive central lateral thalamotomy for chronic neuropathic pain. *Neurosurg Focus.* 2012;32(1):E1.
5. Elias WJ, Huss D, Voss T, et al. A pilot study of focused ultrasound thalamotomy for essential tremor. *N Engl J Med.* 2013;369(7):640–648.
6. McDannold N, Clement GT, Black P, Jolesz F, Hynynen K. Transcranial magnetic resonance imaging- guided focused ultrasound surgery of brain tumors: initial findings in 3 patients. *Neurosurgery.* 2010;66(2):323–332. discussion 332.
7. Yoo SS, Bystritsky A, Lee JH, et al. Focused ultrasound modulates region-specific brain activity. *Neuroimage.* 2011;56(3):1267–1275.
8. Kim H, Chiu A, Lee SD, Fischer K, Yoo SS. Focused ultrasound-mediated non-invasive brain stimulation: examination of sonication parameters. *Brain Stimul.* 2014;7(5):748–756.
9. Lee W, Kim H, Jung Y, Song IU, Chung YA, Yoo SS. Image-guided transcranial focused ultrasound stimulates human primary somatosensory cortex. *Sci Rep.* 2015;5:8743.
10. Tufail Y, Matyushov A, Baldwin N, et al. Transcranial pulsed ultrasound stimulates intact brain circuits. *Neuron.* 2010;66(5):681–694.
11. Tufail Y, Yoshihiro A, Pati S, Li MM, Tyler WJ. Ultrasonic neuromodulation by brain stimulation with transcranial ultrasound. *Nat Protoc.* 2011;6(9):1453–1470.
12. Ginhoux R, Renaud P, Zorn L, et al. A custom robot for Transcranial Magnetic Stimulation: first assessment on healthy subjects. *Conf Proc IEEE Eng Med Biol Soc.* 2013;2013:5352–5355.
13. Grau C, Ginhoux R, Riera A, et al. Conscious brain-to-brain communication in humans using non-invasive technologies. *PLoS One.* 2014;9(8):e105225.
14. Sawaki L, Butler AJ, Leng X, et al. Constraint-induced movement therapy results in increased motor map area in subjects 3 to 9 months after stroke. *Neurorehabil Neural Repair.* 2008;22(5):505–513.
15. Corneal SF, Butler AJ, Wolf SL. Intra- and intersubject reliability of abductor pollicis brevis muscle motor map characteristics with transcranial magnetic stimulation. *Arch Phys Med Rehabil.* 2005;86(8):1670–1675.
16. Kirton A. Modeling developmental plasticity after perinatal stroke: defining central therapeutic targets in cerebral palsy. *Pediatr Neurol.* 2013;48(2):81–94.
17. Daly JJ, Wolpaw JR. Brain-computer interfaces in neurological rehabilitation. *Lancet Neurol.* 2008;7(11):1032–1043.

18. Breshears JD, Gaona CM, Roland JL, et al. Decoding motor signals from the pediatric cortex: implications for brain-computer interfaces in children. *Pediatrics*. 2011;128(1):e160–168.

19. Daly I, Billinger M, Laparra-Hernandez J, et al. On the control of brain-computer interfaces by users with cerebral palsy. *Clin Neurophysiol*. 2013;124(9):1787–1797.

20. Daly I, Faller J, Scherer R, et al. Exploration of the neural correlates of cerebral palsy for sensorimotor BCI control. *Front Neuroeng*. 2014;7:20.

21. Wei P, He W, Zhou Y, Wang L. Performance of motor imagery brain-computer interface based on anodal transcranial direct current stimulation modulation. *IEEE Trans Neural Syst Rehabil Eng*. 2013;21(3):404–415.

22. He W, Wei P, Zhou Y, Wang L. Modulation effect of transcranial direct current stimulation on phase synchronization in motor imagery brain-computer interface. In: *Paper Presented at: Engineering in Medicine and Biology Society (EMBC)*; 2014. 36th Annual International Conference of the IEEE 2014.

23. Kasashima Y, Fujiwara T, Matsushika Y, et al. Modulation of event-related desynchronization during motor imagery with transcranial direct current stimulation (tDCS) in patients with chronic hemiparetic stroke. *Exp Brain Res*. 2012;221(3):263–268.

24. Hong YH, Wu SW, Pedapati EV, et al. Safety and tolerability of theta burst stimulation versus single and paired pulse transcranial magnetic stimulation: a comparative study of 165 pediatric subjects. *Front Hum Neurosci*. 2015;9:29.

25. Wu SW, Gilbert DL, Shahana N, Huddleston DA, Mostofsky SH. Transcranial magnetic stimulation measures in attention-deficit/hyperactivity disorder. *Pediatr Neurol*. 2012;47(3):177–185.

26. Gilbert DL, Isaacs KM, Augusta M, Macneil LK, Mostofsky SH. Motor cortex inhibition: a marker of ADHD behavior and motor development in children. *Neurology*. 2011;76(7):615–621.

27. Enticott PG, Kennedy HA, Rinehart NJ, Tonge BJ, Bradshaw JL, Fitzgerald PB. GABAergic activity in autism spectrum disorders: an investigation of cortical inhibition via transcranial magnetic stimulation. *Neuropharmacology*. 2013;68:202–209.

28. Oberman LM, Pascual-Leone A, Rotenberg A. Modulation of corticospinal excitability by transcranial magnetic stimulation in children and adolescents with autism spectrum disorder. *Front Hum Neurosci*. 2014;8:627.

29. Gilbert DL, Garvey MA, Bansal AS, Lipps T, Zhang J, Wassermann EM. Should transcranial magnetic stimulation research in children be considered minimal risk? *Clin Neurophysiol*. 2004;115(8):1730–1739.

30. Rossi S, Hallett M, Rossini PM, Pascual-Leone A. Safety of TMSCG. Safety, ethical considerations, and application guidelines for the use of transcranial magnetic stimulation in clinical practice and research. *Clin Neurophysiol*. 2009;120(12):2008–2039.

31. Garvey MA, Kaczynski KJ, Becker DA, Bartko JJ. Subjective reactions of children to single-pulse transcranial magnetic stimulation. *J Child Neurol*. 2001;16(12):891–894.

32. Costa RM, Federov NB, Kogan JH, et al. Mechanism for the learning deficits in a mouse model of neurofibromatosis type 1. *Nature*. 2002;415(6871):526–530.

33. Mainberger F, Jung NH, Zenker M, et al. Lovastatin improves impaired synaptic plasticity and phasic alertness in patients with neurofibromatosis type 1. *BMC Neurol*. 2013;13:131.

34. Gilbert DL, Sallee FR, Zhang J, Lipps TD, Wassermann EM. TMS-evoked cortical inhibition: a consistent marker of ADHD scores in Tourette Syndrome. *Biol Psychiatry*. 2005;57:1597–1600.

35. Damji O, Keess J, Kirton A. Evaluating developmental motor plasticity with paired afferent stimulation. *Dev Med Child Neurol*. 2015;57(6):548–555.

36. Oberman L, Ifert-Miller F, Najib U, et al. Transcranial magnetic stimulation provides means to assess cortical plasticity and excitability in humans with fragile x syndrome and autism spectrum disorder. *Front Synaptic Neurosci*. 2010;2:26.

37. Silverman JL, Pride MC, Hayes JE, et al. GABAB receptor agonist r-baclofen reverses social deficits and reduces repetitive behavior in two mouse models of autism. *Neuropsychopharmacology*. 2015;40(9):2228–2239.

38. Buzzi A, Wu Y, Frantseva MV, et al. Succinic semialdehyde dehydrogenase deficiency: GABAB receptor-mediated function. *Brain Res*. 2006;1090(1):15–22.

39. Reis J, Cohen LG, Pearl PL, et al. GABAB-ergic motor cortex dysfunction in SSADH deficiency. *Neurology*. 2012;79(1):47–54.

40. Coxon JP, Stinear CM, Byblow WD. Intracortical inhibition during volitional inhibition of prepared action. *J Neurophysiol*. 2006;95(6):3371–3383.

41. MacDonald HJ, Coxon JP, Stinear CM, Byblow WD. The fall and rise of corticomotor excitability with cancellation and reinitiation of prepared action. *J Neurophysiol*. 2014;112(11):2707–2717.

42. Stinear CM, Coxon JP, Byblow WD. Primary motor cortex and movement prevention: Where Stop meets Go. *Neurosci Biobehav Rev*. 2009;33(5):662–673.

43. Hoegl T, Heinrich H, Barth W, Losel F, Moll GH, Kratz O. Time course analysis of motor excitability in a response inhibition task according to the level of hyperactivity and impulsivity in children with ADHD. *PLoS One*. 2012;7(9):e46066.

44. Enticott PG, Kennedy HA, Rinehart NJ, et al. Interpersonal motor resonance in autism spectrum disorder: evidence against a global "mirror system" deficit. *Front Hum Neurosci*. 2013;7:218.

45. Enticott PG, Kennedy HA, Rinehart NJ, et al. Mirror neuron activity associated with social impairments but not age in autism spectrum disorder. *Biol Psychiatry*. 2012;71(5):427–433.

46. Kapogiannis D, Campion P, Grafman J, Wassermann EM. Reward-related activity in the human motor cortex. *Eur J Neurosci*. 2008;27(7):1836–1842.

47. Thabit MN, Nakatsuka M, Koganemaru S, Fawi G, Fukuyama H, Mima T. Momentary reward induce changes in excitability of primary motor cortex. *Clin Neurophysiol*. 2011;122(9):1764–1770.

48. Klein PA, Olivier E, Duque J. Influence of reward on corticospinal excitability during movement preparation. *J Neurosci*. 2012;32(50):18124–18136.

49. Gupta N, Aron AR. Urges for food and money spill over into motor system excitability before action is taken. *Eur J Neurosci*. 2011;33(1):183–188.

50. Suzuki M, Kirimoto H, Sugawara K, et al. Motor cortex-evoked activity in reciprocal muscles is modulated by reward probability. *PLoS One*. 2014;9(6):e90773.

51. Kapogiannis D, Mooshagian E, Campion P, et al. Reward processing abnormalities in Parkinson's disease. *Mov Disord*. 2011;26(8):1451–1457.

52. Majid DS, Lewis C, Aron AR. Training voluntary motor suppression with real-time feedback of motor evoked potentials. *J Neurophysiol*. 2015;113(9):3446–3452.

53. Nitsche MA, Fricke K, Henschke U, et al. Pharmacological modulation of cortical excitability shifts induced by transcranial direct current stimulation in humans. *J Physiol*. 2003;553(Pt 1):293–301.

54. Monte-Silva K, Kuo MF, Hessenthaler S, et al. Induction of late LTP-like plasticity in the human motor cortex by repeated non-invasive brain stimulation. *Brain Stimul*. 2013;6(3):424–432.

55. Gillick BT, Kirton A, Carmel JB, Minhas P, Bikson M. Pediatric stroke and transcranial direct current stimulation: methods for rational individualized dose optimization. *Front Hum Neurosci*. 2014;8:739.

56. Kessler SK, Minhas P, Woods AJ, Rosen A, Gorman C, Bikson M. Dosage considerations for transcranial direct current stimulation in children: a computational modeling study. *PLoS One*. 2013;8(9):e76112.

57. Ciechanski P, Zewdie E, Kirton A. Enhancement of motor learning with transcranial direct current stimulation in healthy children. In: *44th National Meeting of the Child Neurology Society; October 9, 2015*. National Harbor, MD; 2015.

58. Chen LL, Madhavan R, Rapoport BI, Anderson WS. Real-time brain oscillation detection and phase-locked stimulation using autoregressive spectral estimation and time-series forward prediction. *IEEE Trans Biomed Eng.* 2013;60(3):753–762.
59. Lee GT, Lee C, Kim KH, Jung KY. Regional and inter-regional theta oscillation during episodic novelty processing. *Brain Cogn.* 2014;90:70–75.
60. Ngo HV, Martinetz T, Born J, Molle M. Auditory closed-loop stimulation of the sleep slow oscillation enhances memory. *Neuron.* 2013;78(3):545–553.
61. Wang X, Jiao Y, Tang T, Wang H, Lu Z. Investigating univariate temporal patterns for intrinsic connectivity networks based on complexity and low-frequency oscillation: a test-retest reliability study. *Neuroscience.* 2013;254:404–426.
62. Wei L, Duan X, Zheng C, et al. Specific frequency bands of amplitude low-frequency oscillation encodes personality. *Hum Brain Mapp.* 2014;35(1):331–339.
63. Thut G, Veniero D, Romei V, Miniussi C, Schyns P, Gross J. Rhythmic TMS causes local entrainment of natural oscillatory signatures. *Curr Biol.* 2011;21(14):1176–1185.
64. Helfrich RF, Schneider TR, Rach S, Trautmann-Lengsfeld SA, Engel AK, Herrmann CS. Entrainment of brain oscillations by transcranial alternating current stimulation. *Curr Biol.* 2014;24(3):333–339.
65. Kar K, Krekelberg B. Transcranial alternating current stimulation attenuates visual motion adaptation. *J Neurosci.* 2014;34(21):7334–7340.
66. Pollok B, Boysen AC, Krause V. The effect of transcranial alternating current stimulation (tACS) at alpha and beta frequency on motor learning. *Behav Brain Res.* 2015;293:234–240.
67. Nitsche MA, Paulus W. Excitability changes induced in the human motor cortex by weak transcranial direct current stimulation. *J Physiol.* 2000;527(Pt 3):633–639.
68. Brunoni AR, Nitsche MA, Bolognini N, et al. Clinical research with transcranial direct current stimulation (tDCS): challenges and future directions. *Brain Stimul.* 2012;5(3): 175–195.
69. Priori A, Berardelli A, Rona S, Accornero N, Manfredi M. Polarization of the human motor cortex through the scalp. *Neuroreport.* 1998;9(10):2257–2260.
70. Paulus W. Transcranial electrical stimulation (tES - tDCS; tRNS, tACS) methods. *Neuropsychol Rehabil.* 2011;21(5):602–617.
71. Feurra M, Paulus W, Walsh V, Kanai R. Frequency specific modulation of human somatosensory cortex. *Front Psychol.* 2011;2:13.
72. Moliadze V, Antal A, Paulus W. Boosting brain excitability by transcranial high frequency stimulation in the ripple range. *J Physiol.* 2010;588(Pt 24):4891–4904.
73. Chaieb L, Antal A, Paulus W. Transcranial alternating current stimulation in the low kHz range increases motor cortex excitability. *Restor Neurol Neurosci.* 2011;29(3):167–175.
74. Terney D, Chaieb L, Moliadze V, Antal A, Paulus W. Increasing human brain excitability by transcranial high-frequency random noise stimulation. *J Neurosci.* 2008;28(52): 14147–14155.
75. Chaieb L, Paulus W, Antal A. Evaluating aftereffects of short-duration transcranial random noise stimulation on cortical excitability. *Neural Plast.* 2011;2011:105927.

Index